STATISTICAL PROBLEMS IN
PARTICLE PHYSICS, ASTROPHYSICS AND COSMOLOGY

Proceedings of PHYSTAT05

STATISTICAL PROBLEMS IN PARTICLE PHYSICS, ASTROPHYSICS AND COSMOLOGY

Proceedings of PHYSTAT05

Oxford, UK 12 – 15 September 2005

EDITORS

LOUIS LYONS
University of Oxford, UK

MÜGE KARAGÖZ ÜNEL
University of Oxford, UK

Imperial College Press

ICP

Published by

Imperial College Press
57 Shelton Street
Covent Garden
London WC2H 9HE

Distributed by

World Scientific Publishing Co. Pte. Ltd.
5 Toh Tuck Link, Singapore 596224
USA office: 27 Warren Street, Suite 401-402, Hackensack, NJ 07601
UK office: 57 Shelton Street, Covent Garden, London WC2H 9HE

British Library Cataloguing-in-Publication Data
A catalogue record for this book is available from the British Library.

STATISTICAL PROBLEMS IN PARTICLE PHYSICS, ASTROPHYSICS AND COSMOLOGY
Proceedings of PHYSTAT05

ISBN 1-86094-649-6

Printed in Singapore by World Scientific Printers (S) Pte Ltd

Preface

The PHYSTAT05 Conference was held in Oxford on September 12th to 15th, 2005. Its theme was "Statistical Problems in Particle Physics, Astrophysics and Cosmology." It was the fifth meeting in a sequence that started with the Confidence Limits Workshops held at CERN and Fermilab in 2000, followed by Conferences in 2002 at the Institute for Particle Physics Phenomenology in Durham, UK, and in 2003 at SLAC. While the first 3 meetings had been attended largely by Particle Physicists, at SLAC and at Oxford there was also involvement of Astrophysicists and Cosmologists. The SLAC and Oxford meetings really benefited from a strong presence of statisticians. They enhanced the usefulness of the Conferences in many ways: they were involved in the planning of the meeting, gave invited and contributed talks, and were simply available to discuss statistical issues with Physicists either during or in the breaks between the sessions.

We were delighted to have the Keynote Address given by Sir David Cox. Other invited talks were divided almost equally between Statisticians and Physicists, and there were parallel sessions for the contributed talks. There was also time for a poster session. A Panel Session was devoted to discussion of questions submitted in advance by PHYSTAT05 participants, and the meeting ended with summary talks given by a Statistician, an Astrophysicist and a Particle Physicist. The transparencies of most of these talks are available at the Conference web-site: http://www.physics.ox.ac.uk/phystat05 .

The Conference would not have been possible without the considerable help and effort of many people. Some of these are acknowledged in the Conference after-Dinner talk, extracts of which appear below. We would also like to thank Beverly Roger: the fact that the Panel Discussion appears in these Proceedings is due to her amazing ability to interpret and give meaning to the audio recording of that session.

We hope that these Proceedings will provide a useful record of most of the talks at the Conference. Thanks are due to members of the international Scientific Committee who reviewed the articles appearing here. In several cases this involved very active participation in the production of the final text.

Final thanks go to all the participants in PHYSTAT05 for keeping the speakers and Panel members on their toes, and helping to make it such a productive and interactive meeting.

Louis Lyons and Müge Karagöz Ünel

Part of talk by Louis Lyons at the Conference Dinner in the Jesus College Dining Hall

First I would like to welcome you all to Jesus. The College was founded by Queen Elizabeth I in 1571, just a few years before the publication of his first paper by our keynote speaker, Sir David Cox.

The Hall is adorned by the portraits of Founders, Benefactors and famous old members. The actual work of founding the College was performed by Hugh Price, whose small portrait is below the very large one of Elizabeth. The Queen however insisted on being 'the first author' on the publication of the foundation document of the College. Her portrait is by Nicholas Hilyard, who was a famous miniature painter. This was his largest miniature portrait.

A famous old member depicted here is Sir Harold Wilson. He was well-known in being a predecessor of Sir David as President of the Royal Statistical Society. He also found time to be Prime Minister.

I feel that on an occasion like this it is appropriate to quote from "The Good Book". For those of you with a different background or education, I should explain that by "The Good Book" I of course mean the one that I wrote on "Statistics for Particle and Nuclear Physicists". On the inside cover, there is a quote overheard at an international conference: "This experiment was inconclusive so we had to use statistics." I didn't hear anyone say this at PHYSTAT05, but we did have some very enlightening talks.

So I really want to thank all our invited speakers, who put so much effort into their talks. The success or otherwise of the Conference is largely due to them. Especial thanks go to Nancy Reid, Andrew Jaffe and Gary Feldman who are undertaking the task of summarising the Conference in the concluding talks tomorrow. We are also very grateful to the Statisticians who devoted time, effort and patience to explain statistical issues to physicists, both in their talks and in the numerous discussions that took place between the sessions.

I would also like to thank the speakers of the contributed talks and those who produced posters. We attracted so many presentations that it was necessary to have parallel sessions. I think we were all sorry that we could not be in more than one place at a time. Thanks are due in advance to all speakers who will send in their contributions for the Proceedings by the deadline. The Chairpersons of all the sessions did an admirable job in keeping speakers and discussions within the time constraints imposed by a very tight programme.

The International Committee (whose members are listed at the end of these Proceedings) was very helpful in the planning of this meeting. I feel very fortunate to have had access to this source of advice. An enormous amount of work was undertaken by the local Committee of Andy Carslaw, John Cobb, Sue Geddes and Pete Gronbech. Especial thanks go to Sue. Apart from taking on the bulk of the organisation, she also dealt with the numerous requests from the Conference participants. I think she had an average of about i^{-i} e-mails per person. She coped with all this with great efficiency and good humour. Many thanks Sue!

I would also like to mention Emily Down who undertook the work of arranging that the talks were ready for presentation at all the sessions. With a tight time schedule it was important that this was efficiently done, and Emily certainly did that.

The Conference would not have been possible without the financial support of PPARC, the Institute of Physics, the Royal Statistical Society, the Oxford Statistics Department and the Astor Fund – we appreciate their help. Oxford Particle Physics provided great logistic support. Thanks are especially due to Professors Brian Foster and Susan Cooper, who were very encouraging and supportive of the idea of having this Conference here in Oxford. Brian was also responsible for arranging for the musicians for the excellent Concert in the Holywell Music Room.

I am deeply grateful to my wife Elaine for her ideas of the visit to Bletchley Park and of having a concert in the Holywell Music Room, and even more so for putting up with me during the period when I was arranging the Conference. I promise her not to arrange another Conference, at least in the period until I retire.

Finally I thank all of you for attending the Conference and contributing to its success. I hope you are finding it stimulating and productive.

Contents

---

*Invited Speaker
†Speaker

BAYES/FREQUENTIST

FREQUENTIST AND BAYESIAN STATISTICS: A CRITIQUE (KEYNOTE ADDRESS)

D.R. COX

Nuffield College, Oxford OX1 1NF, UK
E-mail: david.cox@nuf.ox.ac.uk

The broad distinctions between the frequentist and Bayesian approaches to statistical inference are outlined and some brief historical background given. The advantages and disadvantages of the frequentist discussion are sketched and then two contrasting Bayesian views given. The difficulties with the notion of a flat or uninformative prior distribution are discussed.

1. Introduction

There are two broad approaches to formal statistical inference taken as concerned with the development of methods for analysing noisy empirical data and in particular as the attaching of measures of uncertainty to conclusions. The object of this paper is to summarize what is involved.

The issue is this. We have data represented collectively by y and taken to be the observed value of a vector random variable Y having a distribution determined by unknown parameters $\theta = (\psi, \lambda)$. Here ψ is a parameter of interest, often corresponding to a signal whereas λ represents such features as aspects of the data-capture procedure, background noise and so on. In this, probability is an (idealized) representation of the stability of long-run frequencies, whereas ψ aims to encapsulate important underlying physical parameters that are free from the accidents of the specific data under analysis.

How should we estimate ψ and how should we express our uncertainties about ψ?

In the following discussion we assume that the probability model correctly represents the underlying physics. This means that issues of model criticism and possible model reformulation that arise in many other applications of statistical methods can be disregarded.

2. Two Broad Avenues

There are two broad routes to an answer, both with variants.

In the first, the so-called *frequentist* approach, we continue to use probability as representing a long-run frequency. Because ψ is typically an unknown *constant*, it is not in this setting meaningful to consider a probability distribution for ψ. Instead we measure uncertainty via procedures such as confidence limits and significance levels (p-values), whose behaviour is calibrated by their appealing properties under hypothetical repetition. In that the procedure is calibrated by what happens when it is used, it is no different from other measuring devices.

In the second approach, we do aim to attach a probability distribution to the unknown ψ. For this it is essential to extend or change the notion of probability so that it is concerned with uncertainty of knowledge rather than with variability of outcome. Such an approach involves what used to be termed one of inverse probability; it is now generally termed *Bayesian.*

Note that even in those situations where there is a collection of similar parameters that can be regarded as having a probability distribution in the frequency sense it is virtually always necessary to specify their distribution in terms of hyperparameters and a part of the problem of inference is transferred to that for the hyperparameters.

3. A Simple Preliminary

The essence of the Bayesian argument is as follows. Suppose that the possible sets of data that might arise are $\mathcal{D}_1, \mathcal{D}_2, \ldots$ and that the possible explanations are $\mathcal{E}_1, \mathcal{E}_2, \ldots$, and that all events listed have meaningful probabilities. Then

$$\begin{aligned} P(\mathcal{E}_k \mid \mathcal{D}_j) &= P(\mathcal{E}_k \cap \mathcal{D}_j)/P(\mathcal{D}_j) \\ &= P(\mathcal{D}_j \mid \mathcal{E}_k)P(\mathcal{E}_k)/P(\mathcal{D}_j) \\ &\propto P(\mathcal{D}_j \mid \mathcal{E}_k)P(\mathcal{E}_k). \end{aligned}$$

The proportionality is taken as the explanations vary for specified data and the relation has the normalizing constant $1/P(\mathcal{E}_j)$. The words used for the three terms in this last equation, which has the form of an inversion equation, are respectively posterior probability, likelihood and prior probability.

Essentially the same relation holds for probability distributions and parameters in the form

$$f_{\Theta|Y}(\theta \mid y) \propto f_{Y|\Theta}(y \mid \theta) f_{\Theta}(\theta).$$

To obtain the posterior density of the parameter of interest we integrate out with respect to λ.

4. A Brief History

These issues have a long history. Laplace used what are now called Bayesian arguments with a flat prior, whereas Gauss, especially in his work on the optimality properties of the method of least squares, used frequentist concepts. The Irish algebraist Boole strongly criticized flat priors as representations of ignorance or indifference and similar points were made later in the 19th century by Venn. By the end of the 19th century inverse probability was widely regarded as unsatisfactory for inferential purposes.

Pioneering papers on the formulation of statistical inference by the geneticist and statistician R.A.Fisher, especially a major paper in 1922, laid the foundations for a frequentist theory. Some years later Neyman, then in Warsaw, and E.S.Pearson in London began a systematic development designed originally to clarify Fisher's ideas. Later, only partly because of personal friction between Fisher and Neyman, the differences became accentuated and two broad but rather ill-defined schools of frequentist inference can be discerned.

The view that probability is rational degree of belief following on from Laplace was studied in detail in a thesis by the economist John Maynard Keynes. The main work on this theme was done over the late 1920's and 1930's by the geophysicist H. Jeffreys and set out, in particular in a highly influential book *The theory of probability* in 1939. Discussion of how prior distributions might be determined in the absence of evidence have continued, the most notable work being that of J.M.Bernardo. A contrasting view of probability as a degree of belief emphasizes its personalistic character, in particular its link with personal decision making. An early influential contribution was by F.P. Ramsey. Independent major systematic developments were by de Finetti and L.J. Savage.

5. Outline of Frequentist Approach

A summary of the frequentist approach is as follows. In most situations a directly frequency-based concept of probability cannot be applied directly to the unknown of interest, ψ. Instead we introduce measures of security, p-values and confidence limits, whose interpretation is calibrated, as are other measuring instruments, by their properties when used. In this particular context, use is assessed by considering hypothetically how they perform when used repeatedly under the same conditions. The performance may be studied analytically or by computer simulation.

In particular a confidence set specifies all those values of ψ reasonably consistent with the data up to a specified level. In extreme cases, such sets may be the whole space or null, as when the data appear inconsistent with any possible value of ψ.

6. Critique of Frequentist Approach

Major advantages of the approach are that it provides a systematic approach to a wide range of statistical methods and one not requiring additional specification beyond that of the probabilistic representation of the data-generating process. It also gives a route to assessing methods that may have been suggested on relatively informal grounds.

A key problem in principle in frequentist formulations is that of ensuring that the long-run used in calibration is relevant to the analysis of the specific data being analysed. A more immediate issue in applying the ideas is that technically exact solutions are available only for a relatively limited class of situations. Usually, approximations have to be used based on asymptotic analysis and often implemented by computer simulation.

As an example of the last point, suppose that Y has a Poisson distribution with mean $(\gamma + \lambda)t_S$ and that independently Y_B has a Poisson distribution with mean λt_B, correspondingly to observation of first signal plus background and secondly to background alone. Then if interest lies in $\psi = \gamma/\lambda$ exact efficient estimation is possible based on the binomial distribution of Y given $Y + Y_B$ which is a binomial distribution with parameter

$$\frac{t_S(1+\psi)}{t_S(1+\psi) + t_B}.$$

But if interest lies in $\psi^* = \gamma$ itself no formally exact solution is available and we have to use an approximation, typically based on an asymptotic expansion. As with asymptotic expansions in other areas, some care is needed to ensure that the expansions yield good answers in the specific instance.

For example, if the amount of information on background is relatively large, that is the corresponding errors in estimating λ relatively small, the following approximate argument can be used.

For given y, let $p(y, \psi_0; \lambda)$ be the p-value for testing $\psi = \psi_0$, assuming λ is known. Let $\tilde{\lambda}$ be an unbiased estimate for λ with small variance $v(\lambda)$, all conditionally on y. Then a close approximation to the significance level adjusted for errors of estimation of λ is $p(y, \psi_0; \lambda^*)$, where

$$\lambda^* = \tilde{\lambda} - \frac{v(\tilde{\lambda}_0)\partial^2 p/\partial\lambda^2}{2\partial p/\partial\lambda}.$$

The final term has a direct generalization if λ is a vector and may be evaluated at $\lambda = \tilde{\lambda}$.

In particular for the above application with $y = 0$, the p-value for testing $\psi = \psi_0$ leading to an upper confidence limit for ψ is

$$\exp\{-\psi_0 - y_B/t_B + y_B/(2t_B^2)\}.$$

7. Critique of Bayesian Methods

To use Bayesian methods we have to extend the notion of probability so that we can specify a prior distribution for the unknown constant θ. That is we regard probability as measuring a degree of belief in an uncertain event or proposition. There are two radically different ways of doing this.

The first approach is personalistic in which $P(\mathcal{E} \mid \mathcal{I})$ denotes the degree of belief in $\mathcal{E}$ held by a specific individual, conventionally denoted by You, given information $\mathcal{I}$. There is no suggestion that two different people with the same background information will have the same probability. The emphasis is on trying to achieve self-consistency, so-called coherency, in Your probability assessments. The second approach involves a notion of rational degree of belief and, commonly although not necessarily, an attempt, following Laplace, to address the question of assessing the evidence in a specific set of data by using a prior expressing a notion of indifference or ignorance in order to focus attention on the data.

These are to be regarded as two very different approaches and the following comments address them separately.

8. Personalistic Theory

This approach has the ambitious aim, in particular, of introducing into the quantitative discussion uncertain information of a more general kind than is represented by statistical data in the narrow sense. In theoretical discussion it is usually set out as part of a theory of personal decision making. Suppose, in order to simplify the discussion, that there is available a source able to produce events with any specified probability p. Then Your probability of $\mathcal{E}$ is a value of p such that you are indifferent as between

- a valuable prize if $\mathcal{E}$ is true and zero if $\mathcal{E}$ is false
- the same prize if an event with agreed probability p occurs and zero otherwise

A certain kind of consistency of behaviour can be shown to imply that the laws of probability theory hold. Note though that this is not a theory of empirical behaviour based on what people actually do but rather a specification of how they would have to behave to be self-consistent.

A very major difficulty with this as a basis for the public discussion of scientific evidence is that it treats personal intuition as on the same basis as evidence from hard data. More explicitly it treats all probabilities of, say, 0.5 as on an equal footing, whether they are based on careful stable measurements of frequency or on the most transitory of personal judgements. In some situations prior distributions based on a careful summary of expert judgement may be used quantitatively, but then scrutiny of their evidence-base is crucial.

This is not to deny the relevance of personal judgement for the individual decision-maker.

9. Probability as Rational Degree of Belief

While the notion of rational degree of belief can certainly be taken more broadly, for the most part it is associated with the use of priors that are in some sense flat, which aim to represent little or no prior information and which therefore induce posterior distributions having the same objective as frequentist methods, i.e. of summarizing what it is reasonable to learn from data plus assumptions about the structure of the data-generating process.

It is generally accepted from various philosophical standpoints that the notion of representing ignorance as such by a flat prior is treacherous, although in some fields the use of relatively flat priors as noncommittal is quite widespread. The following points arise

- if θ has a flat, i.e. effectively uniform, prior then e^{θ} has an exponential distribution, so that choice of functional form of parameters would be important
- for a one dimensional parameter the Jeffreys prior, essentially uniform in a parameterization for which the Fisher information is constant, leads to a posterior distribution having very good frequentist properties
- flat priors for parameters with a large number of dimensions may give clearly unacceptable answers.

J.M. Bernardo has developed a systematic theory of reference priors. This is based on the notion of finding a prior weighting function that maximizes the expected discrepancy between prior knowledge and prefect knowledge obtained by a specified type of replication of the system. When the parameter space is finite it produces the maximum entropy prior of E.T. Jaynes and for a one-dimensional parameter the Jeffreys prior. Some difficulties are that when there are nuisance parameters

- finding the prior weight is often complicated
- the nuisance parameters have to be arranged in sequence of importance, even though none of them is of intrinsic interest
- if the parameter of interest changes the whole prior structure may change
- if the sampling rule or design changes the prior will in general change
- it is emphasized that the prior weights are not to be thought of as prior probabilities, raising a question-mark over the interpretation of the posterior
- many of the formal simplifications arising from all calculations being probabilistic are lost.

In general reference priors have some good frequentist properties but except in one-dimensional problems it is unclear that they have any special merit in that regard.

10. Concluding Remarks

In conclusion, the following points arise:

- formal inferential aspects are often a relatively small part of statistical analysis
- carefully used, the frequentist approach yields broadly applicable if sometimes clumsy answers
- in simple problems specially chosen prior distributions yield essentially the same answer
- in multiparameter problems flat priors can yield very bad answers
- injection of further information quantitatively through an informative prior may be helpful but scrutiny of the evidence base is essential.

These issues have a very extensive literature. Traditional accounts of the two frequentist viewpoints are by Fisher[1] and Neyman and Pearson[2] and of the two Bayesian approaches by Jeffreys[3] and Savage[4]. An introductory comparative account is by Barnett[5] and a systematic discussion by Cox and Hinkley[6] and Cox[7]. The notion of reference priors is developed in detail by Bernardo[8].

References

1. Fisher, R.A., *The logic of scientific inference.* Edinburgh: Oliver and Boyd (1956).
2. Neyman, J. and Pearson, E.S., *Joint statistical papers of J.Neyman and E.S.Pearson.* Cambridge University Press on behalf of Biometrika Trustees (1967).
3. Jeffreys, H., (1939 and subsequent editions). *The theory of probability.* Oxford University Press (1939).
4. Savage, L.J., *Foundations of statistics.* New York: Wiley (1964).
5. Barnett, V.D., *Comparative statistical inference.* 3rd edition. Chichester: Wiley (1999).
6. Cox, D.R. and Hinkley, D.V., *Theoretical statistics.* London: Chapman and Hall (1974).
7. Cox, D.R., *Principles of statistical inference.* To appear (2006).
8. Bernardo, J.M., Reference priors. To appear (2006).

GENERALIZED FREQUENTIST METHODS FOR CALCULATING p-VALUES AND CONFIDENCE INTERVALS

LUC DEMORTIER

The Rockefeller University, 1230 York Avenue, New York, NY 10021, USA
E-mail: luc@fnal.gov

Generalized frequentism addresses problems that are not exactly solvable using conventional frequentism. Such problems include the calculation of p-values and confidence intervals when nuisance parameters are present, or when interest is focused on a complicated function of the parameters of the model under consideration. Although generalized frequentist methods are based on *exact* probability statements, they do not necessarily yield coverage in the conventional sense. However, simulation studies indicate that these methods tend to overcover, and often surpass other available methods in terms of test power or interval length.

1 Introduction

An often challenging component of frequentist calculations is the elimination of nuisance parameters. There seems to be no method that is generally applicable and at the same time theoretically guaranteed to preserve exact coverage in all cases. However, a couple of likelihood-based methods are known to behave reasonably well in many situations. In the first method, called profiling, the likelihood is *maximized* with respect to the nuisance parameters. The second method, marginalization, *integrates* the likelihood over these parameters. Whichever technique is chosen, its coverage properties for the problem at hand must then be verified a posteriori.

This paper aims to present a third approach, known as generalized frequentism.[1,2] Its strategy is to *extend* the conventional definitions of p-values and confidence intervals in such a way that statistical problems involving nuisance parameters can be solved "exactly", i.e. using exact probability statements. The resulting *generalized* p-values and confidence intervals tend to behave well with respect to the usual frequentist definitions, hence their interest.

2 Generalized p-Values

Let X be a random variable with density $f(x \mid \theta, \nu)$, where θ is the parameter of interest and ν is a nuisance parameter. We are interested in testing:

$$H_0 : \theta \leq \theta_0 \text{ versus } H_1 : \theta > \theta_0.$$

The usual way of solving this problem is to find a *test statistic* $T(X)$, defined as a function of the data X which does not depend on unknown parameters, whose distribution is free of unknown nuisance parameters, and which is stochastically increasing with θ, i.e. such that the probability $\mathbb{P}\mathrm{r}(T(X) \geq t \mid \theta)$ increases with θ for all t. One then calculates the p-value:

$$p = \mathbb{P}\mathrm{r}\big[T(X) \geq T(x) \mid H_0\big],$$

where x is the observed value of X. A small p-value indicates that the observed x does not support H_0.

There are many problems for which test statistics as defined above simply do not exist. In these cases a solution can be found by extending the definition of test statistic to that of a *generalized test variable*, which is a function $T(X, x, \theta, \nu)$ of the random variable X, its observed value x (treated as a constant), and the parameters θ and ν, such that the following requirements are satisfied:

1. $T(x, x, \theta, \nu)$ does not depend on θ or ν;
2. The distribution of $T(X, x, \theta_0, \nu)$ under H_0 is free of ν;
3. Given x and ν, $\mathbb{P}\mathrm{r}\big[T(X, x, \theta, \nu) \geq t \mid \theta\big]$ is a monotonic function of θ.

The generalized p-value based on $T(X, x, \theta, \nu)$ is defined similarly to a conventional p-value:

$$p = \mathbb{P}\mathrm{r}\big[T(X, x, \theta, \nu) \geq T(x, x, \theta, \nu) \mid H_0\big].$$

We emphasize that in this probability statement, only X is considered as a random variable, whereas the observed value x is held constant. Because of the way $T(X, x, \theta, \nu)$ is defined, this p-value is free of unknown parameters and allows the desired interpretation that small p corresponds to lack of support for H_0. However, although p is based on an exact probability statement, the coverage probability $\mathbb{P}\mathrm{r}(p \leq \alpha)$ may depend on nuisance parameters and needs to be checked explicitly.

There exists no general method that will systematically yield all possible generalized test variables for a given problem. However, an easy and useful recipe is available.[3,4] To formulate it we consider a slightly more general problem involving k unknown parameters $\alpha_1, \alpha_2, \ldots, \alpha_k$, and where the parameter of interest θ is a function of the α_i. We make the following assumptions:

1. There exists a set of observable statistics, $(X_1, X_2, \ldots, X_k)$, that is equal in number to the number of unknown parameters α_i.
2. There exists a set of invertible pivots[a], $(V_1, V_2, \ldots, V_k)$, relating the statistics (X_i) to the unknown parameters (α_i).

The recipe is then as follows:

1. By writing the parameter of interest, θ, in terms of the parameters α_i, express θ in terms of the statistics X_i and the pivots V_i.
2. Replace the X_i by their observed values x_i and subtract the result from θ.

For a simple application of this recipe, consider a sample $\{Y_1, Y_2, \ldots, Y_n\}$ drawn from Gauss(μ, σ), a Gaussian distribution with mean μ and width σ, both unknown. We are interested in the ratio $\theta \equiv \sigma/\mu$. The sample mean and standard deviation are a set of minimal sufficient statistics for μ and σ:

$$X_1 \equiv \frac{1}{n}\sum_{i=1}^{n} Y_i \quad \text{and} \quad X_2 \equiv \sqrt{\frac{1}{n-1}\sum_{i=1}^{n}(Y_i - X_1)^2}.$$

The random variables

$$V_1 \equiv \frac{X_1 - \mu}{\sigma/\sqrt{n}} \quad \text{and} \quad V_2 \equiv \frac{n\,X_2^2}{\sigma^2}$$

relate the statistics (X_1, X_2) to (μ, σ), and have distributions free of unknown parameters:

$$V_1 \sim \text{Gauss}(0,1) \quad \text{and} \quad V_2 \sim \chi^2_{n-1}.$$

Applying the recipe yields a generalized test variable, which can be written in terms of (V_1, V_2) or (X_1, X_2):

$$T \equiv \theta - \frac{\sqrt{n}\,x_2}{x_1\sqrt{V_2} - x_2\,V_1} = \theta - \frac{\sigma}{x_1 X_2/x_2 + \mu - X_1}.$$

The first expression for T shows that its distribution under H_0 is free of unknown parameters (the observed values x_1 and x_2 being treated as constants), whereas the second expression shows that the observed value of T is zero. The property of stochastic monotonicity is somewhat harder to verify.

2.1 *Application to Poisson Significance Tests*

For a slightly more complex application we turn to a common problem in high-energy physics. Consider a Poisson process consisting of a background with strength b superimposed on a signal with strength s:

$$f_N(n; b+s) = \frac{(b+s)^n}{n!}\, e^{-b-s}.$$

The nuisance parameter b is determined from a Gaussian measurement x:

$$f_X(x; b) = \frac{e^{-\frac{1}{2}\left(\frac{x-b}{\Delta b}\right)^2}}{\sqrt{2\pi}\,\Delta b}.$$

It is assumed that $b \geq 0$ but that, due to resolution effects, x can take both positive and negative values. We are interested in testing $H_0 : s = 0$ versus $H_1 : s > 0$. This problem has two parameters, b and s, two statistics, N and X, and two pivots:

$$V_1 = \frac{X - b}{\Delta b} \quad \text{and} \quad V_2 = F_N(N; b+s),$$

where $F_N(N; b+s)$ is the cumulative Poisson distribution with mean $b+s$. The pivot V_1 has a Gaussian distribution with mean 0 and width 1. Due to the discreteness of the Poisson distribution however, V_2 is only an approximate pivot. This can be remedied by introducing a uniform random variable U between 0 and 1, and replacing N by $Y \equiv N + U$ for the purpose of applying the recipe of section 2. This is nothing more than a mathematical artifice that provides us with an invertible pivot involving N. Indeed, the cumulative distribution of Y, say $F_Y^+(y, b+s)$, is an invertible pivot with a uniform distribution between 0 and 1. Let $G^+(Y, V)$ be the inverse of that pivot, i.e. $G^+(y, V) = \mu$ if and only if $V = F_Y^+(y, \mu)$. The generalized test variable is then:

$$T = s + \left(x - V_1\,\Delta b\right) - G^+(n, V_2),$$

and the generalized p-value is:

$$p = \mathbb{P}\mathrm{r}(T \geq 0 \,|\, s = 0).$$

From the definition of T it can be seen that this p-value is simply the probability for the difference between a Gauss$(x, \Delta b)$ and a Gamma$(n, 1)$ random variable to be positive. Analytically, the p-value equals the tail area of a convolution between these

[a] Pivots are random variables V_i that depend on the data X_j and the parameters α_k, but whose joint distribution is free of unknown parameters. They are called invertible if, for fixed values of the X_j, the mapping $(\alpha_k) \to (V_i)$ is invertible.

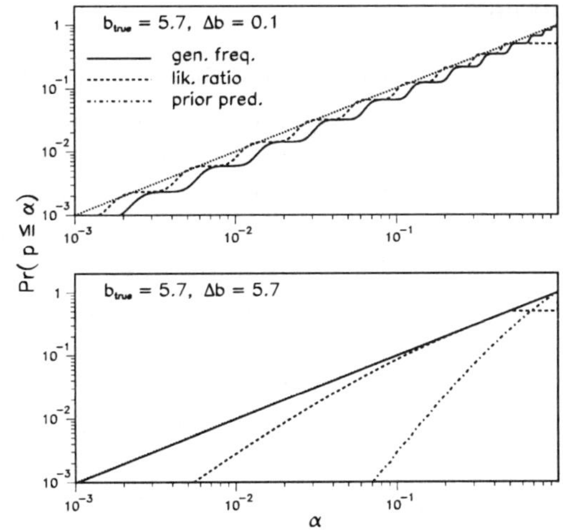

Figure 1. Comparative coverage of p-values. The dotted line represents exact coverage. In the top plot, the coverage of the prior-predictive p-value is indistinguishable from that of the generalized frequentist p-value. In the bottom plot, the coverage of the generalized frequentist p-value is indistinguishable from exact coverage.

random variables; for $n > 0$ it is given by:

$$p = \int_0^{+\infty} dt \, \frac{t^{n-1} \, e^{-t}}{\Gamma(n)} \, \frac{1 + \mathrm{erf}\left(\frac{x-t}{\sqrt{2}\,\Delta b}\right)}{2},$$

and we define p to be 1 when $n = 0$. It is instructive to compare this p-value with two other methods. The first one is quite popular in high-energy physics, and consists in calculating the p-value assuming a fixed value for the nuisance parameter b, and then to average this p-value over $f_X(x; b)$, considered as a prior distribution for b. This yields the so-called "prior-predictive p-value" p_{pp}, which, for $n > 0$, is:

$$p_{pp} = \int_0^{+\infty} dt \, \frac{t^{n-1} \, e^{-t}}{\Gamma(n)} \, \frac{1 + \mathrm{erf}\left(\frac{x-t}{\sqrt{2}\,\Delta b}\right)}{1 + \mathrm{erf}\left(\frac{x}{\sqrt{2}\,\Delta b}\right)}.$$

The second method starts from the likelihood ratio statistic:

$$\lambda = \frac{\sup\limits_{s=0,\, b\geq 0} f_N(n; b+s)\, f_X(x; b)}{\sup\limits_{s\geq 0,\, b\geq 0} f_N(n; b+s)\, f_X(x; b)}$$

For large values of b, the distribution of $-2\ln\lambda$ under H_0 is $\frac{1}{2}\chi_0^2 + \frac{1}{2}\chi_1^2$, i.e. it assigns half a unit of probability to the singleton $\{-2\ln\lambda = 0\}$, whereas the other half is distributed as a chisquared with one degree of freedom over $0 < -2\ln\lambda < +\infty$. We then *define* the likelihood ratio p-value as the appropriate tail area of this distribution. For small values of b this is obviously an approximation, but not a bad one, in the sense that the frequentist validity of the p-value appears to be maintained: $\mathbb{P}\mathrm{r}(p \leq \alpha) \leq \alpha$. Figure 1 compares the coverage probability $\mathbb{P}\mathrm{r}(p \leq \alpha)$ of the three p-values just discussed, as a function of the significance level α, for a simple numerical example. The coverage calculation fluctuates both n and x. For small values of the background uncertainty Δb, the likelihood ratio p-value is somewhat better than the other two, but for large Δb the generalized frequentist p-value is clearly superior.

3 Generalized Confidence Intervals

A standard method for constructing confidence intervals is based on pivots. Let $Q(X, \theta)$ be a pivot for a random variable X with distribution $F_X(x; \theta)$, and let S_α be a subset of the sample space of Q such that

$$\mathbb{P}\mathrm{r}(Q(X, \theta) \in S_\alpha) \;=\; \alpha.$$

Note that the probability in this equation is unambiguously determined since the distribution of Q does not depend on unknown parameters. Given an observed value x for X, a $100\alpha\%$ confidence interval for θ is then:

$$C_\alpha \;=\; \{\theta : \;\; Q(x, \theta) \in S_\alpha\}$$

In problems for which a conventional pivot is not available, one can try to construct a *generalized* pivot, i.e. a function $Q(X, x, \theta, \nu)$ of the random variable X, its observed value x, the parameter of interest θ, and the nuisance parameter ν, such that the following requirements are satisfied:

1. $Q(x, x, \theta, \nu)$ does not depend on ν;
2. The distribution of $Q(X, x, \theta, \nu)$ is free of (θ, ν).

Generalized confidence intervals can then be defined similarly to conventional ones, but using $Q(X, x, \theta, \nu)$ instead of $Q(X, \theta)$.

As with p-values, there is no systematic method for generating all possible generalized pivots for a problem, but a simple recipe is available.[3,4] It is based on the same assumptions as those listed in section 2, and the recipe itself is almost identical to the one used to obtain generalized test variables. The only difference is step 2, which becomes:

2. Replace the X_i by their observed values x_i.

In other words, given a generalized test variable $T(X, x, \theta, \nu)$, the corresponding generalized pivot is obtained as $Q(X, x, \theta, \nu) = \theta - T(X, x, \theta, \nu)$.

3.1 Application to Poisson Upper Limits

Suppose that we observe a Poisson event count X_1 with mean $b + \epsilon\sigma$, where b is a background, ϵ a sensitivity factor, and σ a cross section of interest:

$$X_1 \sim \text{Poisson}(b + \epsilon\,\sigma).$$

Information about b and ϵ are assumed to come from two auxiliary measurements:

$$X_2 \sim \text{Poisson}(c\,b), \qquad X_3 \sim \text{Poisson}(\tau\,\epsilon),$$

where c and τ are known constants. Applying the above recipe yields the following generalized pivot for σ:

$$Q = \frac{\tau\left[G^-(x_1, V_1) - G^-(x_2, V_2)/c\right]}{G^-(x_3, V_3)},$$

where, similarly to the G^+ introduced in section 2.1, G^- is the inverse of the pivot defined by the cumulative distribution of $X - U$, X being a Poisson variate and U a uniform one.[b] The V_i quantities are independent uniform random variables, and the x_i are the observed values of the corresponding X_i.

Suppose now that we wish to calculate upper limits on σ. It is straightforward to verify that the "observed" value of Q is the parameter of interest σ. Therefore, upper limits on σ are obtained by calculating the corresponding quantiles of the distribution of Q. A numerical example of the coverage of these upper limits is shown in Figure 2, together with a reference Bayes calculation. There is slight undercoverage at high σ values.

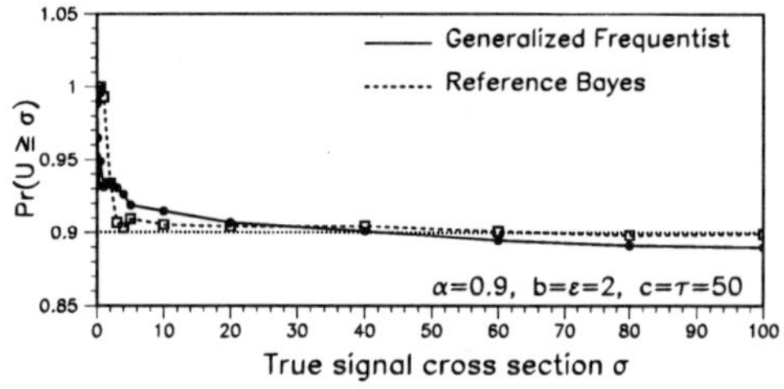

Figure 2. Coverage of upper limits U on the cross section of a signal process, as a function of the true value σ of that cross section. The nominal uncertainties on the background b and the efficiency ϵ are 10%. Solid: generalized frequentist; dashes: reference Bayes.

4 Summary

Generalized frequentist methods allow one to calculate significances and confidence intervals in a wide variety of situations involving nuisance parameters.

In problems with continuous sample spaces, these methods are based on exact probability statements but do not have a conventional frequency interpretation. Nevertheless, their conventional frequentist properties appear to be very good. In fact, Hannig *et al.*[4] have shown that under some general conditions, generalized confidence intervals for scalar or vector parameters have proper frequentist coverage, at least asymptotically.

Although the current literature on generalized frequentism does not appear to treat problems with discrete sample spaces, we have described how these can be solved by introducing a randomization scheme.

Using a simple Poisson example, we have shown that generalized frequentist methods compare favorably to other methods of eliminating nuisance parameters, such as likelihood ratio and Bayes.

References

1. Kam-Wah Tsui and Samaradasa Weerahandi, "Generalized *p*-values in significance testing of hypotheses in the presence of nuisance parameters," J. Amer. Statist. Assoc. **84**, 602 (1989). Erratum: ibid. **86**, 256 (1991).
2. Samaradasa Weerahandi, "Generalized confidence intervals," J. Amer. Statist. Assoc. **88**, 899 (1993). Erratum: ibid. **89**, 726 (1994).
3. Hari K. Iyer and Paul D. Patterson, "A recipe for constructing generalized pivotal quantities and generalized confidence intervals," Colorado State University Department of Statistics Technical Report 2002/10; also at `http://www.stat.colostate.edu/research/2002_10.pdf`.
4. Jan Hannig, Hari Iyer, and Paul L. Patterson, "On fiducial generalized confidence intervals," Colorado State University Department of Statistics Technical Report 2004/12; also at `http://www.stat.colostate.edu/~hari/fiducial/fgpq.pdf`.

[b] When applying generalized frequentist methods to discrete distributions, the results depend slightly on the randomization scheme. The use of G^+ in section 2.1 was dictated by the desire to maintain coverage, even though $G^+(x, V)$ is not defined when $x = 0$. In section 3.1 it seems more important to use a function that *is* defined at $x = 0$, which is the case for $G^-(x, V)$.

BAYESIAN REFERENCE ANALYSIS

LUC DEMORTIER
The Rockefeller University, 1230 York Avenue, New York, NY 10021, USA
E-mail: luc@fnal.gov

As a carefully thought-out attempt to develop the objective side of Bayesian inference, reference analysis provides procedures for point and interval estimation, hypothesis testing, and the construction of objective posterior distributions. For physicists, the interest of these procedures lies in their very general applicability, their invariance under reparametrization, their coherence, and their good performance under repeated sampling.

1 Introduction

One aspect that distinguishes experimental inference in physics from that in other sciences is the objective randomness of quantum processes. As a result, statistical models for quantum phenomena are exact, supporting a strict frequentist analysis of their measurement. Nevertheless, Caves *et al.*[1] have brilliantly motivated a subjective Bayesian interpretation of quantum probabilities, whose form depends on the information available to the observer but is otherwise fully prescribed by a fundamental law. When dealing with actual measurements however, no fundamental law constrains their analysis, summary and report, so that some other objective method must be found.

Ideally, such a method should be very general, applicable to all kinds of measurements regardless of the number and type of parameters and data involved. It should make use of *all* available information, and coherently so, in the sense that if there is more than one way to extract all relevant information from data, the final result will not depend on the chosen way. The desiderata of generality, exhaustiveness and coherence are satisfied by Bayesian procedures, but that of objectivity is more problematic due to the Bayesian requirement of specifying prior probabilities in terms of degrees of belief. Reference analysis[2], an objective Bayesian method developed over the past twenty-five years, solves this problem by replacing the question "what is our prior degree of belief?" by "what would our posterior degree of belief be, if our prior knowledge had a minimal effect, relative to the data, on the final inference?"

In addition to an objective method for specifying priors, reference analysis provides techniques to summarize posterior distributions in terms of point estimates and intervals, and to test precise hypotheses against vague alternatives, a notoriously subtle problem. All these techniques are based on information theory, and in particular on the central concept of intrinsic discrepancy between two probability distributions. This concept is introduced in section 2 and applied to the definition of reference priors in section 3. Section 4 describes the extraction of intrinsic point and interval estimates from posterior distributions.

Due to space limitations, the development of the paper is rather conceptual, with few details in the calculations. The interested reader is encouraged to consult the references, especially Bernardo[2].

2 Intrinsic Discrepancy and Missing Information

The intrinsic discrepancy between two probability densities p_1 and p_2 is defined as:

$$\delta\{p_1,p_2\} \;=\; \min\{\,\kappa\{p_1\,|\,p_2\},\;\kappa\{p_2\,|\,p_1\}\,\}, \qquad (1)$$

$$\text{where}\quad \kappa\{p_i\,|\,p_j\} \;\equiv\; \int dx\; p_j(x)\,\ln\frac{p_j(x)}{p_i(x)} \qquad (2)$$

is the Kullback-Leibler divergence between p_i and p_j. The intrinsic discrepancy $\delta\{p_1,p_2\}$ is symmetric, non-negative, and vanishes if and only if $p_1(x) = p_2(x)$ almost everywhere. It is invariant under one-to-one transformations of x, and is information-additive: the discrepancy for a set of n independent observations is n times the discrepancy for one observation. A simple interpretation of $\delta\{p_1,p_2\}$ is as a measure, in natural information units, of the minimum amount of information that one observation may be expected to provide in order to discriminate between p_1 and p_2. Another interpretation is as the minimum expected log-likelihood ratio in favor of the probability model that generates the data.

Suppose now that we have a parametric model

for some data x:

$$\mathcal{M} \equiv \{p(x \mid \theta),\ x \in \mathcal{X},\ \theta \in \Theta\},$$

and consider the joint probability density of x and θ, $p(x, \theta) = p(x \mid \theta)\, p(\theta)$, where $p(\theta)$ is a prior for θ. Relative to the product of marginals $p(x)p(\theta)$, the joint density captures in some sense the information carried by x about θ. This suggests defining the expected intrinsic information $I\{p(\theta) \mid \mathcal{M}\}$, from one observation of $\mathcal{M}$ about the value of θ when the prior density is $p(\theta)$, as:

$$I\{p(\theta) \mid \mathcal{M}\} \;=\; \delta\{p(x, \theta),\ p(x)\, p(\theta)\}, \tag{3}$$

where $p(x) = \int d\theta\, p(x \mid \theta)\, p(\theta)$. According to this definition, the stronger the prior knowledge described by $p(\theta)$, the smaller the information the data may be expected to provide, and vice-versa. In the limit where $p(\theta)$ is a delta function, $I\{p(\theta) \mid \mathcal{M}\} = 0$

Next, consider the intrinsic information about θ, $I\{p(\theta), \mathcal{M}^k\}$, which could be expected from making k independent observations from $\mathcal{M}$. As k increases, the true value of θ would become precisely known. Thus, as $k \to \infty$, $I\{p(\theta), \mathcal{M}^k\}$ measures the amount of *missing information* about θ which corresponds to the prior $p(\theta)$.

3 Reference Priors

Let $\mathcal{P}$ be a class of sufficiently regular priors that are compatible with whatever initial information is available about the value of θ. The reference prior is defined to be that prior function $\pi(\theta) = \pi(\theta \mid \mathcal{M}, \mathcal{P})$ which maximizes the missing information about the value of θ within the class $\mathcal{P}$. The limiting procedure used to define the missing information requires some care in the calculation of $\pi(\theta)$. Formally, one introduces an increasing sequence of subsets Θ_i of the parameter space Θ, such that $\lim_{i\to\infty} \Theta_i = \Theta$ and $\int_{\Theta_i} \pi(\theta)\; d\theta < \infty$. The reference prior $\pi(\theta)$ is then defined as satisfying:

$$\lim_{k\to\infty} \left[I\{\pi_i \mid \mathcal{M}^k\} - I\{p_i \mid \mathcal{M}^k\} \right] \;\geq\; 0 \qquad \text{for all } \Theta_i, \text{ for all } p \in \mathcal{P}, \tag{4}$$

where $\pi_i(\theta)$ and $p_i(\theta)$ are the renormalized restrictions of $\pi(\theta)$ and $p(\theta)$ to Θ_i.

If the parameter space is finite and discrete, $\Theta = \{\theta_1, \ldots, \theta_m\}$, the missing information is simply the entropy of the prior distribution, $-\sum_{i=1}^m p(\theta_i) \ln p(\theta_i)$, and one recovers the prior proposed by Jaynes for this case. If the parameter is continuous and one-dimensional, and regularity conditions that guarantee asymptotic normality are satisfied, then the reference prior is Jeffreys' prior:

$$\pi(\theta) \propto i(\theta)^{1/2},$$

$$\text{where} \quad i(\theta) = -\int_{\mathcal{X}} dx\; p(x \mid \theta)\, \frac{\partial^2}{\partial \theta^2} \ln p(x \mid \theta). \tag{5}$$

Note that in the definition of reference priors, the limit $k \to \infty$ is *not* an approximation, but an essential part of the definition, since the reference prior maximizes the *missing* information, which is the expected discrepancy between prior knowledge and *perfect* knowledge. A practical advantage of this limiting procedure is that it ensures that reference priors only depend on the asymptotic behavior of the model, thereby greatly simplifying their derivation.

It can be shown that reference priors are independent of sample size, compatible with sufficient statistics (meaning that their form does not depend on whether the model is or is not expressed in terms of sufficient statistics), and consistent under reparametrization (i.e. if ϕ is a one-to-one transformation of θ, then their reference posterior densities are related by $\pi(\phi \mid x) = \pi(\theta \mid x)\, |d\theta/d\phi|$).

Finally, it is important to emphasize that reference priors do not represent subjective belief and should not be interpreted as prior probability distributions (in fact, they are often improper). Only reference *posteriors* have a probability interpretation.

3.1 Treatment of Nuisance Parameters

Suppose the statistical model is $p(x \mid \theta, \lambda)$, with θ the parameter of interest and λ a nuisance parameter. We now need a joint reference prior $\pi(\theta, \lambda)$. The algorithm is sequential:

1. Hold θ constant and apply the one-parameter reference algorithm to obtain the conditional reference prior $\pi(\lambda \mid \theta)$.
2. Derive the one-parameter integrated model:
$$p(x \mid \theta) \;=\; \int_\Lambda d\lambda\; p(x \mid \theta, \lambda)\; \pi(\lambda \mid \theta),$$
where Λ is the parameter space for λ.
3. Apply the one-parameter reference algorithm again, this time to $p(x \mid \theta)$, and obtain the marginal reference prior $\pi(\theta)$.
4. Set $\pi(\theta, \lambda) \;=\; \pi(\lambda \mid \theta)\; \pi(\theta)$.

Note that step 2 will not work if $\pi(\lambda \mid \theta)$ is improper ($p(x \mid \theta)$ will not be normalizable). The solution is to introduce a sequence $\{\Lambda_i\}_{i=1}^{\infty}$ of subsets of Λ that converges to Λ and such that $\pi(\lambda \mid \theta)$ is integrable over each Λ_i. The integration at step 2 is then performed over Λ_i instead of Λ. This procedure results in a sequence of posteriors $\{\pi_i(\theta \mid x)\}_{i=1}^{\infty}$ which converges to the desired reference posterior.

The above algorithm is easily generalized to any number of parameters. However, its sequential character requires that the parameters be ordered. In most applications the order does not affect the result, but there are exceptions. Different orderings may then be used as part of a robustness analysis.

Within a *single* model it is in principle possible to have as many reference priors as there are potential parameters of interest. Indeed, there is no reason for a setup that maximizes the missing information about a parameter θ to be identical to a setup that maximizes the missing information about a parameter η, unless η is a one-to-one function of θ.

3.2 *Example: a Cross Section Measurement*

We illustrate the construction of reference priors with a common problem in high energy physics, that of extracting a cross section σ from an observed number of events n. The latter is assumed to have a Poisson distribution with a mean of the form $b + \epsilon\sigma$, where the sensitivity factor ϵ and the background b are nuisance parameters. The model is:

$$p(n \mid \sigma, \epsilon, b) \;=\; \frac{(b+\epsilon\sigma)^n}{n!}\, e^{-b-\epsilon\sigma}. \tag{6}$$

Note that σ, ϵ, and b are not identifiable from a given n. This problem is usually addressed by using information from calibration data or simulation studies to form a proper, subjective prior for ϵ and b, say $\pi(\epsilon, b)$. We must therefore find the conditional reference prior $\pi(\sigma \mid \epsilon, b)$. If ϵ and b were exactly known, the reference prior for σ would simply be Jeffreys' prior. From the Fisher information for σ:

$$\Sigma_{\sigma\sigma} \;=\; \mathrm{E}\left[-\frac{\partial^2}{\partial\sigma^2}\ln p(n \mid \sigma, \epsilon, b)\right] \;=\; \frac{\epsilon^2}{b+\epsilon\sigma}, \tag{7}$$

this Jeffreys' prior is calculated to be:

$$\pi_{\mathcal{J}}(\sigma \mid \epsilon, b) \;\propto\; \frac{\epsilon}{\sqrt{b+\epsilon\sigma}}. \tag{8}$$

However, this is *not* the reference prior for this problem, i.e. the prior that would be obtained by strict application of equation (4). Although the σ dependence of $\pi_{\mathcal{J}}$ is correct, its ϵ dependence is not, and this matters because $\pi_{\mathcal{J}}$ is improper and ϵ is an unknown parameter. As shown in Sun and Berger[3], the correct reference prior is obtained by renormalizing the above prior using a sequence of nested compact sets for σ. A natural choice for these sets is $[0, u]$, with $u > 0$. Normalizing the above prior over such a set yields:

$$\pi_u(\sigma \mid \epsilon, b) \;=\; \frac{\epsilon}{\sqrt{b+\epsilon\sigma}}\;\frac{\mathbf{1}(u \geq \sigma)}{2\sqrt{b+\epsilon u} \;-\; 2\sqrt{b}},$$

where $\mathbf{1}(u \geq \sigma)$ is 1 if $u \geq \sigma$ and 0 otherwise. The correct conditional reference prior is then:

$$\pi(\sigma \mid \epsilon, b) \;=\; \lim_{u\to\infty}\frac{\pi_u(\sigma \mid \epsilon, b)}{\pi_u(\sigma_0 \mid \epsilon_0, b_0)} \;\propto\; \sqrt{\frac{\epsilon}{b+\epsilon\sigma}},$$

with $(\sigma_0, \epsilon_0, b_0)$ any fixed point. Although this prior is still improper, its ϵ dependence is different from that of equation (8).

We can now write down the reference posterior when σ is the parameter of interest:

$$\pi(\sigma \mid n) \propto \int_0^\infty d\epsilon \int_0^\infty db\; \frac{(b+\epsilon\sigma)^n\, e^{-b-\epsilon\sigma}}{n!}\;\frac{\sqrt{\epsilon}\,\pi(\epsilon, b)}{\sqrt{b+\epsilon\sigma}}. \tag{9}$$

An important aspect of reference posteriors is their behavior under repeated sampling. To test this, we calculate an upper limit U on σ, assuming a product of gamma densities for the subjective prior $\pi(\epsilon, b)$:

$$\pi(\epsilon, b) \;=\; \frac{\tau(\tau\epsilon)^{x-1/2}\, e^{-\tau\epsilon}}{\Gamma(x+1/2)}\;\frac{c(cb)^{y-1/2}\, e^{-cb}}{\Gamma(y+1/2)}. \tag{10}$$

As we are dealing with a mixture of subjective and objective priors, some care is needed in specifying the ensemble with respect to which the coverage of U is to be calculated. Datta and Sweeting[4] suggest to *average* the coverage with respect to the subjective components of the prior. An example of calculation based on this prescription is shown in Figure 1.

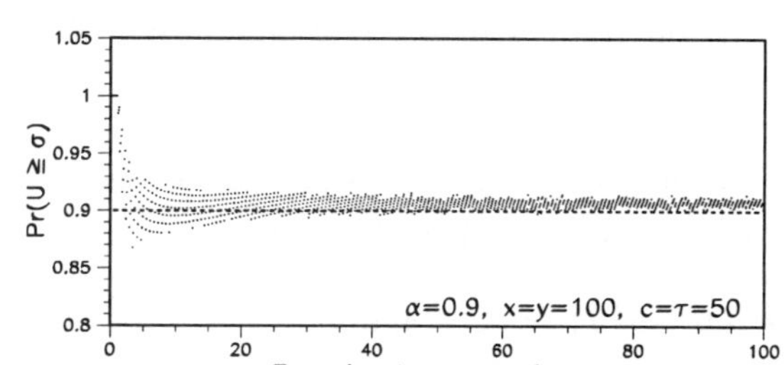

Figure 1. Coverage of 90% credibility level reference Bayes upper limits on a signal cross section σ, as a function of the true value of that cross section.

The coverage appears to converge asymptotically towards the credibility level. Although this behaviour is typical of all sufficiently regular priors, in many cases the convergence is faster when a reference prior is used.

4 Intrinsic Estimation and Testing

It is well known that the Bayesian outcome of a problem of inference is precisely the full posterior distribution for the parameter of interest. However, it is often useful and sometimes even necessary to *summarize* the posterior distribution by providing a measure of location and quoting regions of given posterior probability content.

The typical Bayesian approach formulates point and interval estimation as decision problems. Suppose that $\hat{\theta}$ is an estimate of the parameter θ, whose true value θ_t is unknown. One specifies a loss function $\ell(\hat{\theta}, \theta_t)$, which measures the consequence of using the model $p(x \,|\, \hat{\theta})$ instead of the true model $p(x \,|\, \theta_t)$. The Bayes estimator θ_b of θ minimizes the corresponding posterior loss:

$$\theta_b(x) \;=\; \arg \min_{\hat{\theta}\in\Theta} \int_\Theta d\theta \; \ell(\hat{\theta}, \theta) \; p(\theta \,|\, x).$$

In physics, interest usually focuses on the actual mechanism that governs the data. Therefore we need point and interval estimates that are invariant under one-to-one transformations of the parameter and the data (including reduction to sufficient statistics). A loss function that will deliver such an estimate is the intrinsic discrepancy: $\ell(\hat{\theta}, \theta_t) = \delta\{p(x \,|\, \hat{\theta}), p(x \,|\, \theta_t)\}$. Its reference posterior expectation is:

$$d(\hat{\theta} \,|\, x) \;=\; \int_\Theta d\theta \; \delta\{p(x \,|\, \hat{\theta}), p(x \,|\, \theta)\} \; \pi_\delta(\theta \,|\, x), \quad (11)$$

where $\pi_\delta(\theta \,|\, x)$ is the reference posterior when the intrinsic discrepancy is the parameter of interest.

The *intrinsic estimator* of θ minimizes $d(\hat{\theta} \,|\, x)$:

$$\theta^\star(x) \;=\; \arg \min_{\hat{\theta}\in\Theta} d(\hat{\theta} \,|\, x), \quad (12)$$

and an intrinsic α-credible region for θ is a subset $R^\star_\alpha$ of the parameter space Θ such that:

$$\int_{R^\star_\alpha} d\theta \; \pi(\theta \,|\, x) \;=\; \alpha, \quad \text{and}$$

$$\text{for all } \theta \in R^\star_\alpha,\; \theta' \notin R^\star_\alpha : d(\theta \,|\, x) \leq d(\theta' \,|\, x). \quad (13)$$

Although the concepts of intrinsic estimator and credible region have been defined here for *reference* problems, they can also be used in situations where proper, subjective prior information is available.

Finally, in hypothesis testing, a typical problem is to decide whether a precise value θ_0 may be used as a "proxy" for the unknown value of θ. The reference approach is to use $d(\theta_0 \,|\, x)$ from equation (11), with θ_0 replacing $\hat{\theta}$, as an intrinsic test statistic. Its magnitude is a direct measure of the evidence against the null hypothesis $\theta = \theta_0$.

5 Summary

Noninformative priors have been studied for a long time and most of them have been found defective in more than one way. Reference analysis arose from this study as the only *general* method that produces priors that have the required *invariance* properties, deal successfully with the *marginalization* paradoxes, and have consistent *sampling* properties.

Reference priors should not be interpreted as probability distributions expressing subjective degree of belief; instead, they help answer the question of what could be said about the quantity of interest if one's prior knowledge were dominated by the data.

Reference analysis also provides methods for summarizing the posterior density of a measurement. Intrinsic point estimates, credible intervals, and hypothesis tests have invariance properties that are essential for *scientific* inference.

References

1. Carlton M. Caves, Christopher A. Fuchs, and Rüdiger Schack, "Quantum probabilities as Bayesian probabilities," Phys. Rev. **A** 65, 022305 (2002).
2. José M. Bernardo, "Reference Analysis," Handbook of Statistics **25** (D. Dipak and C.R. Rao, eds.) Amsterdam: Elsevier, 2005. See also `http://www.uv.es/~bernardo/publications.html`.
3. D. Sun and J. O. Berger "Reference priors with partial information," Biometrika **85**, 55 (1998).
4. Gauri Sankar Datta and Trevor J. Sweeting, "Probability matching priors," Research Report No. 252, Department of Statistical Science, University College London (March 2005); also at `http://www.ucl.ac.uk/Stats/research/Resrprts/psfiles/rr252.pdf`.

COSMOLOGICAL BAYESIAN MODEL SELECTION

ROBERTO TROTTA
Oxford University, Astrophysics, Denys Wilkinson Building, Keble Road, OX1 3RH, United Kingdom
E-mail: rxt@astro.ox.ac.uk

Bayesian model comparison can be used to decide whether the introduction of a new parameter is warranted by data. I focus on the Savage-Dickey density ratio as a method to compute the Bayes factor of nested models without carrying out a computationally demanding multi-dimensional integration. I present a new procedure (called "PPOD") to forecast the Bayes factor of a future observation. As an illustration, I consider a few central quantities in the current cosmological concordance model.

1. Introduction

After a decade of intense observational efforts, cosmologists have now at their disposal a wealth of data to guide them in their quest for a cosmological standard model. One important problem which is often encountered is deciding whether or not cosmological data support the introduction of a new quantity in our model. It is generally agreed that a core of 6 parameters is sufficient to describe and in reasonable agreement with most of current cosmological observations[1]. These parameters are the baryon, the cold dark matter and cosmological constant densities, the Hubble parameter, the optical depth to reionization, the scalar spectral index and the amplitude of the primordial (adiabatic) density fluctuations. The status of additional parameters is less certain, as often sampling (frequentist) statistics significance tests do not allow them to be ruled out with high confidence. A wide literature[2] addresses the difficulties arising from the use of p-values (significance level) in assessing the need for a new parameter. Many weaknesses of significance tests are clarified, and some even overcome, by adopting a Bayesian approach to testing. In this work, we take the viewpoint of Bayesian model selection to determine whether a parameter is needed in the light of the data at hand.

The key quantity for Bayesian model comparison is the marginal likelihood, or evidence, whose calculation and interpretation is attracting increasing attention in cosmology and astrophysics[3]. The marginal likelihood has proved useful in other contexts, as well, for instance consistency checks between data sets[4], the detection of galaxy clusters via the Sunayev-Zel'dovich effect[5] and neutrino emissions from type II supernovae[6]. In this paper we use the Savage-Dickey density ratio for an efficient computation of marginal likelihoods ratios (Bayes factor), and we present a new method to forecast the Bayes factor probability distribution of a future observation, called *PPOD* (for "Predictive Posterior Odds Distribution"). We then illustrate applications to some important parameters of current cosmological model building.

2. Bayesian model comparison

Bayesian inference (see e.g. Refs. 7) is based on Bayes' theorem, which is a consequence of the product rule of probability theory:

$$p(\boldsymbol{\theta}|\mathbf{d},\mathcal{M}) = p(\mathbf{d}|\boldsymbol{\theta},\mathcal{M})\pi(\boldsymbol{\theta}|\mathcal{M})/p(\mathbf{d}|\mathcal{M}). \quad (1)$$

On the left-hand side, the posterior probability for the parameters $\boldsymbol{\theta}$ given the data $\mathbf{d}$ under a model $\mathcal{M}$ is proportional to the likelihood $p(\mathbf{d}|\boldsymbol{\theta},\mathcal{M})$, which we will denote in the following by $\mathcal{L}(\boldsymbol{\theta})$, times the prior probability distribution function (pdf), $\pi(\boldsymbol{\theta}|\mathcal{M})$, which encodes our state of knowledge before seeing the data. In the context of model comparison it is more useful to think of $\pi(\boldsymbol{\theta}|\mathcal{M})$ as defining the prior available parameter space under the model $\mathcal{M}$. The normalization constant, independent of the parameters, is the *marginal likelihood for the model* $\mathcal{M}$ (sometimes also called the "evidence") given by

$$p(\mathbf{d}|\mathcal{M}) = \int_\Omega \mathcal{L}(\boldsymbol{\theta})\pi(\boldsymbol{\theta}|\mathcal{M})\mathrm{d}\boldsymbol{\theta} \quad (2)$$

where Ω designates the parameter space under model $\mathcal{M}$.

Consider two competing models $\mathcal{M}_0$ and $\mathcal{M}_1$ and ask what is the posterior probability of each model given the data $\mathbf{d}$. By Bayes' theorem we have

$$p(\mathcal{M}_i|\mathbf{d}) \propto p(\mathbf{d}|\mathcal{M}_i)\pi(\mathcal{M}_i) \quad (i=0,1), \quad (3)$$

where $p(\mathbf{d}|\mathcal{M}_i)$ is the marginal likelihood for $\mathcal{M}_i$ and $\pi(\mathcal{M}_i)$ is the prior probability of the ith model before we see the data. The ratio of the likelihoods for the two competing models is called the *Bayes factor*:

$$B_{01} \equiv \frac{p(\mathbf{d}|\mathcal{M}_0)}{p(\mathbf{d}|\mathcal{M}_1)}, \tag{4}$$

which is the same as the ratio of the posterior probabilities of the two models in the usual case when the prior is presumed to be noncommittal about the alternatives and therefore $\pi(\mathcal{M}_0) = \pi(\mathcal{M}_1) = 1/2$. The Bayes factor can be interpreted as an automatic Occam's razor, which disfavors complex models involving many parameters (see e.g. [8] for details). A Bayes factor $B_{01} > 1$ favors model $\mathcal{M}_0$ and in terms of betting odds it would prefer $\mathcal{M}_0$ over $\mathcal{M}_1$ with odds of B_{01} against 1. It is usual to consider the logarithm of the Bayes factor, for which the rule of thumb[9] is that a positive (strong) preference for one of the competing models requires at least $|\ln B_{01}| \gtrsim 1$ ($\gtrsim 3$).

Evaluating the marginal likelihood integral (2) is in general a computationally demanding task for multi-dimensional parameter spaces. Here we focus instead on the Savage-Dickey density ratio (SDDR, see [10] for references), whose estimation is very promising in terms of reducing the computational effort needed to calculate the Bayes factor of two nested models. Suppose we wish to compare a two-parameter model $\mathcal{M}_1$ with a restricted submodel $\mathcal{M}_0$ with only one free parameter, ψ, and with fixed $\omega = \omega_\star$ (for simplicity of notation we take a two-parameter case, but the calculations carry over trivially in the multi-dimensional case). Assume further that the prior is separable (which is usually the case in cosmology), i.e. that

$$\pi(\omega, \psi|\mathcal{M}_1) = \pi(\omega|\mathcal{M}_1)\pi(\psi|\mathcal{M}_0). \tag{5}$$

Then the Bayes factor B_{01} of Eq. (4) can be written as

$$B_{01} = \frac{p(\omega_\star|\mathbf{d}, \mathcal{M}_1)}{\pi(\omega_\star|\mathcal{M}_1)} \quad \text{(SDDR)}. \tag{6}$$

Thus the evaluation of the Bayes factor of two nested models only requires the properly normalized value of the marginal posterior at $\omega = \omega_\star$ under the extended model $\mathcal{M}_1$, which is a by-product of parameter inference. From (6) it is also clear that the prior available range for ω under $\mathcal{M}_1$ must be carefully assessed, since it is a central ingredient of the model comparison result.

3. Applications to cosmological parameters

We now apply the Bayesian model selection approach outlined above to a few important parameters of the current cosmological concordance model. The first quantity is the scale dependence of primordial density fluctuations, as described by the spectral tilt n_S. Many inflationary models predict a scale-invariant spectrum, i.e. that $n_S = 1$. The second parameter is the spatial curvature of the Universe, Ω_κ; inflation generically predicts that the Universe is flat and thus $\Omega_\kappa = 1$. According to single-field inflation, the initial conditions for primordial fluctuations are adiabatic, but some models (e.g., the curvaton model) allow for the presence of correlated isocurvature modes, as well. We therefore compare a purely adiabatic model, where the isocurvature fraction is $f_{\rm iso} = 0$, to a less restrictive mixture of adiabatic and totally (anti)correlated isocurvature initial conditions. It should be noted that the model comparison result in this case is strongly dependent on how one parameterizes the isocurvature fraction, i.e. which variable one assigns a flat prior to. Finally, we compare a model where the cosmic neutrino background presents primordial anisotropies as predicted by the standard Big Bang theory, parameterized by $c^2_{\rm vis} = 1/3$, to a generic model of non-zero neutrino coupling[11]. We combine cosmic microwave background anisotropies data, observations of the galaxy distribution, the measurement of the Hubble parameter and supernovae type Ia data. For more details on the data used and the priors on the models, see [10, 11].

Table 1 summarizes our results for the Bayes factor, along with the information content of the data I, defined as

$$I \equiv \log_{10}(\sigma_\pi/\sigma_p), \tag{7}$$

where σ_π, σ_p are the standard deviations of the prior and posterior on ω, respectively (both taken to be Gaussians). Thus the quantity I describes the order of magnitude by which our prior knowledge on ω has improved after the arrival of the data. Another characterizing quantity is $\lambda \equiv |\omega_\star - \bar{\omega}|/\sigma_p$, which represents the "number of sigmas" discrepancy between the predicted value under $\mathcal{M}_0$, $\omega_\star$, and the posterior mean, $\bar{\omega}$.

Figure 1 is a useful way of visualizing the outcome of the model comparison procedure for

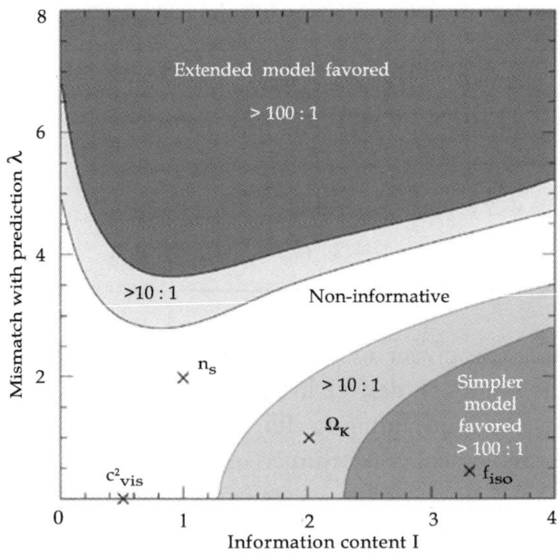

Fig. 1. Informative regions (shaded) where one of the competing models has posterior odds larger than 10 (light shaded regions) or 100 (dark shaded region) against 1. The white region corresponds to a non-informative limbo. The informative regions are computed using the SDDR, Eq. (6), and assuming a Gaussian posterior for the parameter of interest and a Gaussian, separable prior. The location of the parameters discussed in the text is shown by crosses.

Gaussian prior and likelihood, by plotting the status of the parameters under consideration in the $I - \lambda$ plane. For large values of λ, the usual result of sampling theory is recovered, namely that $\mathcal{M}_0$ is highly disfavored. However, as the information content of the data increases (larger I) for moderate values of λ, say $\lambda \lesssim 3$, then the simpler model is favored by the Occam's razor argument implicit in the Bayes factor. As the data becomes better and better, strongly disfavoring a simpler model requires a larger discrepancy between the parameter's measurement and the prediction of the simpler model. In particular, a $2-\sigma$ discrepancy (so frequent in cosmology) from the predicted value $\omega_\star$ is by no means sufficient to strongly disfavor $\mathcal{M}_0$; on the contrary, for high quality data this result does support the view that the extra complexity of the extended model is not justified.

4. Bayes factor forecast

In designing a new observation, it is interesting to assess its potential in terms of its power to address model comparison questions (for a survey optimization approach, see e.g. [12]). To this end, we introduce a new technique which combines a Fisher information matrix forecast with the SDDR formula to obtain a forecast for the Bayes factor of a future observation. The result is a PPOD (for *"Predictive Posterior Odds Distribution"*) for the future model comparison results.

Given the posterior pdf from present-day data, we draw a series of N independent samples (for instance using Markov Chain Monte Carlo techniques). At each sample value $\boldsymbol{\theta}_i, i = 1, \ldots, N$ we perform a Fisher Matrix analysis assuming $\boldsymbol{\theta}_i$ as a fiducial model, which yields a forecast for the likelihood function of the future measurement in the form of its covariance matrix $\mathbf{C}_i$ in parameter space. Writing as before $\boldsymbol{\theta} = (\omega, \boldsymbol{\Psi})$, the forecasted marginalized posterior for the parameter of interest, ω, is a Gaussian pdf centered on ω_i of width $\sigma_i = \sqrt{(\mathbf{C}_i)_{11}}$, which we denote by $\mathcal{N}(\omega, \omega_i, \sigma_i)$. This holds for separable, flat priors along all directions, and assuming that the prior range is much larger than the width of the posterior. Using the SDDR, Eq. (6), we obtain the corresponding Bayes factor comparing the two models $\mathcal{M}_0 : \omega = \omega_\star$ against $\mathcal{M}_1 : \omega \neq \omega_\star$,

$$(B_{01})_i = \mathcal{N}(\omega_\star, \omega_i, \sigma_i)/\pi(\omega_\star|\mathcal{M}_1), \qquad (8)$$

This is a forecast for the Bayes factor *assuming that ω_i is the correct value* for ω, as implicit in the fact that we have taken $\boldsymbol{\theta}_i$ as a fiducial model for the Fisher matrix forecast. In order to obtain $p(B_{01}|\mathbf{d}, \boldsymbol{e})$, the probability of obtaining B_{01} from a future measurement with observational parameters $\boldsymbol{e}$ (encoding e.g. the sensitivity and sky coverage of the experiment) we need to weight the expected Bayes factor by our *present* posterior belief that $\boldsymbol{\theta}_i$ is the true value, obtaining

$$p(B_{01}|\mathbf{d}, \boldsymbol{e}) = \int p(B_{01}|\boldsymbol{\theta}, \mathbf{d}, \boldsymbol{e}) p(\boldsymbol{\theta}|\mathbf{d}) d\boldsymbol{\theta}. \qquad (9)$$

Since the $\boldsymbol{\theta}_i$'s are samples from $p(\boldsymbol{\theta}|\mathbf{d})$, the probability distribution (9) can be obtained by plotting a histogram of the values of $(B_{01})_i, i = 1, \ldots, N$ as given by Eq. (8). This derivation assumes that the marginalized posterior for ω from future data is accurately described by a Gaussian, which is likely to break down in the tails of the distribution. Nevertheless, we can still conclude that fiducial models which have $|\omega_\star - \omega_i|/\sigma_i \gg 1$ strongly disfavor $\mathcal{M}_0$ under future data, even though we cannot attach a precise value to the expected odds. This is why we present PPOD results in five broad bins only (as in Fig. 2), since a finer discrimination would require a better approximation for the forecasted posterior.

Table 1. Summary of model comparison results for the four quantities described in the text.

Quantity	$\mathcal{M}_0$	$\mathcal{M}_1$	I	λ	$\ln B_{01}$	Odds, $\mathcal{M}_0$ vs $\mathcal{M}_1$	Evidence
n_S	$n_S = 1.0$	$0.8 \leq n_S \leq 1.2$	1.0	2.0	~ 0.0	$\sim 1:1$	undecided
Ω_κ	$\Omega_\kappa = 1.0$	$-1.0 < \Omega_\kappa < 1.0$	2.0	1.0	~ 2.6	$\sim 14:1$	positive
$f_{\rm iso}$	$f_{\rm iso} = 0.0$	$-100 \leq f_{\rm iso} \leq 100$	3.3	0.5	~ 7.5	$\sim 1800:1$	strong
$c^2_{\rm vis}$	$c^2_{\rm vis} = 1/3$	$0 \leq c^2_{\rm vis} < 1/3$	0.5	0.0	~ 0.7	$\sim 2:1$	undecided

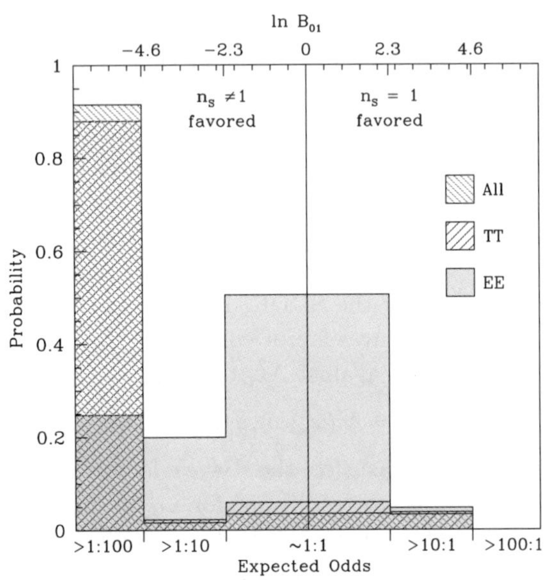

Fig. 2. PPOD for the Planck satellite, comparing a model with ($n_S \neq 1$) and without ($n_S = 1$) a spectral tilt. There is about 90% probability that temperature (TT), E-polarization (EE) and TE-correlation measurements together (All) will strongly disfavor $n_S = 1$ (with odds larger than 1 : 100).

As an example, Fig. 2 shows the PPOD for the cosmic microwave background anisotropies measurements of the Planck satellite, in view of discriminating between a scale invariant $n_S = 1.0$ spectrum versus a more general model with a Gaussian prior distribution with a width $\Delta n_S = 0.2$. We find that there is about 90% probability that the combined temperature and polarisation data will yield a strong preference (with odds larger than 100 : 1) for the non-scale invariant model.

5. Conclusions

Bayesian model comparison tools offer valuable insight into the plausibility of theoretical speculations regarding cosmological parameters in view of the data at hand. Bayes factor forecast techniques can assess the power of future experiments in terms of their ability to deliver high-odds model selection results, thus providing useful guidance in the quest of a cosmological concordance model.

Acknowledgments

It is a pleasure to thank T.J. Loredo for many helpful comments. The author is supported by the Royal Astronomical Society through the Sir Norman Lockyer Fellowship.

References

1. M. Tegmark *et al.* [SDSS Collaboration], Phys. Rev. D **69**, 103501 (2004).
2. A good starting point is the collection of references available from the website of David R. Anderson, Department of Fishery and Wildlife Biology, Colorado State University.
3. P. S. Drell, T. J. Loredo and I. Wasserman, Astrophys. J. **530**, 593 (2000); A. Slosar *et al.*, Mon. Not. Roy. Astron. Soc. **341**, L29 (2003); G. Lazarides, R. R. de Austri and R. Trotta, Phys. Rev. D **70**, 123527 (2004); M. Beltran *et al.*, Phys. Rev. D **71**, 063532 (2005); T. D. Saini, J. Weller and S. L. Bridle, Mon. Not. Roy. Astron. Soc. **348**, 603 (2004).
4. P. Marshall, N. Rajguru and A. Slosar, preprint: astro-ph/0412535.
5. M. P. Hobson and C. McLachlan, Mon. Not. Roy. Astron. Soc. **338**, 765 (2003).
6. T. J. Loredo and D. Q. Lamb, Phys. Rev. D **65** (2002) 063002.
7. E. T. Jaynes, *Probability Theory. The logic of science*, ed. G. L. Bretthorst, (CUP, 2003); P. C. Gregory, *Bayesian Logical Data Analysis for the Physical Sciences* (CUP, 2005).
8. D. J. MacKay, *Information theory, inference, and learning algorithms* (CUP, 2003).
9. R. E. Kass and A. E. Raftery, J. Amer. Stat. Assoc. **90**, 773 (1995).
10. R. Trotta, preprint: astro-ph/0504022.
11. R. Trotta and A. Melchiorri, Phys. Rev. Lett. **95**, 011305 (2005).
12. B. A. Bassett, Phys. Rev. D **71** (2005) 083517.

TOWARDS RECONCILIATION BETWEEN BAYESIAN AND FREQUENTIST REASONING

TOMAŽ PODOBNIK[1, 2] AND TOMI ŽIVKO[2]
[1] *Physics Department, University of Ljubljana, Slovenia*
[2] *"Jožef Stefan" Institute, Ljubljana, Slovenia*
E-mail: tomaz.podobnik@ijs.si , tomi.zivko@ijs.si

A theory of quantitative inference about the parameters of sampling distributions is constructed deductively by following very general rules, referred to as the Cox-Pólya-Jaynes Desiderata. The inferences are made in terms of probability distributions that are assigned to the parameters. The Desiderata, focusing primarily on consistency of plausible reasoning, lead to unique assignments of these probabilities in the case of sampling distributions that are invariant under Lie groups. In the scalar cases, e.g. in the case of inferring a single location or scale parameter, the requirement for logical consistency is equivalent to the requirement for calibration: the consistent probability distributions are automatically also the ones with the exact calibration and *vice versa*. This equivalence speaks in favour of reconciliation between the Bayesian and the frequentist schools of reasoning.

1. Introduction

A theory of quantitative inference about the parameters of sampling distributions is formulated with special attention being paid to the consistency of the theory and to its ability to make verifiable predictions. In the present article only basic concepts of the theory and their most important applications are presented while details can be found elsewhere[1].

Let $p(x_1|\theta I)$ be the probability for a random variate x to take the value x_1 (to take a value in an interval $(x_1, x_1 + dx)$ in the case of a continuous variate), given the family I of sampling distributions, and the value θ of the parameter that specifies a unique distribution within the family (for example, a sampling distribution from the exponential family I, $\tau^{-1}\exp\{-x/\tau\}$, is uniquely determined by the value of the parameter τ). An inference about the parameter is made by specifying a real number, called (degree of) plausibility, $(\theta|x_1 x_2 \ldots I)$, to represent our degree of belief in the value of the (continuous) parameter to be within an interval $(\theta, \theta + d\theta)$. Every such plausibility is conditioned upon the information that consists of measured value(s) x_1, x_2, ... of the sampling variate and of the specified family I of sampling distributions.

We assume all considered plausibilities to be subjects to very general requirements, referred to as the Cox-Pólya-Jaynes (CPJ) Desiderata[1, 2], focusing primarily on consistency of the plausible reasoning. The requirement of consistency can be regarded as the first of the requirements to be satisfied by every theoretical system, be it empirical or non-empirical. As for an empirical system, however, besides being consistent, it must also be falsifiable[3]. We therefore added a Desideratum to CPJ Desiderata, requiring that the predictions of the theory must be verifiable so that, in principle, they may be refuted.

It should be stressed that in this way the list of basic rules is completed. That is, the entire theory of inference about the parameters is built deductively from the aforementioned Desiderata: in order not to jeopardize the consistency of the theory no additional *ad hoc* principles are invoked.

2. Cox's and Bayes' Theorems

Richard Cox showed[4] that a system for manipulating plausibilities is either isomorphic to the probability system or inconsistent (i.e., in contradiction with CPJ Desiderata). Without any loss of generality, we therefore once and for all choose probabilities $p(\theta|x_1 I)$ among all possible plausibility functions $(\theta|x_1 I)$ to represent our degree of belief in particular values of inferred parameters. In this way the so-called inverse probabilities, $p(\theta|x_1 I)$, and the so-called direct (or sampling) probabilities $p(x_1|\theta I)$, become subjects to identical rules.

Transformations of probability distributions that are induced by variate transformations are also uniquely determined by the Desiderata. Let $f(x|\theta I)$ be the probability density function (pdf) for a continuous random variate x so that its probability distribution is expressible as

$$p(x|\theta I) = f(x|\theta I)\, dx \; .$$

Then, if the variate x is subject to a one-to-one transformation $x \longrightarrow y = g(x)$, the pdf for y reads:

$$f(y|\theta I') = f(x|\theta I) \left|\frac{dy}{dx}\right|^{-1} \tag{1}$$

(by using the symbol I' instead of I on the left-hand side of (1) it is stressed that the above transformations may in general alter the form of the sampling distribution). Since the direct and the inverse probabilities are subjects to the same rules, the transformation of the pdf for the inferred parameter, $f(\theta|xI)$, under a one-to-one transformation $\theta \longrightarrow \nu = \bar{g}(\theta)$ is analogous to the transformation of the sampling pdf:

$$f(\nu|xI) = f(\theta|xI) \left|\frac{d\nu}{d\theta}\right|^{-1} . \tag{2}$$

Once the probabilities are chosen, the usual product and sum rules[2] become the fundamental equations for manipulating the probabilities, while many other equations follow from the repeated applications of the two. In this way, for example, Bayes' Theorem for updating the probabilities can be obtained:

$$f(\theta|x_1x_2I) = \frac{f(\theta|x_1I)\, p(x_2|\theta x_1I)}{\int f(\theta'|x_1I)\, p(x_2|\theta' x_1I)\, d\theta'} . \tag{3}$$

Here $f(\theta|x_1I)$ denotes the pdf for θ based on x_1 and I only (i.e., prior to taking datum x_2 into account), $p(x_2|\theta x_1I)$ is the probability for x_2 (the so-called likelihood) given values θ and x_1, while the integral in the denominator on the right-hand side ensures appropriate normalization of the updated pdf $f(\theta|x_1x_2I)$ for θ (i.e., the pdf for θ posterior to taking x_2 into account).

Bayes' Theorem (3) allows only for updating pdf's $f(\theta|x_1I)$ that were already assigned prior to their updating. Consequently, the existing applications of our basic rules must be extended in order to allow for assignment of probability distributions to the parameters, with such assignments representing natural and indispensable starting points in every sequential updating of probability distributions.

3. Consistency Theorem

According to the CPJ Desiderata, the pdf for θ should be invariant under reversing the order of taking into account two independent measurements of the sampling variate x. This is true if and only if the pdf that is assigned to θ on the basis of a single measurement of x, is directly proportional to the likelihood for that measurement,

$$f(\theta|xI) = \frac{\pi(\theta)\, p(x|\theta I)}{\int \pi(\theta')\, p(x|\theta' I)\, d\theta'} , \tag{4}$$

where $\pi(\theta)$ is the consistency factor while the integral in the denominator on the right-hand side of (4) again ensures correct normalization of $f(\theta|xI)$.

There is a remarkable similarity between the Bayes' Theorem (3), applicable for *updating* the probabilities, and the Consistency Theorem (4), applicable for *assigning* the probability distributions to the values of the inferred parameters, but there is also a fundamental and very important difference between the two. While $f(\theta|x_1I)$ in the former represents the pdf for θ prior to taking datum x_2 into account, $\pi(\theta)$ in the latter is (by construction of the Consistency Theorem[1]) just a proportionality coefficient between the pdf for θ and the appropriate likelihood $p(x|\theta I)$, so that no probabilistic inference is ever to be made on the consistency factor alone, nor can $\pi(\theta)$ be subject to the normalization requirement that is otherwise perfectly legitimate in the case of prior pdf's.

The form of the consistency factor depends on the only relevant information that we posses before the first datum is collected, i.e., it depends on the specified sampling model. Consequently, when assigning probability distributions to the parameters of the sampling distributions from the same family I, this must be made according to the Consistency Theorem by using the consistency factors of the forms that are identical up to (irrelevant) multiplication constants.

4. Consistency Factor

According to (2) and (4) combined, the consistency factors $\pi(\theta)$ for θ and $\tilde{\pi}(\bar{g}(\theta))$ for the transformed parameter $\bar{g}(\theta)$ are related as

$$\tilde{\pi}(\bar{g}(\theta)) = k\, \pi(\theta) \left|\bar{g}'(\theta)\right|^{-1} , \tag{5}$$

where k is an arbitrary constant (i.e., its value is independent of either x or θ), while $\bar{g}'(\theta)$ denotes the derivative of $\bar{g}(\theta)$ with respect to θ. However, for the parameters of sampling distributions with the form I that is invariant under simultaneous transformations $g_a(x)$ and $\bar{g}_a(\theta)$ of the sample and the parameter space,

$$f(g_a(x)|\bar{g}_a(\theta)I') = f(x|\theta I) \left|g_a'(x)\right|^{-1} = f(g_a(x)|\bar{g}_a(\theta)I)$$

(i.e., when $I' = I$), $\widetilde{\pi}$ and π must be identical functions up to a multiplication constant, so that (5) reads:

$$\pi\big(\bar{g}_a(\theta)\big) = k(a)\,\pi(\theta)\,\big|\bar{g}'_a(\theta)\big|^{-1} . \qquad (6)$$

Index a in the above expressions indicates parameters of the transformations and k, in general, can be a function of a. In the case of multi-parametric transformation groups the derivative $\bar{g}'_a(\theta)$ is to be substituted by the appropriate Jacobian.

The above functional equation has a unique solution for the transformations $\bar{g}_a(\theta)$ with the continuous range of admissible values a, i.e., if the set of admissible transformations $\bar{g}_a(\theta)$ forms a Lie group. If a sampling distribution for x is invariant under a Lie group, then it is necessarily reducible (by separate one-to-one transformations of the sampling variate $x \to y$ and of the parameter $\theta \to \mu$) to a sampling distribution that can be expressed as a function of a single variable $y - \mu$, $f(y|\mu I) = \phi(y - \mu)$. Sampling distributions of the form $\sigma^{-1}\psi(x/\sigma)$ are examples of such distributions: by substitutions $y = \ln x$ and $\mu = \ln\sigma$ they transform into $\phi(y-\mu) = \exp\{y-\mu\}\,\psi\big(\exp\{y-\mu\}\big)$ (the scale parameters σ are reduced to location parameters μ).

It is therefore sufficient to determine the form of consistency factors for the location parameter μ since we can always make use of (5) to transform $\widetilde{\pi}\big(\mu = \bar{g}(\theta)\big)$ into the appropriate consistency factor $\pi(\theta)$ for the original parameter θ. Sampling distributions of the form $\phi(x - \mu)$ are invariant under simultaneous translations $x \to x+a$ and $\mu \to \mu+a$; $\forall a \in (-\infty, \infty)$, and the functional equation (6) in the case of the translation group reads

$$\pi(\mu + a) = k(a)\,\pi(\mu) ,$$

implying the consistency factor for the location parameters to be $\pi(\mu) \propto \exp\{-q\mu\}$, with q being an arbitrary constant. Accordingly, $\pi(\sigma) \propto \sigma^{-(q+1)}$ is the appropriate form of the consistency factor for the scale parameters.

The value of q is then uniquely determined by recognizing the fact that sampling distributions of the forms $\phi(x - \mu)$ and $\sigma^{-1}\psi(x/\sigma)$ are just special cases of two-parametric sampling distributions

$$f(x|\mu\sigma I) = \frac{1}{\sigma}\,\phi\Big(\frac{x-\mu}{\sigma}\Big) , \qquad (7)$$

with σ being fixed to unity and with μ being fixed to zero, respectively. The consistency factor $\pi(\mu)$ therefore corresponds to assigning pdf's $f(\mu|\sigma x I)$ while $\pi(\sigma)$ is to be used when assigning $f(\sigma|\mu x I)$. When neither σ nor μ is fixed, however, the pdf (7) is invariant under a two-parametric group of transformations, $x \to ax + b$, $\mu \to a\mu + b$ and $\sigma \to a\sigma$; $\forall a \in (0, \infty)$ and $\forall b \in (-\infty, \infty)$, and the functional equation (6) for the consistency factor $\pi(\mu, \sigma)$ for assigning $f(\mu\sigma|xI)$ reads

$$\pi(a\mu + b, a\sigma) = \frac{k(a,b)}{a^2}\,\pi(\mu, \sigma) ,$$

so that $\pi(\mu, \sigma)$ is to be proportional to σ^{-r}, r being an arbitrary constant. According to the product rule, $f(\mu\sigma|xI)$ can be factorized as

$$\begin{aligned} f(\mu\sigma|xI) &= f(\mu|\sigma x I)\, f(\sigma|xI) \\ &= f(\sigma|\mu x I)\, f(\mu|xI) , \end{aligned} \qquad (8)$$

where $f(\sigma|xI)$ and $f(\mu|xI)$ are the marginal pdf's, e.g.

$$f(\sigma|xI) = \int f(\mu'\sigma|xI)\, d\mu' . \qquad (9)$$

The equalities (8) are achieved if and only if $q = 0$ and $r = 1$, i.e., if the three consistency factors, determined uniquely up to arbitrary multiplication constants, read:

$$\pi(\mu) = 1 \;\text{ and }\; \pi(\sigma) = \pi(\mu, \sigma) = \sigma^{-1} . \qquad (10)$$

5. Calibration

In order to exceed the level of a mere speculation, the theory of probabilistic inference about the parameters must be able to make predictions that can be verified (or falsified) by experiments. Therefore, let a random variate x be subject to a family of sampling distributions I and let several independent values x_i of the variate be recorded. The predictions of the theory are made in probabilities

$$P\big(\theta \in (\theta_{i,1}, \theta_{i,2})|x_i I\big) = \int_{\theta_{i,1}}^{\theta_{i,2}} f(\theta'|x_i I)\, d\theta' = \delta \quad (11)$$

that given measured value x_i of the sampling variate, an interval $(\theta_{i,1}, \theta_{i,2})$ contains the actual value of the parameter θ of the sampling distribution. For the sake of simplicity, the intervals are chosen in such a way that the probabilities δ are equal in each of the assignments. The predictions are then verifiable at long term relative frequencies: our probability judgments (11) are said to be calibrated if the fraction of inferences with the specified intervals containing the actual value of the parameter, coincides with δ.

An exact calibration of an inference about a parameter θ is ensured if the assigned pdf $f(\theta|xI)$ is related to the (cumulative) distribution function $F(x, \theta)$ of the sampling variate as

$$f(\theta|xI) = \left|\frac{\partial}{\partial\theta}F(x,\theta)\right| , \tag{12}$$

and the consistency factors $\pi(\mu)$ and $\pi(\sigma)$ (10) do meet the above requirement. Furthermore, if besides being calibrated (12), the pdf for θ is to be assigned according to the Consistency Theorem (4), the distribution of the sampling variate x is necessarily reducible to a distribution of the form $\phi(y-\mu)$ [5]. But exactly the same necessary condition was obtained by requiring invariance of the sampling distribution under a Lie group, with such an invariance being indispensable for determination of consistency factors solely by imposing consistency to the assignment of pdf's. Imposing logical consistency to the theory is thus equivalent to imposing calibration to its predictions: every probabilistic inference about a parameter of a sampling distribution that we are sure is consistent will thus at the same time also be calibrated and, *vice versa*, every calibrated inference, based on a posterior pdf that is factorized according to (4), will simultaneously be logically consistent, too. The equivalence of the two requirements speaks in favour of reconciliation between the (objective) Bayesian and the frequentist schools of reasoning, the former paying attention primarily to logical consistency and the latter stressing the importance of verifiable predictions.

6. Consistency Lost and Regained

Numerous examples can be found with the sampling distributions lacking invariance under Lie groups: there are sampling distributions for continuous random variates (e.g. the Weibull distribution) that are not invariant under continuous groups of transformations, the symmetry can be broken by imposing constraints to parameter spaces of otherwise invariant sampling distributions, or the sampling space may be discrete (e.g. in counting experiments), just to name three of the most common ones. No consistent qualitative parameter inference is possible in such cases, but under very general conditions the remedy is just to collect more data relevant to the estimated parameters. Then, according to the Central Limit Theorem, the discrete sampling distributions approach their dense (Gaussian) limits, the constraints of the parameter spaces become more and more irrelevant, and the sampling distributions of the maximum likelihood estimates of the inferred parameter θ gain Gaussian shapes with θ being the location parameters of the latter, so the ability of making consistent inferences is regained.

7. Consistency Preserved

Consistency factors are determined exclusively by utilizing the tools such as the product rule (8) and marginalization (9), that are deducible directly from the basic Desiderata: in order to preserve consistency of inference it is crucial to refrain from using *ad hoc* shortcuts on the course of inference. For regardless how close to our intuitive reasoning these *ad hoc* procedures may be, how well they may have performed in some other previous inferences, and how respectable their names may sound (e.g. the principle of insufficient reason or its sophisticated version - the principle of maximum entropy, the principle of group invariance, the principle of maximum likelihood, and the principle of reduction), they are all found in general to lead to inferences that are neither consistent nor calibrated.

Acknowledgments

We thank Prof. Louis Lyons for making it possible to present our research at PHYSTAT05 and to Prof. Aleš Stanovnik for proofreading a preliminary draft of this report and correcting several inaccuracies.

References

1. T. Podobnik, T. Živko, *On Consistent and Calibrated Inference about the Parameters of Sampling Distributions*, *physics/0508017* (2005).
2. E. T. Jaynes, *Probability Theory - The Logic of Science*, Cambridge University Press (2003).
3. K. R. Popper, *The Logic of Scientific Discovery*, Hutchinson & Co. Publishers, London (1959).
4. R. T. Cox, *Am. J. Phys.* **14**, 1 (1946).
5. D. V. Lindley, *J. Roy. Stat. Soc.* **B 20**, 102 (1958).

GOODNESS OF FIT

GOODNESS OF FIT — WITH A VIEW TOWARDS PARTICLE PHYSICS

S. L. LAURITZEN
Department of Statistics, University of Oxford
Oxford, United Kingdom
E-mail: steffen@stats.ox.ac.uk

This article reviews aspects of significance testing. Problems of detection of a specific signal with background noise from observed Poisson counts of events is used as a basic example throughout. In particular we discuss issues of using alternative test-statistics, unbinned likelihood fits, and comparing unweighted and weighted histograms. We point at the possibility of adding simultaneous confidence intervals to the statistical toolbox normally used by particle physicists.

Keywords: Borel scales, Cournot's principle, power-divergence statistic, non-identifiability, simultaneous inference.

1. Significance testing

1.1. *General issues*

Significance testing is a well-trodden area of theoretical statistics and it seems just about impossible to say anything about this topic which has not been said many times before. For an excellent discussion of almost every corner, see for example Ref. 1 or Ref. 2. Still, significance testing is causing much controversy between statisticians and it can be hard to find two statisticians who would be in complete agreement.

In the present article we give some brief remarks which primarily serve to set the scene and identify which corner is to be explored. It also briefly indicates the multitudes of issues involved, thus explaining why it may not even be helpful to treat these different situations in a completely unified way, hence giving some rationale for the persistence of disagreement.

Decision vs. evidence There are two related but different types of situation which may be approached by significance testing.

In the first of these, procedures for accepting or rejecting a hypothesis are established with the purpose of using them automatically and repeatedly in a number of similar if not virtually identical situations. Such cases occur for example in industrial quality control. A decision-theoretic framework[3, 4] describes this situation well, the formal Neyman–Pearson theory of significance testing is both appropriate and convincing. Within this theory a linear combination of the probabilities of taking an incorrect decision (type I and II errors) is minimized, often by holding one of these fixed at a given *level of significance.*

The second situation, which forms the basis of this article, pertains to the case where a scientific theory needs to be examined in the light of a single or few related but different experimental results. The decision-theoretic approach seems here less appropriate as the acceptance or rejection of a scientific theory rarely will be a consequence of the experiment under study, but will involve numerous other ways of gaining and incorporating scientific knowledge about the phenomenon. This situation is closer to the Fisherian way of thinking about significance tests and would rather lead to an attempt to quantify the *evidence* in the experimental result for or against the validity of a specific theory, typically in the form of a so-called *p-value* or *significance probability.*

Much of the controversy[5, 6] between Fisher and Neyman on issues of significance testing was centered around these contrasting situations. The difference has probably been exaggerated in the sometimes very heated debate between the two. Most researchers would agree that it would be untenable to quantify the evidence in a given, unique situation in a way that would not have reasonable properties if used repeatedly in conceptual or similar situations. Indeed, the approaches of Neyman and Fisher appear to be less different than what first meets the eye[7].

In any case, the point of view adopted here to analyse problems of goodness of fit is closest to the Fisherian as this seems to have more direct bearing on the context under discussion.

Exploration vs. confirmation The way a significance test is used depends very much on the stage of scientific investigation. In an exploratory phase of a scientific enquiry, significance tests can play an important role in searching for abnormalities in an experimental result, the primary aim being to *identify potentially interesting phenomena* for future exploration and the planning of further experiments. Such cases seem to need a treatment quite different from those in a confirmatory phase, where the issue is to establish conclusive evidence for a given theory which is also *convincing to others.*

Refutation vs. validation Significance tests are used for a variety of different purposes. In some cases they are used in a Popperian quest for refuting a scientific theory, thereby paving the way for establishing alternative and improved theories. In other cases, the objective of the significance test is to validate a certain aspect of a model, to justify assumptions needed for further analysis.

1.2. *Paradigm*

The (largely Fisherian) paradigm of significance testing used in the present article is outlined below:

- A null *hypothesis* H_0 or theory is entertained or proposed and data X collected;
- A *test statistic* $T = t(X)$ is constructed (possibly with an alternative theory in mind) in such a way that large values of T indicate deviations from H_0;
- The *p-value* $p = P(T \geq t_{\text{obs}} \mid H_0)$ is calculated, approximately or exactly;
- The p-value is interpreted by the fundamental principle:

Events of small probability do not happen.

This fundamental principle for relating probabilities to the real world has been termed *Cournot's principle*[8, 9]. Hence, *if p is sufficiently small,* say $p \leq \varepsilon$, H_0 *is untenable.* Emile Borel[10, 11] used the term "the single law of chance" for Cournot's principle and set the following scales for probabilities to be small:

- l'échelle humaine: $\varepsilon \sim 10^{-6}$
- l'échelle terrestre: $\varepsilon \sim 10^{-15}$
- l'échelle cosmique: $\varepsilon \sim 10^{-50}$

Modern statistical practice tends to use $\varepsilon \sim 10^{-1}$, but Particle Physics may well need different scales to allow for scientific progress and simultaneously prevent too many false discoveries.

Although the general issue of significance testing has a strong frequentist flavour, rules such as Cournot's principle are also needed for subjectivist Bayesian probability to make a bridge to observable phenomena in the real world[12].

1.3. *Goodness of Fit*

This term is used to describe particular types of significance tests, but it is used in many different ways and contexts[13, 14], for example:

- Is a given distribution of a specified type?
- Any significance test without explicit specification of an alternative hypothesis;
- Any significance test used to validate, justify, or refute a postulated model.

To avoid the discussion to be too narrow, we will mostly adopt the latter, which conforms well with the application of Cournot's principle.

2. Basic example

To avoid discussing the problems out of context, we will focus on variants of the following problem and setup, describing problems of detection of signal events in the presence of noise in the form of background events. More precisely, we consider the following:

- $X_i = x_i, i = 1, \ldots, n$ are 'binned' counts of independent Poisson events, the i-th bin corresponding to events of mass or energy around m_i.
- The Poisson intensity ν_i in bin i is given as

$$\nu_i = \nu_i(\theta) = \beta_i + \frac{\alpha}{\sigma}\phi\left(\frac{m_i - \mu}{\sigma}\right), \tag{1}$$

where ϕ denotes the standard Gaussian density.

Here β_i is the intensity of *background* events whereas the second term is the intensity of the interesting *signal* events. The background intensity may depend on one or several unknown parameters η so $\beta_i = \beta(\eta_i)$ and $\theta = (\eta, \alpha, \mu, \sigma)$ denotes the vector of all of these parameters. The signal intensity may well be absent, corresponding to $\alpha = 0$ and often the main issue of interest is to infer whether apparent signal events are just random artifacts.

It is quite critical what the exact status is concerning prior knowledge about the background intensity. Can the background intensity be assumed known from other experiments and theory or must it be estimated? How can the background reasonably be modelled? Can the measurement error σ be considered as known or unknown? Is the position of the signal peak μ known? The complexity of problems vary greatly according to circumstance as outlined above.

2.1. *Standard practice*

Standard practice[15] for tackling the situation can be briefly described as follows:

- Fit model to background intensity;
- Calculate goodness of fit statistics using either *the likelihood ratio statistic* G^2

$$\begin{aligned} G^2 &= -2 \log L(\hat{\theta}) \\ &= 2 \sum_{i=1}^{n} \left\{ \nu_i(\hat{\theta}) - X_i + X_i \log \frac{x_i}{\nu_i(\hat{\theta})} \right\} \end{aligned}$$

  or its approximation, known as *Pearson's* χ^2

$$C^2 = \sum_{i=1}^{n} \frac{\{\nu_i(\hat{\theta}) - X_i\}^2}{\nu_i(\hat{\theta})}.$$

  In some cases the latter is substituted with the *Wald statistic*

$$W^2 = \sum_{i=1}^{n} \frac{\{\nu_i(\hat{\theta}) - X_i\}^2}{X_i},$$

  which can be computationally more convenient.
- Calculate p-values approximately or by Monte-Carlo methods.

3. Issues to be considered

The setup described raises a number of issues:

- Is one of the test statistics to be preferred?
- When is the χ^2 distribution appropriate for calculating p-values?
- When calculating p-values using a χ^2-distribution, what is the appropriate number of degrees of freedom?
- If one fits the model with or without the signal component, can the difference between the two test statistics be used and what is its distribution?

Partial answers to these and other questions will be attempted in the following.

3.1. *Power divergence statistics*

It can be helpful to consider the one-parameter family of *power-divergence* statistics[16] given by

$$I_\lambda(X) = \frac{2}{\lambda(\lambda+1)} \sum_{i=1}^{n} X_i \left[\left\{ \frac{X_i}{\nu_i(\hat{\theta})} \right\}^\lambda - 1 \right]$$

for $-\infty < \lambda < \infty$. Provided $\sum_i X_i = \sum_i \nu_i(\hat{\theta})$, it follows that

$$I_1(X) = C^2, \quad \lim_{\lambda \to 0} I_\lambda(X) = G^2,$$

so the commonly used statistics mentioned above are special cases. Ref. 14 recommends $\lambda = 2/3$, which is 'between' C^2 and G^2.

For $\lambda = -1/2$, I_λ becomes the *Freeman-Tukey statistic* F^2

$$F^2 = 4 \sum_i \left\{ \sqrt{X_i} - \sqrt{\nu_i(\hat{\theta})} \right\}^2.$$

The Freeman–Tukey statistic[17] is obviously based on the idea that for a Poisson variable with large mean ν, $\sqrt{X}$ is approximately normally distributed:

$$\sqrt{X} \sim \mathcal{N}(\sqrt{\nu}, 1/4).$$

These statistics all have the same asymptotic χ^2 distribution under the null hypothesis, and each is optimal in some sense. My personal preference would be the likelihood ratio statistic G^2, as it is constructed to have maximal power at the most likely alternative, but it may well be a matter of taste.

For important issues it could be reasonable to calculate I_λ and the associated p-value for a range of different values of λ. It would not be desirable if the interpretation of an experiment depends critically on λ, so if the p-value is on different sides of the threshold for small probabilities as λ varies, the experiment may be considered inconclusive.

The use of W^2 is mostly motivated by the convenience of computation, because its minimization is a direct weighted least squares, whereas the others might be computationally less easy to minimize. The statistic W^2 is potentially less powerful than C^2 against large deviations from the hypothesis, as a large and explicit signal with $X_i > \nu_i(\hat{\theta})$ will yield

$$W_i^2 = \frac{\{\nu_i(\hat{\theta}) - X_i\}^2}{X_i} < \frac{\{\nu_i(\hat{\theta}) - X_i\}^2}{\nu_i(\hat{\theta})} = C_i^2.$$

There is some ambiguity about which of the above statistics is 'best'. Much effort has been used to discuss which of them has a distribution closest to the χ^2-distribution. Much of this depends both on the specific circumstances considered and how closeness is measured. Personally I would be less worried about getting an accurate calculation of the p-value than not detecting a signal because the test statistic is less powerful. Also, because effective Monte-Carlo methods are rapidly being developed, the use of the χ^2-approximation is losing importance.

3.2. *Is the χ^2 distribution appropriate?*

The derivation of the χ^2 distribution is based on the following two elements:

- For ν_i large, X_i are approximately Gaussian $\mathcal{N}(\nu_i, \nu_i)$;
- For ν_i large, the model for the intensity $\nu_i(\theta)$ is approximately linear in the unknown parameters within the likely area of variation of X_i. In particular, the fitting of θ is approximately a linear least squares problem.

In the following some cases where there is trouble will be discussed.

3.2.1. *Unbinned fit*

If k unknown parameters have been fitted based on unbinned data and G^2 is calculated from binned data, the asymptotic distribution of G^2 (or any of the other statistics) is *not* χ^2 with $n-k-1$ degrees of freedom.

Fortunately, its correct asymptotic distribution is well understood. It approximately holds[18] that

$$G^2 = A^2 + \sum_{j=1}^{k} \zeta_j B_j^2,$$

where A^2 is $\chi^2(n-k-1)$ and independent of $B_j^2, j = 1, \ldots, k$, with each B_j^2 distributed as $\chi^2(1)$ and $0 \leq \zeta_j \leq 1$. In particular it holds (approximately) that

$$A^2 < G^2 < A^2 + \sum_{j=1}^{k} B_j^2,$$

where the lower bound is $\chi^2(n-k-1)$ and the upper bound is $\chi^2(n-1)$.

This yields a simple practical way of guarding against problems of this kind: Asymptotically *the correct p-value is between those based on $\chi^2(n-1)$ and $\chi^2(n-k-1)$*. One can just calculate each of them and this will usually be precise enough to identify whether the correct p-value is extremely small.

This result also holds for the other test statistics in the power divergence family[14] and for W^2.

3.2.2. *Parameter singularity*

One specific example where the difficulty in using the χ^2 approximation is due to intrinsic non-linearity of the testing problem is exactly in the case of signal with background noise, as in (1). If the location μ of the peak or the measurement uncertainty σ are not known, a singularity arises because under the null hypothesis $\alpha = 0$, μ and σ do not make sense.

The following method to tackle this problem has been developed by Ref. 19. First proceed as if μ and σ were known, and calculate the usual test statistic for the hypothesis $\alpha = 0$. When μ and σ are known, the hypothesis is a simple, linear hypothesis. Denote the corresponding test statistic as

$$T_{\mu,\sigma} = t_{\mu,\sigma}(X).$$

Each of these follows a χ^2 distribution under the null hypothesis. We now use the test statistic

$$T^* = \sup_{(\mu,\sigma)\in R} T_{\mu,\sigma}$$

where R is a *plausible region* for (μ, σ).

The approximate distribution of T^* is that of the maximum of related χ^2 statistics. The corresponding p-value is not known exactly, but approximate Monte-Carlo methods using the χ^2 distribution for the individual statistics have been developed[19].

The method is somewhat involved, but not unusable, in particular because in many cases, μ is known and σ is approximately known, so the plausible region R can be quite small.

Recently, Ref. 20 has extended and refined the method so that it becomes more accurate and more generally usable. It seems worthwhile to explore the possibility of exploiting this method.

3.3. *Validating the model*

The χ^2-distribution used in the case just discussed would typically be the *difference* between G^2 assuming only background and G^2 when also the peak is fitted.

For the χ^2 distribution to be valid it is important that the model is properly established, in particular that the background intensity is not incorrectly specified.

Thus it must at *least have a non-significant* G^2 *value when the peak is fitted*, to document that the data indeed can be explained in terms of background plus peak.

In addition a careful *residual analysis* should be made to detect systematic or too large deviations from the model (1).

4. Comparing weighted and unweighted histograms

In some cases the information about the background intensity β_i is obtained from an independent experiment with Poisson counts Y_i with intensities $c\rho_i\beta_i$, where ρ_i are known factors and c is a constant determining the total intensity of events in the auxiliary experiments. In other words, the auxiliary experiment has only background events, but may not have the same background rates.

It is then common to form a *weighted histogram* with weights W_i in the ith bin, where

$$W_i = Y_i/\rho_i, \quad i = 1, \dots, n$$

and compare the histogram so obtained with the histogram based on X_i, containing a potential signal peak.

The exact distribution of associated test statistics, calculated as if the weighted events were indeed proper events, cannot be described in simple terms and the asymptotic results cannot be immediately applied to this more complex situation. An alternative would be to compare the histograms with a proper significance test as follows.

Under the null hypothesis $H_0 : \alpha = 0$, the likelihood function in terms of the original observations X_i and Y_i is

$$L(c,\beta) \propto \prod_{i=1}^{n} \beta_i^{x_i+y_i} c^{y_i} e^{-\beta_i(1+c\rho_i)}$$
$$= c^{\sum y_i} e^{-\sum \beta_i - c\sum \rho_i\beta_i} \prod_{i=1}^{n} \beta_i^{z_i},$$

where we have let $\beta = (\beta_1, \dots, \beta_n)$ be the unknown background intensities and $Z_i = X_i + Y_i$ the combined number of events in bin i.

Under H_0, Z_i and the total number of events $T = \sum Y_i$ in the auxiliary experiment are sufficient statistics and the likelihood function is maximized by solving the system of equations which equate their observed values to their expectations:

$$t = \sum_{i=1}^{n} y_i = c\sum_{i=1}^{n} \rho_i\beta_i$$
$$z_i = \beta_i(1+c\rho_i), \quad i = 1, \dots n.$$

These equations can be solved iteratively, for example by using starting values $c = \beta_1 = \dots = \beta_n = 1$ and repeating

$$c \leftarrow \sum_{i=1}^{n} \rho_i\beta_i/t$$
$$\beta_i \leftarrow (1+c\rho_i)/z_i, \quad i = 1, \dots n.$$

This iteration is convergent as it can be seen to be a special instance of the algorithm known as *Iterative Proportional Scaling* or *Iterative Proportional Fitting*[21]. It provides maximum likelihood estimates $\hat{c}$ and $\hat{\beta}_i$ under the null hypothesis. The log-likelihood ratio statistic becomes

$$D = -2\log\frac{L(\hat{c},\hat{\beta})}{L(\hat{\nu})}$$

where $\hat{\nu} = (\hat{\nu}_1, \dots, \hat{\nu}_n)$ is the maximum likelihood estimate under an alternative hypothesis, but many other reasonable test statistics could be used, for example the analogue of C^2

$$\tilde{D} = \sum_i \frac{(x_i - \hat{\beta}_i)^2}{\hat{\beta}_i} + \sum_i \frac{(y_i - \hat{c}\rho_i\hat{\beta}_i)^2}{\hat{c}\rho_i\hat{\beta}_i},$$

or any other statistic from the power-divergence family. The p-value associated with any of these or other statistics can be calculated on the basis of the conditional distribution of the number of events, given the statistic which is sufficient under H_0, as the unknown parameters c and β_i do not enter into that distribution.

This distribution is very easy to simulate using the following Monte-Carlo procedure, which is a variant of Patefield's algorithm for simulating two-way contingency tables, conditional on the marginal totals[22].

A simple argument shows that the conditional distribution of (X_k, Y_k), given $Z_i, i = 1, \dots, n$, $\sum_i Y_i = t$, and $(X_i, Y_i) = (x_i, y_i), i = 1 \dots, k-1$ is given as

$$p(x_k, y_k) = h_k(\rho_k)\rho_k^{y_k} \frac{\binom{s-\sum_1^{k-1} x_i}{x_k}\binom{t-\sum_1^{k-1} y_i}{y_k}}{\binom{s+t-\sum_1^{k-1} z_i}{x_k+y_k}}$$

for $x_k + y_k = z_k$, where $s = \sum_i X_i = \sum_i Z_i - t$ and the expressions in brackets are binomial coefficients. This yields an obvious recursion for simulating from the correct distribution of any test-statistic in cases where the number of events in each bin is limited. For large event numbers, it may be easier to use asymptotic results.

5. Simultaneous confidence intervals

An alternative approach to the problem of assessing whether a peak is indeed present in the model (1) uses the idea of *simultaneous inference*[23]. This approach initially avoids fitting a model altogether and calculates a band within which the true Poisson intensity with high probability must be. If the band is sufficiently narrow, and displays an explicit peak, it might be immediately obvious that the data are inconsistent with any reasonably smooth background model.

Using the fact that the counts in separate bins are independent, it is possible to produce a *simultaneous confidence band* for the Poisson intensity, using that if

$$P(|X_i - \nu_i| > c) = \beta$$

for every bin i, then it follows that

$$P(\max_i |X_i - \nu_i| > c) = 1 - (1 - \beta)^n.$$

Hence, if a $1 - \alpha$ confidence band is desired, we must just choose

$$\beta = 1 - (1 - \alpha)^{1/n}.$$

This now yields a band around the observed histogram within which the proposed background intensity should fit. If this is not possible, then this can be taken as evidence either against the model for background or against absence of the peak.

References

1. D. R. Cox, *Scandinavian Journal of Statistics* **4**, 49 (1977).
2. D. R. Cox and D. V. Hinkley, *Theoretical Statistics* (Chapman and Hall, London, 1974).
3. A. Wald, *Statistical Decision Functions* (John Wiley and Sons, New York, 1950).
4. E. Sverdrup, *Revue de l'Institut International de Statistique* **34**, 309 (1966).
5. R. A. Fisher, *Journal of the Royal Statistical Society, Series B* **17**, 69 (1955).
6. J. Neyman, *Journal of the Operations Research Society of Japan* **3**, 145 (1961).
7. J. O. Berger, *Statistical Science* **18**, 1 (2003).
8. A. A. Cournot, *Exposition de la théorie des chances et des probabilités* (Hachette, Paris, 1843).
9. G. Shafer and V. V. Vovk, *Probability and Finance: It's Only a Game* (John Wiley and Sons, New York, 2001).
10. E. Borel, *Les Probabilités et la Vie* (Presses Universitaires de France, Paris, 1943).
11. E. Borel, *Probabilities and Life* (Dover Publications, New York, 1943). English translation of Borel (1943).
12. A. P. Dawid, *Statistical Science* **19**, 44 (2004).
13. R. B. D'Agostino and M. A. Stephens, *Goodness-of-Fit Techniques* (Marcel Dekker, New York, 1986).
14. T. R. C. Read and N. Cressie, *Goodness-of-Fit Statistics for Discrete Multivariate Data* (Springer-Verlag, New York, 1988).
15. G. Cowan, *Statistical Data Analysis* (Clarendon Press, Oxford, 1998).
16. N. Cressie and T. R. C. Read, *Journal of the Royal Statistical Society, Series B* **46**, 440 (1984).
17. M. F. Freeman and J. W. Tukey, *Annals of Mathematical Statistics* **21**, 607 (1950).
18. H. Chernoff and E. L. Lehmann, *Annals of Mathematical Statistics* **25**, 579 (1954).
19. R. B. Davies, *Biometrika* **74**, 33 (1987).
20. C. Ritz and I. M. Skovgaard, *Biometrika* **92**, 507 (2005).
21. J. N. Darroch and D. Ratcliff, *Annals of Mathematical Statistics* **43**, 1470 (1972).
22. W. M. Patefield, *Applied Statistics* **30**, 91 (1981).
23. R. Miller, *Simultaneous Statistical Inference*, 2nd edn. (Springer-Verlag, 1981).

GOODNESS-OF-FIT FOR SPARSE DISTRIBUTIONS IN HIGH ENERGY PHYSICS

BRUCE YABSLEY

High Energy Physics Department, School of Physics
University of Sydney, NSW 2006 Australia
E-mail: b.yabsley@physics.usyd.edu.au

We consider Pearson's chi-square X^2, the likelihood ratio G^2, and Zelterman's D^2 as goodness-of-fit statistics for high energy physics problems in several dimensions, where the data are sparse. There is a fundamental obstacle in the "ultrasparse" case where all bins have at most one entry ($n_i = 0, 1\ \forall i$). A condition for avoiding this regime is derived; the allowed number of bins k rises faster than the total number of events n: $k_{\max} = 0.4 \times n^{1.4}$. Reasonable binning in many dimensions may thus be possible for modest datasets $n > O(100)$, although special treatment is required to derive p-values. Results for an initial trial problem are encouraging; further studies are underway.

1. Motivation/Historical note

The talk of the Durham meeting was Heinrich's demonstration[1] that the likelihood cannot be used to test goodness-of-fit (g.o.f.) for unbinned maximum likelihood fits (see also Refs 2 and 3). This presents a problem for high energy physics, where the data are often characterised by several variables, leading to the use of unbinned fits to small samples. Due to the importance of such fits at the B-factories, Kay Kinoshita and I both pursued the matter in the following year, considering binning-free tests based on the random walk[4] and the energy test,[5] with inconclusive results. During discussion at PHYSTAT2003, an alternative approach was suggested:[6]

Conventional binned g.o.f. tests rely on results from the asymptotic limit where the number of bins k is fixed, and the number of events $n \to \infty$. This is one reason behind the conventional wisdom that fits should have $n_i \geq 5$ events in each bin. However an alternative limit, where $k \to \infty$ but the ratio of events to bins n/k remains finite, has been studied: see for example Ref. 7. There is considerable statistical literature on g.o.f. in this regime, mostly considering problems in the social sciences (for example, Ref. 8). Here I report the status of an attempt to appropriate this work for use in high energy physics.

2. Adapting a social science example

As a starting point, Zelterman[9] considers a 2D histogram from an employment survey,[10] with $n = 129$ events and $k = 899$ cells: well outside the conventional regime (Fig. 1). The null hypothesis is "independence of the rows [monthly salary] and columns [years since first degree] by using multinomial sampling, conditional on the marginal totals";[9] the alternative hypothesis is that a correlation exists. By inspection, confirmed by linear regression, the data of course *are* correlated. Tests based on

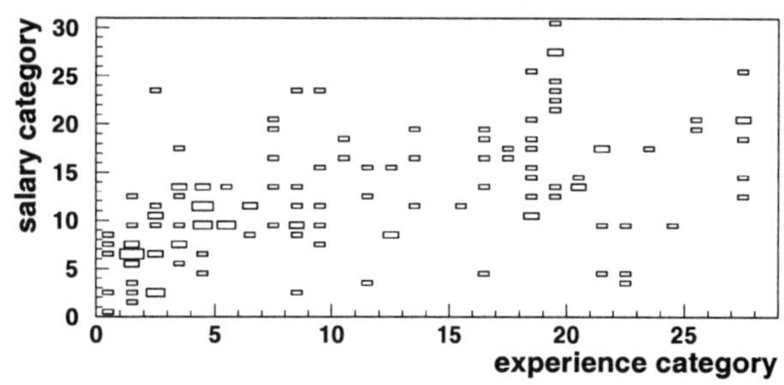

Fig. 1. The sparse histogram used as an example by Zelterman,[9] plotting salary *vs.* years of experience; the smallest squares show one event/bin. The data are taken from an employment survey in Ref. 10.

$$X^2 = \sum_i (n_i - \lambda_i)^2/\lambda_i, \quad (1)$$

$$G^2 = 2\sum_i n_i \log n_i/\lambda_i, \text{ and} \quad (2)$$

$$D^2 = \sum_i \left[(n_i - \lambda_i)^2 - n_i\right]/\lambda_i \quad (3)$$

where n_i is the actual and λ_i the predicted number of events in bin i, find various results: X^2 (Pearson's χ^2) fails to reject the null hypothesis; the likelihood ratio statistic G^2 and D^2 both reject it at extreme signficance. D^2, which is outside the family of Cressie and Read,[11] was introduced by Zelterman for use in the case of sparse data.[9] Both this and G^2 seem suitable to our purpose, based on this example.

No exact mathematical relationship between the quantities in Fig. 1 is expected; they can thus be grouped into categories — binned — according to convenience. Typical high energy physics data are different: the variables are invariant masses, momenta, angles, *etc.*, and the underlying processes are intrinsically simple. Formulae relating various quantities can be derived for some hypotheses, such as decay of a particle with given properties, and for others (*e.g.* combinatorial backgrounds) functions with few parameters are found to fit the data well. It would thus be attractive to choose bins fine enough to distinguish different (possibly correlated) distributions in each quantity: in many dimensions this can lead to arbitrarily small numbers of events-per-bin.

This approach will fail in the limit where $n_i = 0, 1\ \forall$ bins i: a test statistic $I = \sum_i f(n_i, \lambda_i)$ cannot in general distinguish between regions of low and high event density. D^2 collapses to a unique value in this case, for any $\{\lambda_i\}$; all statistics I collapse to the unique value $n \cdot f(1,\lambda) + (k-n)\cdot f(0,\lambda)$ if the distribution is flat, $\lambda_i = \lambda\ \forall i$. (Note that the form given for I is general, including the family of Cressie and Read,[11] D^2, and other statistics.) Since we will typically fit data with floating shape parameters, the limitation is fatal.

To avoid this "ultrasparse" regime, consider the following condition: Let m_j be the number of bins i where $n_i = j$ (so that there are m_1 bins with one entry, *etc.*), and find the number of bins $k = k_{\max}$ such that

$$P\left(m_2 \le \frac{1}{10}n;\ m_j = 0\ \forall j > 2\right) < 0.01 \qquad (4)$$

in the case where the expected bin populations are equal, $\lambda_i = n/k$. If this condition is met, then the majority of datasets will have a significant number of bins with $n_i = 2, 3, \ldots,$ even though the average bin population may be low: at most 1% will be dominated by bins with one entry, $n_i = 1$. (We consider datasets with only a small number of $n_i = 2$ bins $m_2 \le n/10$, and no bins with $n_i > 2$, to be dominated by their single-entry bins.)

For given n and k, the probability of any particular set of counts $(m_0, m_1, m_2, m_3 ...)$ is

$$P(\{m_j\}|n,k) = \frac{n!}{\prod_j (j!)^{m_j}} \cdot \frac{k!}{\prod_j m_j!} \cdot \left(\frac{1}{k}\right)^n, \qquad (5)$$

based on multinomial statistics and simple counting. Using (5) and an arbitrary-precision calculator, it is straightforward to solve (4) for $k_{\max}$. Results are shown in Fig. 2, together with a fit to the power law

$$k_{\max} = 0.4 \times n^{1.4}. \qquad (6)$$

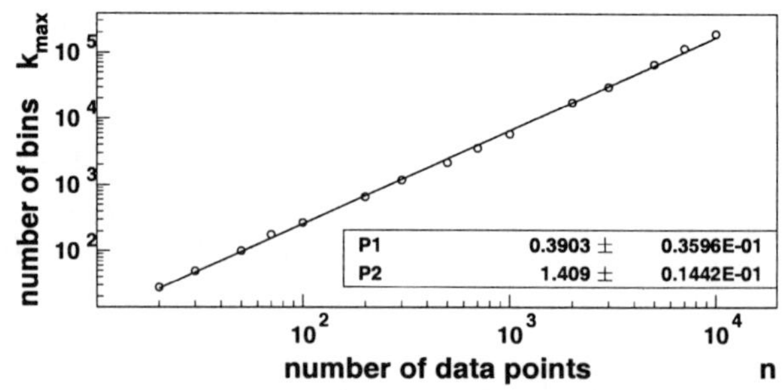

Fig. 2. The largest number of bins $k_{\max}$ satisfying the condition (4), as a function of the number of events n. The results of a fit to the power-law $k_{\max} = \text{P1} \cdot n^{\text{P2}}$ are also shown.

This suggests that binnings with $n/k \ll 1$ may be possible for moderate $n > O(100)$, enabling bins to be chosen in each of many dimensions, as required.

3. Example: $X(3872) \to \pi^+\pi^- J/\psi$

As an example, we consider angular analysis of decays $X(3872) \to \pi^+\pi^- J/\psi$, to determine the quantum numbers J^{PC} of the state; the sample was 58 events including $\approx$ 11 background.[12] Various hypotheses were tested using 1D histograms chosen with typical $n_i \ge 5$, but regions where $\lambda_i \lesssim 1$ on the null hypothesis: see Fig. 2 of Ref. 12. Event counts $n_i \gg 1$ in these bins disfavour the null.

Using toy Monte Carlo (MC) experiments with $n = 50$ and neglecting background, we study the power of tests on (1)–(3) to discriminate against $J^{PC} = 0^{-+}$ using binning in an increasing number of dimensions. Events are generated following Ref. 13, using a complete set of angles $(\theta, \phi, \psi, \chi, \phi_K)$. (Here (θ, ϕ, ψ) are as defined in Ref. 13 for the 0^{-+} case; χ is as defined for 0^{++}; and ϕ_K is the azimuthal angle of the kaon from $B \to KX(3872)$ decay, in the system used to define ϕ.) To bin efficiently, we use non-equidistant bins $[0.0, 0.3]$, $[0.3, 0.6]$, $[0.6, 0.9]$, $[0.9, 1.0]$ in $|\cos\theta|$, where a $\sin^2\theta = 1 - \cos^2\theta$ distribution is expected on the null hypothesis (preserving the small expected population in the last bin,[12] but using fewer bins overall). Fig. 3 shows the value of such binning for discriminating between hypotheses.

Fig. 4 shows the power to reject the null, based on p-values taken from distributions of toy MC experiments. The X^2 and G^2 tests improve noticeably as more dimensions are added, up to case (e) with $k = 128$ bins, comparable to $k_{\max} \approx 95$ (from (6)) for $n = 50$. All tests lose power for binning (f), in the ultrasparse regime ($k = 512$), and inspection of test-

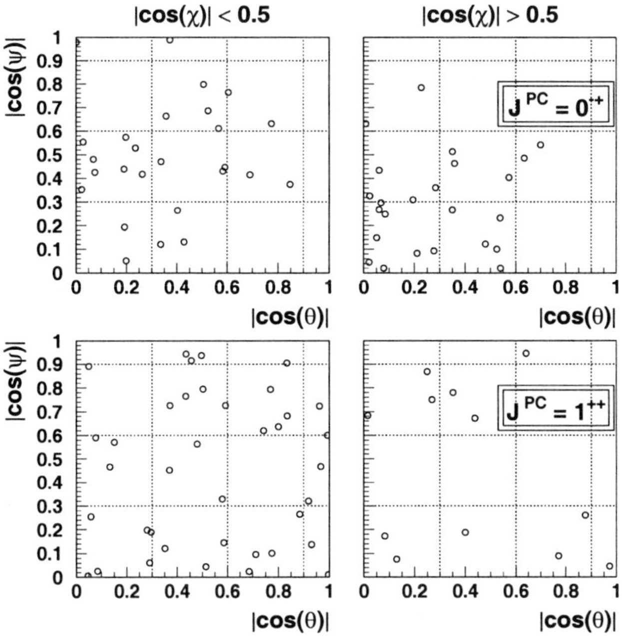

Fig. 3. Data from a single toy MC dataset, $n = 50$, for each of $J^{PC}(X(3872)) = 0^{-+}$ (upper plots) and 1^{++} (lower). Dotted lines show 4×4 binning in $(|\cos\theta|, |\cos\psi|)$; left and right plots show $|\cos\chi| < 0.5$ and ≥ 0.5 bins respectively. For $J^{PC} = 0^{-+}$, we expect $\sin^2\theta\,\sin^2\psi$ dependence and no dependence on $\cos\chi$. For $J^{PC} = 1^{++}$, the distribution $\propto \sin^2\chi$; the dependence on the remaining angles $(\theta, \phi, \psi, \phi_K)$ is nontrivial.

statistic and p-value distributions shows pathologies such as domination by discrete values. X^2 generally shows higher power than G^2, but loses power as $(1-\alpha) \to 1.0$, where α is the significance: the mechanism needs to be studied. D^2 performs poorly in all cases, in marked contrast to the example of section 2.

Also shown is a test based on $\sum_l \ln\mathcal{L}_l$, where $\mathcal{L}_l$ is the likelihood for the lth event: $\propto \sin^2\theta\,\sin^2\psi$ for $J^{PC} = 0^{-+}$. In this case, this test is discriminating, and more powerful than all of the binning-based tests shown. (This result is unlikely to be general, as there are known to be cases where $\mathcal{L}_l$-based tests fail to discriminate against certain alternative hypotheses, even where the null hypothesis is simple; the limitation is related to the failure of these tests to discriminate in the case of unbinned maximum likelihood fits. See Ref. 3.) Since $\mathcal{L}_l$ is a function of $\cos\theta$ and $\cos\psi$ only, it is insensitive to variation in $\phi, \cos\chi$ or ϕ_K, not expected for $J^{PC} = 0^{-+}$ but expected for some other hypotheses, in particular 1^{++} (see Fig. 3). Thus a test combining $\sum_l \ln\mathcal{L}_l$ with appropriate binning in $(\phi, \cos\chi, \phi_K)$ is presumably more powerful still: this remains to be studied.

4. Further work

In addition to further study of the results presented here, the following extensions are planned:

(1.) Varying n in the $X(3872)$ case, to see if Eq. (6) is a reliable guide to the breakdown of tests.

(2.) Application to a basic compound-hypothesis problem: fitting for a possible signal in the presence of background (which may be mismodelled). The prototypical problem of this kind at the B-factories is a search for a rare B-decay.[14]

(3.) A difficult compound-hypothesis problem: angular analysis of $B \to \phi K^*$ or similar decays[15] to determine helicity amplitudes. This is a 3D problem with a few hundred events, and thus combines features of cases (1.) and (2.).

It would be desirable to also apply this method to the analysis in Ref. 16, with $O(100)$ events and the sensitivity due to fine structure in two dimensions. Unfortunately, based on Eq. (6), this is unrealistic.

5. Conclusion

Binned goodness-of-fit tests have been considered for sparse data, where typical bin populations $n_i \ll 5$. Such tests will fail in the "ultrasparse" case where all bins have $n_i = 0, 1$ only; the condition $k_{\max} = 0.4 \times n^{1.4}$ defines a number of bins $k_{\max}$ that avoids this regime, for a given number of events n. For modest sample sizes, $k_{\max}$ is large enough to allow binning in many dimensions. For an angular analysis problem with $n = 50$, substantial improvement in the power of tests is found for careful binning in four dimensions, up to the expected limit $k \approx 100$. Further study using different n, and compound hypotheses, is underway. The (non-χ^2) distribution of test statistics in this regime remains to be studied.

Acknowledgments

I'm indebted to Jan de Leeuw for our discussion at PHYSTAT2003 on the problem of goodness-of-fit statistics for sparse data, and for providing a way into the statistical literature. I would like to thank Nancy Read, Bernard Silverman, and Mike Titterington for their advice and feedback at this meeting; and the conference organisers for providing a setting where these kinds of discussions can take place.

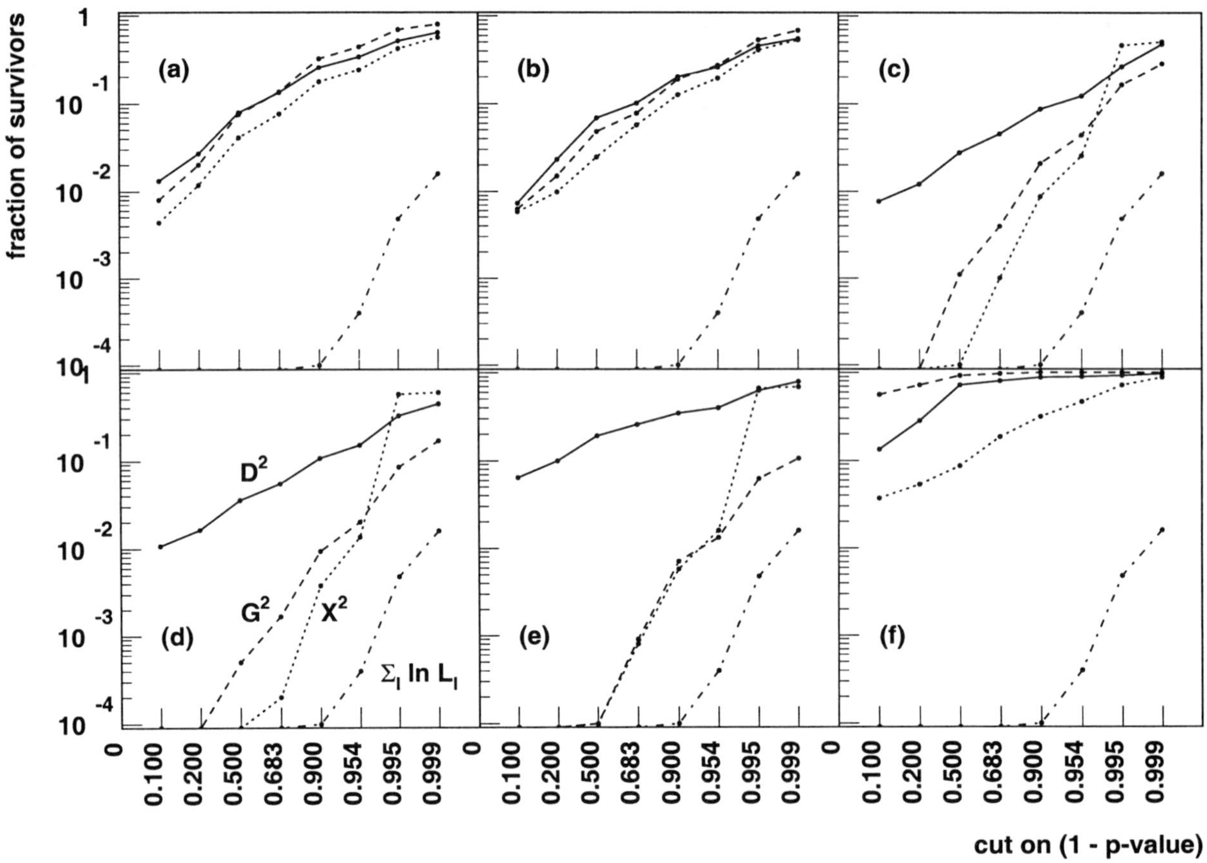

Fig. 4. Angular analysis of $X(3872) \to \pi^+\pi^- J/\psi$ (simulated): Power to reject the null hypothesis $J^{PC} = 0^{-+}$, in the case of 1^{++} data, for hypothesis tests based on D^2 (solid line), G^2 (dashed), X^2 (dotted), and the unbinned log-likelihood $\sum_l \ln \mathcal{L}_l$ (dot-dashed). The fraction of 1^{++} datasets that survive (*i.e.* $(1-\beta)$ where β is the power) is plotted against $(1-\alpha)$, where α is the significance: for tests A and B, A is more powerful if its curve lies below/right of the curve for B. Bins are chosen in a progressively larger number of variables: (a) 10 even bins and (b) 4 variable-width bins in $|\cos\theta|$, (c) 4×4 bins in $(|\cos\theta|, |\cos\psi|)$, (d) $4\times4\times2$ in $(|\cos\theta|, |\cos\psi|, |\cos\chi|)$, (e) $4\times4\times2\times4$ in $(|\cos\theta|, |\cos\psi|, |\cos\chi|, \phi_K)$, and (f) $4\times4\times2\times4\times4$ in $(|\cos\theta|, |\cos\psi|, |\cos\chi|, \phi_K, \phi)$.

References

1. J. Heinrich, CDF Internal Note 5639 (2001). `http://www-cdf.fnal.gov/publications/cdf5639_goodnessoffitv2.ps.gz`
2. K. Kinoshita, in *Proceedings of the Conference on Advanced Statistical Techniques in Particle Physics*, ed. M.R.Whalley, L.Lyons, 176–181 (2002).
3. J. Heinrich, in SLAC–R–703, `eConf C030908` (PHYSTAT2003 Proceedings), 52–55 (2003).
4. K. Kinoshita, in SLAC–R–703, `eConf C030908`, Proceedings), 56–60 (2003). The writeup of my half of the talk is missing from the proceedings.
5. B. Aslan and G. Zech, in SLAC–R–703, `eConf C030908`, (PHYSTAT'03 Proceedings), 97–100 (2003).
6. J. de Leeuw, private communication.
7. C. Morris, *Annals of Statistics* **3**, 165–188 (1975).
8. S.E Fienberg, *J. Royal Stat. Soc. B* **41**, 54–64 (1979).
9. D. Zelterman, *J. Am. Stat. Assoc.* **82**, 624–629 (1987).
10. G. Beatty, "Salary Survey of Mathematicians and Statisticians," in *Proceedings of the Section on Survey Research Methods, American Statistical Association*, pp. 743–747 (1983).
11. N. Cressie and T.R.C. Read, *J. Royal Stat. Soc. B* **46**, 440–464 (1984).
12. K. Abe *et al.*, `arXiv:hep-ex/0505038`.
13. J.L. Rosner, *Phys. Rev. D* **70**, 094023 (2004).
14. Y. Chao, P. Chang *et al.*, *Phys. Rev. Lett.* **94**, 181803 (2005); K. Abe *et al.*, `arXiv:hep-ex/0506080`.
15. K.-F. Chen *et al.*, *Phys. Rev. Lett.* **94**, 221804 (2005).
16. A. Poluektov *et al.*, *Phys. Rev. D.* **70**, 072003 (2004); K. Abe *et al.*, `arXiv:hep-ex/0411049` and `hep-ex/0504013`.

LIKELIHOOD ANALYSIS AND GOODNESS-OF-FIT IN LOW COUNTING RATE EXPERIMENTS

ALDO IANNI
I.N.F.N. Gran Sasso Laboratory, S.S. 17bis km 18+910
Assergi (AQ) 67010, Italy
Email: aldo.ianni@lngs.infn.it

Data analysis in low counting rate experiments require non-standard statistical methods mainly because with a poor data sample the Gaussian approximation is not appropriate. In this paper an example of analysis applied to a sample of data from a low counting rate experiment is presented. Emphasis is put on how to determine uncertainties of physical parameters, on goodness-of-fit and on the application of the Monte Carlo method.

1. Introduction

In experiments with low counting rate, often one has to deal with the problem of how to determine an upper limit or on how to define uncertainties. A complication is that the Gaussian approximation is not valid for a poor data sample, and binning of data can be difficult. Moreover, histograms may have one or more bins without entries. In these cases the very well known χ^2 method should not be used. The maximum likelihood method is best suited for parameter calculation. As far as upper bounds and uncertainties are concerned, there is no well established method (see discussion by G. Feldman in these proceedings).

In this paper we present an example of analysis to determine the activity of ^{85}Kr in a low counting rate detector. In particular, we will consider data from the Counting Test Facility, CTF, of the Borexino solar neutrino detector [1]. The CTF is a 4 ton un-segmented liquid scintillator detector designed to study the radiopurity of organic scintillators in the energy range below 1 MeV. Radioactive contaminants below 1 MeV are ^{238}U, ^{232}Th, ^{85}Kr, ^{39}Ar and ^{210}Pb. In particular, for U and Th the CTF reaches a sensitivity at the level of 10^{-16} g(U,Th)/g. The CTF is equipped with one hundred photomultipliers, an active muon veto and 10^3 m^3 of high purity shielding water. The CTF is located at the underground Gran Sasso Laboratory, Italy. A detailed description of the CTF can be found in [2]. As mentioned above, in the following we will focus on the measurement of ^{85}Kr in the liquid scintillator of the CTF to illustrate statistical methods to be used with a small data sample.

The ^{85}Kr is anthropogenic and it is a beta emitter with an end-point energy equal to 0.687 MeV. With a very small branching ratio (0.43%), it decays (Q_β=0.173 MeV) to ^{85m}Rb which decays to the ground state with τ=1.46μs emitting a photon of 0.514 MeV. This sequence of beta-gamma decays provide a strong signature for the detection of ^{85}Kr. Unfortunately, given the small branching ratio and the low counting rate feature of the detector, the set of collected candidate events is expected to be small in spite of the exposure. In the following, first we will present a sample of data selected in about 555 days. Then we will discuss an analysis method to determine the ^{85}Kr activity, the uncertainty on the measurement and the goodness-of-fit. In particular, we will compare the uncertainties and the goodness-of-fit calculated using the maximum likelihood and a Monte Carlo method, respectively.

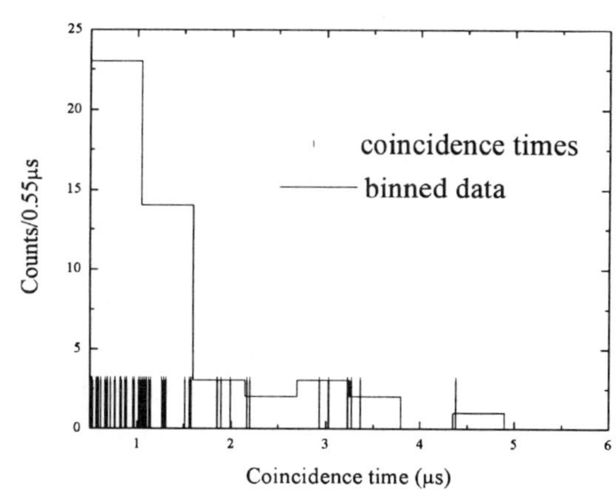

Figure 1. Selected ^{85}Kr events

2. Data

Selected data are reported in Figure 1. In particular, we show the coincidence times, "vertical bars", and a histogram of the data (10 bins). In the whole detector mass (3.73 tons of C_9H_{12}) we found 48 candidates. We expect a fraction of these counts to come from ^{232}Th through the correlated (in space and time) decay sequence: ^{212}Bi(β)->^{212}Po(α,τ=0.435μs).

3. Analysis method

3.1. *Maximum likelihood for parameters estimation*

The method of analysis is based on the so-called un-binned extended maximum likelihood [3]. The data can be described with the following distribution function:

$$\rho(t;a_{Kr},a_{Th},b) = b + \frac{a_{Kr}}{\tau_{Kr}} e^{-t/\tau_{Kr}} + \frac{a_{Th}}{\tau_{Th}} e^{-t/\tau_{Th}} \quad (1)$$

where we have taken into account two sources of background, namely, the ^{212}Bi-^{212}Po decay sequence and a constant contribution from electronics noise. Using Eq. (1), the likelihood function is written:

$$L(a_{Kr},a_{Th},b) = \prod_{i=1}^{N} \frac{\rho(t;a_{Kr},a_{Th},b)}{\int_{t_{\min}}^{t_{\max}} \rho(t;a_{Kr},a_{Th},b)dt} \times$$

$$\left(\int_{t_{\min}}^{t_{\max}} \rho(t;a_{Kr},a_{Th},b)dt\right)^{N} \times e^{-\int_{t_{\min}}^{t_{\max}} \rho(t;a_{Kr},a_{Th},b)dt} = \quad (2)$$

$$e^{-\int_{t_{\min}}^{t_{\max}} \rho(t;a_{Kr},a_{Th},b)dt} \times \prod_{i=1}^{N} \rho(t_i;a_{Kr},a_{Th},b)$$

where N=48 is the number of selected events, and a_{Kr}, a_{Th} and b are unknown parameters. In Eq. (2) t_{min} =0.5μs and t_{max} =6μs. In particular, a_{Kr} is proportional to the ^{85}Kr activity. By searching for the maximum of lnL from Eq. (2), we have determined the best-fit values for the unknown parameters. In the analysis we use implicitly the hypothesis that the decay trend shown by the data in Figure 1 is due to the mean life of Kr and that of Th, as described above. This hypothesis needs to be tested as done below.

Once the best-fit values have been calculated, one needs to determine the uncertainties. A common method is to use the following equation [3,4]:

$$\Delta \ln L \equiv \ln L(\theta) - \ln L_{\max} = \frac{Q}{2}, \quad (3)$$

where lnL(θ) is the profile log-likelihood and Q defines the likelihood ratio error interval (Q=1 for standard interval). We have applied this method: first we have determined the profile log-likelihood for a_{Kr} by maximizing with respect to the other two parameters. Then, we have used Eq. (3) to calculate the standard error interval. It turns out that $a_{Kr} = 30^{+11}_{-10}$. Using the selection cuts efficiency (35%), the Kr activity is calculated to be 36^{+14}_{-12} counts/day in 3.73 tons. We notice that Eq. (3) gives the correct confidence limits if the Gaussian approximation is justified. For the present data sample the log-likelihood is not a Gaussian as can be seen in Figure 2.

We have used a second method in order to determine the "error interval" for a_{Kr}. In particular, using the best-fit values for the parameters as inputs for sampling the distribution in Eq. (1), we have simulated 2000 data sets. For each set we have used Eq. (2) to determine a_{Kr} and the other parameters. The distribution of a_{Kr} is shown in Figure 3. From this distribution we have calculated an interval around the mean value which contains 68.3% of the whole data. It turns out that: $a_{Kr}=29.5^{+9.7}_{-7.5}$ and the Kr activity is 35^{+12}_{-9} counts/day in 3.73 tons.

Once the parameters and uncertainties have been determined, one has to calculate the p-value in order to test the assumed hypothesis. For the data sample considered, we have used two methods: an unbinned test and a chi-squared test combined with Monte Carlo simulation. As un-binned hypothesis test we have considered the Smirnov-Cramer-Von Mises method [5]. In this method a measure of the difference between the data and the model is:

$$w^2 = \int_{-\infty}^{+\infty} [F_{\exp}(x) - F_{th}(x)]^2 f(x)dx. \quad (4)$$

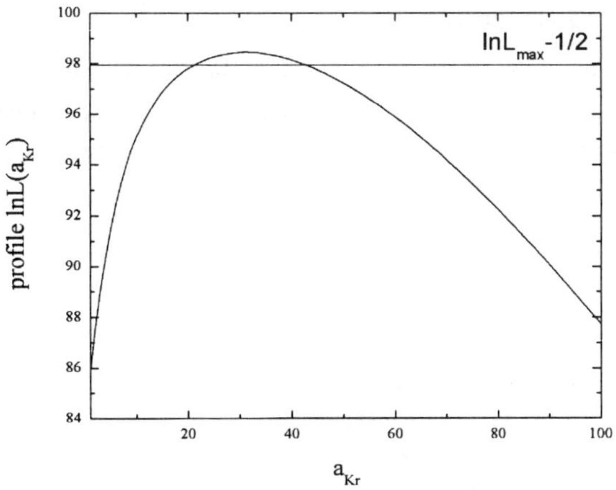

Figure 2. Profile log-likelihood for the parameter a_{Kr}. The horizontal line is used to determine the standard likelihood ratio error interval on the parameter. See text for details.

By writing the data in an ordered form (see Figure 4), Eq. (4) can be written:

$$w^2 = \frac{1}{12N} + \sum_{i=1}^{N} \left[F_{th}(x_i) - \frac{2i-1}{2N} \right]^2. \qquad (5)$$

In Eq. (4) $F_{exp}(x)$ is the experimental cumulative distribution, $F_{th}(x)$ is the model cumulative distribution and f(x) is the model pdf (Eq. (1) defines the pdf of the example presented in the paper). $F_{th}(x)$ is given as a function of the parameters, a_{Kr},a_{Th},b. Best-fit values are used for the parameters in $F_{th}(x)$.

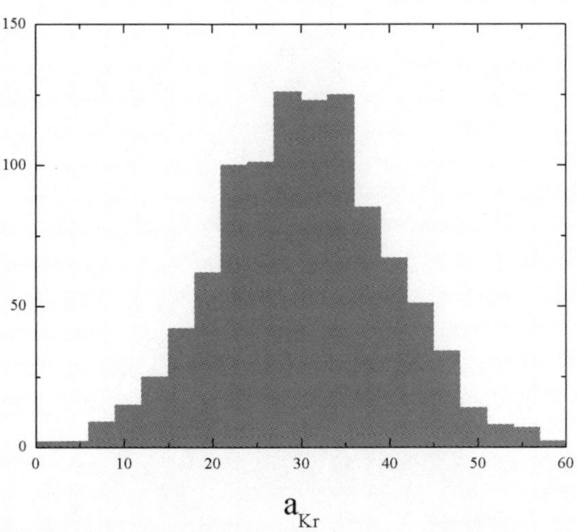

Figure 3. Distribution of the parameter a_{Kr} from the Monte Carlo Method. See text for details.

In Figure 4 we show the experimental cumulative distribution (solid line), and the model cumulative distribution (dashed line). The value determined using Eq. (5) (see Figure 4) corresponds to a p-value of 81% [4].

A second method used to determine the p-value uses binned data and simulates the distribution of the test statistic. In this case the difference between the data and the model is given by:

$$\chi_p^2 = \sum_{i=1}^{N} \left(\frac{N_i^{\exp} - N_i^{th}}{\sqrt{N_i^{th}}} \right)^2, \qquad (6)$$

where N_i^{exp} is the number of counts in the bin i^{th} in the histogram shown in Figure 1, and N_i^{th} is the corresponding expected number of counts. We have used the Monte Carlo method to determine the distribution of the χ_p^2 in Eq. (6) starting from the best-fit values. In Figure 5 we show the result of the simulation. We notice that the distribution of Eq. (6) does not match a χ^2 distribution. This is a general result when one deals with small samples.

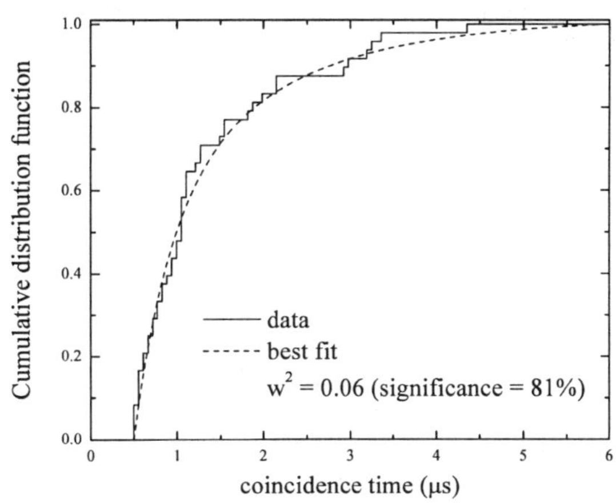

Figure 4. Goodness-of-fit using the Smirnov-Cramer-Von Mises method. See text for details.

The p-value turns out to be equal to 82%.

4. Conclusions

In the paper we have presented a method of analysis to deal with low statistic samples. We have shown how, once the physical parameters are determined with an un-binned maximum likelihood fit, uncertainties and goodness-of-fit can be determined by the Monte Carlo method. However, the methods we used for determining confidence intervals may not give the correct coverage. Un-binned hypothesis tests, such as the Smirnov-Cramer-Von Mises should be used with small data samples to avoid problems with low or zero entries/bin in applying the well-known χ^2 goodness-of-fit method.

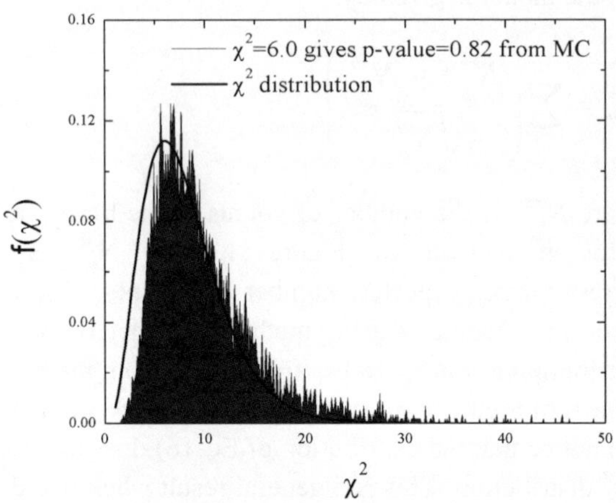

Figure 5. The histogram corresponds to the distribution of Eq. (6) from the Monte Carlo simulation. The solid line shows a χ^2 distribution with the same d.o.f.

Acknowledgments

I wish to thank the organizers of the Conference for the pleasant atmosphere, interesting talks and discussions. I wish also to thank G. Zech for carefully reviewing the draft of the paper.

References

1. Borexino collaboration, G. Alimonti et al., *Astrop. Phys.*, **16**, 205 (2002).
2. Borexino collaboration, G. Alimonti, et al., *Phys. Lett.* **B422**, 349 (1998).
3. G. Cowan, Statistical Data Analysis, *Oxford Science Publications*, Oxford, 1998.
4. Review of Particle Physics, *Phys. Lett.* **B592** (2004).
5. B.P. Roe, Probability and Statistics in Experimental Physics, Springer, Berlin, 1992.

THE "SIEVE" ALGORITHM—SIFTING DATA IN THE REAL WORLD

M. M. BLOCK

Department of Physics and Astronomy, Northwestern University, Evanston, IL, 60201, USA
E-mail: mblock@northwestern.edu

Experimental data are rarely, if ever, distributed as a normal (Gaussian) distribution, in real world applications. A large set of data—such as the cross sections for particle scattering as a function of energy contained in the archives of the Particle Data Group[1]—is a compendium of all published data, and hence, unscreened. For many reasons, these data sets have many outliers—points well beyond what is expected from a normal distribution—thus ruling out the use of conventional χ^2 techniques. We suggest an adaptive algorithm that applies to the data sample a sieve whose mesh is coarse enough to let the background fall through, but fine enough to retain the preponderance of the signal, thus sifting the data. The "Sieve" algorithm gives a robust estimate of the best-fit model parameters in the presence of a noisy background, together with a robust estimate of the model parameter errors, as well as a determination of the goodness-of-fit of the data to the theoretical hypothesis. Computer simulations were carried out to test the algorithm for both its accuracy and stability under varying background conditions.

1. Introduction

Our major assumptions about the experimental data are:

(1) The experimental data can be fitted by a model which successfully describes the data.
(2) The signal data are Gaussianly distributed, with Gaussian errors.
(3) That we have "outliers" only, so that the background consists only of points "far away" from the true signal.
(4) The noise data, *i.e.* the outliers, do not completely swamp the signal data.

2. The Adaptive Sieve Algorithm

2.1. *Algorithmic steps*

We now outline our adaptive Sieve algorithm:

(1) Make a robust fit of *all* of the data (presumed outliers and all) by minimizing Λ_0^2, the tuned Lorentzian squared, defined as

$$\Lambda_0^2(\boldsymbol{\alpha};\boldsymbol{x}) \equiv \sum_{i=1}^{N} \ln\left\{1+0.179\Delta\chi_i^2(x_i;\boldsymbol{\alpha})\right\}, \quad (1)$$

described in detail in Block[2]. The M-dimensional parameter space of the fit is given by $\boldsymbol{\alpha}=\{\alpha_1,\ldots,\alpha_M\}$, $\boldsymbol{x}=\{x_1,\ldots,x_N\}$ represents the abscissas of the N experimental measurements $\boldsymbol{y}=\{y_1,\ldots,y_N\}$ that are being fit and $\Delta\chi_i^2(x_i;\boldsymbol{\alpha}) \equiv \left(\frac{y_i-y(x_i;\boldsymbol{\alpha})}{\sigma_i}\right)^2$, where $y(x_i;\boldsymbol{\alpha})$ is the theoretical value at x_i and σ_i is the experimental error. As discussed in Block[2], minimizing Λ_0^2 gives the same total $\chi^2_{\rm min} \equiv \sum_{i=1}^N \Delta\chi_i^2(x_i;\boldsymbol{\alpha})$ from eq. (1) as that found in a χ^2 fit, as well as rms widths (errors) for the parameters—for Gaussianly distributed data—that are almost the same as those found in a χ^2 fit. The quantitative measure of "far away" from the true signal, *i.e.*, point i is an outlier corresponding to Assumption (3), is the magnitude of its $\Delta\chi_i^2(x_i;\boldsymbol{\alpha}) = \left(\frac{y_i-y(x_i;\boldsymbol{\alpha})}{\sigma_i}\right)^2$.

If $\chi^2_{\rm min}$ is satisfactory, make a conventional χ^2 fit to get the errors and you are finished. If $\chi^2_{\rm min}$ is not satisfactory, proceed to step 2.

(2) Using the above robust Λ_0^2 fit as the initial estimator for the theoretical curve, evaluate $\Delta\chi_i^2(x_i;\boldsymbol{\alpha})$, for the N experimental points.
(3) A largest cut, $\Delta\chi_i^2(x_i;\boldsymbol{\alpha})_{\rm max}$, must now be selected. For example, we might start the process with $\Delta\chi_i^2(x_i;\boldsymbol{\alpha})_{\rm max}=9$. If any of the points have $\Delta\chi_i^2(x_i;\boldsymbol{\alpha}) > \Delta\chi_i^2(x_i;\boldsymbol{\alpha})_{\rm max}$, reject them—they fell through the "Sieve". The choice of $\Delta\chi_i^2(x_i;\boldsymbol{\alpha})_{\rm max}$ is an attempt to pick the largest "Sieve" size (largest $\Delta\chi_i^2(x_i;\boldsymbol{\alpha})_{\rm max}$) that rejects all of the outliers, while minimizing the number of signal points rejected.
(4) Next, make a conventional χ^2 fit to the sifted set—these data points are the ones that have been retained in the "Sieve". This fit is used to estimate $\chi^2_{\rm min}$. Since the data set has been truncated by eliminating the points with $\Delta\chi_i^2(x_i;\boldsymbol{\alpha}) > \Delta\chi_i^2(x_i;\boldsymbol{\alpha})_{\rm max}$, we must slightly

renormalize the χ^2_{min} found to take this into account, by the factor $\mathcal{R}$, whose inverse is shown in Fig. 9a of Block[2].

If the renormalized χ^2_{min}, *i.e.*, $\mathcal{R} \times \chi^2_{\text{min}}$ is acceptable—in the *conventional* sense, using the χ^2 distribution probability function—we consider the fit of the data to the model to be satisfactory and proceed to the next step. If the renormalized χ^2_{min} is not acceptable and $\Delta\chi^2_i(x_i;\boldsymbol{\alpha})_{\text{max}}$ is not too small, we pick a smaller $\Delta\chi^2_i(x_i;\boldsymbol{\alpha})_{\text{max}}$ and go back to step 3. The smallest value of $\Delta\chi^2_i(x_i;\boldsymbol{\alpha})_{\text{max}}$ that makes much sense, in our opinion, is $\Delta\chi^2_i(x_i;\boldsymbol{\alpha})_{\text{max}} = 2$. After all, one of our primary assumptions is that the noise doesn't swamp the signal. If it does, then we must discard the model—we can do nothing further with this model and data set!

(5) From the χ^2 fit that was made to the "sifted" data in the preceding step, evaluate the parameters $\boldsymbol{\alpha}$. Next, evaluate the $M \times M$ covariance (squared error) matrix of the parameter space which was found in the χ^2 fit. We find the new squared error matrix for the Λ^2 fit by multiplying the covariance matrix by the square of the factor r_{χ^2} (for example[2], $r_{\chi^2} \sim 1.02, 1.05$, 1.11 and 1.14 for $\Delta\chi^2_i(x_i;\boldsymbol{\alpha})_{\text{max}} = 9$, 6, 4 and 2, respectively), shown in Fig. 9b of Block[2]. The values of $r_{\chi^2} > 1$ reflect the fact that a χ^2 fit to the *truncated* Gaussian distribution that we obtain—after first making a robust fit—has a rms (root mean square) width which is somewhat greater than the rms width of the χ^2 fit to the same untruncated distribution. Extensive computer simulations[2] demonstrate that this *robust* method of error estimation yields accurate error estimates and error correlations, even in the presence of large backgrounds.

You are now finished. The initial robust Λ^2_0 fit has been used to allow the phenomenologist to find a sifted data set. The subsequent application of a χ^2 fit to the *sifted set* gives stable estimates of the model parameters $\boldsymbol{\alpha}$, as well as a goodness-of-fit of the data to the model when χ^2_{min} is renormalized for the effect of truncation due to the cut $\Delta\chi^2_i(x_i;\boldsymbol{\alpha})_{\text{max}}$. Model parameter errors are found when the covariance (squared error) matrix of the χ^2 fit is multiplied by the appropriate factor $(r_{\chi^2})^2$ for the cut $\Delta\chi^2_i(x_i;\boldsymbol{\alpha})_{\text{max}}$.

It is the *combination* of using both Λ^2_0 (robust) fitting and χ^2 fitting techniques on the sifted set that gives the Sieve algorithm its power to make both a robust estimate of the parameters $\boldsymbol{\alpha}$ as well as a robust estimate of their errors, along with an estimate of the goodness-of-fit.

Using this same sifted data set, you might then try to fit to a *different* theoretical model and find χ^2_{min} for this second model. Now one can compare the probability of each model in a meaningful way, by using the χ^2 probability distribution function of the numbers of degrees of freedom for each of the models. If the second model had a very unlikely χ^2_{min}, it could now be eliminated. In any event, the model maker would now have an objective comparison of the probabilities of the two models.

3. Evaluating the Sieve Algorithm

We will give two separate types of examples which illustrate the Sieve algorithm. In the first type, we use computer-generated data, normally distributed about

- a constant, along with random noise to provide outliers. The advantage here, of course, is that we know which points are signal and which points are noise.

For our real world example, we took eight types of experimental data for elementary particle scattering from the archives of the Particle Data Group[1]. For all energies above 6 GeV, we took total cross sections and ρ-values and made a fit to these data. These were all published data points and the entire sample was used in our fit. We then made separate fits to

- $\bar{p}p$ and pp total cross sections and ρ-values,
- π^-p and π^+p total cross sections σ and ρ-values,

using eqns. (2) and (3) below.

4. Real World Data–$\bar{p}p$ and pp

We will illustrate the Sieve algorithm by simultaneously fitting all of the published experimental data above $\sqrt{s} > 6$ GeV for both the total cross sections σ and ρ values for $\bar{p}p$ and pp scattering, as well as for π^-p and π^+p scattering. The ρ value is the ratio of the real to the imaginary forward scattering amplitude and $\sqrt{s}$ is the cms energy E_{cms}. The data sets used have been taken from the Web site of the Particle Data Group[1] and have not been modified.

4.1. *Testing the Froissart Bound Hypothesis*

Testing the hypothesis that the cross sections rise asymptotically as $\ln^2 s$, as $s \to \infty$, the four functions $\sigma^\pm$ and $\rho^\pm$ that we will *simultaneously* fit for $\sqrt{s} > 6$ GeV are:

$$\sigma^\pm = c_0 + c_1 \ln\left(\frac{\nu}{m}\right) + c_2 \ln^2\left(\frac{\nu}{m}\right) + \beta_{\mathcal{P}'}\left(\frac{\nu}{m}\right)^{\mu-1} \pm \delta\left(\frac{\nu}{m}\right)^{\alpha-1}, \tag{2}$$

$$\rho^\pm = \frac{1}{\sigma^\pm}\left\{\frac{\pi}{2}c_1 + c_2\pi\ln\left(\frac{\nu}{m}\right) - \beta_{\mathcal{P}'}\cot\left(\frac{\pi\mu}{2}\right)\left(\frac{\nu}{m}\right)^{\mu-1} + \frac{4\pi}{\nu}f_+(0) \pm\delta\tan\left(\frac{\pi\alpha}{2}\right)\left(\frac{\nu}{m}\right)^{\alpha-1}\right\}, \tag{3}$$

where the upper sign is for pp (π^+p) and the lower sign is for $\bar{p}p$ (π^-p) scattering[3]. The laboratory energy is given by ν and m is the proton (pion) mass. The exponents μ and α are real, as are the 6 constants c_0, c_1, c_2, $\beta_{\mathcal{P}'}$, δ and the dispersion relation subtraction constant $f_+(0)$. We set $\mu = 0.5$, appropriate for a Regge-descending trajectory, leaving us 7 parameters. We then require the fit to be anchored by the experimental values of $\sigma_{\bar{p}p}$ and σ_{pp} (σ_{π^-p} and σ_{π^+p}), as well as their slopes, $\frac{d\sigma^\pm}{d(@}$, at $\sqrt{s} = 4$ GeV for nucleon scattering and $\sqrt{s} = 2.6$ GeV for pion scattering. This in turn imposes 4 conditions on the above equations and we thus have three free parameters to fit: c_1, c_2 and $f_+(0)$.

4.2. *$\bar{p}p$ and pp raw scattering data*

The raw experimental data for $\bar{p}p$ and pp scattering for $E_{\text{cms}} > 6$ GeV were taken from the Particle Data Group[1]. There are a total of 218 points in these 4 data sets. We fit these 4 data sets *simultaneously* using eq. (2) and eq. (3). Before we applied the Sieve, we obtained $\chi^2_{\text{min}} = 1185.6$, whereas we expected 215. Clearly, either the model doesn't work or there are a substantial number of outliers giving very large $\Delta\chi^2_i$ contributions. The Sieve technique shows the latter to be the case.

4.3. *The results of the Sieve algorithm*

We now study the effectiveness and stability of the Sieve. Table 1 contains the fitted results for $\bar{p}p$ and pp scattering using 2 different choices of the cut-off, $\Delta\chi^2_{i\,\text{max}} = 4$ and 6. It tabulates the fitted parameters from the χ^2 fit together with the errors found in the χ^2 fit, the total χ^2_{min}, ν, the number of degrees of freedom (d.f.) after the data have been sifted by the indicated $\Delta\chi^2_i$ cut-off and the renormalized $\chi^2/d.f.$

To get robust errors, the errors quoted in Table 1 for each parameter should be multiplied by the common factor r_{χ^2}=1.05, using the cut $\Delta = 6$. See Block[2] for details.

Table 1. The results for a 3-parameter fit to Eqns. 2 and 3. The renormalized χ^2_{min}/ν, taking into account the effects of the $\Delta\chi^2_{i\,\text{max}}$ cut, is given in the row labeled $\mathcal{R} \times \chi^2_{\text{min}}/\nu$.

Fitted Parameters	$\Delta\chi^2_{i\,\text{max}}$ = 4	6
c_1 (mb)	-1.452 ± 0.066	-1.448 ± 0.066
c_2 (mb)	0.2828 ± 0.0061	0.2825 ± 0.0060
$f(0)$ (mbGeV)	-0.065 ± 0.56	-0.020 ± 0.56
χ^2_{min}	142.8	182.8
ν (d.f).	182	190
$\mathcal{R} \times \chi^2_{\text{min}}/\nu$	1.014	1.040

We note that for $\Delta\chi^2_{i\,\text{max}} = 6$, the number of retained data points is 193, whereas we started with 218, giving a background of $\sim 13\%$. We have rejected 25 outlier points (5 σ_{pp}, 5 $\sigma_{\bar{p}p}$, 15 ρ_{pp} and no $\rho_{\bar{p}p}$ points) with χ^2_{min} changing from 1185.6 to 182.8. We find $\chi^2_{\text{min}}/\nu = 0.96$, which when renormalized for $\Delta = 6$ becomes $\mathcal{R} \times \chi^2_{\text{min}}/\nu = 1.04$, a very likely value with a probability of 0.34.

Obviously, we have cleaned up the sample—we have rejected 25 datum points which had an average $\Delta\chi^2_i \sim 40$! We have demonstrated that the goodness-of-fit of the model is excellent and that we had very large $\Delta\chi^2_i$ contributions from the outliers that we were able to Sieve out. These outliers, in addition to giving a huge χ^2_{min}/ν, severely distort the parameters found in a χ^2 minimization, whereas they were easily handled by a robust fit which minimized Λ^2_0, followed by a χ^2 fit to the sifted data. Inspection of Table 1 shows that the parameter values c_1, c_2 and $f_+(0)$ effectively do not depend on $\Delta\chi^2_{i\,\text{max}}$, our cut-off choice, having only very small changes compared to the predicted parameter errors. Figure 1 shows the result of the fit of eq. (2) to the sieved data sample of $\bar{p}p$ and pp cross sections. Clearly, this is an excellent fit. Its prediction at the LHC is $\sigma_{pp} = 107.6 \pm 0.1$ mb.

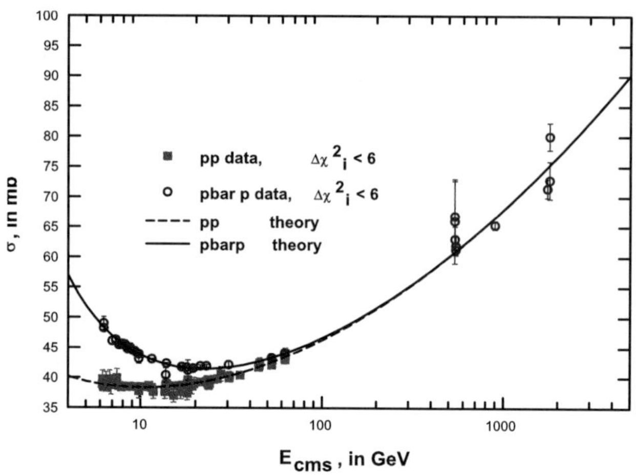

Fig. 1. A plot of $\sigma_{\bar{p}p}$ and σ_{pp}, in mb *vs.* $\mathrm{E_{cms}}$, the center of mass system energy , in GeV. The data points shown are the result of screening *all* of the cross section points for those points with $\Delta\chi_i^2 < 6$. The open circles are $\sigma_{\bar{p}p}$ and the squares are σ_{pp}. The solid line is the theoretical fit to $\sigma_{\bar{p}p}$ and the dashed line is the theoretical fit to σ_{pp}.

Due to space limitations, similarly good fits to the ρ values using eq. (3), as well as $\sigma_{\pi p}$ and $\rho_{\pi p}$ fits, are not shown—see ref. 2 for complete details.

5. Comments and Conclusions

Computer simulations[2] have shown the Sieve algorithm works well in the case of backgrounds in the range of 0 to $\sim 40\%$. Extensive computer data were generated about a straight line, as well as about a constant. It also works well for the $\sim 13\%$ to 19% contamination for the eight real-world data sets taken from the Particle Data Group[1]. However, the Sieve algorithm is clearly inapplicable in the situation where the outliers (noise) swamp the signal. In that case, nothing can be done. See ref. 2 for computer simulation results.

Our particular choice of minimizing the Lorentzian squared in order to extract the robust parameters needed to apply our Sieve technique seems to be a sensible one for both artificial computer-generated noisy distributions, as well as for real-world experimental data. The choice of filtering out all points with $\Delta\chi_i^2 > \Delta\chi_{i\,\max}^2$—where $\Delta\chi_{i\,\max}^2$ is as large as possible—is optimal in both minimizing the loss of good data and maximizing the loss of outliers.

The utilization of the "Sieved" sample with $\Delta\chi_i^2 < \Delta\chi_{i\,\max}^2$ allows one to:

(1) use the *unbiased* parameter values found in a χ^2 fit to the truncated sample for the cut $\Delta\chi_i^2(x_i;\boldsymbol{\alpha})_{\max}$, even in the presence of considerable background.
(2) find the renormalized $\chi^2_{\min}/\nu$, *i.e.*, $\mathcal{R} \times \chi^2_{\min}/\nu$.
(3) use the renormalized $\chi^2_{\min}/\nu$ to estimate the goodness-of-fit of the model employing the standard χ^2 probability distribution function. We thus estimate the probability that the data set fits the model, allowing one to decide whether to accept or reject the model.
(4) make a robust evaluation of the parameter errors and their correlations, by multiplying the standard covariance matrix C found in the χ^2 fit by the appropriate value of $(r_{\chi^2})^2$ for the cut $\Delta\chi_{i\,\max}^2$.

In conclusion, the " Sieve" algorithm gains its strength from the combination of making first a Λ_0^2 fit to get rid of the outliers and then a χ^2 fit to the sifted data set. By varying the $\Delta\chi_i^2(x_i;\boldsymbol{\alpha})_{\max}$ to suit the data set needs, we easily adapt to the different contaminations of outliers that can be present in real-world experimental data samples. Not only do we now have a robust goodness-of-fit estimate, but we also have also a robust estimate of the parameters and, equally important, a *robust* estimate of their errors and correlations. The phenomenologist can now eliminate the use of possible personal bias and guesswork in "cleaning up" a large data set.

6. Acknowledgements

I would like to thank Professor Steven Block of Stanford University for valuable criticism and contributions to this manuscript and Professor Louis Lyons of Oxford University for many valuable discussions. Further, I would like to acknowledge the hospitality of the Aspen Center for Physics.

References

1. K. Hagiwara *et al.* (Particle Data Group), Phys. Rev. D **66**, 010001 (2002).
2. M. M. Block, **arXiv** physics/0506010 (2005).
3. In deriving these equations, we have employed real analytic amplitudes derived using unitarity, analyticity, crossing symmetry, Regge theory and the Froissart bound.

χ^2 TEST FOR THE COMPARISON OF WEIGHTED AND UNWEIGHTED HISTOGRAMS

N.D. GAGUNASHVILI

University of Akureyri, Faculty of Information Technology, Borgir, v/Nordursló̀d, IS-600 Akureyri, Iceland
E-mail: nikolai@unak.is

The widely used χ^2 homogeneity test for comparing histograms(unweighted) is modified for cases involving unweighted and weighted histograms. Numerical examples illustrate an application of the method for the case of histograms with a small statistics of events and also for large statistics of events. This method can be used for the comparison of simulated data histograms against experimental data histograms.

1 Introduction

The χ^2 criteria of homogeneity [1] which is used to compare two or more histograms is well established. Without limiting the general nature of the discussion, we consider two experimental histograms with the same binning and the number of bins equal to r. Let us denote the number of events in the ith bin in the first histogram as n_i and as m_i in the second one. The total number of events in the first histogram is equal to $N = \sum_{i=1}^{r} n_i$, and $M = \sum_{i=1}^{r} m_i$ in the second histogram.

The hypothesis of homogeneity is that the two histograms represent random values with identical distributions. This is equivalent to there existing r constants $p_1, ..., p_r$, such that $\sum_{i=1}^{r} p_i = 1$, and the probability of belonging to the ith bin for some measured value in both experiments is equal to p_i. If the hypothesis of homogeneity is valid, then $p_i, i = 1, ..., r$, can be estimated from the data as

$$\hat{p}_i = \frac{n_i + m_i}{N + M}\ , \tag{1}$$

and then

$$X^2 = \sum_{i=1}^{r} \frac{(n_i - N\hat{p}_i)^2}{N\hat{p}_i} + \sum_{i=1}^{r} \frac{(m_i - M\hat{p}_i)^2}{M\hat{p}_i} \tag{2}$$

has approximately a $\chi^2_{(r-1)}$ distribution [1].

2 The test

A simple modification of the ideas described above can be used for the comparison of unweighted and weighted histograms. Let us formulate the hypothesis of identity of an unweighted histogram to a weighted histogram so that there exist r constants $p_1, ..., p_r$, such that $\sum_{i=1}^{r} p_i = 1$, and for any ith bin the following equations are valid:

$$n_i = Np_i + \delta(n_i), \qquad w_i = Wp_i + \delta(w_i)\,. \tag{3}$$

Here w_i is the weight of the contents of an ith bin, $W = \sum_i w_i$ is the common weight of the weighted histogram; $\delta(n_i)$, $\delta(w_i), i = 1, ..., r$, are the random residuals with expectation values $\mathrm{E}\,\delta(n_i) = \mathrm{E}\,\delta(w_i) = 0$ and variances $\mathrm{Var}\,\delta(n_i) = En_i$, $\mathrm{Var}\,\delta(w_i) = \sigma_i^2$. If we replace the variance $\mathrm{Var}\,\delta(n_i)$ with the estimate n_i, the variance $\mathrm{Var}\,\delta(w_i)$ with estimate s_i^2 (sum of squares of weights of events in the ith bin) and the hypothesis of identity is valid, then $p_i, i = 1, ..., r$, can be estimated from the data by the Least Squares Method [2]

$$\hat{p}_i = \frac{N + w_i W/s_i^2}{N^2/n_i + W^2/s_i^2}\,. \tag{4}$$

We may then use the test statistic

$$X^2 = \sum_{i=1}^{r} \frac{(n_i - N\hat{p}_i)^2}{n_i} + \sum_{i=1}^{r} \frac{(w_i - W\hat{p}_i)^2}{s_i^2} \tag{5}$$

and it is plausible that this has approximately a $\chi^2_{(r-1)}$ distribution.

This method, as well as the original one [1], has a restriction on the number of events in a bin. The number of events recommended for the proposed method is more than 25. In the case of a weighted histogram if the number of events is unknown, then we can apply this recommendation for the equivalent number of events as $n_i^{equiv} = w_i^2/s_i^2$.

The studentised residuals

$$R_i = \frac{n_i - N\hat{p}_i}{\sqrt{n_i}\sqrt{1 - 1/(1 + W^2 n_i/N^2 s_i^2)}} \tag{6}$$

have approximately a normal distribution with mean equal to 0 and standard deviation equal to 1 [2]. Analysis of the residuals can be useful for the identification of bins that are outliers, or bins that have a big influence on X^2.

3 Numerical example

The method described herein is now illustrated with an example. We take a distribution

$$\phi(x) = \frac{2}{(x-10)^2+1} + \frac{1}{(x-14)^2+1} \qquad (7)$$

defined on the interval $[4, 16]$. Events distributed according to the formula (7) are simulated to create the unweighted histogram. Uniformly distributed events are simulated for the weighted histogram with weights calculated by formula (7). Each histogram has 20 bins. Fig. 1 shows the result of comparison of the unweighted histogram with 2500 events and the weighted one with 500 events.

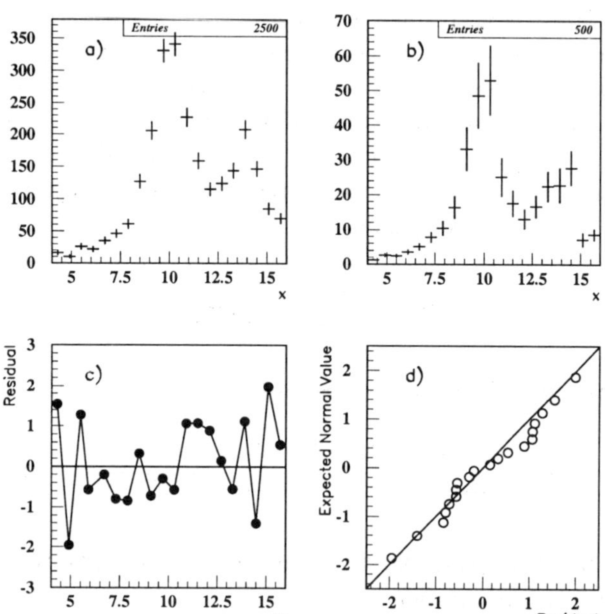

Figure 1. An example of comparison of the unweighted histogram with 2500 events and the weighted histogram with 500 events: a) unweighted histogram; b) weighted histogram; c) studentised residuals plot; c) normal Q-Q plot of residuals.

The value of test statistic X^2 is equal to 21.36 with p-value equal to 0.31, so the hypothesis of identity of the two distributions can be accepted. The behavior of the studentised residuals plot (see Fig. 1c) and the normal Q-Q plot (see Fig. 1d) of residuals are regular and we cannot identify the outliers or bins with a big influence on X^2.

To investigate the dependence of the distribution of the test statistics on the number of events, three cases were considered. The first case is the unweighted histogram with 1000 events and weighted with 200 events; the second case is 2500 events in unweighted histogram and 500 events in weighted; and the third case has 10000 and 2000 events respectively. In each case 10000 pairs of histograms were simulated with calculation of X^2 statistics for the each pair. Fig. 2 shows the Chi-square Q-Q plots and the histograms of X^2 statistics. As we can see the real distribution of test statistics obtained for low number of events has a heavier tail than the theoretical χ^2_{19} distribution. It means that the p-value calculated with the theoretical χ^2_{19} distribution is lower than the real p-value and any decision about rejecting the hypothesis of identity of the two distributions is conservative. The distribution of test statistics for the second case is close to the theoretical distribution and confirms that the greater than 25 entries in a bin is reasonable for the application of the method.

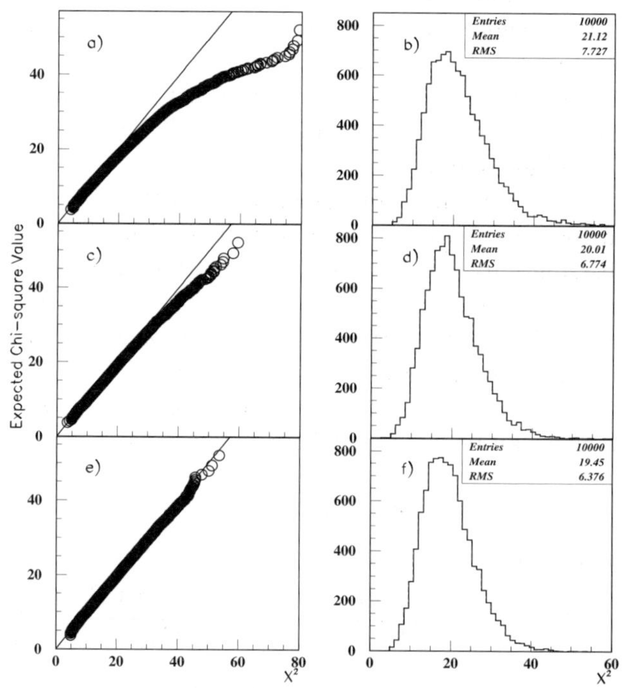

Figure 2. Chi-square Q-Q plots and histograms of X^2 statistics for: a),b) unweighted histograms with 1000 events and weighted with 200; c),d) unweighted histograms with 2500 events and weighted with 500; e),f) unweighted histograms with 10000 events and weighted with 2000.

References

1. H. Cramer, *Mathematical methods of statistics* (Princeton University Press, 1946).
2. G.A.F. Seber, *Linear Regression Analysis* (John Wiley & Sons Inc, 2003).

LIKELIHOOD/PARAMETER ESTIMATION

REDUCTION OF THE NUMBER OF VARIABLES IN PARAMETER INFERENCE

G. ZECH

Universitaet Siegen, D-57068 Siegen, Germany
E-mail: zech@physik.uni-siegen.de

Whenever acceptance losses and measurement uncertainties are to be corrected by Monte Carlo simulation, simple parameter inference with the Maximum Likelihood method suffers from statistical difficulties. MLE is especially problematic when the p.d.f. is a function of several variables. We present two methods to solve the problem. Method 1 reduces the number of variables by an optimized variable transformation. It is efficient if the number of parameters is lower than the number of variables. Method 2 fits the data to an analytic distribution which approximates the distorted distribution. The corresponding approximate likelihood estimates and their errors are then corrected by a Monte Carlo simulation. In cases where resolution and acceptance effects are very large, a crude correction is implemented to avoid a loss of precision.

Introduction

Experiments in particle physics determine interesting parameters usually through a comparison of experimental histograms to a theoretical prediction. This is problematic if the number of observed events is small and the p.d.f. depends on several variables leading to multidimensional histograms with few entries per bin. The obvious solution is to construct the likelihood function of individual events and to compute the MLE. However, the need to take into account acceptance and resolution effects prohibits this approach in most cases. The data have to be compared to a Monte Carlo simulation.

A typical historical example for this kind of problem is the determination of the vector versus axial vector coupling in the decay of the τ particle observed at the e^+e^- storage rings around 1975. At the PLUTO experiment at DESY a few dozen events of the type $e^+e^- \to \tau^+\tau^- \to l^+l^-(\nu_\tau\bar{\nu}_\tau\nu_l\bar{\nu}_l)$ had been observed. The p.d.f. is a function of the momentum vectors of the leptons and depends linearly on the parameter of interest. One of the six variables can easily be eliminated using symmetry arguments but the five remaining variables are still too many for a comparison of an experimental five dimensional histogram to the theoretical prediction which is only available in the form of a Monte Carlo simulation.

In the following section we show how in many cases the number of variables can be reduced to a single variable without loss of precision. The method has been described partially in a previous publication[1]. It was re-invented later[2] and called *the optimal observable method.* In Section 2 we present another more general method which is preferable when the p.d.f. depends on several parameters or when its function of a parameter is very non-linear.

1. Reduction of variables by variable transformation

We start with the simple case where the p.d.f. depends linearly on one parameter and two variables:

$$f(x, y, |\theta) = f_1(x, y) + \theta f_2(x, y).$$

With the substitution $v = f_1(x, y)$, $u = f_2(x, y)/f_1(x, y)$ we obtain the new distribution

$$g(u, v|\theta) = v(1 + \theta u)\frac{\partial(x, y)}{\partial(u, v)}$$

where inference of θ requires only the distribution of the *single* variable u as becomes obvious when we write down the likelihood as a function of θ,

$$\ln L(\theta) = \sum_i \ln(1 + \theta u_i) + const. \qquad (1)$$

where the sum runs over all events $u_i(x_i, y_i)$. Note that for a likelihood fit based on the very simple likelihood function (1) we would not need to know the analytic form of $g(u, v|\theta)$ which usually is a complicated function of u and v.

In most cases we have to take into account acceptance losses and resolution effects and cannot use (1) directly to fit θ. We have to simulate the marginal distribution $g'_u(u'|\theta)$ of the observed variable $u'(x', y')$ and fit θ to the observed data set u'_{obs}. Again we do not need the analytic form of $g_u(u'|\theta)$. The Monte Carlo simulation based on $f(x, y, |\theta)$ provides events (x', y') and consequently distributions of $u'(x', y')$.

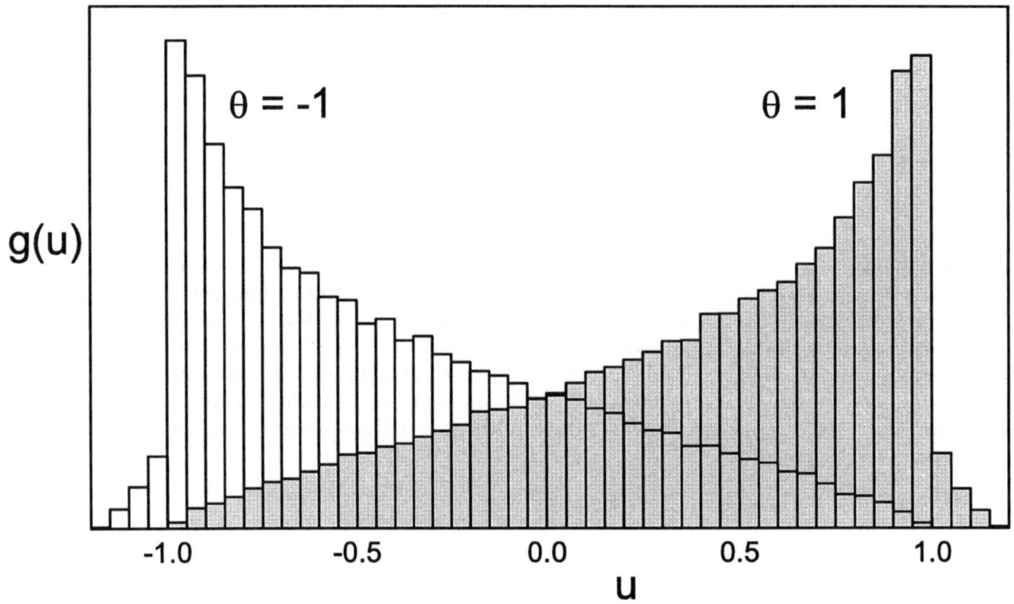

Fig. 1. Distribution of the optimum variable u.

The method is easily generalized to more than two variables. We illustrate it in a simple example. For the p.d.f.

$$f(x,y,z|\theta) = \frac{1}{\pi}\left[(x^2+y^2+z^2)^{1/2} + (x+y^3)\theta\right],$$
$$x^2+y^2+z^2 \le 1$$

the interesting variable is

$$u = \frac{x+y^3}{(x^2+y^2+z^2)^{1/2}}, \ |u| \le \sqrt{2}.$$

Figure 1 shows Monte Carlo simulations for $\theta = \pm 1$. Experimental data would have to be compared to a superposition of those two distributions. For low event numbers a likelihood fit could be applied.

Often, p.d.f.s will not depend linearly on a single parameter. We solve the problem by expanding the p.d.f. in a Taylor series at first rough estimates $\theta_{10}, \theta_{20}, ..$ of the components of the parameter vector θ:

$$\begin{aligned} f(\mathbf{x}|\theta) &= f(\mathbf{x}|\theta_0) + \frac{\partial f}{\partial \theta_1}|_{\theta_0}\,\Delta\theta_1 + \frac{\partial f}{\partial \theta_2}|_{\theta_0}\,\Delta\theta_2 + ... \\ &= v(1 + u_1\Delta\theta_1 + u_2\Delta\theta_2 + ...), \\ u_i &= \frac{1}{f(\mathbf{x}|\theta_0)}\frac{\partial f}{\partial \theta_i}|_{\theta_0} \end{aligned}$$

We determine the deviations $\boldsymbol{\Delta\theta}$ of the parameters from the initial estimates. If necessary the procedure can be iterated.

Now, one new variable is required for each parameter. Thus our method reduces the number of components of our variable vector $\mathbf{x}$ only if this number is larger than the number of parameters which have to be determined. But independent of the fitting procedure it will be instructive to investigate the individual marginal u-distributions.

The variable transformation not only simplifies fitting but also helps to visualize the parameter dependence and to apply goodness-of-fit tests to sensible variables.

2. Monte Carlo correction of approximate estimators

We can reduce the experimental information even further to a single constant per parameter. A simple example illustrates our approach. Let us consider a lifetime distribution which suffers from acceptance losses and resolution effects. In an ideal experiment the mean value of the observed times is a sufficient statistic. (This is even true when the time interval is restricted.)

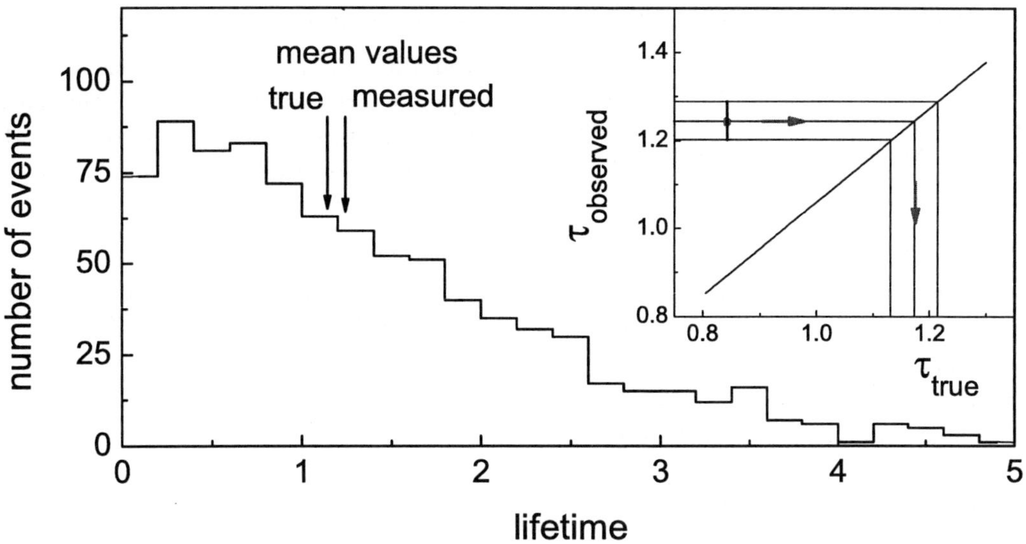

Fig. 2. Monte Carlo correction of mean life. The insert shows the correction function.

Now, the mean value $\bar{t}_{observed}$ of the lifetimes observed in the real experiment still contains most of the relevant information. The estimate $\hat{\tau}$ is obtained by a Monte Carlo simulation which provides the relation between τ_{true} and $\bar{t}_{observed}$. A schematic example where a bias due to resolution effects is corrected is shown in Figure 2.

In the general case, we compute the maximum likelihood estimate $\hat{\theta}_{observed}$ of the parameters using the undistorted p.d.f. $f(\mathbf{x}'|\theta)$ with the experimental variables $\mathbf{x}'$. The relation $\theta(\hat{\theta}_{observed})$ is obtained by a Monte Carlo simulation. The simulation of the experiment is performed with a fixed parameter set θ_0. By re-weighting the events by $f(\mathbf{x}|\theta)/f(\mathbf{x}|\theta_0)$ the full function $\theta(\hat{\theta}_{observed})$ is generated.

When the experimental resolution is extremely poor, it might happen that $f(\mathbf{x}'|\theta)$ is undefined for some values of the measured variable vector $\mathbf{x}'$. Then we either have to exclude the corresponding events or better to scale the variables in such a way that all events are within the physically allowed range.

The methods described in Section 1 and so far in Section 2 work well whenever acceptance and resolution effects are not too large. Then the loss in precision due to the applied approximations is negligible. We now consider large acceptance and resolution effects. We first include *acceptance losses* $\alpha(\mathbf{x})$. The modified p.d.f. is

$$h(\mathbf{x},\theta) = \frac{\alpha(\mathbf{x})f(\mathbf{x}|\theta)}{\int \alpha(\mathbf{x})f(\mathbf{x}|\theta)\mathbf{dx}}$$

With the abbreviation $A(\theta) = \int \alpha(\mathbf{x})f(\mathbf{x},\theta)\mathbf{dx}$, the log-likelihood function for N events is

$$\ln L(\theta|\mathbf{x}) = \sum_{i=1}^{N} \ln f(\mathbf{x}_i|\theta) - N\ln A(\theta) + N\ln\alpha(\mathbf{x}) \tag{2}$$

where we can drop the last term. The function $A(\theta)$ is proportional to the event rate for a given luminosity. Whenever it does not depend on θ it can be neglected and the problem is reduced to the previous case. Usually $A(\theta)$ will depend only weakly on θ. We obtain it either from a Monte Carlo simulation or approximate it by an analytic estimate. The MLE from eqn. 2 and the observed data provide $\hat{\theta}_{observed}$. The transition to $\hat{\theta}$ is performed as described above. Since the simulation takes care of all experimental effects including the estimation method, all approximations are corrected. Therefore, a crude estimate of $A(\theta)$ is sufficient.

In the rare cases where resolution effects strongly depend on the parameters, we have also to correct for the convolution. Standard unfolding methods require histogramming of events and thus are not applicable since the ML method requires individual events. Binning free unfolding[3] can be used but this is a rather involved approach. Usually, it will be simpler to convolute the p.d.f. Since the convoluted p.d.f. is used in the likelihood fit where it has to be repeated with each parameter change, the convolution cannot be performed in an elaborate Monte Carlo simulation but has to be approximated by a simple smearing algorithm. Again, a crude approximation is sufficient since all approximations will be corrected automatically.

Acknowledgments

I would like to thank Louis Lyons and his team for the perfect organization of a very stimulating PHYSTAT Conference. I am grateful to Roger Barlow and Louis Lyons for useful comments.

References

1. G. Zech, *Nucl. Instr. and Meth.* **137**, 551 (1987).
2. M. Diehl and O. Nachtmann *Z. f. Phys.* C **62**, 397 (1994).
3. B. Aslan and G. Zech, *Nucl. Instr. and Meth.* A **537**, 626 (2005).

ERRORS FROM THE LIKELIHOOD FUNCTION

ROGER BARLOW

School of Physics and Astronomy, The University of Manchester, Manchester UK

The 'errors' on a result are often evaluated by finding the points at which the log likelihood falls by $\frac{1}{2}$ from its peak value. This is examined for two cases: a lifetime measurement and a Poisson measurement. Results are compared with the exact central Neyman construction and with the Bartlett approximation. It is shown that the agreement of the log likelihood method with the central Neyman construction is poor, and the Bartlett approach explains why.

1. Errors from the Likelihood

The 'errors' on a result are generally interpreted as the 68% confidence interval about the central value[1]. If the number of measurements N is large, the quoted error $\sigma_{\hat{a}}$ of the maximum likelihood estimator $\hat{a}$ of a parameter a is read off the parabolic log likelihood curve from the points at which $\ln L(a)$ falls by $\frac{1}{2}$ from its peak value $\ln L(\hat{a})$: $\Delta \ln L = -\frac{1}{2}$. For experiments with finite N, which give non-Gaussian results, the statistical error is evaluated by a similar procedure. The values $a_\pm$ below and above $\hat{a}$ for which $\Delta \ln L = \ln L(a_\pm) - \ln L(\hat{a}) = -\frac{1}{2}$ are found, and the 68% (more accurately, 68.27%) interval quoted as $[a_-, a_+]$ or $[\hat{a} - \sigma_-, \hat{a} + \sigma_+]$.

A non-rigorous argument for doing so is that, whatever the shape of the likelihood function, it can (unless it has multiple peaks) be made Gaussian by a suitable transformation of the variable a. The standard procedure applied to this transformed variable will give the $\Delta \ln L = -\frac{1}{2}$ points, which would then be transformed back into those of the original variable. This is convenient, but not justified. Even if a finite-N likelihood function has, by chance or design, a Gaussian form so that the log likelihood has a parabolic shape entirely described by the second derivative, $\frac{d^2 \ln L}{da^2}$, the value of that second derivative will in general, because N is finite, differ from the expectation value $\left\langle \frac{d^2 \ln L}{da^2} \right\rangle$, which gives the variance of the estimate. It is now being increasingly questioned[2–5] and an examination of how well it works in practice is needed to inform the discussion.

We examine two cases: the determination of the lifetime of an unstable state decaying according to the radioactive decay law, and the determination of the number of events produced by a Poisson process, typical cases in particle physics. We use the 68% central region, but the techniques can be applied to central or one-sided regions with any probability content. A version of this note is already published[6].

Bayesian statistics can also be used to give confidence intervals. This is not considered here: we compare the exact Neyman confidence intervals with two other frequentist-motivated methods.

2. An Exact Interval

If $\hat{a}$ is continuous the central Neyman confidence interval can be found[7] from the values satisfying:

$$\int_0^{\hat{a}} P(\hat{a}'; a_+) d\hat{a}' = 0.16 \qquad \int_{\hat{a}}^{\infty} P(\hat{a}'; a_-) d\hat{a}' = 0.16$$

(depicted graphically in Figure 1) where $P(\hat{a}; a)$ is the probability density for a true value a giving an estimate $\hat{a}$. These equations define the confidence belt such that the probability of a measurement lying within the region is, by construction, 68%, and the probability of an under- or over-estimate is 16%. If $\hat{a}$ is discrete then the integrals become sums and the equations are inequalities, so that the central probability is at least 68% and each tail at most 16%.

3. The Bartlett Approximation

An alternative approximation technique is due to Bartlett[8–10]. If $\hat{a}$ is unbiassed then for any N

$$\frac{d \ln L}{da} = \sum_1^N \frac{d \ln P(x_i; a)}{da}$$

is distributed with mean zero and variance $-\left\langle \frac{d^2 \ln L}{da^2} \right\rangle$. For large N the Central Limit Theorem states that its distribution is Gaussian.

Within the limits of this Gaussian approximation (refinements can correct for the non-Gaussian finite N behaviour, but these lie outside our scope)

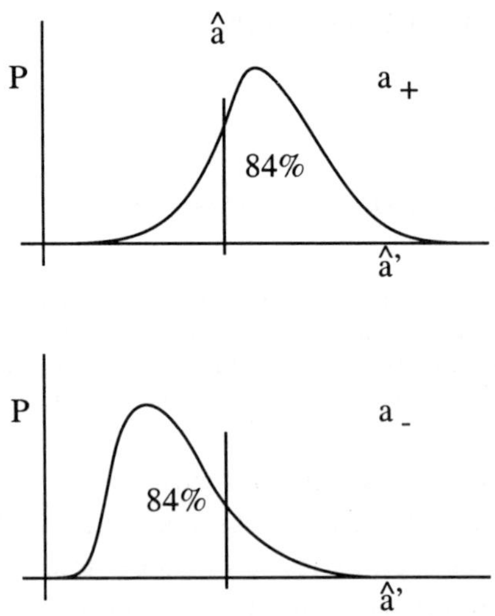

Fig. 1. If the true value is a_+ then the probability of getting a result $\hat{a}$ or less is at most 16%, and similarly for a_-

this gives probability regions for $\frac{d \ln L}{da}$. If this quantity can be expressed in terms of $\hat{a}-a$ then these can be interpreted as confidence regions for a. This will be illustrated in the two examples.

4. Example 1: Measuring a Lifetime

The probability for a state with mean lifetime τ to decay after an observed time t is given by

$$P(t;\tau) = \frac{1}{\tau} e^{-t/\tau}.$$

The log likelihood for N measurements $t_1 \dots t_N$ is

$$\ln L = -N\frac{\bar{t}}{\tau} - N \ln \tau$$

where $\bar{t} = \frac{1}{N}\sum t_i$. Differentiation to find the maximum gives $\hat{\tau} = \bar{t}$ and $\ln L(\hat{\tau}) = -N(1 + \ln \bar{t})$.

The probability distribution for $\bar{t}$ is given[3] by convoluting the $P(t)$ with itself $N-1$ times:

$$P(\bar{t};\tau) = \frac{N^N \bar{t}^{N-1}}{\tau^N (N-1)!} e^{-N\bar{t}/\tau}.$$

For the exact Neyman region we require the integral of this from zero to the measured value, which is to be 16% for the upper limit $\tau_+ \equiv \bar{t} + \sigma_+$ and 84% for the lower limit $\tau_- \equiv \bar{t} - \sigma_-$. This is given by

$$\int_0^{\bar{t}} P(\bar{t}';\tau)\, d\bar{t}' = 1 - e^{-N\bar{t}/\tau} \sum_{j=0}^{N-1} \frac{\bar{t}^j N^j}{j! \tau^j}.$$

Thus for a given N and $\bar{t}$ one has to adjust the value of τ until the desired probability content (here 0.84 or 0.16 for lower or upper limits) is achieved.

The region obtained, expressed as differences from the measured $\bar{t}$, is shown in the columns 2 and 3 of Table 1, for various values of N. (Everything scales with $\bar{t}$, so the errors are given in terms of that value.)

Table 1. Error bars for 68% confidence regions obtained by the 3 methods for a Lifetime measurement

N	Exact		$\Delta \ln L$	$= -\frac{1}{2}$	Bartlett	
	$\sigma_-/\bar{t}$	$\sigma_+/\bar{t}$	$\sigma_-/\bar{t}$	$\sigma_+/\bar{t}$	$\sigma_-/\bar{t}$	$\sigma_+/\bar{t}$
1	0.457	4.787	0.576	2.314	0.500	∞
2	0.394	1.824	0.469	1.228	0.414	2.414
3	0.353	1.194	0.410	0.894	0.366	1.366
4	0.324	0.918	0.370	0.725	0.333	1.000
5	0.302	0.760	0.340	0.621	0.309	0.809
6	0.284	0.657	0.318	0.550	0.290	0.690
7	0.270	0.584	0.299	0.497	0.274	0.608
8	0.257	0.529	0.284	0.456	0.261	0.547
9	0.247	0.486	0.271	0.423	0.250	0.500
10	0.237	0.451	0.260	0.396	0.240	0.463
15	0.203	0.343	0.219	0.310	0.205	0.348
20	0.182	0.285	0.194	0.261	0.183	0.288
25	0.166	0.248	0.176	0.230	0.167	0.250
50	0.124	0.164	0.129	0.156	0.124	0.165
100	0.0908	0.1109	0.0937	0.1070	0.0909	0.1111
250	0.0594	0.0675	0.0607	0.0660	0.0595	0.0675
500	0.0428	0.0468	0.0434	0.0461	0.0428	0.0468
1000	0.0306	0.0326	0.0310	0.0323	0.0307	0.0327

The $\Delta \ln L = -\frac{1}{2}$ points are found by interpolating to find the values of τ for which $N - N/\tau - N \ln \tau = -\frac{1}{2}$, and shown in columns 4 and 5.

For the Bartlett approximation, we have

$$\frac{d \ln L}{d\tau} = \frac{N}{\tau^2}(\bar{t} - \tau)$$

which clearly has a mean of zero: differentiating again and taking the expectation value gives its variance as $\frac{N}{\tau^2}$. So, for any τ, $\bar{t}$ has mean τ and standard deviation $\tau/\sqrt{N}$. This is exact. We then – this is the approximation – take this as being Gaussian and use it in the Neyman prescription, accordingly requiring that $\bar{t}$ lie one standard deviation above $\tau_- \equiv \bar{t} - \sigma_-$ and one standard deviation below $\tau_+ \equiv \bar{t} + \sigma_+$

$$\bar{t} = \tau_+ - \frac{\tau_+}{\sqrt{N}} \qquad \bar{t} = \tau_- + \frac{\tau_-}{\sqrt{N}}$$

i.e. $\sigma_- = \frac{\bar{t}}{\sqrt{N+1}}$ and $\sigma_+ = \frac{\bar{t}}{\sqrt{N-1}}$.

These are shown in the final two columns of Table 1. The results are also presented in Figure 2.

The Bartlett error bars are surprisingly close to the central Neyman ones. The $\Delta \ln L = -\frac{1}{2}$ error bars are surprisingly different: about ten events are needed for them to be within 10%, and hundreds for agreement at the few per cent level.

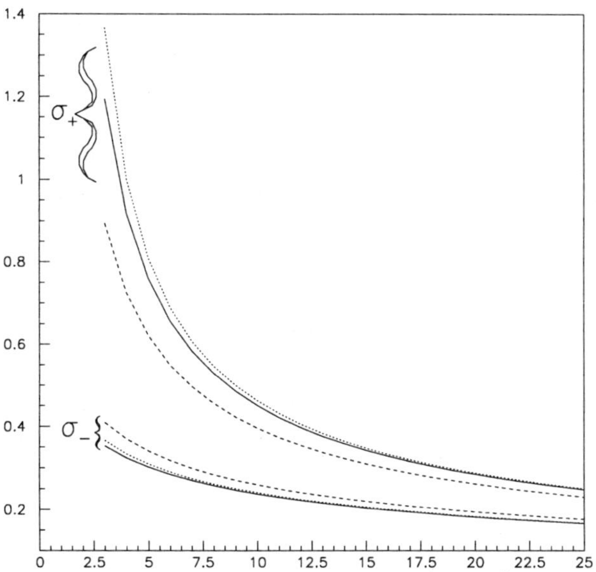

Fig. 2. Upper and lower limits on the 68% central confidence interval for a lifetime measurement with various numbers of data values, showing the exact construction (solid), the Bartlett approximation (dotted) and the $\Delta \ln L$ approximation (dashed). The limits are shown as fractions, $\sigma/\bar{t}$.

5. Example 2: A Poisson Counting Experiment

If N events are seen from a Poisson process of true mean λ, the maximum likelihood estimate $\hat{\lambda}$ is just N. The Neyman limits are found from

$$\sum_{r=0}^{N} e^{-\lambda_+} \frac{\lambda_+^r}{r!} = 0.16 \qquad \sum_{r=0}^{N-1} e^{-\lambda_-} \frac{\lambda_-^r}{r!} = 0.84.$$

The resulting σ values are shown in columns 2 and 3 of Table 2. The $\Delta \ln L = -\frac{1}{2}$ errors are read off from $N - \lambda + N \ln(\lambda/N)$. These are shown in columns 4 and 5.

The Bartlett method gives the familiar fact that the variance of $N - \lambda$ is just λ. This suggests that

$$N - \lambda_- = \sqrt{\lambda_-} \qquad \lambda_+ - N = \sqrt{\lambda_+}.$$

However $P(N; \lambda)$ is defined for integer N only. To use this set of discrete spikes as a function requires us to replace it by a histogram defined as $e^{-\lambda}\lambda^N/N!$ for values of the continuous variable between $N - \frac{1}{2}$ and $N + \frac{1}{2}$. This requires us to add $\frac{1}{2}$ to each of the ranges, giving

$$\sigma_- = \sqrt{N + \frac{1}{4}} \qquad \sigma_+ = \sqrt{N + \frac{1}{4}} + 1$$

These are shown in columns 6 and 7 of Table 2. The data are shown graphically in Figure 3.

Table 2. Error bars for 68% confidence regions obtained by the 3 methods for a Poisson measurement

N	Exact		$\Delta \ln L$	$= -\frac{1}{2}$	Bartlett	
	σ_-	σ_+	σ_-	σ_+	σ_-	σ_+
1	0.827	2.299	0.698	1.358	1.118	2.118
2	1.292	2.637	1.102	1.765	1.500	2.500
3	1.633	2.918	1.416	2.080	1.803	2.803
4	1.914	3.162	1.682	2.346	2.062	3.062
5	2.159	3.382	1.916	2.581	2.291	3.291
6	2.380	3.583	2.128	2.794	2.500	3.500
7	2.581	3.770	2.323	2.989	2.693	3.693
8	2.768	3.944	2.505	3.171	2.872	3.872
9	2.943	4.110	2.676	3.342	3.041	4.041
10	3.108	4.266	2.838	3.504	3.202	4.202
15	3.829	4.958	3.547	4.213	3.905	4.905
20	4.434	5.546	4.145	4.811	4.500	5.500
25	4.966	6.066	4.672	5.339	5.025	6.025
50	7.046	8.117	6.742	7.408	7.089	8.089
100	9.982	11.03	9.669	10.34	10.01	11.01
250	15.80	16.83	15.48	16.15	15.82	16.82
500	22.35	23.37	22.03	22.70	22.37	23.37
1000	31.61	32.63	31.29	31.96	31.63	32.63

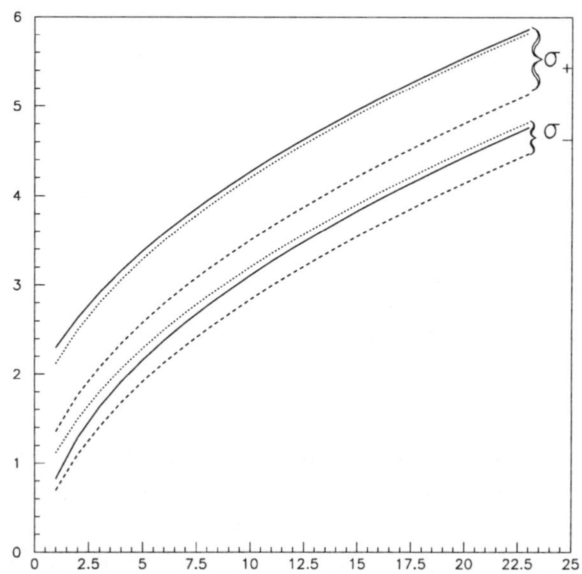

Fig. 3. Upper and lower σ values on the 68% central confidence interval for a Poisson measurement, as a function of the measured value N, showing the exact construction (solid), the Bartlett approximation (dotted) and the $\Delta \ln L$ approximation (dashed)

Again, the Bartlett error bars are close to the central Neyman ones, and the $\ln L$ error bars very different, with tens of events required for 'fair' and hundreds for 'good' agreement. Furthermore, in this case it underestimates both errors. Adding 0.5 to each limit, to account for the discrete binning, would improve the agreement though, unlike the Bartlett case where this is forced on us, this is an *ad hoc* adjustment not done in practice. Anyway inspection of the table shows that even after adjusting in this way, agreement with the exact values would still be worse than that from the Bartlett approximation.

6. Coverage

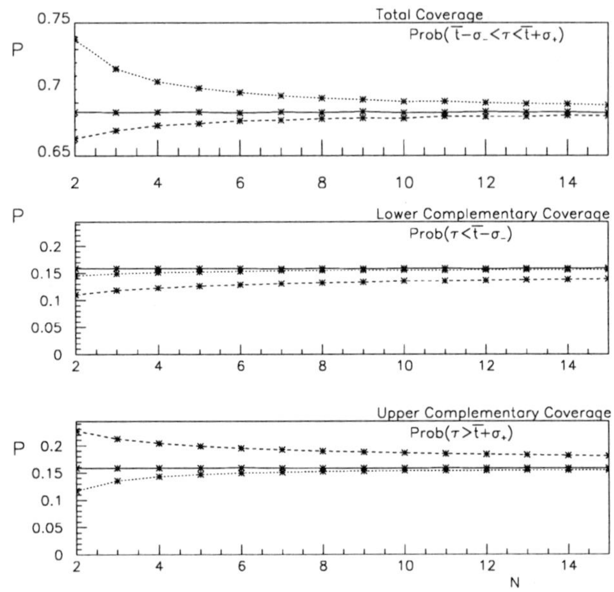

Fig. 4. Coverage for the Lifetime measurement with different N, for the exact method (solid), the $\Delta \ln L$ approximation (dashed) and the Bartlett approximation (dotted)

The probability integrals of section 4 scale with $\bar{t}/\tau$, so the coverage - the probability that a particular value τ will generate a value of $\bar{t}$ such that the interval $[\bar{t}(1-\sigma_-/\bar{t}), \bar{t}(1+\sigma_+/\bar{t})]$ encloses it - is, for any chosen pair of values $(\sigma_-/\bar{t}, \sigma_+/\bar{t})$, independent of τ, although it does depend on N.

The top plot of Figure 4 shows the coverage for the three methods. The exact method, by construction, has 68% coverage for all N. The Bartlett method (dotted line) overcovers slightly and the $\Delta \ln L$ method (dashed line) undercovers slightly. For the strict frequentist overcoverage is, though to be avoided if possible, permissible, whereas any undercoverage invalidates the method. Even so, this small undercoverage is probably admissible.

The figure also shows the probabilities that the interval lie above the true value (second plot) or below it (third plot). Again, the exact method gives 16% for all N. The Bartlett method's overcoverage (in these plots a point below the 16% line shows a low probability of true value lying outside the interval, i.e. overcoverage) is similar above and below. But the $\Delta \ln L$ method overcovers in the lower region and undercovers in the upper region. (Its values of σ_- in Table 1 are greater than the exact values whereas its values of σ_+ are less.) The 'admissible' coverage in the top plot is achieved through a cancellation in the lower two. Although the $\Delta \ln L$ interval is a good approximation to a 68% confidence interval, it is not nearly so good an approximation to a central 68% interval.

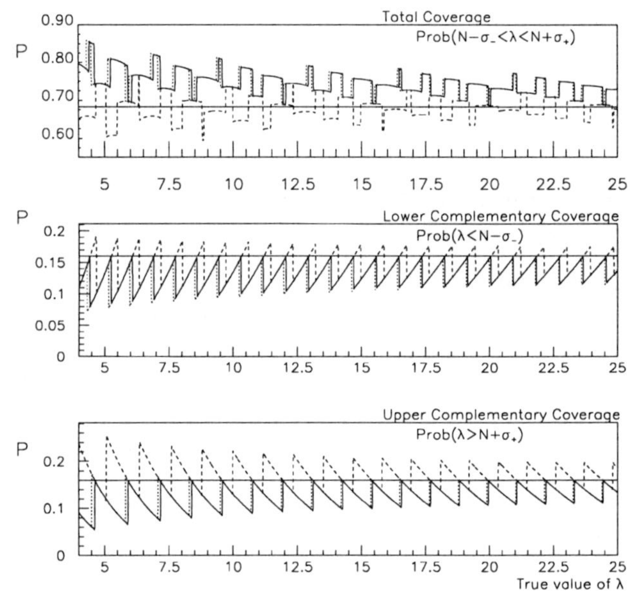

Fig. 5. Coverage for the Poisson measurement with the exact method (solid), the $\Delta \ln L$ approximation (dashed) and the Bartlett approximation (dotted)

For the Poisson measurement we do have to consider coverage as a function of λ. For the exact interval this is shown by the solid line in the top plot of Figure 5. N (unlike $\bar{t}$) is discrete so the probability content cannot in general be adjusted to 68%, and one has settle for a larger value. The interval overcovers. The overcoverage effect for a central interval is greater than for a single-sided upper limit, as the excess comes from both upper and lower ends. This is shown by the solid line in the lower plots, which again display the probabilities of the lower edge of the interval being greater than the true value, and the upper edge being less. As λ increases the probability of the lower limit exceeding it increases until it gets to 16%, at which point the allowed range of N increases by 1, removing this probability from the sum. At the upper limit, increasing λ lowers the probability of a result in the lower tail, until the lowest N value in the allowed range can be removed from it, giving a step increase in probability. The combination of the two sawtooth plots gives the crenellated effect in the total.

The coverage from the Bartlett method (dotted line) is, as would be expected from Table 2, a very close match to that of the exact interval. The $\Delta \ln L$ interval (dashed line) overcovers for some λ and undercovers for others, with the upper value typically undercovering and the lower typically overcovering, to give a combined result within 10% of what it purports to be. It can be argued[11, 12] that overcoverage is undesirable, and the $\Delta \ln L$ approximation is preferable as averaging over many cases the over- and under-coverage tend to cancel each other out. But if this is done it should be a conscious decision, not an unintentional consequence of the method. A discussion of coverage in Poisson measurements, for various approximations, can be found in[13].

7. Discussion

If reported errors obtained by $\Delta \ln L = -\frac{1}{2}$ method are regarded, by an author or a reader, as an approximation to the central Neyman interval, then the inaccurary is appreciable. Even for values of N of order 100, generally considered 'large', it is typically wrong in the second significant figure, and often grossly wrong. And yet values obtained by this method are frequently quoted to considerable precision by experiments.

Only two measurements have been considered. Yet they are typical in particle physics and (in more complicated form) cover a wide range of practice.

As used here the Bartlett method is just like the familiar Gaussian. The probability is

$$P(\hat{a}; a) = \frac{1}{\sqrt{2\pi}\sigma(a)} e^{-(a-\hat{a})^2/2\sigma(a)^2}$$

where the a dependence of σ is written in explicitly. Confidence regions are read off as usual (68% is 1σ, 95% is 2σ, and so on) except that σ is a function of the parameter being investigated. The 68% limits thus correspond to a fall of $\frac{1}{2}$ in the log likelihood from the exponential. But the total log likelihood also changes due to the denominator. The $\Delta \ln L = -\frac{1}{2}$ method includes this $-\ln \sigma(a)$ term, thereby making it differ from the remarkably accurate Bartlett approximation.

The $\Delta \ln L = -\frac{1}{2}$ method is widespread and there is no prospect that people will abandon it. Indeed, it does convey basic properties of the Likelihood function in a compact way, and the coverage is generally fair. This study does not advocate that practitioners stop using it. But discretion must be exercised in using its results by interpreting them as equivalent to a central Neyman confidence interval.

References

1. L. Alvarez-Gaumé et al *Review of Particle Properties, Phys. Lett.* B **592**, 1 (2004) page 14.
2. R. J. Barlow *Introduction to Statistical Issues in Particle Physics,* in PHYSTAT2003 (Ed. L. Lyons), Stanford, Sept. 2003. SLAC-R-703 (2004).
3. A. Bukin:*A Comparison of Methods for Confidence Intervals* in PHYSTAT2003 (Ed. L. Lyons), Stanford, Sept. 2003. SLAC-R-703 (2004).
4. G. Zech, Private communication.
5. G. D'Agostini:*Bayesian Reasoning in Data Analysis - a Critical Introduction,* World Scientific Publications, 2003.
6. R. J. Barlow, *Nucl. Instrum. Methods Phys. Res.* A **550**, 392 (2005).
7. M.G. Kendall and A. Stuart: *The Advanced Theory of Statistics,* Charles Griffin & Co., Vol II, 4th Edition (1979).
8. M.S. Bartlett: *On the Statistical Estimation of Mean Lifetimes,* Phil. Mag. **44** 244 (1953).
9. M.S. Bartlett: *Estimation of Mean Lifetimes from Multiple Plate Cloud Chamber Tracks,* Phil. Mag. **44** 1407 (1953).
10. B. P. Roe:*Probability and Statistics in Experimental Physics,* 2nd Edition, Springer, 2001. See pp. 202 ff.
11. F. Porter: *Interval Estimation using the Likelihood Function,* Nucl. Instrum. Methods Phys. Res. **A368** 783-803 (1996).
12. F. Porter *Statistical Issues in Particle Physics - a View from BaBar,* in PHYSTAT2003 (Ed. L. Lyons), Stanford, Sept. 2003. SLAC-R-703 (2004).
13. J. Heinrich: *Coverage of Error Bars for Poisson Data,* CDF statistics memo 6438, on `http://www-cdf.fnal.gov/publications/cdf6438_coverage.pdf`

ASYMMETRIC STATISTICAL ERRORS

ROGER J. BARLOW
School of Physics and Astronomy, Manchester University, Manchester, UK

We investigate ways of parametrising non-Gaussian likelihood functions when the only information given is the location of the peak and the errors σ_- and σ_+. We show that the traditional method leads to bias, and improvement is possible.

1. Introduction: Likelihood Functions

In general the likelihood function is not symmetric about its maximum, and the quoted errors σ_- and σ_+, estimated using the $\Delta \ln L = -\frac{1}{2}$ method[1], are different. For simple (large N) results this difference is small and the log likelihood can be treated as a parabola. For a complex result (and a conscientious investigator) the full likelihood function is reported explicitly. But for many results the only information on the likelihood is the peak location $\hat{a}$ and the points $\hat{a} + \sigma_+$, $\hat{a} - \sigma_-$, at which lnL falls by $\frac{1}{2}$.

The combination of statistical errors and the combination of results require the complete likelihood functions[2]. So these have to be modelled. We examine some models and their performance in typical cases, and apply them to some toy experiments.

2. Drawing a Curve Through 3 Points

We seek to parametrise $F(a) = lnL(\vec{x}; a)$ given $F'(\hat{a}) = 0 \qquad F(\hat{a} + \sigma_+) = F(\hat{a} - \sigma_-) = F(\hat{a}) - \frac{1}{2}$

The function has 3 parameters, obtained from the 3 input values $\hat{a}, \sigma_-$ and σ_+. (A 4th parameter corresponding to the actual value of $\ln L(\hat{a})$ is of no interest.) This is not a well-formed problem. There are infinitely many curves that go through the 3 points and satisfy these equations. But it has to be asked, as combination of asymmetric errors is going on in many experiments. We cannot hope to find the 'correct' answer as there is none: instead we seek something sensible and consistent and easy to use We illustrate this with two examples. One is a Poisson measurement with $n = 5$. The likelihood function is $L(5; a) = e^{-a} a^5/5!$. Its peak is at $\hat{a} = 5$ and $\sigma_+ = 2.581$, $\sigma_- = 1.916$. We take these values and see how well the parametrisations reproduce $L(5; a)$. The second is a Gaussian measurement of mean 8.0 and standard deviation 3.0, for which the logarithm is taken. $\hat{a} = ln(8) = 2.0744$, $\sigma_+ = ln(11) - ln(8) = 0.3184$, $\sigma_- = ln(8) - ln(5) = 0.4700$. (This is not the logNormal distribution, just the equivalent of plotting a normal distribution on log-linear graph paper.) We investigate how well the parametrisations reproduce $L(8; a) = \frac{1}{3\sqrt{2\pi}} exp(\frac{-(e^a - 8)^2}{18})$.

2.1. *Traditional Method*

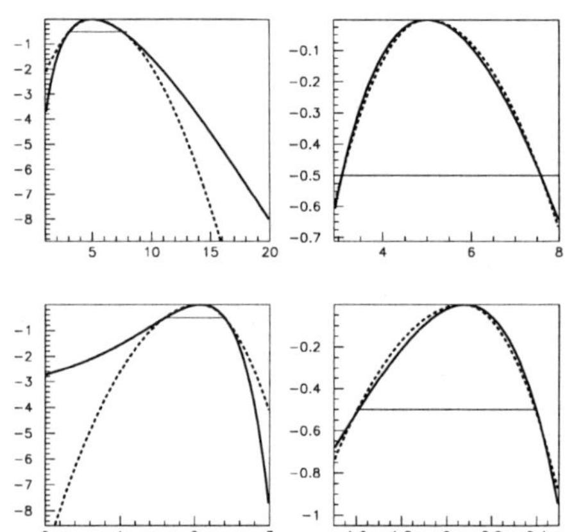

Fig. 1. Traditional Method. For explanation see text.

The usual way of combining asymmetric errors is to add the positive and negative errors separately in quadrature. This amounts to using a split Gaussian with σ_+ above the peak, σ_- below.

This is shown in Fig. 1. In all 4 plots the solid lines show the true likelihood function, dashed lines the parametrisation. The upper plots show this for the first (Poisson) example, the lower plots for the second (log-linear Gaussian). The right hand plots are close-ups of the central region. The horizontal line is $\frac{1}{2}$ below the peak. All the fits agree, by construction, with the true curves where they cross this line. Plots for later methods use the same system.

For this method, the agreement in the central region appears to be fair in both cases, whereas further away it is not very good.

2.2. *PDG Method*

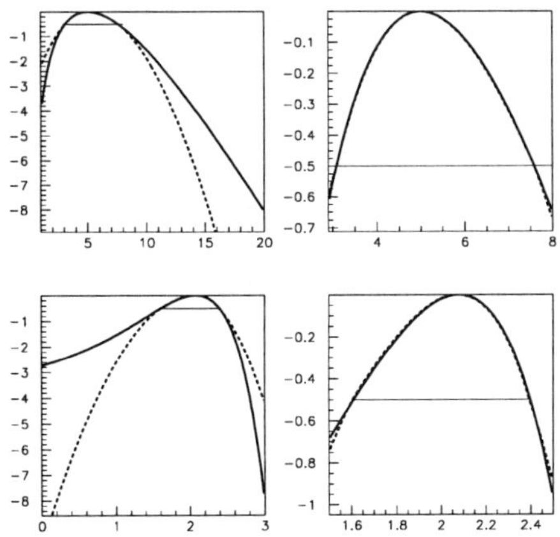

Fig. 2. PDG Method. For explanation see text of 2.1.

The PDG[3], in combining results from different experiments, adapts this method to avoid the step in the second derivative. Between $[\hat{a}-\sigma_-, \hat{a}+\sigma_+]$ it uses a Gaussian interpolation between $\sigma = \sigma_-$ and $\sigma = \sigma_+$. Agreement (Fig. 2) within the central region is improved, but the model outside is unchanged.

2.3. *Cubic*

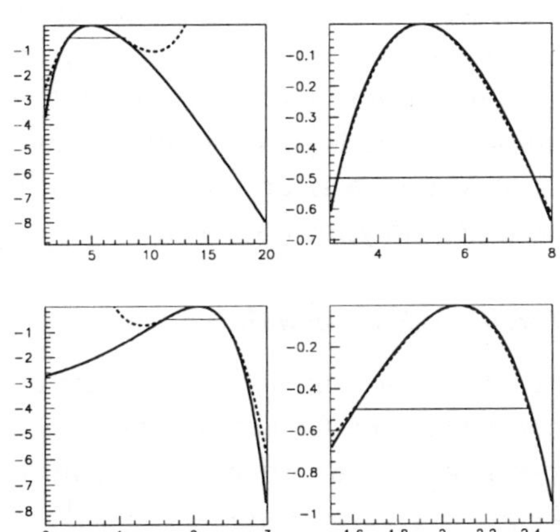

Fig. 3. Cubic Method. For explanation see text of 2.1.

A small cubic term gives $F(a) = \alpha a^2 + \beta a^3$ with $\alpha = -\frac{1}{2}\frac{\sigma_-^3+\sigma_+^3}{\sigma_+^2\sigma_-^2(\sigma_-+\sigma_+)}$ and $\beta = -\frac{1}{2}\frac{\sigma_-^2-\sigma_+^2}{\sigma_+^2\sigma_-^2(\sigma_-+\sigma_+)}$ (Here and later we take $\hat{a}=0$ for simplicity.) But a small term does not stay small, and although the central region is fine, it rapidly becomes bad outside (see Fig. 3). $F(a)$ tends to $+\infty$ rather than $-\infty$ for either large a or large $-a$. This approach is unworkable.

2.4. *Restricted Quartic*

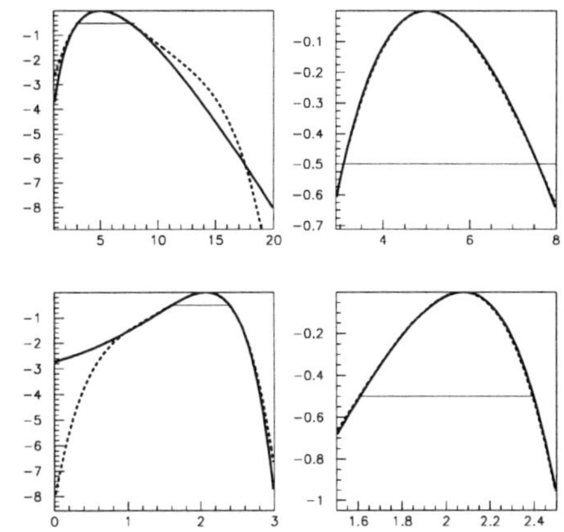

Fig. 4. Quartic Method. For explanation see text of 2.1.

A quartic $F(a) = -\left(\frac{\alpha^2 a^2}{2} + \frac{\alpha\beta a^3}{3} + \frac{\beta^2 a^4}{12}\right)$ has a single peak as the second derivative is negative definite. β and α are found exactly, with expressions too complicated to be given here. Agreement (Fig. 4) is good in the central region, and outside for the second example but not the first.

2.5. *Logarithmic*

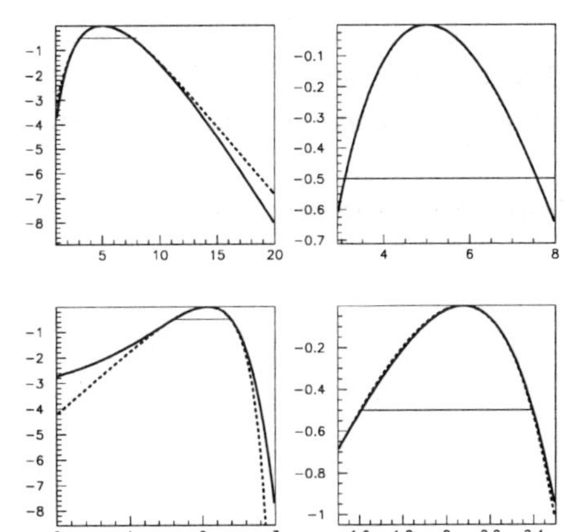

Fig. 5. Logarithmic Method. For explanation see text of 2.1.

Another method for an asymmetry is to scale the x-axis proportionately to the distance from the peak. This leads to $F(a) = -\frac{1}{2}\left(\frac{ln(1+\gamma a)}{ln\beta}\right)^2$ with $\beta = \sigma_+/\sigma_-, \gamma = \frac{\sigma_+-\sigma_-}{\sigma_-\sigma_+}$. Agreement (Fig. 5) is reasonable. However the form is badly behaved when $1+\gamma a$ goes negative, and does not give a parabola for $\sigma_- = \sigma_+$.

2.6. *Generalised Poisson*

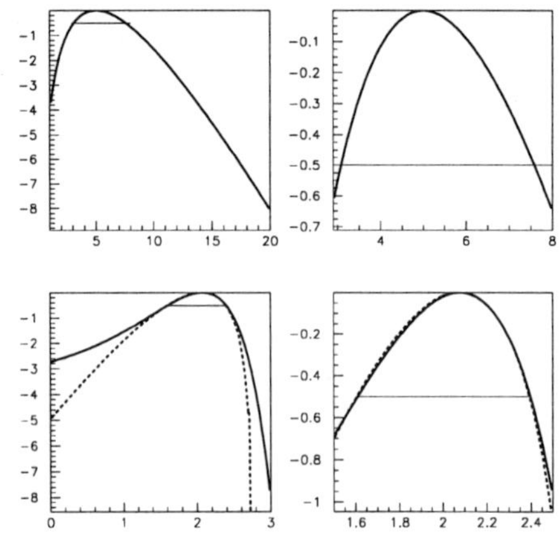

Fig. 6. Poisson Method. For explanation see text of 2.1.

The Poisson likelihood for a true mean a and N observed events is (apart from a constant term) $lnL(a) = N \ln a - a$. We can use this as a function of a with N as a real (as opposed to integer) parameter - denote it ν to indicate this. Increasing ν increases the location of the peak, increases the width, and decreases the skew. The first two can be accounted for by introducing separate location and scale parameters, giving a form $F(a) = -\alpha a + \nu ln(1 + \frac{\alpha a}{\nu})$ Determining α and ν requires numerical solution of an equation (again, details are complicated). Performance (Fig. 6) is perfect for the Poisson, as expected. It is poor for the second example, on the upper side.

2.7. *Gaussian - Linear Sigma*

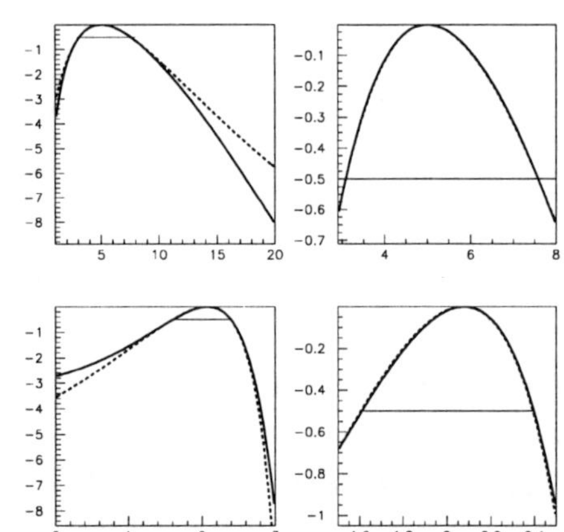

Fig. 7. Linear σ Method. For explanation see text of 2.1.

Consideration of the Bartlett approximation suggests[1] a Gaussian-like form, with the standard deviation a function of the variable being determined $F(a) = -a^2/2\sigma(a)^2$. The simplest variation is linear, giving $F(a) = -\frac{1}{2}(\frac{a^2}{\sigma_0+\sigma' a})^2$ with $\sigma_0 = \frac{2\sigma_+\sigma_-}{\sigma_+ + \sigma_-}$ $\sigma' = \frac{\sigma_+ - \sigma_-}{\sigma_+ + \sigma_-}$. This performs (Fig 7) excellently on the second example, and quite acceptably on the first.

2.8. *Gaussian - Linear Variance*

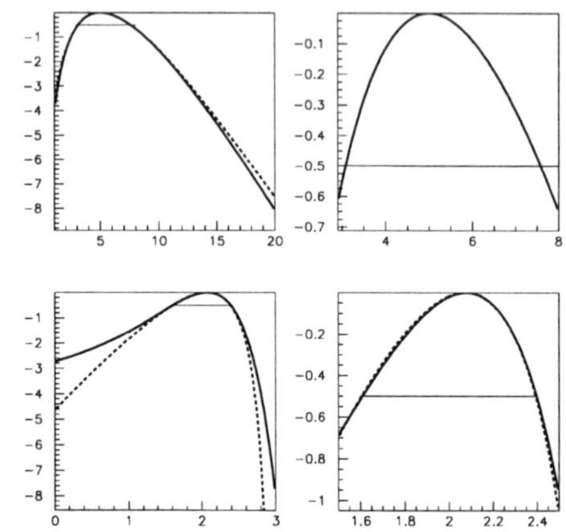

Fig. 8. Linear V Method. For explanation see text.

A modification of this is to make it linear for the variance rather than the standard deviation, giving $F(a) = -\frac{1}{2}\frac{a^2}{V_0+V'a}$ where $V_0 = \sigma_-\sigma_+$, and $V' = \sigma_+ - \sigma_-$. This performs (Fig 8) even better for the first example (which is understandable, as with the Poisson the Variance is proportional to the mean) and quite acceptably for the second.

2.9. *Model 9: Edgeworth*

The distribution for samples from a finite number N of sources, where the Gaussian is not yet attained, can be described by the Edgeworth expansion[4]

$$\frac{1}{\sqrt{2\pi}}e^{-x^2/2}\left[1 + \frac{\gamma_1}{6\sqrt{N}}H_3(x) + \frac{1}{N}(\ldots)\right]$$

where the H_i are the Hermite polynomials. This suggests a parametrisation

$$F(a) = -\frac{1}{2}\frac{a^2}{\sigma^2} + ln(1 + C\left((a/\sigma)^3 - 3(a/\sigma)\right))$$

Solution for the 3 parameters can be reduced to two simultaneous equations which have to be solved numerically. However there are no solutions if the asymmetry is above about 14%. Above this, increasing C merely increases the size of the second part of the expression, which dominates and produces a fairly symmetric peak at an increasing offset.

3. Toy Experiments

We investigate how the 8 parametrisations perform in realistic model experiments in which two values are combined. We have (privileged) knowledge of the full likelihoods and know what the result should be, and compare it with results obtained from quoting the asymmetric errors of the two results, finding the model parameters, combining the parametrised likelihoods and maximising.

In the first model a Poisson measurement is performed twice, under identical conditions. The correct procedure is a simple average, but only as we know the experiments are identical. We simulate this 10,000 times, comparing this average with the results obtained using the parametrisations.

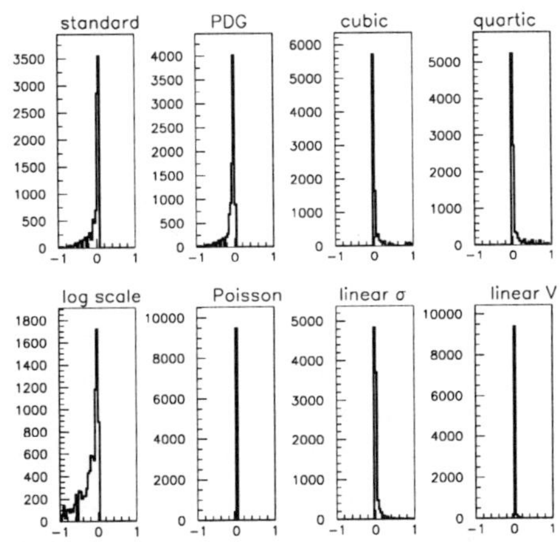

Fig. 9. Deviations for Model 1

Fig. 9 shows the deviation between the privileged answer and those from the models. Methods 1 and 2 spread to smaller values. Methods 3 and 4 do better. Model 5 is awful, 7 is good, 6 and 8 are excellent. One would expect 6 to do well as a Poisson is modelling a Poisson, but 8 is virtually as good.

In the second model one Poisson has a true mean of 10.0 and the other (representing an experiment that ran for half the time) has a mean 5.0 and is scaled up by 2. The performances (Fig. 10) are similar. Methods 1 and 2 are even worse!

Given two measurements of a lifetime, each based on 5 events, the correct (privileged) answer is again the simple average. For a particular result $\bar{t}$ the errors[1] $\sigma_+ = 0.621\bar{t}, \sigma_- = 0.340\bar{t}$ were used in the models. The logarithmic model had technical problems so we omit it and the cubic. Again (Fig. 11) the Poisson and linear Variance Gaussian do best.

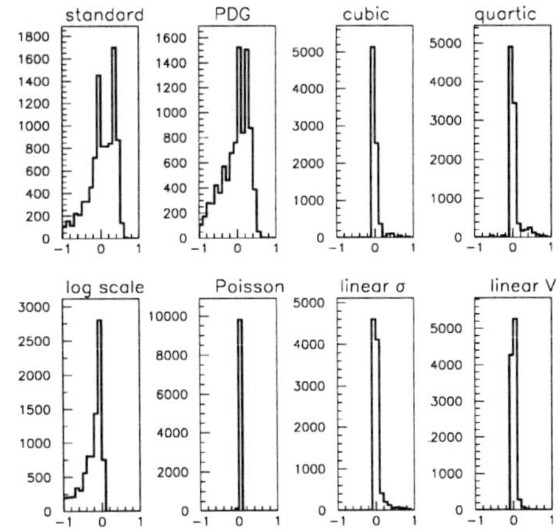

Fig. 10. Deviations for Model 2

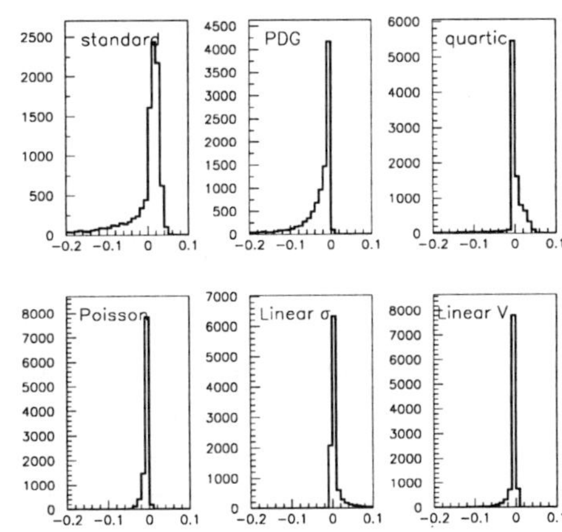

Fig. 11. Deviations for Model 3

4. Conclusions

More study is needed, but provisionally it appears that we can do better than traditional or PDG methods. The Generalised Poisson and Gaussian - linear V methods perform similarly well, but the latter's parameters are much easier to determine. *Its use should be recommended.* This may be acceptable to the community as it is just the PDG method, but continuing the interpolation outside the $[-\sigma_-, \sigma_+]$ region, and using σ^2 rather than σ.

References

1. R. J. Barlow, *Errors from the Likelihood Function*, these proceedings.
2. R. J. Barlow, *Asymmetric Errors*, Proc. PHYSTAT2003 (Ed. L. Lyons.) SLAC-R-703 (2003) p. 250.
3. S. Eidelman *et al.*, Phys. Lett. **B592** 1 (2004)
4. D. R. Cox and D. V. Hinkley *Theoretical Statistics*, Chapman & Hall (1974) p464.

BIAS-FREE ESTIMATION IN MULTICOMPONENT MAXIMUM LIKELIHOOD FITS WITH COMPONENT-DEPENDENT TEMPLATES

P. CATASTINI

I.N.F.N.-Sezione di Pisa, Universitá di Siena, Italy
E-mail: pierluigi.catastini@pi.infn.it

G. PUNZI

I.N.F.N.-Sezione di Pisa, Scuola Normale Superiore, Pisa, Italy
E-mail: giovanni.punzi@pi.infn.it

The possibility of strong biases in a multicomponent Maximum Likelihood fits with component-dependent templates has been demonstrated in some toy problems. We discuss here in detail a problem of practical interest, particle identification based on time-of-flight or dE/dx information. We show that large biases can occur in estimating particle fractions in a sample if differences between the momentum spectra of particles are ignored, and we present a more robust fit technique, allowing bias-free estimation even when the particle spectra in the sample are unknown.

1. Introduction

It has been shown in some toy problems[1] that strong biases may occur in a multicomponent Maximum Likelihood fit whenever the templates, i.e. the functions, used to parameterize the probability distributions used in the fit are not fixed but depend on event observables. An interesting example of such a problem in the practice of experimental High Energy Physics is the statistical separation of different kinds of particles on the basis of limited–precision measurements of particle–dependent quantities, like Time–of–Flight or energy loss (dE/dx).

2. Particle Fractions Estimation

Consider a sample of particles generated by a certain physical process in our experiment. We know that the given sample is a mixture of known particle types, for example *Pions*, *Kaons* and *Protons*, but unfortunately we don't know the fractions of each type, respectively indicated by f_π, f_K, f_P. Let's assume that our experimental apparatus includes a Particle Identification (PID) device, providing the measurement of some quantities whose distribution depends on the particle type. Using this PID information we want to estimate f_π, f_K and f_P, by means of an Unbinned Maximum Likelihood fit of our data sample.

The above problem is very common in particle physics, for example it occurs in separating different decay modes of a given particle[3] (same final state multiplicity and topology but different final state particle types), in studies of fragmentation of heavy quarks[2], or in optimizing the performances of algorithms for tagging the flavor of B mesons[2].

We will consider two common methods for particle identification: one is based on the measurement of energy loss of charged particles due to the ionization of a gas or of a semiconductor (often the same device used to measure particle momentum), the so called dE/dx measurement; the other is based on the measurement of the Time-of-Flight (TOF) of the particle. A common feature of PID devices based on the above principles is that the separation power between different particles is not a constant, but strongly depends on the momentum of the given, unknown, particle. A clear example of this feature is shown in Fig 1 where the dE/dx mean response of different particles is plotted as a function of momentum in the drift chamber of a typical High-Energy Physics experiment. Assuming that the resolution of the measurement is constant, the separation power dramatical changes in a short momentum range. As a consequence of the dependence of the mean value of the PID response on the particle momentum, the templates describing the PID variable's $p.d.f.$ are not fixed but depend, on an event-by-event base, on the momentum of the particle: we clearly are in the situation described in[1] where the templates of the fit depend on a component of the fit itself.

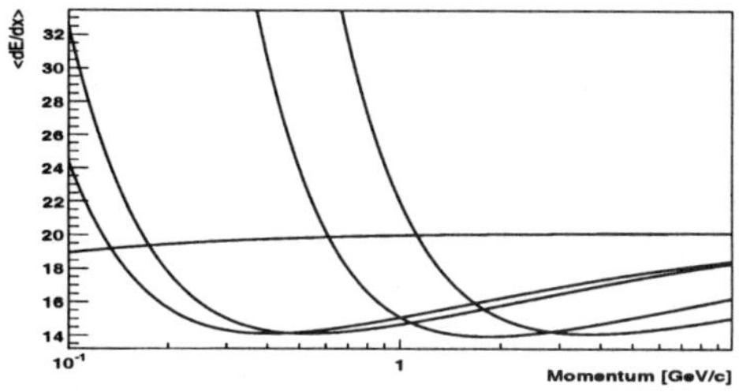

Fig. 1. The mean value of the energy loss of charged particles as a function of the momentum in a typical experiment.

2.1. *The Likelihood expression*

Consider, for simplicity, only the PID information provided by a dE/dx measurement. Our observables are then the dE/dx (*pid*) and the momentum of the track (*mom*).We will indicate as *type* the particular particle hypothesis. Unfortunately, we cannot simply write the Likelihood function as:

$$L(f_j) = \prod_i \Big(\sum_{j=\pi,K,P} f_j P(pid_i|mom_i, type_j)\Big). \quad (1)$$

Using expression (1) may give a strongly biased result if our additional variable, the momentum, has different distributions depending on the particle type (see next section). As discussed in[1], whenever the templates used in a multi–component fit depend on additional observables, to avoid the bias it is necessary to use the correct, complete Likelihood expression, including the explicit distributions of all observables for all classes of events. In our case, the above implies that we need to include in our Likelihood the momentum distributions of each particle type. We should also notice that in practice those distributions are almost always different.

We then write the correct Likelihood function as:

$$\begin{aligned} L(f_j) &= \prod_i \Big(\sum_{j=\pi,K,P} f_j P(pid_i, mom_i|type_j) \quad (2) \\ &= \prod_i \Big(\sum_{j=\pi,K,P} f_j P(pid_i|mom_i, type_j) \\ &\qquad \times P(mom_i|type_j)\Big), \end{aligned}$$

with the condition:

$$\sum_{j=\pi,K,P} f_j = 1. \quad (3)$$

3. A Toy Study

We generated a sample of different particle types with known composition as follow:

- PID variable is distributed, for each particle, according to a typical resolution function (i.e. the template used in the fit) defined as:

$$PID_{\text{measured}} - PID_{\text{expected}}(mom) \quad (4)$$

Note the dependence on momentum of the expected PID.

It is important to note that we have chosen typical realistic values for all needed parameters.

This distribution represents:

$$P(pid_i|mom_i, type_j) \quad (5)$$

in Eq. (2).

- Momenta of the particles are distributed according a Gaussian $N(\mu_j, \sigma_j)$, where $j = \pi, K, P$ and:

$$\mu_\pi = 1.00, \quad \mu_K = 1.25, \quad \mu_P = 1.25,$$

$$\sigma_\pi = \sigma_K = \sigma_P = 0.50.$$

Those distributions obviously represent:

$$P(mom_i|type_j) \quad (6)$$

of equation 2.

- Particle fractions where fixed to:

$$f_\pi = 50\%, \quad f_K = 35\%, \quad f_P = 15\%.$$

We then used an unbinned Maximum Likelihood fit to estimate the particle fractions of the sample using the Likelihood function described in Eq. (2) where:

$$P(mom_i|type_j) = N(\mu_j, \sigma_j). \quad (7)$$

In Fig. 2 (upper plot) the distribution of the estimators for f_π and f_P are shown for thirty toy samples of ten thousand particles each. As expected, the fractions returned by the fit are well centered on the true values given by the input.

Conversely, the same distributions obtained with the incomplete Likelihood function of Eq. (1) (Fig. 2, lower plot) are affected by a bias much larger than the nominal statistical uncertainty of those measurements, due to the difference in the momentum distribution of each particle type. This demonstrates that the effect predicted in [1] is actually very significant in real–life problems of Particle Identification.

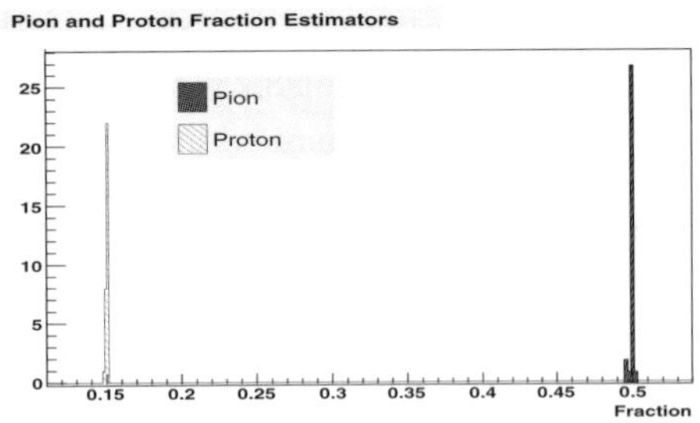

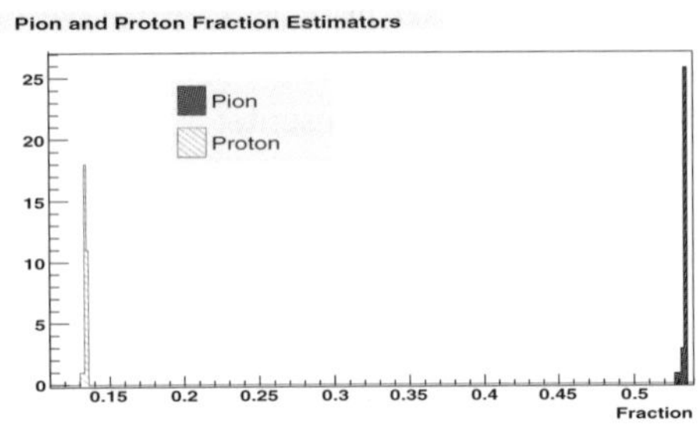

Fig. 2. The Pion and Proton fraction estimator distributions when the complete (top) and incomplete (bottom) Likelihood expression is used.

3.1. *The case of unknown momentum distributions*

Writing the complete Likelihood function considering the distribution of all the observables used in the fit is relatively straightforward in principle.

On the other hand, in practice, we often have poor information about those distributions; sometimes they are completely unknown. It is the case, for example, of the particle fractions produced during the fragmentation of heavy quarks where the corresponding momentum distributions are unknown and no functional hypothesis can be made.

Considering what was shown in the previous section, we now wonder how to avoid the bias and write the complete Likelihood if the additional observable distributions are unknown.

If no specific functional form can be assumed, we may want to use a general one, e.g. we could consider a Series Expansion as a description of the distributions with the expansion coefficients left as free parameters to be determined by the fit.

We then write the momentum term of the Likelihood function (2):

$$P(mom_i, type_j) = \sum_m a_{mj} U_m(mom_i) \qquad (8)$$

where m is the order and U_m are the basis vectors used for the series expansion.

Coming back to our toy sample, we considered Orthogonal Polynomials as a basis for the expansion. Amongst a number of possibilities, we selected Second Type Chebyshev Polynomials (denoted by U_m).

We then replaced in expression (2) the term Eq. (7) with Eq. (8) and we performed again the unbinned Maximum Likelihood Fit, this time by fitting also the parameters of the polynomial expansion. As shown in Fig. 3, now the bias is brought back to zero, as it was when we assumed perfect knowledge of the individual momentum distributions of each particle type. We have been able to avoid the bias in

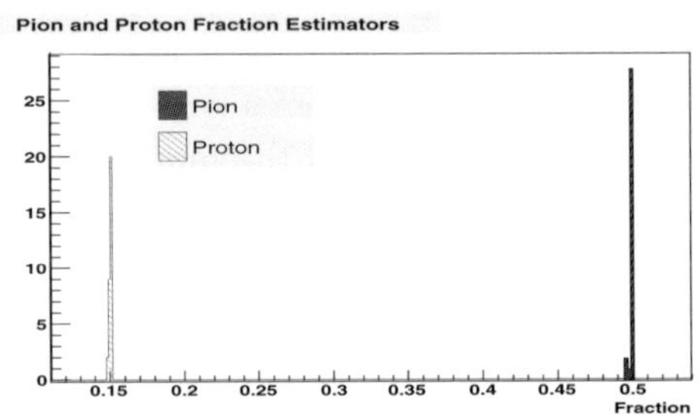

Fig. 3. The Pion and Proton fraction estimator distributions using a Series Expansion as a parameterization of the momentum distribution.

the fraction fit, without any particular assumption on the functional form of the momentum distributions. In such a way we simulated the practical case where no information is known about the additional observable distributions. Please notice also that just the first seven terms of the Second Type Chebyshev Expansion were needed in order to parametrize each particle type momentum distribution. Another interesting aspect is that comparing Fig. 3 to Fig. 2 no significant degradation in the resolution of the estimator is observed, although the number of parameters is increased. In Fig. 4 the projections of the fit to the toy sample are shown.

3.2. *A more complicated case: Time of flight*

Suppose that our PID information is obtained by the measurement of the Time of Flight. The expression of the expected TOF is a function of two observables:

$$TOF_{\text{expected}}(mom, L) = \frac{L/c}{\sqrt{1 + (m_j/mom)^2}} \qquad (9)$$

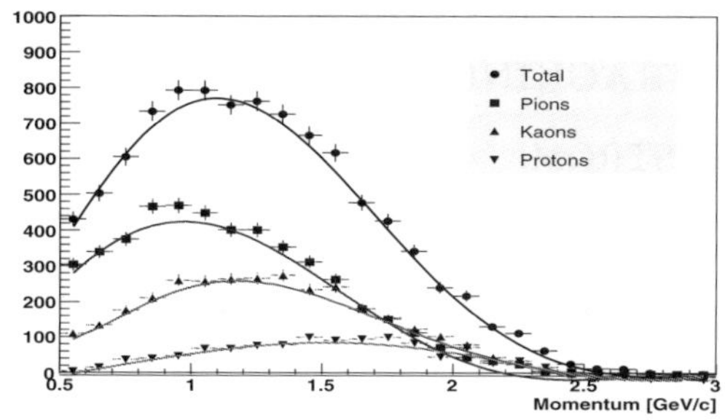

Fig. 4. The momentum projections for each particle type superimposed on the corresponding generated distributions.

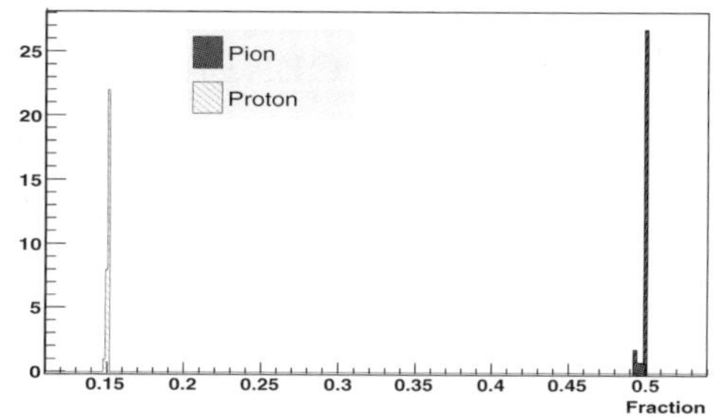

Fig. 5. The Pion and Proton fraction estimator distributions using two Series Expansions as a parameterization of the momentum and the arclength distributions.

where L is the length travelled by the particle during its time measurement (arclength) and it is a functon of the production angle of the particle (in the cylindrical geometry of the TOF detector), c is the speed of light, m_j is the mass of the particle hypothesis j and mom is again the momentum. Both the momentum and the arclength distributions could be different for each particle type, i.e., both observables could be source of bias in the particle fractions estimation. Assuming no correlations between the momentum and the arclength, we have to modify the expression (2) to be:

$$\begin{aligned} L(f_j) &= \prod_i \Big(\sum_{j=\pi,K,P} f_j P(pid_i mom_i, arc_i|type_j) \qquad (10) \\ &= \prod_i \Big(\sum_{j=\pi,K,P} f_j P(pid_i|mom_i, type_j) \\ &\qquad \times P(mom_i|type_j) \\ &\qquad \times P(arc_i|type_j)\Big). \end{aligned}$$

We then added the simulation of the arclength in our toy sample according to a normal distribution $N(\mu_j, \sigma_j)$ using the values:

$$\mu_\pi = 90, \quad \mu_K = 100, \quad \mu_P = 110,$$

$$\sigma_\pi = \sigma_K = \sigma_P = 25.$$

Considering again the case where no information is available about the distributions of each particle type, we used the same technique of the Series Expansion for both variables. We repeated our fit on thirty toy samples and also in this case, as shown in Fig. 5, no bias was observed for our estimator. It is also interesting to observe that we used just three terms of the Chebyshev Expansion for the arclength parameterization, that results in an approximate description of data (see arclength projections in Fig. 6) but it doesn't affect the results of the fit.

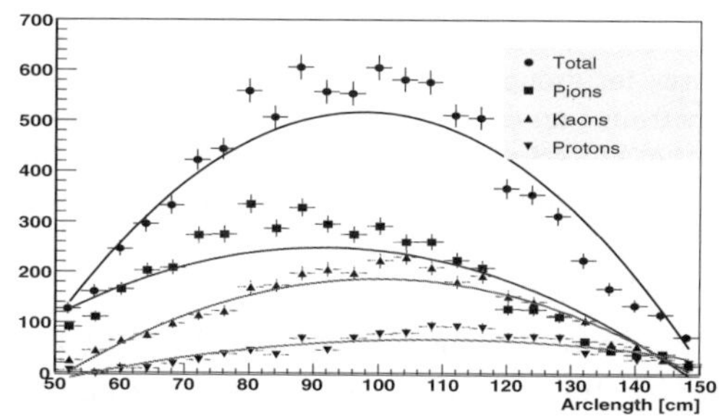

Fig. 6. The arclength projections for each particle type superimposed on the corresponding generated distributions.

4. Conclusions

In this short paper we focused on a practical and common problem of particle physics: the estimation of the particle type fractions using Particle Identification information. We showed that a significant bias can arise from the use of an incomplete expression of the Likelihood under realistic conditions. We also considered a practical problem where no information was assumed about an observable. We eliminated the bias by using Series Expansions of the unknown distributions in orthogonal polynomials, where the coefficients of the expansions are free parameters determined by the fit. We also considered a more complicated example where two relevant observables have unknown distributions, and also in this case the Series Expansion was successful in avoiding biases in determining the fractions of each component.

References

1. G. Punzi, *physics/0401045*.
2. K. Anikeev *et al.*, *hep-ph/0201071*.
3. G. Punzi, in *Proceedings of 32nd International Conference on High-Energy Physics (ICHEP04), hep-ex/0504045*.

LEAST SQUARES APPROACH TO THE ALIGNMENT OF THE GENERIC HIGH PRECISION TRACKING SYSTEM

PAWEL BRÜCKMAN DE RENSTROM

University of Oxford, DWB, Keble Road, Oxford OX1 3RH, UK

and

Institute of Nuclear Physics PAN, ul. Rdzikowskiego 152, Kraków 31-342, PL

E-mail: p.bruckman1@physics.ox.ac.uk

STEPHEN HAYWOOD

Rutherford Appleton Laboratory, Chilton Didcot OX11 0QX, UK

E-mail: S.Haywood@ral.ac.uk

A least squares method to solve a generic alignment problem of a high granularity tracking system is presented. The algorithm is based on an analytical linear expansion and allows for multiple nested fits, e.g. imposing a common vertex for groups of particle tracks is of particular interest. We present a consistent and complete recipe to impose constraints on either implicit or explicit parameters. The method has been applied to the full simulation of a subset of the ATLAS silicon tracking system. The ultimate goal is to determine ≈35,000 degrees of freedom (DoF's). We present a limited scale exercise exploring various aspects of the solution.

1 Introduction

The ultimate alignment precision of the modern HEP tracking systems can be achieved by means of a global χ^2 fit of the alignment parameters to trajectories of real particles reconstructed in the detector. The advantage of the method is that it uses all the available information, and potentially can correct all possible misalignments without the need for iteration. However, in common with any algorithm based on reconstructed tracks, there are certain distortions of the detector which are difficult to correct. These arise from distortions which displace detector hits such that they still lie on helices. Among the most common ones are well known sagitta distortions (global: $R\delta\phi = \alpha + \beta R + \gamma R^2$, ϕ dependent: $\delta X = a + bR + cR^2$ and θ dependent: $\delta\phi = \kappa R\cot(\theta)$), so-called "telescope" ($\delta Z = e + fR$) and various radial deformations (e.g. elliptical). These global distortions of the detector geometry, so called "weak modes", can lead to significant biases in the reconstructed track parameters. In this paper, we present the generic formalism to solve the least squares alignment problem as well as discuss various extensions leading to better control of the weak modes.

2 The Formalism

The alignment algorithm is based on the minimisation of the "global χ^2" defined as:

$$\chi^2 = \sum_{tracks} r^T V^{-1} r \quad \text{where} \quad r_i \equiv (\vec{m}_i - \vec{e}_i(\pi, a)).\hat{k} \tag{1}$$

Here $\vec{e}_i$ denotes the i'th intersection point of the extrapolated track with a sensor plane and $\vec{m}_i$ is the position of the associated detector hit. $\hat{k}$ is the unit vector defining the measurement direction for the sensor plane[a]. The intersection point depends on the track parameters (π) as well as on the subset of alignment parameters related to the intersected module (a). V is the covariance matrix of hit position measurements.

2.1 The Basic Least Squares Linear Expansion

The minimisation condition requires:

$$\frac{d\chi^2}{da} = 0 \implies \sum_{tracks} \frac{dr^T}{da} V^{-1} r = 0 \tag{2}$$

The assumption about the corrections being small allows us to use a linear expansion throughout. In particular, all second order derivatives are neglected. The expansion reads:

$$\sum_{tracks} \frac{dr^T}{da_0} V^{-1} \left(r_0 + \frac{dr}{da_0} \delta a \right) = 0 \tag{3}$$

from which the generic solution can be obtained:

$$\delta a = -\left(\sum_{tracks} \frac{dr^T}{da_0} V^{-1} \frac{dr}{da_0} \right)^{-1} \sum_{tracks} \frac{dr^T}{da_0} V^{-1} r_0 \tag{4}$$

where r_0 is the initial residual and $\frac{dr}{da_0} \equiv \frac{dr}{da}\big|_{\pi=\pi_0,\ a=a_0}$. Because r is a function of π and a

[a] Recall that for a pixel detector, each physical hit corresponds to two distinct measurements (2D) and therefore gives rise to two residuals along the two measurement directions.

the full derivative from Eq. 4 can be written as:

$$\frac{dr}{da} = \frac{\partial r}{\partial a} + \frac{\partial r}{\partial \pi}\frac{d\pi}{da} \tag{5}$$

where $\frac{d\pi}{da}$ is obtained by differentiating the solution from a single track fit. In exact analogy to 4, we get:

$$\delta\pi = -\underbrace{\left(\frac{\partial r^T}{\partial \pi_0} V^{-1} \frac{\partial r}{\partial \pi_0}\right)}_{J}^{-1} \frac{\partial r^T}{\partial \pi_0} V^{-1} r(\pi_0, a) \tag{6}$$

Combining 4, 5 and 6 yields the solutions for the alignment parameters alone:

$$\delta a = -\underbrace{\left(\sum_{tracks} \frac{\partial r^T}{\partial a_0} W \frac{\partial r}{\partial a_0}\right)}_{\mathcal{M}}^{-1} \underbrace{\sum_{tracks} \frac{\partial r^T}{\partial a_0} W r(\pi_0, a_0)}_{\mathcal{V}} \tag{7}$$

where

$$W \equiv V^{-1} - V^{-1}\mathrm{E}J^{-1}\mathrm{E}^T V^{-1}, \quad \mathrm{E} \equiv \frac{\partial r}{\partial \pi}|_{\pi=\pi_0} \tag{8}$$

$\mathcal{M}$ is a symmetric $n \times n$ matrix and $\mathcal{V}$ a vector of size n, where n is the number of alignment DoF's. Formula 7 can be shown to be equivalent to the one obtained using purely matrix manipulation methods [1].

2.2 Fitting a Common Event Vertex

In order to include a vertex fit into the formalism, we have to redefine the track parameterisation. Only three perigee parameters survive ($\pi = \pi(\phi,\ \cot\theta,\ Q/p_{\mathrm{T}})$). Impact parameters are replaced by the common vertex for the event ($b = (x_b, y_b, z_b)$). Definition of the residuals (Eq. 1) takes the new form:

$$r_i \equiv (\vec{m}_i - \vec{e}_i(\pi, b, a)).\hat{k} \tag{9}$$

The generic solution from Eq. 4 still holds, however, the full derivative takes a more complicated form:

$$\begin{aligned} \frac{dr}{da} &= \frac{\partial r}{\partial a} + \mathrm{E}\frac{d\pi}{da} + \mathrm{F}\frac{db}{da} \\ \frac{d\pi}{da} &= -J^{-1}\mathrm{E}^T V^{-1}(\frac{\partial r}{\partial a} + \mathrm{F}\frac{db}{da}) \\ \frac{db}{da} &= -\underbrace{\left(\sum_{tracks}^{ev} \mathrm{F}^T W \mathrm{F}\right)}_{M_b}^{-1} \left(\sum_{tracks}^{ev} \mathrm{F}^T W \frac{\partial r}{\partial a}\right) \end{aligned} \tag{10}$$

where we additionally defined $\mathrm{F} \equiv \frac{\partial r}{\partial b}$. Despite the above complexity, the final solution can be written in a compact form:

$$\delta a = -\underbrace{\left(\sum_{tracks} \frac{\partial r^T}{\partial a_0} X \frac{\partial r}{\partial a_0}\right)}_{\mathcal{M}}^{-1} \underbrace{\sum_{tracks} \frac{\partial r^T}{\partial a_0} X r(\pi_0, b_0, a_0)}_{\mathcal{V}} \tag{11}$$

where we used the relation:

$$\frac{dr}{da} = X\frac{\partial r}{\partial a}, \quad X \equiv W - W\mathrm{F}M_b^{-1}\left(\sum_{tracks}^{ev} \mathrm{F}^T W\right) \tag{12}$$

2.3 Adding External Constraints

Consider an example of constraints on track parameters. In general, they may be non-linear, however they have to be linearised before they can enter the formalism. Constraints appear as extra bi-linear terms in the expression for χ^2:

$$\chi^2 = \sum_{tracks} \left(r^T V^{-1} r + (\pi - x)^T S^{-1} (\pi - x)\right) \tag{13}$$

where vector x and covariance matrix S define the constraint on π. The solution for the track parameters is given by:

$$\begin{aligned} \delta\pi &= -J^{-1}\left(\mathrm{E}^T V^{-1} r(\pi_0, a) + S^{-1}(\pi_0 - x)\right) \\ J &\equiv \mathrm{E}^T V^{-1}\mathrm{E} + S^{-1} \end{aligned} \tag{14}$$

The solution for the alignment parameters can be derived in the usual way and yields:

$$\delta a = -\mathcal{M}^{-1} \sum_{tracks} \left(\frac{\partial r^T}{\partial a_0} W r_0 - \frac{\partial r^T}{\partial a_0} V^{-1}\mathrm{E}J^{-1}S^{-1}(\pi_0 - x)\right) \tag{15}$$

where $\mathcal{M}$ is as in Eq. 11, but with J as in Eq. 14.

3 Example Tests

The above formalism was implemented in prototype code in order to test the alignment of the ATLAS Silicon Tracking System [2]. The entire system consists of 1744 pixel modules (2D readout, $14 \times 115\mu$m resolution) and 4088 double-layer strip detector modules with two sides rotated by 40 mrad stereo angle ($16 \times 580\mu$m resolution). Given 6 DoF's of every module, there are 34,992 parameters to be determined. Solution for the entire system presents a substantial numerical challenge and as such is beyond the scope of this report. Here we present results from a test setup consisting of a region of $0 < \eta < 1$. The selected subset of the system contains 1030 silicon modules (both pixel and strip) and corresponds to over 1/6 of the entire tracking system (6180 DoF's). Only a limited data sample of 450,000 muon tracks ($2 < p_T < 20$ GeV/c) was used, so the results do not represent the ultimate precision.

3.1 The Baseline Algorithm

Diagonalisation of the matrix $\mathcal{M}$ obtained from Eq. 7 yields in the eigenvalue spectrum shown in Figure 1.

First four values are zero (up to the numerical accuracy) and correspond to unresolved translations and rotation w.r.t. the Z axis of the entire system.[b] To obtain meaningful results reciprocals of these four eigenvalues are set to zero which is equivalent to freezing these modes. The "weak modes" correspond to the lowest (finite) eigenvalues and consequently dominate the overall error on the alignment parameters.[c] More importantly, these global shape deformations lead directly to biases on fitted track parameters. Figure 2 shows pulls of the alignment corrections as determined from the perfectly aligned detector. The distribution is nicely Gaussian, centred at zero and the scatter plot does not reveal any suspicious structures. To further test the algorithm we collectively shifted all pixel detectors by $\Delta X = 200\mu$m, $\Delta Y = 100\mu$m, $\Delta Z = 400\mu$m in the ATLAS global frame (Z axis is parallel to the beam line). Tracks were refitted to the modified geometry and the alignment algorithm run. We observed no outstanding deformations to cylinders. In order to make the discussion more quantitative, we projected alignment parameters on rigid cylinders using the Jacobian transformation: $dr/dA_l = (dr/da_k)(\partial a_k/\partial A_l)$, with A_l being the 7×6 DoF's of the seven rigid cylinders. Note that this simple technique may prove very useful as a day-0 solution or a genuine method to reduce number of DoF's. Results are given in Table 1. The solution settled on a minor "telescope" mode which is one of the weakest and most difficult to control. Otherwise, corrections in the orthogonal plane are consistent with the imposed misalignment within the statistical error.

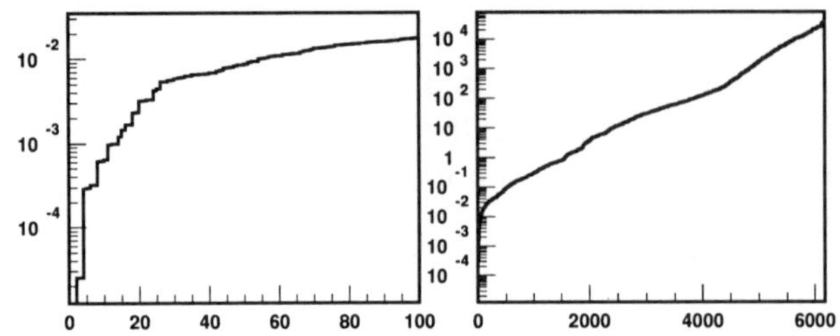

Figure 1. Eigen-spectrum of the matrix $\mathcal{M}$. Left plot zooms on the 100 weakest modes.

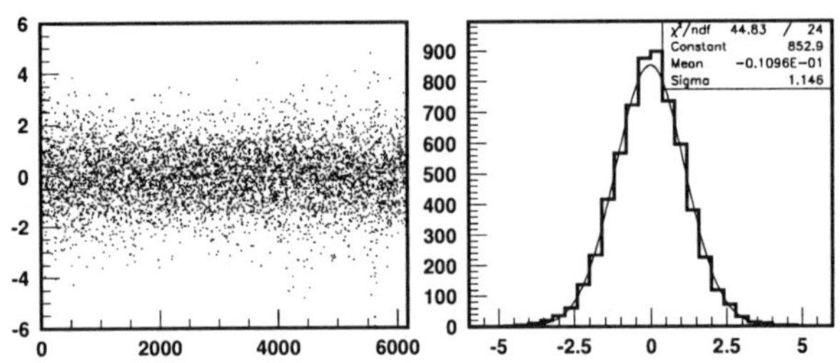

Figure 2. Pulls of the alignment parameters in the diagonal space as determined for the perfectly aligned detector.

[b]The other two rotations do not result in singular modes due to the defined and fixed direction of the magnetic field.
[c]Recall that the error is proportional to square root of the reciprocal of the eigenvalue.

Table 1. Corrections (μm) to rigid cylinders.

cylinder	ΔX	ΔY	ΔZ
PIX b-layer	-198 ± 5	-105 ± 5	-450 ± 29
PIX layer 1	-199 ± 4	-102 ± 4	-445 ± 27
PIX layer 2	-200 ± 3	-101 ± 3	-440 ± 25
SCT barrel 3	-2 ± 3	0 ± 3	-22 ± 15
SCT barrel 4	-2 ± 2	0 ± 2	-16 ± 10
SCT barrel 5	-1 ± 1	0 ± 1	-2 ± 5
SCT barrel 6	0 ± 0	0 ± 0	0 ± 0

3.2 The Common Vertex Constraint

Applying the common vertex constraint fit of Eq. 11 (there are $\approx$10 muons per event in our data sample) yielded qualitatively similar results but the absolute error on the pixel module positions (close to the interaction point) was reduced by a factor of two. Figure 4a shows the difference in the eigenvalues of the weak modes (first 100) after applying the vertex fit.

3.3 Constraints on Track Parameters

Starting from a perfect detector we imposed specific constraints on all track parameters: $\cot\theta' = \cot\theta - 0.001$, $\sigma = 0.0001$ and $Q/p_T' = Q/p_T - 0.01$, $\sigma = 0.001$ (GeV/c)$^{-1}$. These particular constraints were chosen as they directly correspond to well known weak modes, namely the "telescope" mode and the "sagitta" distortion. Alignment solution of Eq. 15 was determined with the above constraints imposed.

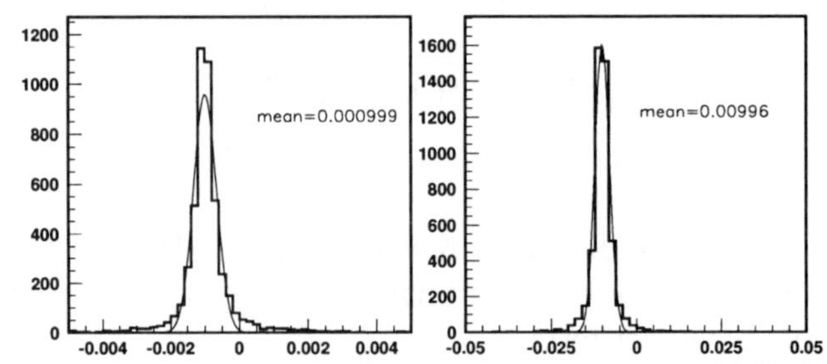

Figure 3. Change to track parameters after the refit to distorted detector geometry. See section 3.3 for more details.

Then, an independent track sample was refitted to the modified detector geometry. Figure 3 shows the resulting shifts to the track parameters. The deformation to the detector geometry led precisely to the required change of track parameters.

3.4 Constraint on the Mass of a Resonance

The idea of the constraints on track parameters can be extended to the constraint on the mass of a known resonance (e.g. $Z \to \mu^+\mu^-$, $J/\Psi \to \mu^+\mu^-$). All that is needed is an extra term in the χ^2 expression:

$$\chi'^2 = \chi^2 + (m_i - M)^T \frac{1}{\sigma_i^2}(m_i - M) \qquad (16)$$

where M is the known mass of the resonance and σ_i its assumed width accounting for experimental resolution. The solution for the constrained fit is obtained using $dm/da = (\partial m/\partial \pi)(d\pi/da)$:

$$\mathcal{J}' = \mathcal{J} + \frac{\partial m^T}{\partial \pi}\frac{1}{\sigma_i^2}\frac{\partial m}{\partial \pi}, \quad \mathcal{V}' = \mathcal{V} + \frac{dm_i^T}{da}\frac{1}{\sigma_i^2}(m_i^0 - M) \qquad (17)$$

The idea was tested in a very naive way using the muon event sample. Tracks with $p_T > 5$ GeV/c were combined into pseudo-resonances if the resulting mass was 5 GeV/c^2 or larger. The initial mass of the pair was used for the M value in each case. σ was set to 0.1 GeV/c^2 for all pairs. The improvement of the sensitivity to weak modes is shown in Figure 4*b*. Results are encouraging but the method clearly deserves proper validation using true Z and J/Ψ samples.

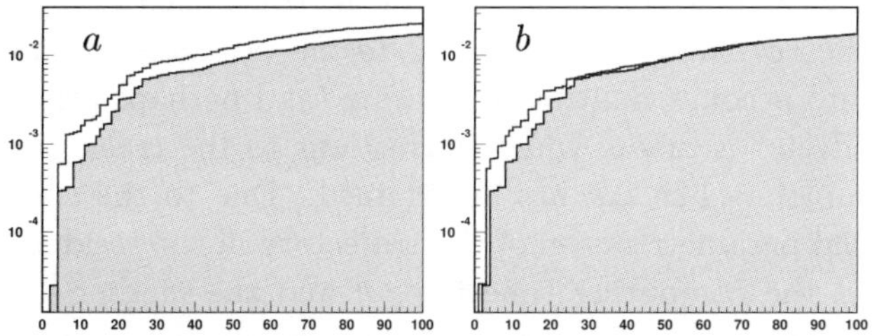

Figure 4. Eigen-spectrum for the baseline algorithm (shaded histogram) with (*a*) the vertex constraint and (*b*) mass constraint superimposed. Only 100 weakest modes shown.

3.5 External Constraints on the Geometry

External constraints may result from various mechanical considerations, actual hardware monitoring of the deformations of the support structure, etc. Whatever the source, they rarely determine positions of the individual modules. Instead, they give constraints in terms of arbitrary shape functions F_k. The extra term takes the form:

$$\chi'^2 = \chi^2 + p_k \frac{1}{\sigma_k^2} p_k \quad \text{with} \quad p_k = F_{ki}a^i - A_k \qquad (18)$$

where A_k is the requested amplitude of the k'th shape function and σ_k is the corresponding error. The constraint results in the following extra contributions to the final big matrix $\mathcal{M}$ and vector $\mathcal{V}$:

$$\mathcal{M}' = \mathcal{M} + \frac{1}{\sigma_k^2}\underbrace{(\hat{F}_k \hat{F}_k^T)}_{\text{tensor}}, \quad \mathcal{V}' = \mathcal{V} - \frac{1}{\sigma_k^2} A^k \hat{F}_k \qquad (19)$$

We tested the idea using directly two known weak modes, namely an "elliptical" and a "telescope" distortion. Table 2 shows the imposed constraint and the resulting amplitude of the corresponding mode after realignment. It was found that all other modes were unchanged relative to the unconstrained solution of section 3.1.

Table 2. Imposed vs reconstructed constraints on the geometry (arbitrary units.)

mode	"elliptical"	"telescope"
constrained A	1.0000	0.00000
constrained $\sigma(A)$	0.0100	0.00100
reconstructed A	0.9870	0.00007
reconstructed $\sigma(A)$	0.0099	0.00100

4 Summary

The least squares solution to the alignment of large HEP tracking systems has been presented. It has the potential to incorporate various extra constraints to improve its sensitivity to weak modes. Preliminary tests of the proposed extensions using the ATLAS silicon tracking system have been presented.

Acknowledgements

We would like to thank the ATLAS Inner Detector Alignment Group for helpful discussions. In particular, we acknowledge valuable discussions and cross-checks provided by Adlene Hicheur and thank him for his collaboration. The datasets used were generated using ATLAS software, for which we thank the ATLAS Software Team. This work had been inspired by the ideas used by ALEPH: ALEPH-97-116, A.Bonissent et al.

References

1. V. Blobel, C. Kleinwort, "A new method for the high-precision alignment of track detectors", Proceedings of the Conference on Advanced Statistical Techniques in Particle Physics, University of Durham, UK March 18-22, 2002.
2. ATLAS Collaboration, ATLAS Detector and Physics Performance TDR, ATLAS TDR 15, CERN/LHCC/99-15, 25 May 1999.

A NEW FAST TRACK-FIT ALGORITHM BASED ON BROKEN LINES

V. BLOBEL

Institut für Experimentalphysik, Universität Hamburg, Germany
E-mail: volker.blobel@desy.de

The determination of the particle momentum in HEP experiments requires a fit of a parametrization to the points measured in a tracking chamber. A new non-recursive track-fit algorithm based on broken lines allows the reconstruction of the particle trajectory taking into account details of the multiple scattering. It provides optimal parameters and their covariance matrices at track start and end, and optimal values at each measured point along the trajectory including the variances. The parametrization of the trajectory allows the use of sparse-matrix techniques with a total execution time $\mathcal{O}(n)$, and the new algorithm is, under test conditions, a factor six faster than the Kalman filter.

1. Track measurement in particle physics

For a HEP tracking detector with a homogeneous magnetic field $\boldsymbol{B}_z$ (in z-direction) the *ideal track parametrization* is a *helix* with five parameters: the curvature κ (inverse radius, signed), the distance d_{ca} and the angle ϕ_0 at the point of closest approach to the axis, the intercept z_0 and the slope parameter $\tan\lambda \equiv \cot\vartheta$. Various effects can result in deviations to the ideal helix curve, and the track fit with a pure helix parametrization is not optimal. Multiple scattering deflections will influence all downstream measurement in a correlated way, and delimit the accuracy of momentum measurement at low momenta. There are effects of the field inhomogeneity and continuous energy loss along the trajectory (radiation in case of electrons).

Different methods of track fitting[2] exist which are able to take the effects mentioned above into account. In global methods with a computing time $\propto n^3$, where n is the number of data points, the track parameters are determined in a single step. In the *matrix method* all effects of multiple scattering are included in the covariance matrix of the measured points, which becomes non-diagonal; in the *break-point method* a certain number of scattering planes is defined, increasing the number of parameters, while the covariance matrix of the measured points remains diagonal. In the *progressive method*[3] the track is followed by incorporating measurement after measurement with update of the parameter vector and covariance matrix, starting from the outer detector. The method is equivalent to the *Kalman filter*, which became the standard method of track fitting; multiple scattering is introduced as process noise and optimal track parameters are determined at both ends of the track by smoothing in the direction opposite to the filter. These methods have a computing time $\propto n$ and are faster by a large factor compared with the global methods mentioned before.

The method proposed here can be considered as a global method too. First approximate track parameters are determined from simple 2D fits of a circle

$$\frac{1}{2}\kappa\left(x_i^2 + y_i^2 + d_{\mathrm{ca}}^2\right) - \left(1 + \kappa d_{\mathrm{ca}}\right)\left(x_i \sin\phi_0 - y_i \cos\phi_0\right) + d_{\mathrm{ca}} = 0$$

and of a straight line $z_i = z_0 + (\tan\lambda)\cdot s_i$ to the data. Then residuals w.r.t. the circle and the straight line are calculated as a function of the track length s_i (in the $r\phi$-plane); a *detailed fit to the residuals* taking into account multiple scattering (and perhaps other effects) is made, where corrections to the track parameters like $\Delta\kappa$ are determined. Due to the special parametrization of the trajectory in the residual fit the computing time is $\propto n$ and the method is, under test conditions, faster by a factor of six, compared with the Kalman filter and smoothing. Results for the track parameters are almost identical for the global and Kalman methods. An example for a track fit is shown in Figure 3. A circle fit is shown on the left, and the residuals to the circle are shown on the right as a function of the track length; the fit of the residuals is discussed in later sections.

2. Multiple scattering

A charged particle traversing material will make a large number of small angle collisions, called multiple scattering, which is dominated by Coulomb scattering off the nuclei. Multiple scattering is parametrized by two mutually orthogonal, uncorrelated angles.

The Review of Particle Properties PDG[1] quotes the formula

$$V[\theta] = \theta_0^2 = \left(\frac{13.6 \text{ MeV}}{\beta pc}\right)^2 t \; [1 + 0.038 \ln t]^2 \quad (1)$$

for the variance of the deflection angle θ of a singly charged particle with momentum p and velocity β. The quantity t is the thickness of the material in units of the radiation length X_0, thus $t = \Delta s / X_0$.

The trajectory of a charged particle traversing a *homogeneous* medium of thickness Δs between two detector planes is shown in Figure 1. The effect of multiple scattering after traversal of a homogeneous medium of thickness Δs can be described by two parameters, e.g. the *deflection angle* θ_{plane}, or just θ, and the *angle* ψ, which is the angle between the original particle direction and the straight line between the two intersection points (circles). The two angles are statistically correlated. In an ideal detector

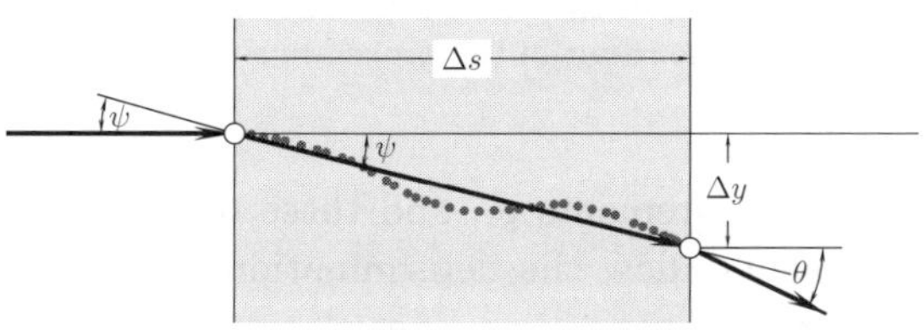

Fig. 1. Quantities used to describe multiple scattering.

the intersections could be measured with high precision and the straight line between the two intersections gives the complete information on the particle trajectory, which is available from the measurement. The angle between the direction of this line and the true particle direction is $\psi_{\text{left}} \equiv \psi$ on the left and $\psi_{\text{right}} \equiv \theta - \psi$ on the right of the medium. Expectation and variance of these two angles are identical with

$$E \begin{bmatrix} \psi_{\text{left}} \\ \psi_{\text{right}} \end{bmatrix} = \begin{pmatrix} 0 \\ 0 \end{pmatrix} \quad V \begin{bmatrix} \psi_{\text{left}} \\ \psi_{\text{right}} \end{bmatrix} = \begin{pmatrix} 1/3 & 1/6 \\ 1/6 & 1/3 \end{pmatrix} \theta_0^2 \quad (2)$$

for a homogeneous medium between the two detector planes. Usually the material distribution between two detector planes is inhomogeneous and the covariance matrix has to be calculated from the geometry of the material distribution within the layer i:

$$\boldsymbol{V} \begin{bmatrix} \psi_{\text{left}} \\ \psi_{\text{right}} \end{bmatrix}_i = \begin{pmatrix} V_{L,i} & V_{LR,i} \\ V_{LR,i} & V_{R,i} \end{pmatrix}, \qquad (3)$$

where the matrix elements are proportional to the value of θ_0^2, calculated from t by the formula (1).

3. Tracking in the sz-plane

The approximate value of the momentum p determined in the simple 2D fits allows the calculation of the multiple scattering variances $V[\theta]$. In the fit of the residuals in the sz-plane corrections to the parameters z_0 and $\tan\lambda$ are determined, taking multiple scattering into account.

Figure 2 shows the trajectory of a charged particle with multiple scattering, and the intersection points of the trajectory with the detector planes. The coordinates y_i, with standard deviation σ_i, represent the residuals of the measurements at the detector planes with coordinates s_i, for $i = 1, 2, \ldots n$; they are transverse to the average track, and are uncorrelated. For an improved fit taking into account the multiple scattering effects, the new track-fit method developed here uses two phases in the track reconstruction:

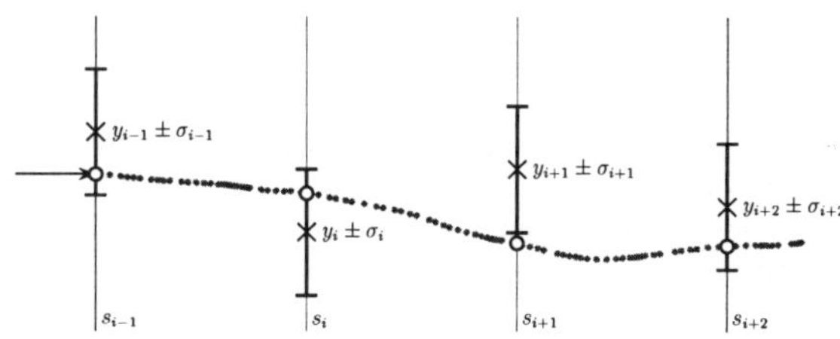

Fig. 2. Particle trajectory and measured residuals y_i.

Reconstruction of the trajectory: The trajectory, represented by the intersection points of the trajectory with the detector planes, is determined in a least squares fit; the estimates of the intersection points are denoted by u_i.

Track parameter determination: From the fitted u_i-values the two track parameters intercept and slope, required for the physics analysis, are determined at both sides of the track.

The least squares fit thus has n parameters u_i to be determined, one for each measured value y_i. The sum of squares to be minimized includes the sum of $(y_i - u_i)^2/\sigma_i^2$. The intersection points u_i to be fitted are shown in Figure 4. Each pair of adjacent points are connected by a straight line, and due to multiple scattering there is a kink angle,

$$\beta_i = \psi_{\text{right},i-1} - \psi_{\text{left},i}$$

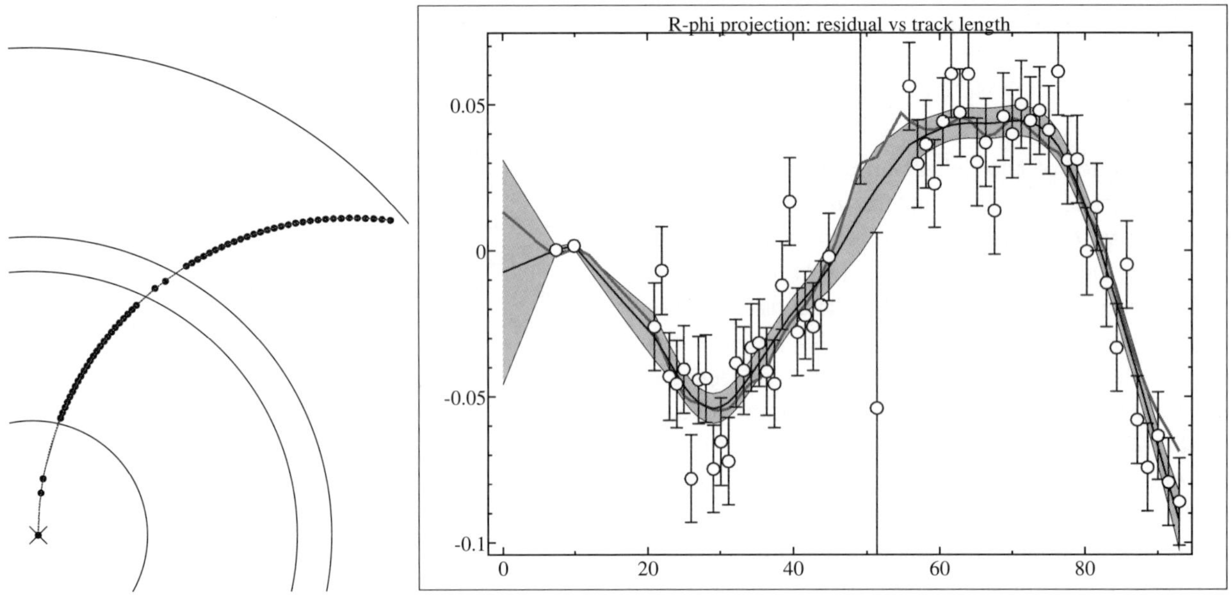

Fig. 3. A 200 MeV/c track in a detector, similar to the H1 detector. On the left the hits in the $r\phi$-plane (perpendicular to the magnetic-field direction) are show together with the result of a circle fit. The residuals of the hits w.r.t. the circle fit are shown on the right as a function of the track length. The lines and the band representing the result of the broken-line fit are explained in section 4.

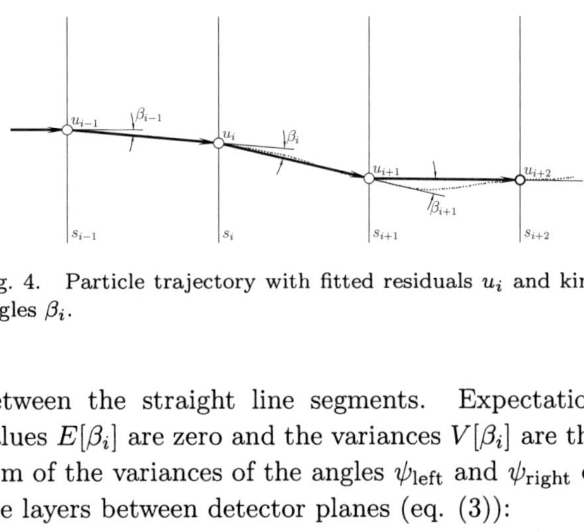

Fig. 4. Particle trajectory with fitted residuals u_i and kink angles β_i.

between the straight line segments. Expectation values $E[\beta_i]$ are zero and the variances $V[\beta_i]$ are the sum of the variances of the angles ψ_{left} and ψ_{right} of the layers between detector planes (eq. (3)):

$$V[\beta_i] = V[\psi_{\text{right},i-1}] + V[\psi_{\text{left},i}] \ .$$

The approximation of the true trajectory is given by the points (s_i, u_i). There are $(n-2)$ kink angles β_i, which are (to good approximation) linear functions of the values u_{i-1}, u_i and u_{i+1},

$$\beta_i = [u_{i-1}\,\delta_{i-1} - u_i\,(\delta_{i-1} + \delta_i) + u_{i+1}\,\delta_i] \qquad (4)$$

with the definition $\delta_i = 1/(s_{i+1} - s_i)$. The angles β_i, which all have an expectation value of zero and a variance given by the multiple scattering theory, can be considered as $(n-2)$ measurements in addition to the n measurements y_i, and these total $(2n-2)$ measurements allow the determination of estimates of the n true points u_i in a linear least squares fit by minimizing the function

$$\boxed{S(\boldsymbol{u}) = \sum_{i=1}^{n} \frac{(y_i - u_i)^2}{\sigma_i^2} + \sum_{i=2}^{n-1} \frac{\beta_i^2}{\sigma_{\beta,i}^2}} \qquad (5)$$

with respect to the values u_i; no explicit parametrization of the trajectory is defined.

The vector $\boldsymbol{u}$ that minimizes the sum-expression $S(\boldsymbol{u})$ is given by the solution of the standard normal equations

$$\boldsymbol{C}_u \boldsymbol{u} = \boldsymbol{r}; \; \boldsymbol{C}_u = \begin{pmatrix} C_{11} & C_{12} & C_{13} & & & \\ C_{21} & C_{22} & C_{23} & C_{24} & & \\ C_{31} & C_{32} & C_{33} & C_{34} & C_{35} & \\ & C_{42} & C_{43} & C_{44} & C_{45} & \cdots \\ & & C_{53} & C_{54} & C_{55} & \cdots \\ & & & & & \ddots \end{pmatrix} \qquad (6)$$

of linear least squares. The matrix $\boldsymbol{C}_u$ is a symmetric band matrix, where all elements outside the narrow band (bandwidth $m = 2$) vanish. A fast solution with computing time $\propto n$ is based on the (Cholesky) decomposition of the matrix $\boldsymbol{C}_u$ according to $\boldsymbol{C}_u = \boldsymbol{L}\boldsymbol{D}\boldsymbol{L}^{\mathrm{T}}$, where the matrix $\boldsymbol{L}$ is a left unit triangular matrix (diagonal elements are 1) and $\boldsymbol{D}$ is a diagonal matrix; the band structure is kept in this decomposition. The vector $\boldsymbol{u}$ is determined in the steps

$$\begin{array}{lrl} \text{decompose} & \boldsymbol{C}_u = \boldsymbol{L}\boldsymbol{D}\boldsymbol{L}^{\mathrm{T}} & (6n) \\ \text{solve} & \boldsymbol{L}\boldsymbol{v} = \boldsymbol{r}_u & (2n) \\ \text{solve} & \boldsymbol{L}^{\mathrm{T}}\boldsymbol{u} = \boldsymbol{D}^{-1}\boldsymbol{v} & (3n)\,. \end{array}$$

The number of operations (multiplication, division) per step is indicated in the equations; in total $11n$ operations are needed.

After the reconstruction of the trajectory the corrections Δz_0 and $\Delta(\tan\lambda)$ at track start are calculated from the two first $\boldsymbol{u}$-values u_1 and u_2 from the fitted trajectory and added to the initial approximations. In order to calculate the covariance matrix of the track parameters a few elements of the covariance matrix $\boldsymbol{V}_u \equiv \boldsymbol{C}_u^{-1}$ are required. A special method[4] can be used to calculate those elements of the inverse matrix which are in the band of the original matrix, in a computation time linear in n, using the decomposition $\boldsymbol{L}\boldsymbol{D}\boldsymbol{L}^{\mathrm{T}}$; for the bandwidth of $m = 2$ there are only $6n$ operations.

4. Tracking in the $r\phi$-plane

Corrections to the parameters κ, d_{ca} and ϕ_0 are determined in a fit of the residuals in the $r\phi$-plane; in addition to the parameters of section 3 there is a curvature correction $\Delta\kappa$. Corrections Δd_{ca} and $\Delta\phi_0$ are calculated from the first two $\boldsymbol{u}$-values u_1 and u_2. The mean value of the kink angle β_i, as defined in equation (4), is now different from zero, due to the magnetic deflection. The magnetic deflection is taken into account by the *re-definition* of the kink angle

$$\beta_i = [\ldots] \quad + (a_{i-1} + a_i)\cdot\Delta\kappa/2 \tag{7}$$

(compare equation (4); a_i is the distance between the points i and $i+1$) in the expression of equation (5), with $E[\beta_i] = 0$, and this has to be used in the function $S(\boldsymbol{u}, \Delta\kappa)$ to be minimized, which now depends on the additional parameter $\Delta\kappa$. The solution of the minimization problem is only slightly more complicated. The linear least squares expression $S(\boldsymbol{u}, \Delta\kappa)$ is minimized by the solution of the matrix equation:

$$\left(\begin{array}{c|c} C_\kappa & \boldsymbol{c}^{\mathrm{T}} \\ \hline \boldsymbol{c} & \boldsymbol{C}_u \end{array}\right) \left(\begin{array}{c} \Delta\kappa \\ \hline \boldsymbol{u} \end{array}\right) = \left(\begin{array}{c} r_\kappa \\ \hline \boldsymbol{r}_u \end{array}\right), \tag{8}$$

where $\boldsymbol{C}_u$ is as before in section 3, C_κ is a scalar and $\boldsymbol{c}$ is a vector. The solution with the steps

$$\begin{array}{rll} \boldsymbol{C}_u = \boldsymbol{L}\boldsymbol{D}\boldsymbol{L}^{\mathrm{T}} & & (6n) \\ \boldsymbol{C}_u\boldsymbol{z} = \boldsymbol{c} & & (5n) \\ \boldsymbol{B}_\kappa = \left(\boldsymbol{C}_\kappa - \boldsymbol{c}^{\mathrm{T}}\boldsymbol{z}\right)^{-1} & & (n+1) \\ \Delta\kappa = \boldsymbol{B}_\kappa\left(\boldsymbol{r}_\kappa - \boldsymbol{z}^{\mathrm{T}}\boldsymbol{r}_u\right) & & (n+1) \\ \boldsymbol{C}_u\widetilde{\boldsymbol{u}} = \boldsymbol{r}_u & & (5n) \\ \boldsymbol{u} = \widetilde{\boldsymbol{u}} - \boldsymbol{z}\Delta\kappa & & (n) \end{array}$$

requires again a number of operations with is proportional to n. The submatrix $\boldsymbol{V}_u$ at the position of the matrix $\boldsymbol{C}_u$ in the inverse matrix is $\boldsymbol{C}_u^{-1} + \boldsymbol{z}B_\kappa\boldsymbol{z}^{\mathrm{T}}$, which again allows the calculation of the covariance matrix of the parameters in a number of operations proportional to n. Figure 3 shows on the right the true simulated and the fitted trajectory, together with the $\pm 1\sigma$ band around the fitted trajectory.

Summary

The proposed algorithm allows fast track fits, fully taking into account multiple scattering; extensions to include energy loss and magnetic-field inhomogeneity are possible. The algorithm gives the full information on every measured point: the fitted value with propagated error, and pulls of position and kink angle. The new algorithm is faster by a factor of six in comparison with the Kalman filter (under test conditions), and gives the result without iterations or recursion.

References

1. Particle Data Group, Review of Particle Physics, *Phys. Lett.* B **592**, 1 (2004).
2. R. Mankel, Pattern recognition and event reconstruction in particle physics, *Rep. Prog. Phys.* **67**, 553 (2004).
3. P. Billoir, Track Fitting with Multiple Scattering: A New Method, *Nucl. Instr. Meth. A* **225**, 352 (1984).
4. K. Takahashi, J. Fagan and M. Chin, Formation of a sparse bus impedance matrix and its applications to short circuit study, Proceedings 8th PICA Conference (1973), Minneapolis, Minnesota.

NUISANCE PARAMETERS/LIMITS/DISCOVERY

TREATMENT OF NUISANCE PARAMETERS IN HIGH ENERGY PHYSICS, AND POSSIBLE JUSTIFICATIONS AND IMPROVEMENTS IN THE STATISTICS LITERATURE

ROBERT D. COUSINS

Dept. of Physics and Astronomy, University of California, Los Angeles, CA 90095, USA
E-mail: cousins@physics.ucla.edu

Nuisance parameters are common in high energy physics (HEP), and various methods are used for incorporating their effects into measurements of physics interest. A survey of some of the professional statistics literature provides justification for and insight into most of the methods commonly used in HEP. There are extensions and refinements whose usefulness could still be explored, especially in higher-dimensional problems.

Keywords: nuisance parameters; systematic uncertainties

1. Introduction

Nuisance parameters appear in virtually every physics measurement of interest because the measuring apparatus must be calibrated, and for all but the simplest apparatus, the calibration technique involves unknowns that are not directly of physical interest. In high energy physics (HEP), the primary measurement nearly always involves counting particle physics interactions of interest known as "events". The numbers of events with various characteristics are used to make inferences about underlying processes, typically Poisson. For example, suppose that n events from a Poisson interaction process are observed during a measured time interval t, and one wishes to make inferences about the mean interaction rate per unit time, Γ. By a variety of methods discussed in the next section, an interval pertaining to the unknown Poisson mean μ from which n is sampled can be constructed. If the time interval t is known with negligible uncertainty, then an interval pertaining to Γ is obtained by dividing the endpoints of the interval for μ by t.

If the measurement of t itself has non-negligible uncertainty, then t (or some surrogate) becomes a nuisance parameter, and the question arises as to how to incorporate the uncertainty in t into the interval for Γ. Already in this simple example, there is much food for thought. When one adds the common complication of "background" events that mimic the "signal" events of interest, any uncertainty in the mean rate of background events adds another nuisance parameter, and the possibilities proliferate further.

The uncertainties in nuisance parameters often correspond to what we call "systematic uncertainties" in HEP. At the last PhyStat, Sinervo[52] presented a more careful discussion of this correspondence while advocating a more precise set of definitions of three classes of systematic errors.

In this paper, I survey some representative literature from both the high energy physics and professional statistical communities, and compare and contrast the respective approaches for dealing with nuisance parameters. The ease with which one can follow citations and download papers on the Web resulted in a collection that has a lot of stimulating articles. But while I have taken at least a cursory look at all papers cited and have read a number of them, my study was tightly constrained by the decreasing time I find available to devote to my statistics hobby. Therefore, much of this paper is an annotated bibliography, and I hope that others will be able to pursue these leads.

For context and definiteness, I use the construction of an interval used to characterize the uncertainty in a single unknown parameter of interest. Such intervals are nearly always quoted in experimental HEP papers. (On the other hand, to go beyond intervals, e.g., to explicit decision theory, is rarely if ever done in a formal manner in HEP publications.) I emphasize to statisticians that physicists do not interpret confidence intervals rigidly according to the caricature of "rejecting" or "accepting" the hypothesis, but generally find confidence intervals useful as a way of conveying the results of experiments.

Table 1. 68% C.L. intervals for the mean μ of a Poisson distribution, based on the single observation $n_0 = 3$, calculated by various methods. Only the frequentist intervals avoid under-coverage for all values of μ. The boldface numbers highlight the fact that the frequentist central interval shares the right endpoint with the Bayesian interval with uniform prior, and the left endpoint with the Bayesian interval with $1/\mu$ prior, explaining why neither set of Bayesian intervals covers for all values of μ.

Method	Prior	Interval	Length
rms deviation	–	(1.27, 4.73)	3.46
Bayesian central	1	(2.09, **5.92**)	3.83
Bayesian shortest	1	(1.55, 5.15)	3.60
Bayesian central	$1/\mu$	(**1.37**, 4.64)	3.27
Bayesian shortest	$1/\mu$	(0.86, 3.85)	2.99
Likelihood ratio	–	(1.58, 5.08)	3.50
Frequentist central	–	(**1.37, 5.92**)	4.55
Frequentist shortest	–	(1.29, 5.25)	3.96
Frequentist LR ordering	–	(1.10, 5.30)	4.20

In Sec. 2, I start with a simple problem with no nuisance parameters in order to foreshadow the proliferation of methods and illustrate the first correspondence between frequentist and Bayesian methods. In Sec. 3, I discuss the role of conditioning, which is a concept that I believe deserves more awareness within HEP. In Secs. 4, 5 and 6, I describe methods for incorporating nuisance parameters in, respectively, Bayesian credible intervals, likelihood intervals, and explicitly constructed confidence intervals. I conclude in Sec. 7 with some recommendations given existing tools, and some areas yet to be explored.

2. Intervals for a Poisson Mean

To set the stage, we first recall some ways to construct an interval corresponding to a confidence level (or analog) of 68.27% for an unknown Poisson mean μ after a single observation of n events. We take $n = 3$ for definiteness. Table 1, taken from Refs. 21 and 25, gives intervals that we can identify as:

(1) *estimate of mean and rms deviation:* $n \pm \sqrt{n}$. This is just a crude estimate at small n, and I do not consider it further.
(2) *credible intervals* constructed by assigning the indicated prior $P(\mu)$ and constructing a Bayesian credible interval, using an auxiliary condition as noted.
(3) *likelihood intervals* constructed from likelihood *ratios*, with no integration or other reference to a metric on μ.
(4) *confidence intervals* constructed from Neyman's construction, with auxiliary conditions specified as noted.

The frequentist coverage probability of these types of intervals as a function of μ can be easily studied; only the confidence intervals give exact or higher coverage for all values of μ. (Ref. 62 examines the coverage of likelihood intervals.) As noted in Ref. 21, traditionally high energy physicists are rather strict about coverage, in contrast to the attitude expressed by statisticians in Refs. 42 and 63.

By far the most common Bayesian prior for a Poisson mean in HEP is the uniform prior, for which the right endpoint of a central credible interval coincides with that of a frequentist central confidence interval; this makes upper limits identical. On the other hand, left endpoints, and hence *lower* limits, are identical for the $1/\mu$ prior that actually has some motivation in terms of scale invariance. It was advocated by Jeffreys, although the rule for "Jeffreys' Priors" yields the prior $1/\sqrt{\mu}$. See Refs. 21 and Reid's respondent's talk[72] for further discussion. Reid draws attention to the $1/\sqrt{\mu}$ prior as the more fundamental "matching prior", and regards the exact matching of the endpoints in Table 1 as essentially an artifact of discreteness.

To such a diverse set of starting points, we add a variety of techniques for coping with nuisance parameters. The recent professional statistics literature seems to be mainly concerned with likelihood and Bayesian methods, while at least some of us in HEP are still interested in confidence intervals that give correct coverage by construction. The various points of view inform each other. The further confidence intervals stray from conditioning or its extreme, the likelihood principle, the more susceptible they are to being deemed irrelevant to the data set at hand; likewise, credible intervals with poor frequentist behavior place the prior under increased scrutiny. In both HEP and some professional statistics literature, performing a Bayesian-style integration over nuisance parameters in an otherwise non-Bayesian method is considered a reasonable thing to try; I think that the ultimate justification (or lack thereof) comes from studying the frequentist properties of the results (although this interpretation can be problematic for some uncertainties). This is the point of view taken by Linnemann in his interesting study[51] of various measures of significance at the previous PhyStat.

3. The Role of Conditioning (or Absence thereof) in HEP

In HEP, it is common to calculate coverage probabilities for confidence intervals (or Type I and II error probabilities) by Monte Carlo simulation using an ensemble of pseudo-experiments that includes all possible data sets that might be obtained according to the experimental procedure. Since our usual procedure is to take data for an amount of "live time" that is well defined (though usually not exactly specified in advance), the number of events obtained in each pseudo-experiment fluctuates according to a Poisson distribution. Consider, however, a situation in which the intrinsic *uncertainty* on the measurement of a parameter θ depends on the total number of events, but in which the number of events itself carries no information about θ. One can argue that the result of a particular experiment should be a confidence interval in which the ensemble used to calculate coverage should consist of pseudo-experiments that all have the same number of events as was actually observed. The argument goes back to Fisher and *conditioning* on an ancillary statistic.

As reviewed by Reid[20] and references therein (including notable work by Cox, also speaking at this conference), conditioning on some aspect of the data actually observed has a variety of justifications, including elimination of nuisance parameters. In HEP, conscious conditioning seems to be considered only rarely. To the extent that Bayesian-inspired techniques observe the likelihood principle (as in the case for pure subjective Bayesians), the extreme of conditioning on the actual set of data observed is built in, but I do not know how widespread this is recognized in HEP. Although I have attempted to read some fraction of the vast statistical literature on this topic (including the reviews by Reid in 1995[20], by Fraser in 2004[64], and the discussion in Ref. 30) I still find myself in the state of "a little knowledge is a dangerous thing". Therefore I will confine my remarks to examples of personal interest, and some pointers to the literature; see also Sec. 5 below. Demortier[50], another high energy physicist, gave his perspective at the last PhysStat.

3.1. *Ratio of Poisson Means*

In an example from HEP, an experiment observes x events of one type from Poisson X with unknown mean μ, and observes y events of another type from (independent) Poisson Y with unknown mean ν. Suppose the physics of interest is in the *ratio* of Poisson means, the single parameter $\lambda = \mu/\nu$. Then either of the individual means, or the sum, can be taken as a nuisance parameter, and we wish to obtain a confidence interval for λ from the data (x, y) in the presence of unknown nuisance parameter. The product of Poisson probabilities can be rewritten as the product of a single Poisson probability with mean $\tau = \mu + \nu$ for the total number of events $Z = X + Y$, and the binomial probability that this total is divided as such with the binomial parameter $\rho = \lambda/(1 + \lambda)$:

$$
\begin{aligned}
P(x,y) &= \left(\frac{\mathrm{e}^{-\mu}\mu^x}{x!}\right) \times \left(\frac{\mathrm{e}^{-\nu}\nu^y}{y!}\right) \\
&= \left(\frac{\mathrm{e}^{-(\mu+\nu)}\,(\mu+\nu)^z}{z!}\right) \\
&\quad \times \left(\frac{z!}{x!(z-x)!}\,\rho^x\,(1-\rho)^{(z-x)}\right). \qquad (1)
\end{aligned}
$$

That is, rewriting in terms of observables (X, Z) and parameters (λ, τ):

$$
P\;(x, y; \mu, \nu) = P(z; \mu + \nu)\;P(x|z; \rho) \qquad (2)
$$

$$
\begin{aligned}
P\;(x, z - x; \lambda\tau/(\lambda+1), \tau/(\lambda+1)) \\
= P(z;\tau)\;P(x|z;\lambda/(1+\lambda)). \qquad (3)
\end{aligned}
$$

In this form, all the information about λ is in the *conditional* binomial probability for the observed "successes" x, *given* the observed total number of events z. In the words of Reid[20], "...it is intuitively obvious that there is no information on the ratio of rates from the total count...". The same conclusion was reached in our community by James and Roos[5]. Therefore one simply uses x and z to look up a standard confidence interval for ρ, and rewrites it in terms of λ.

3.1.1. *Inference about the Total Mean: Marginalization*

Suppose that the parameter of interest and the nuisance parameter are reversed: one desires inference about sum of means $\tau = \mu + \nu$, and the ratio λ is the nuisance parameter! As discussed by Reid[20], it is no longer conditioning that is appropriate, but rather *marginalization*, i.e., integrating over a sub-space of the *sample* space. This can be seen from Eq. 2; if we sum over observed x, then the inference on τ is made from the resulting Poisson $P(z;\tau)$.

Thus, this example illustrates the use of both conditioning and marginalization. Both these concepts return repeatedly in modifications to the profile likelihood discussed in Sec. 5. I find it hard to understand, however, how one would be able to develop a general algorithm based on one concept or the other, when this simple example alternates between concepts depending on the parameter of interest.

3.1.2. *Epilogue on the Ratio of Poisson Means*

Many years ago while teaching a seminar on data analysis, I studied the coverage of the confidence intervals in Ref. 5, and found that they not only typically over-covered (as do confidence intervals for a Poisson mean), but that they *always* over-covered by a finite amount! There were *no* combinations of μ and ν for which the set of confidence intervals had coverage even close to the nominal confidence level. This convinced me that there must exist proper subsets of the James/Roos intervals that still covered. A literature search revealed that ratio-of-Poisson-means intervals were derived in an astounding variety of contexts, but that everyone obtained the same intervals, and there was even a theorem by Lehmann and Scheffé to justify the intuitive use of the above factorization. Nonetheless, after playing around with Neyman-like constructions, I found some "improved" intervals, and wrote up the story with all the references[26]. It was clear that the discreteness of the problem evaded the theorem (as Lehmann had warned).

A problem with some aspects in common (2×2 contingency tables) has been argued about for over 50 years in the statistics literature, with most people coming down on the side of enforcing strict conditioning. Whether or not my intervals (which still over-cover and are shorter by any metric since they are proper subsets of the standard ones) are "improved" or not is a matter of some debate. I tend to conclude that using the statistical fluctuations in the total number of events is a natural and effective way to average out the discreteness, especially in light of the willingness of statisticians to average over discreteness in what seems to me to be a more arbitrary way[42, 63]. I come back to the construction I used in Sec. 6 below.

3.2. *Non-Standard Conditioning in HEP on the Observed Constraint on the Number of Background Events*

In HEP, it has become common in one context to use non-standard conditioning that, as far as I know, has no foundation in the statistics literature. While not requiring a nuisance parameter, I mention it here for completeness, and because the generalization common in HEP does have a nuisance parameter. X and Y are random Poisson variables for (experimentally indistinguishable) signal and background, respectively, and one observes $z = x + y$ from the sum $Z = X + Y$. The mean b of the background Y is known, and one desires a confidence interval on the unknown mean μ of X. This problem has a long history including the paper by Feldman and myself[25] that constructs frequentist confidence intervals using the likelihood-ratio ordering in Ref. 30. These intervals cover by construction for the ensemble of all experiments, but they have been criticized for badly violating the likelihood principle[34]. The most blatant case is when $z = 0$ is observed, in which case one *knows* that for the experiment at hand, there are no background events ($y = 0$). In general, whenever z is observed, one knows that $y \leq z$.

In 1989, Zech[12] calculated upper limits on μ by calculating probabilities conditioned on $y \leq z$; this has been commonly used and extended in other contexts[22, 29]. (For further perspective on the evolving point of view of Zech on this and other methods, see Ref. 47.) This conditioning on a inequality was proposed independently in 1999 in a modification to Ref. 25 by Roe and Woodroofe (RW)[28]. However, Zech's original paper was criticized by Highland[23], and RW was criticized by me[39]. (Subsequently RW advocated a different technique[41].) That Zech and RW were using the same conditioning escaped me for some time, but I have explained it in detail in Ref. 37, along with Highland's objections, with which I tend to agree. The conditioning has properties that some find desirable, in particular for upper limits. But for two-sided intervals it leads to a situation in which the intervals cover for the restricted ensemble but not for the unconditional ensemble[39].

Read, one of the advocates of a generalized version of this conditioning[22], recommended using it for upper limits[46], and using Ref. 25 when there is a clear signal and there is no issue of interpretation.

It is notable that while most of the vast literature on conditioning seems not to have found its way into HEP, a non-standard way with no apparent formal justification was invented in HEP and gained a large following in HEP. Given the shaky foundation, caution should be used in any new application.

3.3. *Conditioning in Comparing Simple Hypotheses*

Berger et al.[18] showed that for testing a simple hypothesis against a simple alternative, the Bayesian posterior for equal prior probabilities has a nice frequentist interpretation in terms of error probabilities conditioned on the value of the likelihood ratio statistic actually observed in the data. Dass and Berger[58] generalized this to certain composite hypotheses. Neither paper seems to be cited much in the statistical literature (except by Berger himself), and a recent review in Ref. 57 is accompanied by spirited and on the whole rather unsympathetic commentary from statisticians.

I actually found Ref. 18 to be somewhat appealing (in the admittedly rare special cases in which we have simple hypotheses), and Ref. 58 to be intriguing. Given the apparent usefulness of conditioning, and the apparent difficulties of conditioning in many of our frequentist techniques in HEP, it would be interesting to see if Berger's point of view could provide some useful inspiration. I note, however, that Reid, the respondent to the present paper, cautions me that part of what I find attractive depends on a certain type of "flat" prior and so may not have good properties in general.

4. Nuisance Parameters in Bayesian Intervals

In the Bayesian world, all the difficulties with nuisance parameters are pushed (where else?) into the prior pdfs for the nuisance parameters. It could be that HEP, with its nearly universal usage of uniform priors, has something substantial to gain from the professional literature, in particular by investigating the so-called reference priors of Bernardo and collaborators[14].

As Bayesians are fond of pointing out, once the priors are specified, turning the crank is intuitive and straightforward: one constructs the posterior pdf as usual and integrates out the nuisance parameters to obtain the marginal posterior pdf for the unknown parameter of interest, and proceeds as from there as if there had been no nuisance parameters.

Liseo[17] compares a Bayesian analysis based on reference priors (Berger and Bernardo) with the profile likelihood and its modifications (Sec. 5 below), and concludes that "the frequentist coverage properties of the credible sets derived from the reference priors are shown to be better than those computed from the likelihood approach." (A more extensive update is in Ref. 67.) In the Response to the present paper, Reid informs us that "Liseo's comparison of Bayesian analysis methods is somewhat misleading... as it does not use the more accepted likelihood approach...", with reference to her article on this topic.

Berger, Liseo, and Wolpert[27] review integrating out nuisance parameters from a point of view somewhat detached from the Bayesian motivation, simply studying the performance and practical issues. Their point of view is unambiguous: in response to a suggestion in the discussion that profile likelihoods be compared to integrated likelihoods as a form of sensitivity analysis, the authors respond that it might provide some assurance if they agree, but if they disagree badly the authors would "simply suspect that it is a situation with a 'bad' profile likelihood."

As advocated by Prosper[10, 24], the D0 experiment at Fermilab[78] has been using Bayesian methods for some time, integrating out the nuisance parameters. This practice has now spread to other collaborations. The statistics committee of the CDF[79] collaboration at Fermilab has performed a study[59] of Bayesian elimination of nuisance parameters in upper limit calculations; the associated software is available. Conway, a member of this committee, has separately released a program[70] for combining different experiments, including correlations. Demortier, another member of this committee, has separately studied [45, 60] Bayesian techniques, and gives quite an interesting discussion of the prior pdf and the dangers of improper priors, and his recommended solution. At this conference, he has given a nice overview of reference priors, with much food for thought[74]. Also at this conference, Heinrich[76] has presented an important study of the dangers of uniform priors for multiple background processes.

D'Agostini[35] has also been forcefully advocating a Bayesian approach for some time, with less emphasis on frequentist properties.

5. Nuisance Parameters in Likelihood Intervals

A widely used and appreciated parameter-fitting package in high energy physics is MINUIT[2], written and maintained for several decades by CERN physicist James. The MINUIT manual and the accompanying published paper[6] describe its method of MINOS for obtaining confidence intervals and regions from likelihood ratios (increments in the negative log-likelihood). It uses Wilks's theorem[1] as applied to the profile likelihood, although until recently[40], the name profile likelihood was used rarely in HEP. The profile likelihood maximizes the likelihood over the nuisance parameters, separately for each value of the parameter(s) of interest.

Rolke and Lopez[40] have studied in detail the method of the profile likelihood as applied to the Poisson signal plus background problem in which the background is determined (with some uncertainty). I believe there was some confusion regarding the relationship of this work to MINUIT, that has now been resolved. The paper begins with the formalism of the likelihood ratio test as in Refs. 30, 25, but implements a rather conventional profile likelihood as in the method of MINOS, with an additional patch to improve the performance. Rolke, Lopez, and Conrad[68] have further studied the performance of the profile likelihood in some of HEP's prototype problems, with encouraging results.

Since MINUIT was first written, there has been quite a bit of study in the professional statistics community of cases in which the simple profile likelihood runs into difficulties, and of ways to overcome them. I am not aware of any of this research being applied routinely in HEP. Already in 1970, Kalbfleisch and Sprott[3] surveyed a variety of methods for eliminating parameters from the likelihood function: integrated likelihoods, maximum relative likelihoods, marginal likelihoods, and conditional likelihoods. (The accompanying discussion by a number of luminaries of the day includes this gem from A.W.F. Edwards: "Let me say at once that I can see no reason why it should always be possible to eliminate nuisance parameters. Indeed, one of the many objections to Bayesian inference is that is always permits this elimination.") In 1977, Basu[4] presented an even longer list and reviewed in detail the marginalizing and conditioning methods, and worked on a proper definition of nuisance parameter including the Bayesian view.

Barndorff-Nielson, in 1983[7] and 1986 [8], seems to have triggered a renewed look a the problem from the point of view of speed of asymptotic convergence by studying a "modified profile likelihood" "...with, generally, better inferential properties than the ordinary profile likelihood", and related concepts. He constructed approximate confidence intervals for the parameter of interest that are correct to order $O(n^{-3/2})$.

In 1987, Cox and Reid[9] proposed transforming the nuisance parameters into a set that is (at least locally) orthogonal to the parameters of interest, in the sense that off-diagonal elements of the information matrix vanish. Then the idea is to condition on the observed values of the nuisance parameters. The result is a formula similar to that of Barndorff-Nielson but able to neglect a term due to the orthogonalization (although thereby losing parameterization invariance). In the discussion, G.A. Barnard also takes the point of view (as did Edwards above) that one should not eliminate nuisance parameters "if the data do not permit it." Given that these methods are quite complex, for me the most interesting question was posed by F. Critchley: "Which values of n are sufficiently sub-asymptotic to make the more elaborate procedures worthwhile and yet sufficiently large to retain enough accuracy in the crucial approximation on which rests the key advantage of parameter orthogonality?" The answer to this question affects whether or not it is worth it to us in HEP to attempt to implement something like this in MINUIT, for example. My concern is that, for very small n that we frequently have in HEP, the asymptotic advantages are not yet apparent.

In the ensuing years, Fraser and Reid[11] added additional commentary; McCullagh and Tibshirani[13] proposed yet another "adjustment" to the profile likelihood; and Cox and Reid[15] added further clarification regarding when the "modifications" gives a real improvement over the vanilla profile likelihood. Severini[31] also discusses the relationships among the various modified likelihoods and Bayesian methods. At the last PhysStat, Reid and Fraser[49] provided a useful introduction for non-statisticians, with detailed explanations of examples relevant to HEP.

6. Nuisance Parameters in Frequentist Neyman-like Construction of Confidence Intervals

Traditionally many high energy physicists, including myself, have found confidence intervals to be appealing because probability P is defined in a way we understand and can simulate, and because Neyman taught us how to construct intervals that have the stated coverage (or greater) by construction. There is indeed the issue of educating people that confidence intervals are not the answer to the subjective questions that people want answered, e.g., "How much should I believe the hot new theory given the data in hand, and should I change what I do when I get up in the morning?" I remain optimistic[33] that we can teach people in HEP that $P(\text{data}|\text{theory})$ differs from $P(\text{theory}|\text{data})$, and that decisions require further subjective input about risk tolerance.

In HEP, central confidence intervals and upper confidence limits were for a long time the norm, with the choice of which one to use typically based on the data. It was only in the last decade or so that it became common knowledge in HEP that confidence intervals in general correspond to inverting a hypothesis tests on a parameter, and that the likelihood ratio test is an obvious default test to invert[30]. The application to prototype cases of interest in the absence of nuisance parameters was worked out by Feldman and myself in 1997–98[25], and then we investigated the extension to nuisance parameters, guided by the terse prescription (for an approximate method) in Ref. 30. Except for Feldman's talk at the Fermilab CLW[36], neither this work nor some follow-up work by Feldman has been written up. The initial delay was caused by the realization that one obtains a different answer in the limit the uncertainty on the nuisance parameter goes to zero than that obtained in the absence of a nuisance parameter; this is due to inserting a continuous variable into a discrete problem. Feldman described a patch for this in Ref. 36. As discussed below, for large n where this patch is irrelevant, others have continued and extended this approach.

Fraser, Reid, and Wong[61] argue that the whole approach of confidence intervals is decision-theoretic, and that likelihood-based inference, with ranges of p-values, is a preferred option.

6.1. *Full Multi-Dimensional Neyman Construction*

In principle, a brute-force technique is to consider a fine grid in the entire multi-dimensional parameter space, including nuisance parameters, and for each grid point construct an acceptance region of the desired confidence level in the data space. For this one needs an algorithm for ordering the data. Then, for a particular value of the parameter of interest, one takes the union of all the acceptance regions for that value and all values of the nuisance parameters, and proceeds to find confidence regions as usual. This typically leads to confidence intervals or regions that badly over-cover for any particular set of true values of the parameters, in order to cover for all sets.

In practice, I am aware of only a few cases in HEP where this has been attempted[26, 53, 44]. In the ratio of Poisson means problem described above[26], I played around with the ordering and managed to build acceptance regions that were subsets of the standard acceptance regions based on conditioning. But this is a tough (although fun) game that becomes increasingly harder as the number of nuisance parameters increases. I think that practically speaking, using an approximate method and checking the coverage is generally more productive than using the brute-force construction (in which case one will still probably want to check the coverage, to see how badly it over-covers).

In the previous PhyStat, Cranmer[48] presented a full construction for dealing with the background uncertainty in frequentist hypothesis testing, similar in concept to that in Ref. 26, but using the full generalization of the likelihood ratio ordering in Refs. 25, 30. At this conference[73], he compares this method with other methods. This is important work that should be "required reading" for those working on these issues at CERN's Large Hadron Collider and elsewhere. Related work by Punzi[77] adds further valuable insight into the full Neyman construction.

6.2. *Integrating Nuisance Parameters*

While participating in a number of experiments looking for "new physics" that we did not find, I encountered the simplest example of the problem discussed in the introduction, namely no event found ($n = 0$), and thus needing an upper confidence limit on Γ in the presence of uncertainty in t. (The symbols Γ and t are typically replaced by more general symbols such as those for cross section and luminosity.) In 1990, it was common either to ignore the uncertainty in t or to adjust the upper limit on Γ by adjusting $\hat{t}$ by some factor times σ_t.

Using intuition that would make a Bayesian smile, Highland and I averaged upper limits over the pdf for t centered on the measured $\hat{t}$ and obtained well-behaved results[16]. F. James explained to us that this was Bayesian averaging grafted on to a frequentist upper limit, but we stayed with it, since a purely frequentist solution had behavior that seemed unlikely to be accepted[16, 21]. Indeed, such intuitive averaging had already been used in the CDF experiment and elsewhere[16]. The important qualitative result was that, for uncertainties of 10% or so in t that were common in that day, the practice of ignoring the uncertainty was a better approximation than adjusting the upper limit by 10% or more. A fully Bayesian treatment with a uniform prior for the Poisson mean μ gave the same upper limit (if one did a sensible thing when the denominator neared zero), and in the cases we tested, the method over-covered for reasons that made sense to us.

More comprehensive coverage tests have been done internally in some collaborations and seem always to find that the method yields upper limits that over-cover (except for an incorrect study that found under-coverage). Blocker and the CDF statistics committee[69] find the performance of the algorithm in Ref. 16 to be essentially identical to a fully Bayesian technique for setting upper limits, and prefer the latter.

Barlow[43] has made available a calculator program with which one can explore results calculated in the spirit of Ref. 16 from n, Γ, t, and in addition the background estimate and its uncertainty.

Conrad et al.[54], and Tegenfeldt and Conrad[71], studied the properties of integrating out nuisance parameters for background uncertainty as well as luminosity uncertainty in the context of the intervals of likelihood-ratio ordering construction of Ref. 25. The program for performing the calculation is also published[65] and since updated, including the treatment recommended by Hill[56] for a pathology in the case of fewer than expected background events. Their conclusions are consistent with the observation in other contexts that such a treatment of nuisance parameters leads to over-coverage for any particular value of nuisance parameters.

Lista[66] has integrated out a Gaussian uncertainty on the background in the context of the upper limits from non-standard conditioning described in Sec. 3.2.

Cranmer[73] has explored what happens if one integrates nuisance parameters out to 5σ significance (!). He finds severe undercoverage. At that level, knowing the form of the pdf for the nuisance parameter becomes a real issue.

7. Conclusions

From the extensive and continuing literature on this topic in both the high energy and statistical communities, it seems clear that more work is necessary before a consensus is attained for even a "convention" that everyone agrees on. As the HEP community seems to be increasingly fond of 5σ significance, this places rather extreme demands on any approximate methods. Regarding what can be tried today, I believe it is worth emphasizing the following.

- As the quotes from Edwards and Barnard above indicate, it may not always be fruitful to eliminate nuisance parameters. In cases where the inference depends strongly on the value of the nuisance parameter, the clearest presentation may be simply to enumerate cases.
- In a completely Bayesian analysis, "turning the crank" within the methodology may be straightforward, but specification of priors is fraught with pitfalls (especially in high dimensions), and interpretation of probability "P" can be a challenge if P is not consistently subjective degree of belief in all the inputs.
- It seems to me that the widespread availability of MINUIT, our long tradition of using it in HEP, and the reasonable frequentist performance of its output combine to make it mandatory that one use the method of MINOS (differences in log of the profile likelihood) on one's likelihood function while trying out various options. The contours provided by MINUIT give insight into how sensible it is to eliminate the nuisance parameters.
- Already in 2000, Feldman[36] outlined the way we interpreted Ref. 30's prescription to include nuisance parameters in likelihood-ordered Neyman construction, but with the paucity of examples

outside of our NOMAD collaboration, this did not become widely known. Now that Cranmer[48, 73] and Punzi[77] have discussed the prescription and its more exact generalization (another way to interpret Ref. 30) in more detail, this situation is much improved. Feldman[75] and I believe that the Neyman construction using the approximation he presented in 2000 is a scalable, reasonable approach that deserves more study.

- No matter what method is used, the common practice of exploring the frequentist properties of the result should be strongly encouraged.

In addition, from the references and the talks at this conference, some next steps seem to be apparent for further development:

- The performance of reference priors[14], as discussed by Demortier[74] at this conference, should be explored by those in HEP who advocate a Bayesian approach.
- Conditioning when appropriate should become a part of our conscious thinking, and the pros and cons of restricted and global ensembles should be better understood in our community.
- It would be interesting to explore the consequences of modern modifications to the profile likelihood beyond the examples shown by Reid and Fraser[49] at the previous PhyStat.

Finally, I end on a note of caution that has its roots in recent work in my current collaboration. If the underlying sources of the nuisance parameters are systematic uncertainties that become quite large, one becomes very sensitive to the details of the pdfs for the nuisance parameters, which can be much more poorly specified than the Poisson process that underlies our statistical uncertainties. In that case, one must be vigilant against blind use of a high-powered algorithm that in the end is not robust in this context, especially when one is applying it in extreme tails such as 5σ significance.

Acknowledgments

I owe a great debt to the many people with whom I have discussed these issues, beginning with the late Virgil Highland, and continuing with Gary Feldman, Fred James, Louis Lyons, Günter Zech, and many others. (Of course, we do not always agree, and the opinions in this paper are my own.) Special thanks go to Louis for organizing another stimulating conference, and to the statisticians who so generously and graciously help us physicists to understand their work. I particularly thank Nancy Reid for her enlightening comments on my talk and manuscript during and after the conference.

References

1. S.S. Wilks, "The large-sample distribution of the likelihood ratio for testing composite hypotheses", Annals of Math. Stat. **9** (1938) 60.
2. F. James, "MINUIT. Function Minimization and Error Analysis," wwwas-doc.web.cern.ch/wwwasdoc/minuit/minmain.html
3. J.D. Kalbfleisch and J.D. Sprott, "Application of likelihood methods to models involving large numbers of parameters," Jour. Roy. Stat. Soc. Series B **32**, 175 (1970).
4. D. Basu "On the Elimination of Nuisance Parameters" JASA **72**, 355 (1977)
5. F. James and M. Roos, "Errors on Ratios of Small Numbers of Events", Nuclear Physics **B172** 475 (1980).
6. F. James, "Interpretation of the shape of the likelihood function around its minimum," Comput. Phys. Commun. **20** 29 (1980).
7. O. Barndorff-Nielsen "On a Formula for the Distribution of the Maximum Likelihood Estimates" Biometrika **70**, 343 (1983)
8. O. Barndorff-Nielsen "Inference on Full or Partial Parameters Based on the Standardized Signed Log Likelihood Ratio" Biometrika **73**, 307 (1986)
9. D.R. Cox, N. Reid "Parameter Orthogonality and Approximate Conditional Inference", Jour. Roy. Stat. Soc. Series B **49**, 1 (1987)
10. H. B. Prosper, "Small Signal Analysis In High-Energy Physics: A Bayesian Approach," Phys. Rev. D **37**, 1153 (1988); see also D. A. Williams, "Comment On 'Small Signal Analysis In High-Energy Physics: A Bayesian Approach'," Phys. Rev. D **38**, 3582 (1988), and reply.
11. D.A.S. Fraser, N. Reid, "Adjustments to Profile Likelihood" Biometrika **76**, 477 (1989)
12. G. Zech, "Upper Limits In Experiments With Background Or Measurement Errors," Nucl. Instr. and Meth. **A277** 608 (1989).
13. P. McCullagh, R. Tibshirani, "A Simple Method for the Adjustment of Profile Likelihoods" Jour. Roy. Stat. Soc. Series B **52**, 325 (1990)
14. See references in Ref. 17.
15. D.R. Cox, N. Reid, "A Note on the Difference Between Profile and Modified Profile Likelihood" Biometrika **79**, 408 (1992)
16. R.D. Cousins and V.L. Highland, "Incorporating systematic uncertainties into an upper limit," Nucl. Instrum. Meth. A **320**, 331 (1992).

17. B. Liseo "Elimination of Nuisance Parameters with Reference Priors" Biometrika **80**, 295 (1993); see also
18. J.O. Berger, L.D. Brown, and R.L. Wolpert, "A unified conditional frequentist and Bayesian test for fixed and sequential simple hypothesis-testing", Ann. Stat. **22** 1787 (1994).
19. V. Innocente and L. Lista, "Evaluation of the upper limit to rare processes in the presence of background, and comparison between the Bayesian and classical approaches," Nucl. Instrum. Meth. A **340**, 396 (1994).
20. N. Reid, "The Roles of Conditioning in Inference" Stat. Sci. **10**, 138 (1995). A more recent, very concise summary of likelihood-based inference is in Ref. 38.
21. R. D. Cousins, "Why isn't every physicist a Bayesian?," Am. J. Phys. **63**, 398 (1995).
22. A.L. Read, "Modified Frequentist Analysis of Search Results (The CL_s Method)", Workshop on Confidence Limits, CERN (2000). doc.cern.ch/yellowrep/2000/2000-005/p81.pdf; A.L. Read, "Optimal statistical analysis of search results based on the likelihood ratio and its application to the search for the MSM Higgs boson at sqrt(s) =161 and 172 GeV", DELPHI collaboration note, 97-158 PHYS 737 (1997).
23. V. Highland, Nucl. Instr. and Meth. **A398** 429 (1997), followed by reply by G. Zech.
24. P. C. Bhat, H. B. Prosper and S. S. Snyder, "Bayesian analysis of multi-source data," Phys. Lett. B **407**, 73 (1997).
25. G. J. Feldman and R. D. Cousins, "A Unified approach to the classical statistical analysis of small signals," Phys. Rev. D **57**, 3873 (1998).
26. R. D. Cousins, "Improved central confidence intervals for the ratio of Poisson means," Nucl. Instrum. Meth. A **417**, 391 (1998).
27. J.O. Berger, B. Liseo, R.L. Wolpert "Integrated Likelihood Methods for Eliminating Nuisance Parameters" Stat. Sci. **14**, 1 (1999)
28. B. P. Roe and M. B. Woodroofe, "Improved probability method for estimating signal in the presence of background," Phys. Rev. D **60**, 053009 (1999)
29. T. Junk, "Confidence level computation for combining searches with small statistics", Nucl. Instr. and Meth. **A434** (1999) 435.
30. A. Stuart, K. Ord, and S. Arnold, *Kendall's Advanced Theory of Statistics*, Volume 2A, 6th ed., (London:Arnold, 1999), and earlier editions by Kendall and Stuart.
31. T.A. Severini, "On the Relationship Between Bayesian and Non-Bayesian Elimination of Nuisance Parameters" Statistica Sinica **9**, 713 (1999)
32. J. Conway, "Inclusion of systematic uncertainties in upper limits and hypothesis tests," Workshop on Confidence Limits, CERN (2000).
33. R. Cousins, "Comments on methods for setting confidence limits," Workshop on Confidence Limits, CERN (2000).
34. G. Zech, "Confronting classical and Bayesian confidence limits to examples," Workshop on Confidence Limits, CERN (2000). arXiv:hep-ex/0004011.
35. www-zeus.roma1.infn.it/ agostini/prob+stat.html
36. G. Feldman, "Multiple measurements and parameters in the unified approach," Workshop on Confidence Limits, Fermilab (2000), conferences.fnal.gov/cl2k/copies/feldman2.pdf.
37. R.D. Cousins, "Additional comments on methods for setting confidence limits", Workshop on Confidence Limits, Fermilab (2000), conferences.fnal.gov/cl2k/copies/bcousins2.ps.
38. N. Reid "Likelihood" J. Am. Stat. Assoc. **95**, 1335 (2000)
39. R. D. Cousins, "Comment on [Improved probability method for estimating signal in the presence of background]," Phys. Rev. D **62** (2000) 098301.
40. W. A. Rolke and A. M. Lopez, "Confidence intervals and upper bounds for small signals in the presence of background noise," Nucl. Instrum. Meth. A **458**, 745 (2001), arXiv:hep-ph/0005187.
41. B. P. Roe and M. B. Woodroofe, "Setting confidence belts," Phys. Rev. D **63**, 013009 (2001)
42. L.D. Brown, T.T. Cai, A. DasGupta "Interval Estimation for a Binomial Proportion" Stat. Sci. **16**, 101 (2001)
43. R. Barlow, "A calculator for confidence intervals," Comput. Phys. Commun. **149**, 97 (2002), arXiv:hep-ex/0203002.
44. D. Nicolo, G. Signorelli, "An application of the strong confidence to the Chooz experiment with frequentist inclusion of systematics," Conference on Advanced Statistical Techniques in Particle Physics, Durham, England (2002), www.ippp.dur.ac.uk/Workshops/02/statistics/.
45. L. Demortier, "Bayesian treatments of systematic uncertainties," Conference on Advanced Statistical Techniques in Particle Physics, Durham, England (2002), www.ippp.dur.ac.uk/Workshops/02/statistics/.
46. A.L. Read, "Presentation of Search Results — the CL_s Technique," Conference on Advanced Statistical Techniques in Particle Physics, Durham, England (2002), www.ippp.dur.ac.uk/Workshops/02/statistics/.
47. G. Zech, "Frequentist and Bayesian confidence limits," Eur. Phys. J. C **4**, 12 (2002), arXiv:hep-ex/0106023.
48. K. S. Cranmer, "Frequentist hypothesis testing with background uncertainty," PHYSTAT2003, SLAC (2003). arXiv:physics/0310108.

49. N. Reid, D.A.S. Fraser, "Likelihood inference in the presence of Nuisance parameters," PHYSTAT2003, SLAC (2003). arXiv:physics/0312079.
50. L. Demortier, "Constructing ensembles of pseudo-experiments," PHYSTAT2003, SLAC (2003) arXiv:physics/0312100.
51. J. Linnemann, "Measures of significance in HEP and astrophysics," PHYSTAT2003, SLAC (2003) arXiv:physics/0312059.
52. P. Sinervo, "Definition and treatment of systematic uncertainties in high energy physics and astrophysics," PHYSTAT2003, SLAC (2003).
53. G. Punzi, "Including systematic uncertainties in confidence limits," CDF Statistics note (2003).
54. J. Conrad, O. Botner, A. Hallgren and C. Perez de los Heros, "Including systematic uncertainties in confidence interval construction for Poisson statistics," Phys. Rev. D **67**, 012002 (2003), arXiv:hep-ex/0202013. See also Ref. 55.
55. J. Conrad, O. Botner, A. Hallgren and C. P. de los Heros, "Coverage of confidence intervals for Poisson statistics in presence of systematic uncertainties," arXiv:hep-ex/0206034.
56. G. C. Hill, "Comment on 'Including systematic uncertainties in confidence interval construction for Poisson statistics'," Phys. Rev. D **67**, 118101 (2003), arXiv:physics/0302057.
57. J.O. Berger, "Could Fisher, Jeffreys and Neyman Have Agreed on Testing?" Stat. Sci. **18**, 1 (2003)
58. S.C. Dass and J.O. Berger, "Unified Conditional Frequentist and Bayesian Testing of Composite Hypotheses", Scand. J. Statist. **30** 199 (2003).
59. J. Heinrich, C. Blocker, J. Conway, L. Demortier, L. Lyons, G. Punzi, and P. K. Sinervo, "Interval estimation in the presence of nuisance parameters. 1. Bayesian approach," CDF/MEMO/STATISTICS/PUBLIC/7117 (2004), arXiv:physics/0409129.
60. L. Demortier, "A fully Bayesian computation of upper limits for Poisson processes", CDF/MEMO/STATISTICS/PUBLIC/5928 (2004)
61. D. A. S. Fraser, N. Reid and A. C. M. Wong, "Inference for bounded parameters," Phys. Rev. D **69**, 033002 (2004).
62. R. Barlow, "A note on $\Delta \ln L = -1/2$ Errors," arXiv:physics/0403046.
63. M.J. Bayarri, J.O. Berger "The Interplay of Bayesian and Frequentist Analysis" Stat. Sci. **19**, 58 (2004)
64. D.A.S. Fraser, "Ancillaries and Conditional Inference" Stat. Sci. **19**, 333 (2004); see also T.J. diCiccio and M.E. Thompson, "A Conversation with Donald A.S. Fraser," Stat. Sci. **19**, 370 (2004).
65. J. Conrad, "A program for confidence interval calculations for a Poisson process with background including systematic uncertainties: POLE 1.0," Comput. Phys. Commun. **158**, 117 (2004).
66. L. Lista, "Including Gaussian uncertainty on the background estimate for upper limit calculations using Poissonian sampling," Nucl. Instrum. Meth. A **517**, 360 (2004); see also Ref. 19.
67. B. Liseo, "The Elimination of Nuisance Parameters", geostasto.eco.uniroma1.it/utenti/liseo/pub.htm (2004).
68. W. A. Rolke, A. M. Lopez and J. Conrad, "Limits and Confidence Intervals in the Presence of Nuisance Parameters," arXiv:physics/0403059 (v4, 7 Jul 2005).
69. C. Blocker, "Interval Estimation in the Presence of Nuisance Parameters. 2. Cousins and Highland Method," CDF/MEMO/STATISTICS/PUBLIC/7539 (2005).
70. J. Conway, "Calculation of cross section upper limits combining channels incorporating correlated and uncorrelated systematic uncertainties", CDF/PUB/STATISTICS/PUBLIC/6428 (2005)
71. F. Tegenfeldt and J. Conrad, "On Bayesian treatment of systematic uncertainties in confidence interval calculations," Nucl. Instrum. Meth. A **539**, 407 (2005), arXiv:physics/0408039.
72. N. Reid, respondent to this paper, these proceedings.
73. K. Cranmer, "Statistical Challenges of the LHC", these proceedings.
74. L. Demortier, "Bayesian Reference Analysis", these proceedings.
75. G. Feldman, "Concluding Talk", these proceedings.
76. J. Heinrich, "The Bayesian approach to setting limits: what to avoid", these proceedings.
77. G. Punzi, "Ordering algorithms and Confidence Intervals in the presence of nuisance parameters", these proceedings.
78. D0 collaboration, www-d0.fnal.gov.
79. CDF collaboration, www-cdf.fnal.gov.

RESPONSE (TO COUSINS)

N. REID

Department of Statistics, University of Toronto, 100 St. George St., Toronto Canada M53 3G3
E-mail: reid@utstat.utoronto.ca

The paper by Cousins has provided an excellent overview of many of the problems arising with nuisance parameters. In my view it is easier and clearer to think about confidence limits from the approach using p-values, which is fairly common in the statistical literature but perhaps less so in the physics literature. Assume we have data x from a model $X \sim f(x;\theta)$ where the unknown parameters $\theta = (\psi, \nu)$ are partititoned into parameters of interest ψ and nuisance parameters ν. A p-value for testing $H_0 : \psi = \psi_0$ is given by $\Pr(T \geq t(x); \psi_0)$, where $T = t(X)$, and Pr is computed under model f for x.

Confidence intervals are easily obtained from p-values, by considering the function $p(\psi) = \Pr(T \geq t^{obs}; \psi)$: a $(1-\alpha)$ interval is obtained by finding $\{\psi : \alpha/2 \leq p(\psi) \leq 1-\alpha/2\}$, and a confidence bound can be obtained using $\{\psi : p(\psi) \geq \alpha\}$ or $\{\psi : p(\psi) \leq 1-\alpha\}$. The conversion from the function $p(\psi)$ to limits for ψ can be done exactly, approximately, or by simulation. For example, if $\bar{x} \sim N(\mu, 1)$, the p-value function for μ is $p(\mu) = 1 - \Phi\{\sqrt{n}(\bar{x} - \mu)\}$, and the confidence interval is $\bar{x} \pm \sqrt{n} z_{\alpha/2}$. If the p-value is obtained by using the approximation $-2\ln L \sim \chi^2_1$, then the confidence interval is computed by interpolation, and is sometimes summarized as $\hat{\psi} - \sigma_-, \hat{\psi} + \sigma_+$.

There are two aspects to the definition of p-values: the choice of the summary statistic $t(X)$ to be used in assessing the consistency of the data with H_0, and the calculation of p which may be carried out exactly, by some approximation, or by simulation. I think it is helpful to separate these two aspects. For example we might be able to find a statistic $t(X)$ whose distribution is free of ν. While this distribution might be complicated, basing confidence intervals on it would guarantee coverage for all values of the nuisance parameter. The comparison in Cousins' Table 1 uses approximate calculation of the limits for the likelihood ratio, but exact calculation for the frequentist limits. However the likelihood ratio intervals could be inverted exactly and would then be identical to the frequentist shortest intervals.

There is considerable interest in the statistical literature on so-called 'matching' priors, which are priors for which the Bayesian posterior limit has good coverage. Defining the posterior limit $\theta^{(1-\alpha)}(x)$ by $\Pr\{\theta \geq \theta^{(1-\alpha)}(x) \mid x\} = \alpha$, the matching condition is $\Pr\{\theta^{(1-\alpha)}(X) \leq \theta \mid \theta\} = \alpha + \epsilon$, where $\epsilon = 1/\sqrt{n}$ or $1/n$, etc. Matching to $O(1/\sqrt{n})$ turns out to be too weak; all priors achieve this. Matching to $O(1/n)$ is uniquely achieved for models with just one parameter by Jeffreys' prior $\pi(\theta) \propto i^{1/2}(\theta)$. For the Poisson mean this gives $\pi(\mu) \propto \mu^{-1/2}$, and the corresponding interval for Cousins' Table 1 is (1.72, 5.27). This actually does match the frequentist interval, but the latter must be defined using the mid p-value $\Pr(T > t^{obs}) + (1/2) * \Pr(T = t^{obs})$.

For the ratio of Poisson means, we have a complete factorization of the likelihood $L(\lambda, \nu; x, y) = L_1(\tau)L_2(\lambda)$, where $\tau = \lambda(1+\nu)$, and L_2 is the binomial likelihood for x 'successes' out of $x+y$ events, and probability of 'success' $\lambda/(\lambda+1)$. Using L_2 is equivalent to choosing T to be X given $X+Y$, and has the advantage that distribution is free of τ. It also has the interpretation, perhaps more important, that it directly measures λ. Although it will not usually be the case, in this model the profile likelihood is identical to L_2. However the conditional distribution has fewer points of support than full distribution so it is less likely that we will be able to observe a p-value of exactly 0.05, and it is this discreteness that leads to overcoverage. This is a different phenomenon than the separation of the parameter of interest from the nuisance parameter.

Finally, I appreciate the uncertainty over the 'correct' form of adjustment to profile likelihood. The statistical literature is full of suggestions, but there is no magic bullet here either. One version that seems to improve the usual approximations involves a correction related to the observed information, and is described in my summary paper. Experience with the correction in many examples indicates that it is worth the effort to incorporate it, especially in small n situations. Liseo's[1] comparison of Bayesian analysis to methods based on modified likelihoods is somewhat misleading, as it does not use the more widely accepted likelihood approach based on higher order asymptotic theory.[2]

References

1. B. Liseo, *Biometrika* **80**, 295 (1993).
2. N. Reid, in *Bayesian Statistics V*, Ed. J.M. Bernardo *et al.*, Oxford University Press, 351, 1996.

ORDERING ALGORITHMS AND CONFIDENCE INTERVALS IN THE PRESENCE OF NUISANCE PARAMETERS

GIOVANNI PUNZI

I.N.F.N.-Sezione di Pisa, Largo B. Pontecorvo 3, 56100 Pisa, Italy
E-mail: giovanni.punzi@pi.infn.it

We discuss some issues arising in the evaluation of confidence intervals in the presence of nuisance parameters (systematic uncertainties) by means of direct Neyman construction in multi–dimensional space. While this kind of procedure provides rigorous coverage, it may be affected by large overcoverage, and/or produce results with counterintuitive behavior with respect to the uncertainty on the nuisance parameters, or other undesirable properties. We describe a choice of ordering algorithm that provides results with good general properties, the correct behavior for small uncertainties, and limited overcoverage.

1. Introduction

A conceptually straightforward method to incorporate systematics into Confidence Limits is to apply the usual Neyman construction directly on the complete *pdf* of the problem, including the set of additional parameters ν describing the systematic effects, and then project the solution on the space of parameters of interest μ. Systematic uncertainties may take the form of an allowed range for the ν's, or may be defined by the observables of the problem. Although the method can be applied to a more general situation, we will assume in the following discussion that measurements are available of some ("subsidiary") observable(s) y, whose only purpose is to provide information on the systematic parameters, through the dependence of their *pdf* on ν. In this case, one will consider the overall *pdf*:

$$p((x, y)|(\mu, \nu)) \tag{1}$$

that gives the joint probability of observing the value of the "physics observables" x plus all "systematic measurements" y, given all unknown parameters, physics and systematics. One starts by deriving Confidence Limits in the larger (μ, ν) space from the observed values of (x, y) with the same procedure that could have been used in absence of systematics to derive limits on μ: one simply needs to sample a number of points inside the parameter space and require coverage for each of them. Then, in order to get results containing only the physical parameters, one needs to project the confidence region in (μ, ν) onto the μ space, so as to get rid of unwanted information on the nuisance parameters.

Although the above procedure is general, conceptually simple, and rigorous, other methods have been preferred in the vast majority of problems in physics. This can be ascribed to a few important difficulties with this method. To begin with, the problem of numerical calculation of Confidence Regions (CRs) in multi–dimensional spaces is often quite complex and CPU–consuming. Then there is a non–trivial question of what ordering algorithm to use in the Neyman construction. There is an issue of "efficiency", or power, of the solution, because projecting the band on the μ space effectively means to inflate a limited region in (μ, ν) to an unlimited band in the ν direction, thereby increasing the coverage for all additional points (μ, ν) included. This means that the final limits quoted on μ will almost always *overcover*, and sometimes badly, especially when ν has many dimensions; this is indeed the case with standard choices of ordering[1]. A related additional problem is that the behavior of the limits when the systematic uncertainty approaches zero is in many cases unsatisfying. It often happens that the limit for small systematics is quite different from the result one would quote in absence of that systematic; this problem, however, is not unique to the projection method.

If the above problems could be alleviated, this methodology could find greater use in HEP.

2. A Benchmark problem

Our discussion, although general, will be centered on a specific problem that has been the initial motivation for this work: a Poisson distributed signal in presence of a known background, with a systematic uncertainty on the signal normalization (efficiency).

We have:

$$x \sim Pois(\epsilon\mu + b) \quad , \quad e \sim G(\epsilon, \sigma) \qquad (2)$$

where e is the result of a subsidiary measurement with resolution σ of the unknown efficiency ϵ, which is intended to be a generic "normalization factor", not necessarily smaller than one. In the following we will mostly assume a normal distribution for G for simplicity; the possibility of negative values of the efficiency estimate does not pose any problems to the algorithms discussed in this document. This can actually occur, for instance, when the efficiency measurement implies some sort of background–subtraction procedure[a].

3. Looking for an optimal band

What one would like to accomplish is to find a clever enough rule for constructing the initial Confidence Band, to minimize the amount of unnecessary coverage added when the band is projected onto the "interesting parameters" space. It is not obvious what the minimum is for a particular problem, because the frequentist requirement of minimum coverage for every possible true value of the parameters may imply some minimum amount of overcoverage, which is unavoidable regardless of the algorithm used in the construction, much in the way overcoverage occurs in discrete problems. Therefore, there is no reason for being a–priori discouraged about the capability of the projection method to provide powerful solutions (that is, narrow intervals). A striking demonstration of this is provided by the use of the projection method, with an appropriately designed algorithm for band construction, in producing a more efficient solution to a classical, well–explored problem like the ratio of Poisson means[3].

It is intuitively obvious that in order to obtain an efficient solution, the initial confidence band must extend as far as possible along the direction of the nuisance parameter. This is not trivial to achieve, since the band needs to be built in the (x, e) space, while the objective is to produce a desired shape in the (μ, ϵ) space. A good general requirement to impose is that, given any two sections of the band at two fixed values of the nuisance parameter ϵ, one must be completely included in the other. It is intuitive that a band cannot be optimal if it does not satisfy this requirement, because if one had to take one of the two sections and expand it to completely include the other, the projected confidence region in μ would be unaffected, and conversely one could exploit the coverage gained in this way to trim a part of the exceeding part of the chosen section, thus creating the conditions for a tightening of the projected confidence region.

4. Ordering algorithm

One way to define how to construct the confidence band in the complete space is to derive it from an ordering function $f(x, e; \mu, \epsilon)$, so that the confidence band is defined by the inequality $f(x, e; \mu, \epsilon) > c(\mu, \epsilon)$, where the threshold c is determined for each value of the parameters from the usual Neyman's requirement of coverage:

$$\int_{f(x,e;\mu,\epsilon)>c(\mu,\epsilon)} p(x, e|\mu, \epsilon)dxde \geq CL \qquad (3)$$

where CL is the desired Confidence Level. It is worth noting that this is not the only conceivable way to define a band satisfying the coverage condition[3], but it is attractive for reasons of simplicity. A simple way to implement in an ordering algorithm the requirement of inclusion formulated in the previous section is to impose that $f(x, e; \mu, \epsilon)$ is independent of ϵ: $f(x, e; \mu, \epsilon_1) = f(x, e; \mu, \epsilon_2)$. In this way, sections taken at different ϵ for the same value of μ will only differ in the value of $c(\mu, \epsilon)$, and will therefore be included in one another. This requirement is also very convenient from the point of view of computing, as it implies that the ordering function f need only be calculated once for every μ.

As an additional requirement, we want the projected confidence regions to converge to the results in absence of systematic uncertainty when the size of the uncertainty goes to zero. We do not restrict to a specific ordering (one may want to be able to choose, for instance, between central and upper limits), so we start from a given generic ordering function $f_0(x; \mu)$ in the restricted space. This defines the behavior of the ordering function along the direction of observable x, but careless extension of any such rule to the

[a]This simple and common example has been selected by the CDF statistics committee as a benchmark in performing comparisons between a number of different methods. A minor difference from the current example is that a positive, Poisson–like distribution is assumed for the subsidiary measurement instead of a Gaussian, in order to avoid problems with Bayesian treatment[2].

whole (x, e) space will not work. As an example, extending the trivial ordering used to achieve upper limits ($f_0(x;\mu) = x$) results in substantial overcoverage (see fig.1a). We need additional criteria to ensure proper behavior in the subsidiary observable e. We don't want to give special preference to any values, because this will amount to attempting to extract information on the nuisance parameter, while we want to maximize information on the physical parameter μ. We do this by choosing the following ordering function:

$$f(x, e; \mu) = \int_{f_0(x') < f_0(x)} p(x'|e; \mu, \hat{\epsilon}(e))dx' \qquad (4)$$

where $\hat{\epsilon}(e)$ is the maximum–Likelihood estimate of ϵ for the given e. That implies that the same integrated conditional probability will be contained in the band for each value of e.

We make an exception to the rule of being indifferent to the value of e, ϵ for very unlikely values: we select an interval of values $[e_{min}, e_{max}]$ such that the probability for a measurement to fall outside is $\ll 1 - CL$, and assign lowest rank to all points lying outside this interval. From the above conditions, they will never be reached by the ordering procedure, so they can simply be ignored, which saves computation. This clipping technique has already been advocated as a help in keeping the projections small[4]; in our context however it seemed to have no significant effects beyond saving computation.

5. Results

We have applied the ordering rule of equation (4) to our problem of choice (sec. 2), with an ordering f_0 corresponding to upper limits. Fig. 1b shows that this time very little overcoverage is obtained, except from some discretization–related "ripples". It is interesting to note that these limits are tighter than the limits obtained with other popular methods (compare, for instance, the coverage obtained for the same problem with Bayesian[2] or Cousins–Highland methods[5, 6]), although guaranteed by construction to cover for every possible value of both μ and ϵ. This confirms the capability of the projection method to produce powerful results, when used in conjunction with an appropriate ordering algorithm, as per Eq. (4).

The procedure we have described can be used with any other desired ordering. If we apply it to Unified Intervals[7], we find an interesting fact: because of the Likelihood Ratio theorem, ensuring the independence of the distribution from true parameter values, the ordering algorithm defined by Eq. (4) is approximately equivalent to ordering based on the ratio of profile Likelihoods. That quantity has been suggested as a good intuitive ordering to use in handling systematics since [8], and has been used in neutrino experiments[1, 9] (with a conditional frequentist motivation), and in a problem very similar to ours, the Poisson with uncertainty on background[4]. It reappears here as an approximation of the more general rule defined by Eq. (4). Fig 2 shows that coverage plot for our benchmark problem, which is close to the nominal constant 0.9, indicating that there is very little to be further gained.

6. Continuity

One of our initial goals was to obtain a continuous behavior when $\sigma_{syst} \to 0$. In previous examples, although the limit is approximated much better than with other frequentist methods (see for instance[10]), there is still a slight difference. For instance, the upper limit with the Unified method at 90% for $n = 4, b = 3$ is[7] 5.6 , while our results approach $\simeq 5.47$ when $\sigma \to 0$. More annoyingly, the limit found with systematics is lower. This is a well known problem, tied to the transition between discrete and continuous regime[11], and is pretty much independent of the specific algorithm. However, our method for evaluating limits allows a very simple fix, requiring no alterations to the ordering: all that is needed is to keep the size of the grid used in the numerical calculations from becoming too small in the direction of the nuisance observable. This has a natural justification under the same principles that guided the general design of our algorithm: we are trying to disregard detailed information on the the subsidiary observable, in favor of information on the physics parameter μ. In our problem, by choosing a minimum step $\Delta e = 0.1$ we obtain perfect continuity at zero (fig. 3). A side effect of this limitation is to save some computing time.

7. Systematic uncertainties given as ranges

The approach we have described has wider application than the examples mentioned above. For instance, it can handle in a natural way the important situations in which no subsidiary measurement in available to provide information on the nuisance parameter. This often occurs in real life: the systematic uncertainty may be due to a physical constraint, or related to a choice within a range (discrete or continuous) of theoretical predictions or assumptions, or can otherwise be specified in a way that is not detailed enough to uniquely identify a probability distribution. In these cases, usually the only available information on ϵ is represented by a range of admissible values.

This situation is automatically handled by our approach: one simply has one less observable to worry about, but the rest of the construction works exactly in the same way. In fact, calculations are much faster with the lack of a subsidiary measurement, so that when dealing with small systematics it is actually more convenient to transform any possible nuisance measurement into an appropriate range for the nuisance parameter, and simply use that information as input, in order to save computing time. Again, our tests yielded very limited overcoverage, compatible with what was required simply by the discrete nature of the problem.

It is worth noting that a range of values is not at all equivalent to a uniform distribution, which implies more precise knowledge. For instance, by comparing the limits obtained in the two cases, it is seen that the limits for the range case are looser then in the uniform distribution case, as intuitively expected due to the smaller information content in a statement about a range (see Table 1). This is in contrast with what happens in a Bayesian approach, where a prior function is always required, and a uniform distribution is often chosen to represent lack of information.

In general, treating systematic uncertainties as ranges is a good candidate approach to problems with many nuisance parameters, as it allows big savings in CPU time, in addition to avoiding the trouble of having to worry about the accuracy of the distributions assumed to represent the systematic uncertainties.

8. Conclusions

We have presented a general method to incorporate systematic uncertainties in a limit calculation in a rigorous frequentist way, which is powerful (does not produce large overcoverage), has the right limit for small uncertainties, can be used even with uncertainties given as ranges, and can easily be calculated in practice. This is based on projection of a traditional Neyman construction with an ordering algorithm specified by Eq. (4). We have applied it to the specific problem of Poisson measurement with an uncertainty on the efficiency.

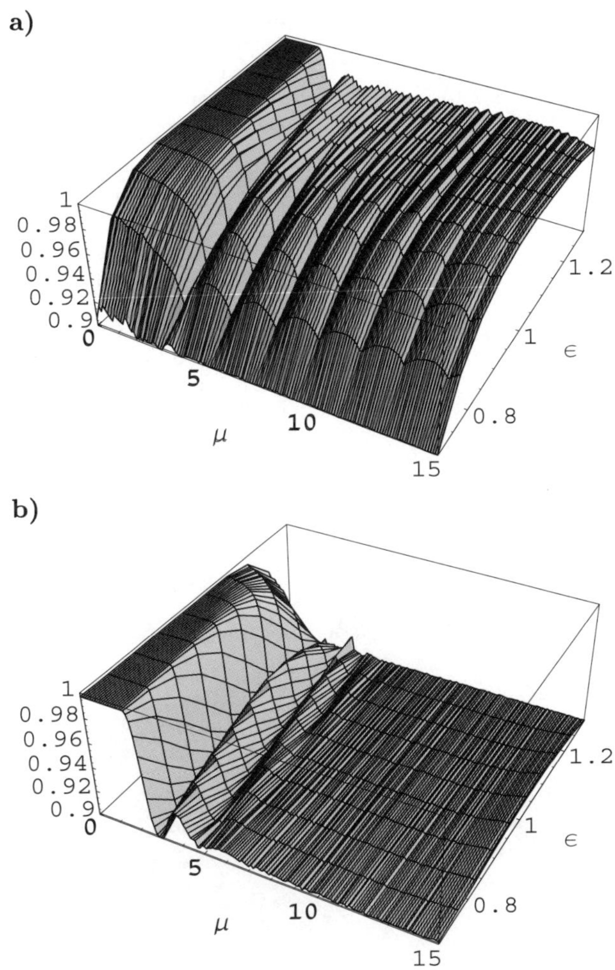

Fig. 1. Coverage plots for upper limits with a naive ordering (a), and with the ordering of eq.(4) (b). The efficiency is measured with a Gaussian uncertainty, $\sigma = 0.1$.

Acknowledgments

The Author is deeply indebted to Louis Lyons for continuing inspirational interaction and for organizing such pleasant and productive meetings; also wishes to thank Robert Cousins for helpful comments, and all members of the CDF statistics committee for many stimulating discussions.

Table 1. Confidence Limits for Poisson+background with systematic uncertainty on the efficiency, obtained by extending Unified Limits through Eq.(4). Results are given for $b = 3$, $e = 1.0$, and various models of uncertainty on ϵ (see text).

n_{obs}	Without systematics	Gaussian $\sigma = 0.1$	Uniform ± 0.15	Range ± 0.15
0	0.00 , 1.08	0.0 , 1.1	0.0 , 0.9	0.0 , 1.0
1	0.00 , 1.88	0,0 , 1.9	0.0 , 1.7	0.0 , 1.9
2	0.00 , 3.04	0.0 , 3.0	0.0 , 2.7	0.0 , 3.0
3	0.00 , 4.42	0.0 , 4.4	0.0 , 4.0	0.0 , 4.5
4	0.00 , 5.60	0.0 , 5.9	0.0 , 5.4	0.0 , 6.0
5	0.00 , 6.99	0.0 , 7.4	0.0 , 6.9	0.0 , 7.4
6	0.15 , 8.47	0.0 , 8.9	0.2 , 8.2	0.1 , 8.9
7	0.89 , 9.53	0.9 , 10.3	1.0 , 9.6	0.8 , 10.4
8	1.51 , 10.99	1.4 , 11.7	1.5 , 10.9	1.3 , 11.8
9	1.88 , 12.30	2.0 , 13.1	2.1 , 12.3	1.9 , 13.1

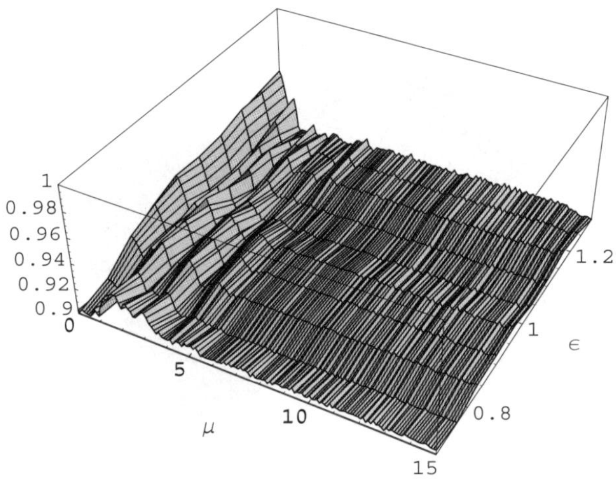

Fig. 2. Coverage plot for Unified limits, Gaussian uncertainty, $b = 3, \sigma = 0.1$.

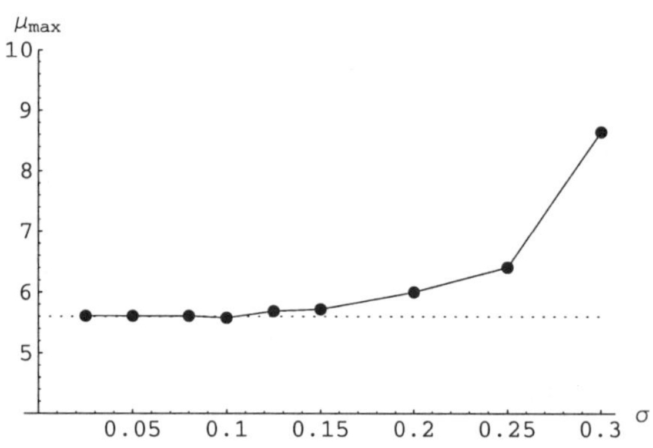

Fig. 3. Behavior of upper limit when $\sigma \to 0$, same problem as in Fig. 2.

References

1. G. Punzi, in *Proceedings of Conference on Advanced Statistical Techniques in Particle Physics, Durham, England, 18-22 Mar 2002* IPPP-02-39, p. 22-27; G. Signorelli and D. Nicolò, *ibidem*, p. 152-156. [http://www.ippp.dur.ac.uk/Workshops/02/ statistics/proceedings.shtml]
2. J. Heinrich et al. [arXiv:physics/0409129].
3. R. Cousins, Nucl. Instrum. Meth. A **417**, 391 (1998).
4. K. S. Cranmer, eConf **C030908**, WEMT004 (2003) [arXiv:physics/0310108]; see also K. S. Cranmer, these proceedings.
5. C. Blocker et al., CDF internal note 7539, FNAL, 3/15/05. (publicly available on the CDF web site).
6. F. Tegenfeldt and J. Conrad, Nucl. Instrum. Meth. A **539**, 407 (2005). [arXiv:physics/0408039].
7. G. J. Feldman and R. D. Cousins, Phys. Rev. D **57**, 3873 (1998) [arXiv:physics/9711021].
8. M. Kendall and A. Stuart, "The Advanced Theory of Statistics", (1961). Vol. 2, Ch. 24.
9. M. Apollonio *et al.* Eur. Phys.J. C **27**, 331 (2003) [arXiv:hep-ex/0301017]
10. W. A. Rolke, A. M. Lopez and J. Conrad, Nucl. Instrum. Meth. A **551**, 493 (2005). [arXiv:physics/0403059].
11. G. Feldman, presentation at Workshop on Confidence Limits, FNAL 3/28/2000. See also: G. Feldman, these proceedings.

LIKELIHOOD RATIO INTERVALS WITH BAYESIAN TREATMENT OF UNCERTAINTIES: COVERAGE, POWER AND COMBINED EXPERIMENTS

J. CONRAD

CERN, PH-EP Dept., CH-1211 Geneva 23, Switzerland
(now at: Royal Institute of Technology (KTH), Particle and Astroparticle Physics,
AlbaNova University Center, SE-10691 Stockholm, Sweden)
E-mail: Jan.Conrad@cern.ch

F. TEGENFELDT

Iowa State University
Ames, IA 5011-3160, USA
E-mail: Fredrik.Tegenfeldt@cern.ch

In this note we present studies of coverage and power for confidence intervals for a Poisson process with known background calculated using the Likelihood ratio (aka Feldman & Cousins) ordering with Bayesian treatment of uncertainties in nuisance parameters. We consider the variant where the Bayesian integration is performed in both the numerator and the denominator, and also the modification where the integration is done only in the numerator whereas in the denominator the likelihood is taken at the maximum likelihood estimate of the parameters. Furthermore we discuss how measurements can be combined in this framework and give an illustration with limits on the branching ratio of a rare B-meson decay recently presented by CDF/D0. A set of C++ classes has been developed which can be used to calculate confidence intervals for single or combining multiple experiments using the above algorithms and considering a variety of parameterizations to describe the uncertainties.

1. Introduction

A popular technique to calculate confidence intervals in recent years is the one suggested by Feldman & Cousins[1]. The method consists of constructing an acceptance region for each possible hypothesis (in the way proposed by Neyman[2]) and fixing the limits of the region by including experimental outcomes according to rank which is given by the likelihood ratio[a]:

$$R(s,n)_{\mathcal{L}} = \frac{\mathcal{L}(n|s+b)}{\mathcal{L}(n|s_{best}+b)} \tag{1}$$

where s is the hypothesis, n the experimental outcome, b the expected background, s_{best} is the hypothesis most compatible with n and $\mathcal{L}$ the Likelihood function. The expected background b is an example of a so-called *nuisance parameter*, i.e. a parameter which is not of primary interest but which still affects the calculated confidence interval. Another example of such a nuisance parameter could be the signal efficiency. In the originally proposed method by Feldman & Cousins, only the presence of background was considered and it was assumed to be exactly known. The question on how to treat uncertainties in nuisance parameters in confidence interval calculation, in particular in context of the frequentist construction, has drawn considerable attention in the recent years. In 1992 Cousins & Highland[3] proposed a method which is based on a Bayesian treatment of the nuisance parameters. The main idea is to use a probability density function (pdf) in which the average is taken over the nuisance parameter:

$$P(n|s,\epsilon) \longrightarrow \int P(n|s,\epsilon')P(\epsilon'|\epsilon)d\,\epsilon' := q(n|s,\epsilon) \tag{2}$$

where ϵ' is the true value of the nuisance parameter, ϵ denotes its estimate and s and n symbolize the signal hypothesis and the experimental outcome respectively.

Cousins & Highland only treated the case of Gaussian uncertainties in the signal efficiency. The method has since been generalized by Conrad et al.[4] to operate with the Feldman & Cousins ordering scheme and taking into account both efficiency and background uncertainties as well as correlations. This generalized method has already been used in a number of particle and astroparticle physics experiments (see references in Tegenfeldt & Conrad[5]). FHC[2] denotes this generalized method in the remain-

[a]Throughout this note we consider Poisson distributions with experimental outcome n, hypothesis parameter s and (possibly not exactly) known background b.

der of this note. If there are significantly less events than expected background, FHC^2 tends to result in confidence intervals which become smaller with increasing uncertainties. Hill[6] therefore proposed a modification where the ordering of the likelihood ratio is defined as:

$$R(s,n)_{\mathcal{L}} = \frac{q(n|s+b)}{\mathcal{L}(\max(0, n_{obs} - \hat{b}) + \hat{b})} \quad (3)$$

here $\hat{b}$ is the maximum likelihood estimate of b given the subsidiary observation of b. MBT ("Modified Bayesian Treatment") denotes this modification in the remainder of this note.

In this contribution, we discuss coverage and power of FHC^2 and MBT as well as the combination of different experiments with and without correlations. We start by introducing the C++ library which has been developed to be able to do the necessary calculations.

2. POLE++

For the coverage studies presented in this paper, a reasonably fast and efficient code is required. Hence, a user-friendly and flexible C++ library of classes was developed based on the FORTRAN routine presented by Conrad[7]. The library is independent of external libraries and consists of two main classes, *Pole* and *Coverage*. The first class takes as input the number of observed events, the efficiency and background with uncertainties and calculates the limits using the method described in this paper. The integrals are solved analytically. *Coverage* generates user-defined pseudo-experiments and calculates the coverage using *Pole*. Presently the library supports Gauss, log-Normal and flat pdf for description of the nuisance parameters. Several experiments with correlated or uncorrelated uncertainties in the nuisance parameters can be combined. The pole++ library can be obtained from http://cern.ch/tegen/statistics.html

3. Coverage and Power

The most crucial property of methods for confidence interval construction is the coverage, which states that a fraction (1-α) of infinitely many repeated experiments should yield confidence intervals that include the true hypothesis irrespective of what the true hypothesis is. For confidence interval construction (according to Neyman) without uncertainties in nuisance parameters this property is fulfilled by construction. In the present case however, we have to test the coverage employing Monte Carlo experiments.

Power on the other hand is a concept which is defined in the context of hypothesis testing: the power of a hypothesis testing method is the probability that it will reject the null hypothesis, s_0, given that the alternative hypothesis s_{true} is true. This concept is rather difficult to generalize to confidence intervals since the alternative hypothesis is not uniquely defined. We use the following definition for power:

$$\Pi(s_{true})_{s_0} = \sum_{n \notin Acc(s_0)} q(n|s_{true}, \epsilon) \quad (4)$$

and view power as a function of s_{true}. $Acc(s_0)$ here denotes the acceptance region of s_0. This seems an intuitively appealing measure: given the choice between different methods, the method which has minimally overlapping acceptance regions should be taken.

Typical examples of the coverage as a function of signal hypothesis are shown in Figure 1. It can be seen that the introduction of a continuous variable leads to a considerable smoothing of the coverage plot. A modest amount of over-coverage is introduced, similarly for the MBT method and the FHC^2 method. For high Gaussian uncertainties in efficiency ($\sim$ 40%) the over-coverage of MBT is less pronounced than that for FHC^2. More detailed coverage studies of the FHC^2 method have been presented by Tegenfeldt & Conrad[5]. The power of the FHC^2 and MBT methods is compared in Figure 1 for 40% uncertainties in the efficiency. FHC^2 has higher power for hypotheses rather far away from the null hypotheses. This is true only for large signals and comparably large uncertainties (and for not too large differences between s_0 and s_{true}), otherwise differences are negligible.

4. Combining Different Experiments

Combination of experiments can be divided into two cases. The simpler case is the one of completely uncorrelated experiments: in this case the pdf used in the construction is given by a multiplication of the pdfs of the single experiments:

$$q(\vec{n}|s) = \prod_{i=1}^{n_{exp}} q(n_i|s, \epsilon_i) \quad (5)$$

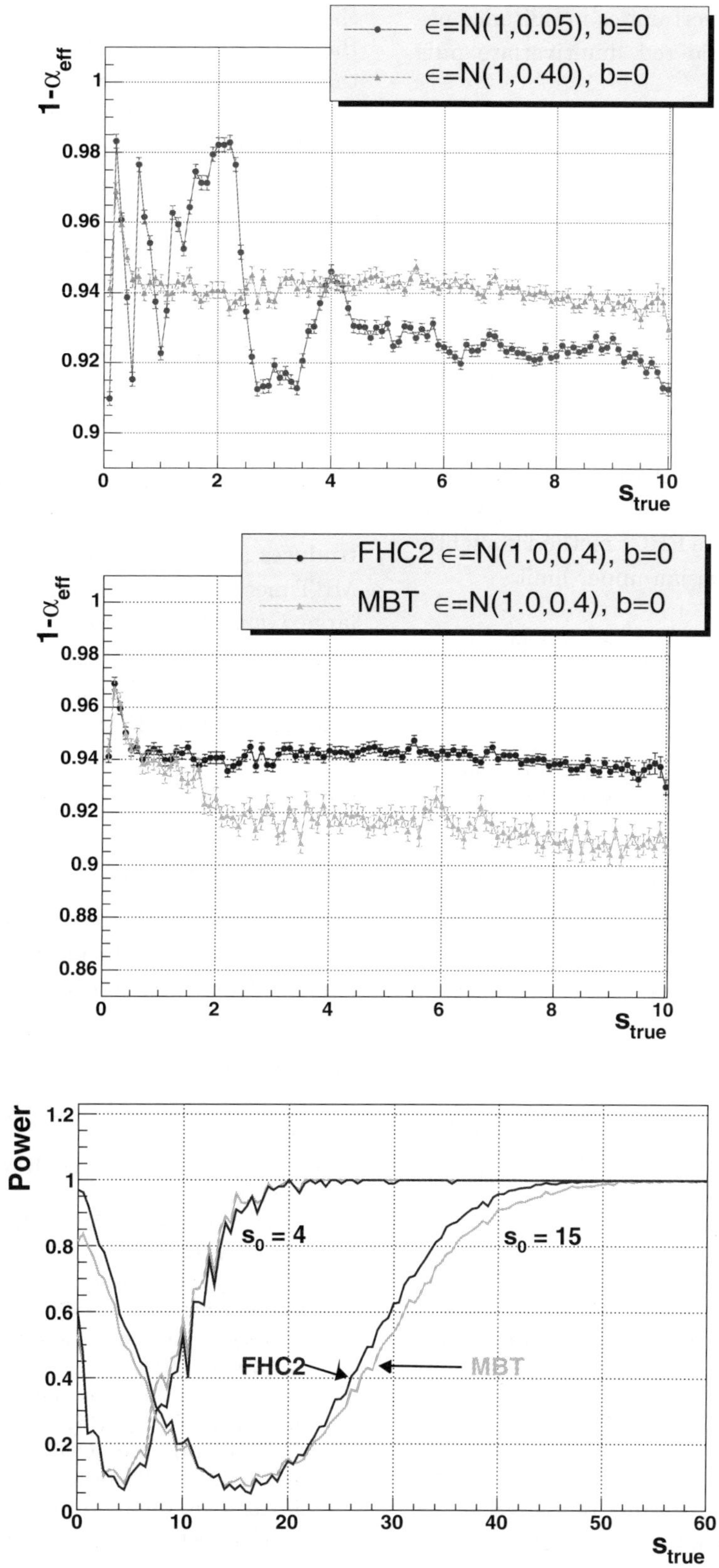

Fig. 1. Examples for the coverage and power of the discussed methods. Uppermost figure: coverage of the FHC^2 method assuming 5% and 40% Gaussian uncertainties in efficiency. Middle figure: the coverage for the FHC^2 method compared to the MBT method for 40% Gaussian efficiency uncertainties. Lowest figure: the power of the two methods compared for 40% Gaussian uncertainties in efficiency.

If correlations between uncertainties in nuisance parameters have to be considered, multivariate pdfs have to be employed:

$$q(\vec{n}|s,\vec{\epsilon}) = \int_0^{\infty} ... \int_0^{\infty} \prod_{i=1}^{n_{exp}} P(n|s,\epsilon_i') P(\vec{\epsilon'}|\vec{\epsilon}) \prod_{i=1}^{n_{exp}} d\epsilon_i' \tag{6}$$

We illustrate the effect of combining different experiments with the example of the CDF limit on the branching ratio for $B_s^0 \to \mu^+\mu^-$, see Table 1. In this case, two CDF data sets are combined with an uncorrelated uncertainty in the background expectation and an uncertainty in the efficiency which can be factorized into a correlated and uncorrelated part[8]. Bernhard et al.[8] presented a fully Bayesian combination, which is included in the table for comparison. The limit obtained using the FHC^2 method is slightly smaller than the fully Bayesian upper limit.

Table 1. The CDF single and combined limits on $B_s^0 \to \mu^+\mu^-$ calculated by FHC^2. CDF1 and CDF2 denote the two different data sets used for single limits. The quoted uncertainties are for the single experiments, the efficiency uncertainties change to 13.1 and 11.1% for the uncorrelated part if experiments are combined. The number in the parentheses is the result of the purely Bayesian calculation[7].

	CDF 1	CDF 2
background uncertainty [%]	14.8	19.7
eff. uncertainty [%]	10.4	11.3
corr. eff. uncertainty.[%]	15.5	
95% CL [10^{-7}]	2.5	4.3
95% combined [10^{-7}]	1.7 (2.0)	

5. Discussion & Conclusion

There are two main caveats when interpreting the presented results: first of all, the methods (more or less implicitly) assume a flat prior probability for the true nuisance parameter. Thus, conclusions on the coverage and power are true only for that prior. This assumption seems particularly harmful in the case of combined experiments, a case for which we did not calculate the coverage. Results presented at this conference by Heinrich[9] indicate that the assumption of a flat prior for nuisance parameters in each channel leads to significant under-coverage for fully Bayesian confidence intervals. Heinrich also shows that this behavior can be remedied with an appropriate choice of prior (in his particular example: $1/\epsilon$). For the methods presented here this might imply that there is under-coverage in the case of several combined experiments. A second caveat is that we test the coverage only for 90% confidence level. At this conference Cranmer[10] presented results that indicate under-coverage for very high confidence levels (> 5 σ) if uncertainties in the background are treated in the Bayesian way. Tests of coverage for high confidence levels and combined experiments are currently under way.

With these caveats in mind, we conclude that the Bayesian treatment of nuisance parameters introduces a moderate amount of over-coverage. The MBT method has less over-coverage for the case with large Gaussian uncertainties in the signal efficiencies. We also compared the power of the two suggested methods. For large uncertainties and large true signals, the FHC^2 method has higher power for hypotheses relatively far away from the null hypothesis.

Acknowledgments

We would like to thank the conference organizers, in particular Louis Lyons for organizing this useful and very enjoyable conference.

References

1. G. J. Feldman and R. D. Cousins, Phys. Rev. **D57**, 3873 (1998).
2. J. Neyman, Phil. Trans. Royal Soc. London **A**, 333 (1937).
3. R. D. Cousins and V. L. Highland, Nucl. Instrum. Meth. A **320**, 331 (1992).
4. J. Conrad, O. Botner, A. Hallgren and C. P. de los Heros, Phys. Rev. **D67**, 012002 (2003).
5. F. Tegenfeldt and J. Conrad, Nucl. Instrum. Meth. A **539**, 407 (2005).
6. G. C. Hill, Phys. Rev. **D67**, 118101 (2003).
7. J. Conrad, Comp. Phys. Comm. **158** 117 (2004).
8. R. Bernhard *et al.* [CDF Collaboration], arXiv:hep-ex/0508058
9. J. Heinrich, these proceedings.
10. K. Cranmer, these proceedings.

LIMITS AND CONFIDENCE INTERVALS IN THE PRESENCE OF NUISANCE PARAMETERS

WOLFGANG A. ROLKE and ANGEL M. LÓPEZ
Department of Mathematics, University of Puerto Rico - Mayagüez, Mayagüez, PR 00681, USA

JAN CONRAD
PH-Department, CERN, CH-1211, Geneva 23, Switzerland

We study the frequentist properties of confidence intervals computed by the method known to statisticians as the profile likelihood. It is seen that the coverage of these intervals is surprisingly good over a wide range of possible parameter values for important classes of problems, in particular whenever there are additional nuisance parameters with statistical or systematic errors. Programs are available for calculating these intervals.

We consider the problem of setting confidence limits for the signal rate in the presence of background which is estimated from data sidebands or Monte Carlo. Specifically we study the situation where the signal can be modeled by a Poisson distribution, a background with either a Poisson or a Gaussian distribution and an efficiency with either a Binomial or a Gaussian distribution. We establish the domain of validity, enabling comparison with other methods. We show that this method, together with some minor adjustments, has very good coverage even in cases when the parameters lie close to or at the physical boundaries.

Although this paper, and the corresponding routines, only deal with the specific problems outlined above, the results show that the method of profile likelihood is a viable technique for dealing with nuisance parameters, and it should be useful for other problems as well.

A stand-alone FORTRAN routine for calculating the limits discussed here is available at http://charma.uprm.edu/~rolke/publications.htm. It is also available as TRolke which is part of the ROOT system. Both routines also allow the calculation of the experimental sensitivity. Finally, at least for the cases where there are more events in the signal region than are expected from background, one could use MINUIT/MINOS to carry out the calculations, though in this case care needs to be taken to set the limits on the parameters correctly.

It is to be hoped that the profile likelihood method yields good results also in situations other than the ones discussed here. Because it is already available as part of MINUIT, its implementation for different problems should be quite straightforward. It needs to be emphasized, though, that the profile likelihood method can not be assumed to yield good results in all cases and that it might require some adjustments to the general method as we have done here. It is therefore strongly recommended that a thorough check of its performance be done whenever it is applied to a new problem. In the case of setting limits, this means a coverage study as described above, at least for the range of likely parameter values.

This work has previously been published in "*Limits and Confidence Intervals in the Presence of Nuisance Parameters*", W.A. Rolke, A.M. López and J.Conrad, Nuclear Instruments and Methods A, 551/2-3, 2005, pp. 493-503. It is available for downloading at http://xxx.lanl.gov/abs/physics/0403059.

THE BAYESIAN APPROACH TO SETTING LIMITS: WHAT TO AVOID

JOEL HEINRICH

University of Pennsylvania, Philadelphia, PA 19104, USA
E-mail: heinrich@fnal.gov

The task of setting limits in situations involving nuisance parameters with uncertainties has proved a difficult one in practice. CDF's Statistics Committee has recently recommended a Bayesian approach to setting limits. While investigating the performance of that approach, one rather restricted scenario was found to result in poor coverage behavior. The scenario is described, the resulting poor coverage behavior is illustrated, and solutions are proposed.

1. Introduction

The CDF Statistics Committee's web site[1] gives recommendations for setting limits[2] that promote a Bayesian method and advocate checking the method's frequentist coverage properties. Following those recommendations, we show in Sec. 2 that flat priors are adequate for the Bayesian method of setting upper limits in the single channel Poisson case. We consider multiple channels in Sec. 3, and find that flat subsidiary priors lead to poor coverage behavior in certain cases. Coverage is restored in Sec. 4 by replacing the flat subsidiary priors.

2. First Test Case: Single Channel

We observe n events from a process with Poisson rate $\epsilon s + b$, where s is cross section, ϵ is acceptance$\times$luminosity, b is background, and obtain the Bayesian posterior for s. Nuisance parameters ϵ and b are determined via Poisson subsidiary measurements, whose posteriors serve as the priors for ϵ and b in the main measurement. The specified Bayesian priors are

- flat prior for $s \geq 0$
- flat (subsidiary) prior for $\epsilon \geq 0$
- flat (subsidiary) prior for $b \geq 0$

We obtain the joint posterior $p(s, \epsilon, b|n)$, and marginalize over ϵ and b. An upper limit for s is obtained by integrating the posterior $p(s|n)$ with respect to s from $s = 0$ to the value of s that yields credibility level β.

The subsidiary measurement for ϵ observes m events with Poisson rate $\kappa\epsilon$, where κ is a known constant. The subsidiary posterior,

$$p(\epsilon|m) = \frac{\kappa(\kappa\epsilon)^m e^{-\kappa\epsilon}}{m!} \tag{1}$$

becomes the prior for ϵ in the main measurement. The mean of $p(\epsilon|m)$ is $(m+1)/\kappa$. (This is a calibration measurement of ϵ.)

The subsidiary measurement for b observes r events with Poisson rate ωb, where ω is a known constant. The subsidiary posterior,

$$p(b|r) = \frac{\omega(\omega b)^r e^{-\omega b}}{r!} \tag{2}$$

becomes the prior for b in the main measurement. The mean of $p(b|r)$ is $(r+1)/\omega$. (This is a sideband determination of b.)

The posterior $p(s|n)$ is calculated analytically, given in Refs. 3, 4. Figure 1 shows a typical posterior p.d.f.

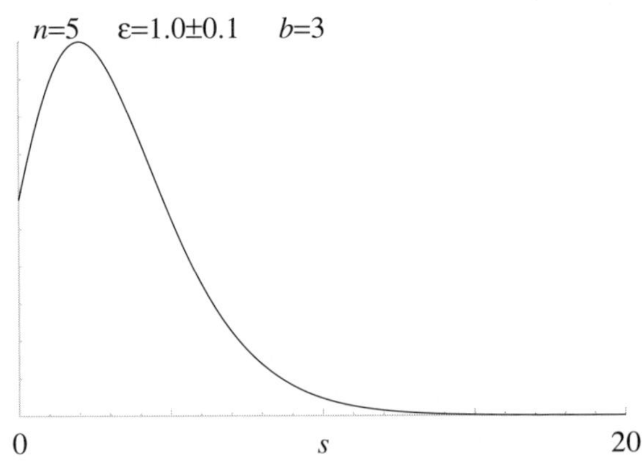

Fig. 1. An example $p(s|n)$ with b fixed ($\kappa = 100$ and $m = 99$).

We employ an objective Bayesian methodology. The priors, which are improper (and not related to personal belief), are evaluated using a frequentist technique. See Ref. 5 for a discussion of this approach.

The frequentist coverage probability C is used as a diagnostic to check the performance of the limit setting scheme. For upper limits on s, C is the probability that, for fixed (true) values of the parameter of interest s and nuisance parameters ϵ and b, the resulting upper limit will be larger than s_{true}. The coverage is calculated by summing over all possible outcomes of the main and subsidiary measurements.

For this single channel case, $C > \beta$ for every combination of s_{true}, ϵ_{true}, and b_{true} tested, with this choice of priors, even when uncertainties on ϵ and b are very large (illustrated in Figs. 2–4). Although opinions differ on whether *any* undercoverage is acceptable, large undercoverage is considered bad. The single channel test case passes this test.

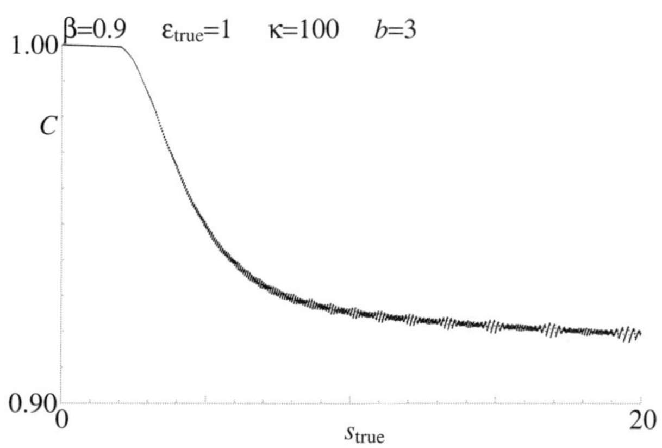

Fig. 2. Typical single channel case. Coverage for 90% credibility level upper limits, acceptance uncertainty = 10%, background uncertainty = zero.

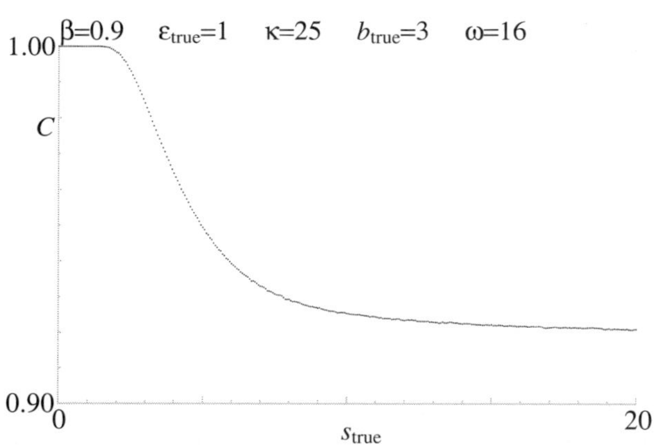

Fig. 3. Typical single channel case. Coverage for 90% credibility level upper limits, acceptance uncertainty = 20%, background uncertainty = 15%. This example is divided into N channels in Sec. 3.

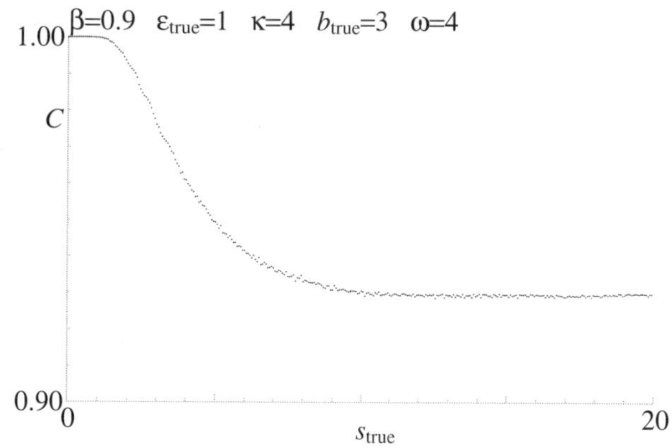

Fig. 4. Extreme single channel case. Coverage for 90% credibility level upper limits, acceptance uncertainty = 50%, background uncertainty = 29%. Larger ϵ and b uncertainties lead to slightly larger C here than in Fig. 3.

3. Second Test Case: Multiple Channels

Given N channels, and n_k observed events in the kth channel, $k = 1, 2, \ldots, N$, the Poisson probability of obtaining the observed result is

$$\prod_{k=1}^{N} \frac{e^{-(s\epsilon_k + b_k)}(s\epsilon_k + b_k)^{n_k}}{n_k!}$$

where s the cross section, and ϵ_k and b_k are the acceptance and expected background for the kth channel, respectively. One multiplies by the prior for s, the $2N$ nuisance priors, and marginalizes.

Reference 6 describes a MC integration approach to calculating the Bayesian posterior for s, given a prior flat in s, but no restrictions on the nuisance priors.

For our test case, we specify that the data of the first test case (both the main measurement and the subsidiary measurements) are divided into N samples that are treated independently, to derive an upper limit on the common parameter s. Flat priors are specified for the $2N$ subsidiary measurements, leading to $2N$ subsidiary posteriors that become the nuisance priors for the main measurement. The prior for s remains flat.

For this Poisson example, we find that, when the size of the initial subsidiary data sets is not large, dividing into N independent channels drives C progressively further down as N increases. A typical example is shown in Figs. 5–7. The coverage of the 2-channel case (Fig. 5) is close to the nominal 90%; it drops to ~87% in the 3-channel case (Fig. 6), and down to ~84% in the 4-channel case (Fig. 7). One can drive the coverage arbitrarily low by simply increasing N.

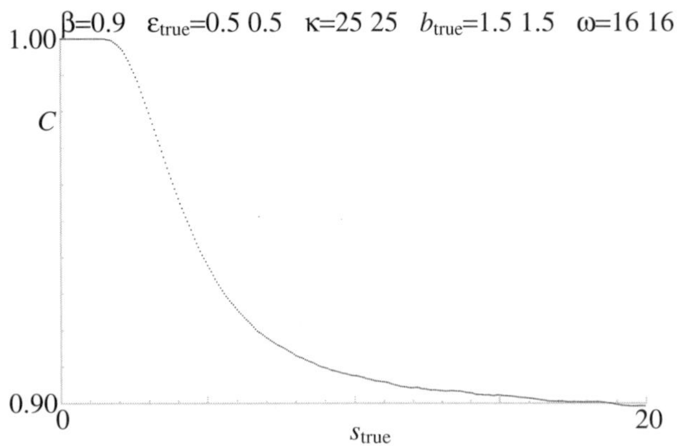

Fig. 5. 2 independent channels. Coverage for 90% credibility level upper limits, acceptance uncertainty = 29%/channel, background uncertainty = 20%/channel.

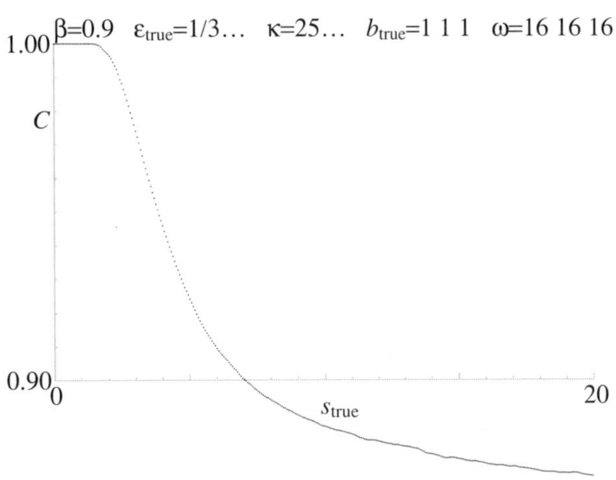

Fig. 6. 3 independent channels. Coverage for 90% credibility level upper limits, acceptance uncertainty = 34%/channel, background uncertainty = 25%/channel.

4. The Solution

The fault is in our choice of priors for the Poisson subsidiary measurements. For example, a flat prior for each channel's ϵ_k subsidiary measurement yields an ϵ^{N-1} prior for the total acceptance, creating a large bias when $N > 2$. (Same bias problem for b.)

With respect to upper limits, a flat prior for s leads to a bias producing overcoverage in simple Poisson cases. This bias in the subsidiary measurements leads to undercoverage in the main measurement,

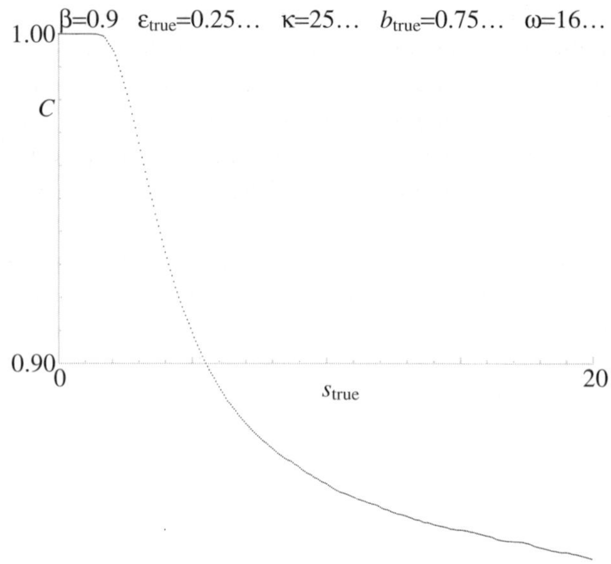

Fig. 7. 4 independent channels. Coverage for 90% credibility level upper limits, acceptance uncertainty = 40%/channel, Background uncertainty = 29%/channel.

since an overestimate of ϵ or b leads to an underestimate for s. In our test case, using a flat prior is "conservative" for s, but "anticonservative" for ϵ and b. When $N = 1$, they roughly balance. When $N > 2$, the subsidiary priors dominate.

For our test case, a "perfect" solution is available: Use $1/\epsilon_k$ and $1/b_k$ priors[7, 8] for the subsidiary measurements. This is illustrated in Fig. 8 for the 4-channel case.

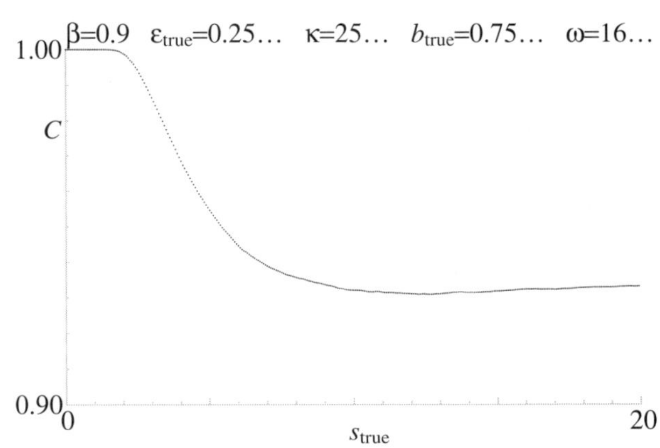

Fig. 8. 4 independent channels. Coverage for 90% credibility level upper limits, acceptance uncertainty = 40%/channel, background uncertainty = 29%/channel. Use of $1/\epsilon_k$ and $1/b_k$ subsidiary priors here restores coverage, compared with Fig. 7.

With this choice of subsidiary priors, the nuisance priors for the kth channel become

$$p(\epsilon_k|m_k) = \frac{\kappa_k(\kappa_k\epsilon_k)^{m_k-1}e^{-\kappa_k\epsilon_k}}{(m_k-1)!}$$

and

$$p(b_k|r_k) = \frac{\omega_k(\omega_k b_k)^{r_k-1}e^{-\omega_k b_k}}{(r_k-1)!}$$

These have the same gamma distribution form as Eqs. 1 and 2, but are shifted downward. The means are m_k/κ_k and r_k/ω_k, respectively, eliminating the bias:

$$\langle m_k/\kappa_k\rangle = \epsilon_{\text{true},k} \qquad \langle r_k/\omega_k\rangle = b_{\text{true},k}$$

That is, the mean of the nuisance prior is now an unbiased estimator of the true value of the nuisance parameter.

5. Conclusions

- The multichannel case involves a multidimensional nuisance prior. In hindsight, this should have led us to distrust a prior flat in multiple dimensions, since this is well known[9] to lead to problems.
- Our example is not entirely realistic, as it specifies unusually low precision calibrations. Also, correlations among the ϵ_k and b_k, which would effectively reduce the dimensionality, are absent. But extreme cases are useful for testing the method.
- Marginalization over nuisance parameters using Bayesian priors is a common feature of many methods for setting limits (e.g. Ref. 10). Using unbiased priors will help avoid pathologies.
- The $1/\epsilon_k$ and $1/b_k$ subsidiary priors are matched to this Poisson case. Other cases will require different solutions.
- In the objective Bayesian approach, the choice of subsidiary priors is just as important as the choice of prior for the parameter of interest in the main measurement. Switching to $1/\epsilon_k$ and $1/b_k$ subsidiary priors to remove the bias in the nuisance priors raised the coverage significantly, and may make use of a $1/\sqrt{s}$ prior in the main measurement more appealing. (Reference 3 shows that a $1/s$ prior yields an unnormalizable posterior in the presence of background, while a $1/\sqrt{s}$ prior combined with a flat subsidiary prior for ϵ leads to a slight undercoverage in the single channel case.)
- Coverage calculations are useful in revealing poor choices of prior in the objective Bayesian approach.

Acknowledgments

I would like to thank the members of the CDF Statistics Committee for helpful discussions, and the organizers of PHYSTAT05 for arranging an excellent conference.

References

1. CDF Statistics Committee, `www-cdf.fnal.gov/physics/statistics/`.
2. CDF Statistics Committee, C. Blocker *et al.*, *Recommendations Concerning Limits*, CDF Internal Note 7739, Fermilab (2005), `www-cdf.fnal.gov/publications/cdf7739_limit_recommendation_2.pdf`.
3. J. Heinrich *et al.*, *Interval estimation in the presence of nuisance parameters. 1. Bayesian approach.*, CDF Internal Note 7117, Fermilab (2004), physics/0409129, `www-cdf.fnal.gov/publications/cdf7117_bayesianlimit.pdf`.
4. J. Heinrich, *User Guide to Bayesian-Limit Software Package*, CDF Internal Note 7232, Fermilab (2004), `www-cdf.fnal.gov/publications/cdf7232_blimitguide.pdf`.
5. M. J. Bayarri and J. O. Berger, *Statistical Science* **19**, p. 58 (2004), `projecteuclid.org/Dienst/UI/1.0/Summarize/euclid.ss/1089808273`, `www.isds.duke.edu/~berger/papers/interplay.html`.
6. J. Heinrich, *Bayesian limit software: multi-channel with correlated backgrounds and efficiencies*, CDF Internal Note 7587, Fermilab (2005), `www-cdf.fnal.gov/publications/cdf7587_genlimit.pdf`.
7. H. Jeffreys, *Theory of Probability, 3rd edn.* (Oxford University Press, 1961), ch. 3, p. 120.
8. R. D. Cousins, these proceedings.
9. D. R. Cox, these proceedings.
10. J. Conrad and F. Tegenfeldt, these proceedings.

STATISTICALLY DUAL DISTRIBUTIONS IN STATISTICAL INFERENCE

S. I. BITYUKOV and V. V. SMIRNOVA

Institute for high energy physics, 142281 Protvino, Russia
E-mail: Serguei.Bitioukov@cern.ch, vera@cub.ihep.su

N. V. KRASNIKOV

Institute for nuclear research RAS, 117312 Moscow, Russia
E-mail: krasniko@ms2.inr.ac.ru

V. A. TAPERECHKINA

Moscow State Academy of Instrument Engineering and Computer Science, Moscow, Russia

The estimation of a parameter of a model by the measurement of a random variable whose distribution depends on this parameter is one of the main tasks of statistics. In this paper the notion of the statistically dual distributions is introduced. An approach, based on the properties of the statistically dual distributions, to resolve the given task is proposed.

1. Introduction

As shown in refs. [1, 2], in the framework of frequentist approach we can construct the probability distribution of the possible magnitudes of the Poisson distribution parameter to give the observed number of events $\hat{n}$ in a Poisson stream of events. This distribution, which can be called a confidence density function of a parameter, is described by a Gamma-distribution with the probability density function which looks like a Poisson distribution of probabilities. This is the reason for naming this pair of distributions as statistically dual distributions. Also, the interrelation between the Poisson and Gamma distributions was used in these papers to reconstruct the confidence density of the Poisson distribution parameter by a unique way and, correspondingly, to construct any confidence interval for the parameter.

According to B. Efron [3] the confidence density is the fiducial [4] distribution of the parameter. This distribution is considered as a genuine *a posteriori* density for the parameter without prior assumptions.

The same relation [5–7], which allows one to reconstruct the confidence density of a parameter in a unique way, exists between several pairs of statistically self-dual distributions (normal and normal, Laplace and Laplace and, as is shown below, Cauchy and Cauchy).

Note that the posterior distribution of the parameter also is used for the definition of conjugate families in the Bayesian approach. The interrelation between the statistically dual distributions and conjugate families is discussed in ref. [7].

2. Statistically dual distributions

Let us define statistically dual distributions.

Definition 1: Let $\phi(x,\theta)$ be a function of two variables. If the same function can be considered both as a family of the probability density functions (pdf) $f(x|\theta)$ of the random variable x with parameter θ and as another family of pdf's $\tilde{f}(\theta|x)$ of the random variable θ with parameter x (i.e. $\phi(x,\theta) = f(x|\theta) = \tilde{f}(\theta|x)$), then this pair of families of distributions can be named as **statistically dual distributions**.

The statistical duality of Poisson and Gamma-distributions follows from simple discourse.

Let us consider the Gamma-distribution with probability density

$$g_x(\beta,\alpha) = \frac{x^{\alpha-1}e^{-\frac{x}{\beta}}}{\beta^\alpha\Gamma(\alpha)}. \quad (1)$$

Changing the standard designations of the Gamma-distribution $\frac{1}{\beta}$, α and x for a, $n+1$ and μ, we get the following formula for the probability density of Gamma-distribution

$$g_n(a,\mu) = \frac{a^{n+1}}{\Gamma(n+1)}e^{-a\mu}\mu^n, \quad (2)$$

where a is a scale parameter and $n+1>0$ is a shape parameter. Suppose $a=1$, then the formula of the probability density of Gamma-distribution $\Gamma_{1,n+1}$ is

$$g_n(\mu)=\frac{\mu^n}{n!}e^{-\mu},\ \mu>0,\ n>-1. \tag{3}$$

It is a common supposition that the probability of observing n events in the experiment is described by a Poisson distribution with parameter μ, i.e.

$$f(n|\mu)=\frac{\mu^n}{n!}e^{-\mu},\ \mu>0,\ n\geq 0. \tag{4}$$

One can see that if the parameter and variable in Eq. (3) and Eq. (4) are exchanged, in other respects the formulae are identical. As a result these distributions (Gamma and Poisson) are **statistically dual distributions**. These distributions are connected by the identity [1] (see, also, this identity in another form in refs. [8, 9, 10])

$$\sum_{i=\hat{n}+1}^{\infty} f(i|\mu_1)+\int_{\mu_1}^{\mu_2} g_{\hat{n}}(\mu)d\mu+\sum_{i=0}^{\hat{n}} f(i|\mu_2)=1, \tag{5}$$

i.e.

$$\sum_{i=\hat{n}+1}^{\infty}\frac{\mu_1^i e^{-\mu_1}}{i!}+\int_{\mu_1}^{\mu_2}\frac{\mu^{\hat{n}}e^{-\mu}}{\hat{n}!}d\mu+\sum_{i=0}^{\hat{n}}\frac{\mu_2^i e^{-\mu_2}}{i!}=1$$

for any real $\mu_1\geq 0$ and $\mu_2\geq 0$ and non-negative integer $\hat{n}$.

The definition of the confidence interval (μ_1,μ_2) for the Poisson distribution parameter μ using [1, 5]

$$P(\mu_1\leq\mu\leq\mu_2|\hat{n})=P(i\leq\hat{n}|\mu_1)-P(i\leq\hat{n}|\mu_2), \tag{6}$$

where $P(i\leq\hat{n}|\mu)=\sum_{i=0}^{\hat{n}}\frac{\mu^i e^{-\mu}}{i!}$, allows one to show that a Gamma-distribution $\Gamma_{1,1+\hat{n}}$ is the probability distribution of different values of μ parameter of Poisson distribution on condition that the observed value of the number of events is equal to $\hat{n}$, i.e. $\Gamma_{1,1+\hat{n}}$ is the confidence density of the parameter μ. This definition is consistent with the identity Eq. (5). Note, if we suppose in Eq. (5) that $\mu_1=\mu_2$ we have a conservation of probability. The right-hand side of Eq. (6) determines the frequentist sense of this definition.

Another example of statistically dual distribution is the Cauchy distribution with unknown parameter θ and known parameter b. Here we also can exchange the parameter θ and variable x while conserving the same formula of the probability density.

The probability density of the Cauchy distribution is

$$C(x|\theta)=\frac{b}{\pi(b^2+(x-\theta)^2)}. \tag{7}$$

The probability density of its statistically dual distribution is also the Cauchy distribution:

$$\tilde{C}(\theta|x)=\frac{b}{\pi(b^2+(x-\theta)^2)}. \tag{8}$$

In such a way the Cauchy distribution can be named as **statistically self-dual distribution**. An identity like Eq. (5) also holds,

$$\int_{\hat{x}}^{\infty} C(x|\theta_1)dx+\int_{\theta_1}^{\theta_2}\tilde{C}(\theta|\hat{x})d\theta+\int_{-\infty}^{\hat{x}} C(x|\theta_2)dx=1, \tag{9}$$

where $\hat{x}$ is the observed value of random variable x and $\tilde{C}(\theta|\hat{x})$ is the confidence density.

3. Statistical duality and estimation of the parameter of a distribution

It is easy to show that the reconstruction of the confidence density is unique if Eqs. (5) or (9) holds [5–7].

As a result we have the **Transform** (both for Poisson-Gamma pair of families of distributions and for statistically self-dual distributions)

$$\int_{\hat{x}}^{\infty} f(x|\theta_1)dx+\int_{\theta_1}^{\theta_2}\tilde{f}(\theta|\hat{x})d\theta+\int_{-\infty}^{\hat{x}} f(x|\theta_2)dx=1 \tag{10}$$

between the space of the realizations $\hat{x}$ of random variable x and the space of the possible values of the parameter θ, i.e.

$$\tilde{f}(\theta|\hat{x})=T_{cd}\hat{x}, \tag{11}$$

where T_{cd} is the operator of the Transform. Here θ_1 and θ_2 are the bounds of the confidence interval for location parameter θ. As is shown above in the case of Gamma- and Poisson distributions, the two integrals are replaced by sums and $-\infty$ is replaced by 0.

The Transform Eq. (10) allows one to use statistical inferences about the random variable for estimation of an unknown parameter.

The simplest examples of this are given by several infinitely divisible distributions.

Definition 2: A distribution F is **infinitely divisible** if for each n there exist a distribution function F_n such that F is the n-fold convolution of F_n.

As known the Poisson, Gamma-, normal and Cauchy distributions are infinitely divisible distributions. The sum of independent and identically distributed random variables, which obey one of the above families of distributions, also obeys the distribution from the same family. Applying the Transform Eq. (10) to this sum allows one to reconstruct the confidence density of the parameter in the case of several observation of the same random variable. It means that we construct the relation

$$\tilde{f}(n\theta|\hat{x}_1+\hat{x}_2+\cdots+\hat{x}_n) = T_{cd}(\hat{x}_1+\hat{x}_2+\cdots+\hat{x}_n), \quad (12)$$

where T_{cd} is the operator of the Transform Eq. (10), the set $\hat{x}_1, \hat{x}_2, \ldots, \hat{x}_n$ are the observed values. Thereafter we reconstruct the confidence density of θ, i.e. $\tilde{f}(\theta|\hat{x}_1, \hat{x}_2, \ldots, \hat{x}_n)$.

The method for construction of the confidence density of the mean value of several random variables which obey the Poisson distribution (sample with size > 1) and the way how to take into account the statistical uncertainty is shown in refs. [2].

The use of the confidence density also can be formulated in Bayesian framework. Let us consider, as an example, the Cauchy distribution. We suppose in our approach that the parameter θ is not a random value and before the measurement we do not prefer any values of this parameter, i.e. possible values of the parameter have equal probability and a prior distribution of θ is $\pi(\theta) = const$. Suppose we observe $\hat{x}_1$ and update our prior via the Transform Eq. (10) to obtain $\tilde{C}(\theta|\hat{x}_1)$, which is the pdf of the Cauchy distribution. This becomes our new prior before observing $\hat{x}_2$. It is easy to show that in the case of observing $\hat{x}_2$ the reconstructed confidence density $\tilde{C}(2\theta|\hat{x}_1 + \hat{x}_2)$ also is the pdf of the Cauchy distribution and, correspondingly, $\tilde{C}(\theta|\hat{x}_1, \hat{x}_2)$ is our next new prior. By induction this argument extends to sequences of any number of observations, i.e. we use the iterative procedure

$$\tilde{C}(\theta|\hat{x}_1, \hat{x}_2, \ldots, \hat{x}_{n-1}, \hat{x}_n) = T_{pd}(\tilde{C}(\theta|\hat{x}_1, \hat{x}_2, \ldots, \hat{x}_{n-1}), \hat{x}_n), \quad (13)$$

where T_{pd} is the operator of the Transform between a priori density and a posteriori density of the parameter.

4. Conclusions

We have formulated the notion of statistically dual distributions in the framework of probabilistic (and, in this sense, frequentist) approach.

We have shown that the statistical duality allows one to connect the estimation of the parameter with the measurement of the random variable of the distribution due to the Transform Eq. (10).

All considered cases of statistically dual distributions belong to conjugate families (which are defined in the framework of Bayesian approach). For example [11], the distributions conjugate to Poisson distributions were built by a Monte Carlo method (i.e. in the frequentist approach). Hypothesis testing confirms that these distributions are Gamma-distributions as expected in this case.

This means that statistical duality gives a clear frequentist sense to the confidence density of the parameter and allows one to construct confidence intervals in an easy way.

Acknowledgments

We are grateful to Louis Lyons, Victor Matveev and Vladimir Obraztsov for interest and support of this work.

We also thank Glen Cowan, Bob Cousins, Yuri Gouz, George Kahrimanis, Alexandre Nikitenko and Claudia Wulz for very useful discussions. S.B. would like to thank the Organizing Committee of PhyStat05 for having organized this interesting Workshop which is the wonderful opportunity for exchange of ideas. This work has been partly supported by grants RFBR 04-01-97227 and RFBR 04-02-16020.

References

1. S.I. Bityukov, N.V. Krasnikov, V.A. Taperechkina, *Preprint IFVE* 2000-61, Protvino, 2000; also, *e-Print:* hep-ex/0108020, 2001.
2. S.I.Bityukov, *JHEP* 09 (2002) 060; also, *e-Print:* hep-ph/0207130, 2002; S.I. Bityukov, N.V. Krasnikov, *Nucl.Inst.&Meth.* **A502**, 795 (2003).

3. B. Efron, *Stat.Sci.* **13** 95 (1998).
4. R.A. Fisher, *Proc. of the Cambridge Philosophical Society* **26**, 528 (1930).
5. S.I. Bityukov, V.A. Taperechkina, V.V. Smirnova, *e-Print:* math.ST/0411462, 2004.
6. E.A. Barkova, S.I. Bityukov, V.A. Taperechkina, *e-Print:* math.ST/0507452, 2005.
7. S.I. Bityukov, N.V. Krasnikov, in Proc. of MaxEnt'05, August 2005, San Jose, CA, USA, AIP Conference Proceedings **803** 398 (2005).
8. E.T. Jaynes: Papers on probability, statistics and statistical physics, Ed. by R.D. Rosenkrantz, D.Reidel Publishing Company, Dordrecht, Holland, 1983.
9. A.G.Frodesen, O.Skjeggestad, H.Toft, *Probability and Statistics in Particle Physics,* UNIVERSITETSFORLAGET, Bergen-Oslo-Tromso, 1979.
10. R.D. Cousins, *Am.J.Phys.* **63** 398 (1995) 398.
11. S.I. Bityukov et al., *Nucl.Instr.&Meth.* **A534**, 228 (2004); also, *e-Print:* physics/0403069, 2004.

PROGRAM FOR EVALUATION OF SIGNIFICANCE, CONFIDENCE INTERVALS AND LIMITS BY DIRECT CALCULATION OF PROBABILITIES

S. I. BITYUKOV

Institute for high energy physics, 142281 Protvino, Russia

S. E. EROFEEVA

Moscow State Academy of Instrument Engineering and Computer Science, Moscow, Russia

N. V. KRASNIKOV

Institute for nuclear research RAS, 117312 Moscow, Russia

A. N. NIKITENKO

Imperial College, London, United Kingdom, on leave from ITEP, Moscow

We propose a program which allows one to estimate significance, confidence intervals and limits, taking into account systematics and statistical uncertainties of variables described by Poisson distributions. The given program can be used for combining searches. The motivation of the direct probabilities calculations is determined by two reasons. Firstly, the tail of a Poisson distribution is heavier than that of a normal distribution. In the case of small probabilities the Gaussian approximation gives the wrong values of estimators. Secondly, the estimators which are constructed on the basis of the likelihood ratio often have poor statistical properties.

1. Introduction

In searches for new particles in high energy physics the tasks of the comparison of the possibilities of different planned experiments and of the setting and combining of the results of running experiments are vitally important. The determination of the significance of the results and the setting of confidence limits are related to these tasks.

There are many reviews about using of the notion of a significance in various areas of physics (for example, [1–3]). According to ref. [4] "Common practice is to express the significance of an enhancement by quoting the number of standard deviations". In the case of asymmetric distributions this statement is not clear. Then the significance can be quoted in terms of equivalent standard deviations of the normal distribution.

The definition chosen for significance depends on the goal. The simplest significance $S_1 = \dfrac{s}{\sqrt{b}}$ (or Z_{sb} in the notation of refs. [2]) takes into account only the fluctuation of the background with the assumption that the background obeys the normal distribution with mean and variance equals to b. The S_1 corresponds the case when we observed $s + b$ events. The significance $S_2 = \dfrac{s}{\sqrt{s+b}}$ takes into account the fluctuation of the normal distributed random value *signal+background* with the expected background b. The significance $S_{12} = \sqrt{s+b} - \sqrt{b}$ [5] takes into account the fluctuations of the signal and of the background with the assumption that the signal and the background obey the normal distribution with mean and variance which equal s for signal and b for background.

In the case of the asymmetrically distributed signal and background these significances can be used only as approximations [6]. Thus for Poisson distributions, the significance S_{cP} [7] was proposed as the analogy of S_1. The significance S_{cP} is determined by direct calculations of probabilities.

The correct choice of the confidence interval for the parameter of the distribution under study also is a hot problem of data analysis. This matter was discussed in many Workshops and Conferences [8]. There are many methods of constructing intervals: credible, tolerant, fiducial. Physicists [9] formulated the problem of the specialist in data analysis "the only remaining problem: make a choice ... chosen method should be as simple as possible, but not wrong". In many cases the reconstruction of the confidence density [10] of the parameter (by direct calculation of probabilities) is the solution of this problem. The knowledge of the confidence density allows to determine any confidence intervals (central interval, shortest interval, interval with optimal coverage and so on) by the unique way and to take into account the systematics and statistical uncertainties.

2. Significance

The significance S_{cP} is the probability from Poisson distribution with mean b to observe $b+s$ or more events, converted to the equivalent number of sigmas of a Gaussian distribution, i.e.

$$\beta = \frac{1}{\sqrt{2\pi}} \int_{S_{cP}}^{\infty} e^{-\frac{x^2}{2}} dx, \text{ where } \beta = \sum_{i=s+b}^{\infty} \frac{b^i e^{-b}}{i!}. \quad (1)$$

The program ScP allows one to calculate the significance defined in a such way. The background uncertainties are incorporated into the program. The program takes into account two types of uncertainties: the systematic uncertainty with statistical properties (the normal distribution with mean which equals 0 and the variance – σ_b^2) and the uncertainty without statistical properties [5] (theoretical uncertainty with bias in the background $b + \delta_b$ and conserving the scale, i.e. $s + b$ without bias). Also, the program allows one to combine several observed values of significance. The values of S_{cP} by definition have the irregular behavior for non-integer $s + b$. The program has the option which allows the smoothing of the result by the using of "the continuous Poisson distribution" [11].

3. Confidence Limits

The knowledge of the confidence density allows one to determine any confidence limits for the parameter of the distribution. The program *Limsb* is the realization of this idea. The Gamma-distributions $\Gamma_{1,\hat{n}+1}$ is statistically dual to the Poisson distribution in the case of $\hat{n}$ observed events from the Poisson flow of events [10, 11]. It means that the direct calculations of the probability density of the Gamma-distribution $\Gamma_{1,\hat{n}+1}$ are the reconstruction of the confidence density of the unknown parameter of the Poisson distribution. The confidence intervals produced by the program *Limsb* coincide with the corresponding confidence intervals calculated in the framework of the Bayesian approach with a uniform prior.

4. Conclusions

All significances described in Sec. 1 relate to the measurement of a random variable. However, significance S_{cP} can be generalized for the determination of the significance of the estimated parameter by the use of confidence densities. We plan to include the presented approach for determination of the significance and confidence limits into the ROOT system [12]. The programs *ScP* and *Limsb* can be found in Web page http://cmsdoc.cern.ch/~bityukov.

Acknowledgments

The authors are grateful to V.B. Gavrilov, V.A. Kachanov and A. De Roeck for the interest and useful comments, R. Cousins, A. Lanyov, G. Quast and S. Shmatov for fruitful discussions. This work has been supported by grants RFBR 04-02-16020 and RFBR 05-07-90072.

References

1. P. Sinervo, Proc. of Int. Conf. *Advanced Statistical Techniques in Particle Physics*, March 18-22, 2002, eds. L. Lyons, M. Whalley, Durham, UK, p.64.
2. J.T. Linnemann, Proc. of Int. Conf. *Statistical Problems in Particle Physics, Astrophysics and Cosmology*, Sept. 8-11, 2003, eds. L. Lyons, R. Mount, R. Reitmeyer, SLAC, Stanford, CA, USA, p.35.
3. V.Bartsch and G.Quast, *CMS Note 2005/004*. Aug., 2003; R. Cousins, J. Mumford, V. Valuev, *CMS Note 2005/003*. Feb., 2005.
4. A.G.Frodesen, O.Skjeggestad, H.Tøft, *Probability and Statistics in Particle Physics,* UNIVERSITETS-FORLAGET, Bergen-Oslo-Tromsø, 1979.
5. S.I.Bityukov and N.V.Krasnikov, *Modern Physics Letters* **A13** (1998) 3235.
6. S.I.Bityukov and N.V.Krasnikov, *Nucl.Instr.&Meth.* **A452** (2000) 518.
7. I.Narsky, *Nucl. Instrum. Meth.* **A450** (2000) 444.
8. Proc. of Workshop *Confidence Limits*, eds. F. James, L. Lyons, Y. Perrin, Jan. 17-18, 2000, CERN, Geneva, Switzerland; also, Conferences in refs [1, 2].
9. G. Quast, CMS Physics Analysis Days, May 9, 2005.
10. S.I. Bityukov *et al., "Statistically dual distributions in statistical inference"*, in these Proceedings.
11. S.I. Bityukov *et al., Preprint IFVE 2000-61*, Protvino, 2000; hep-ex/0108020, 2001.
12. R. Brun and A. Kreshuk, *"Statistics in ROOT"*, in these Proceedings.

EXAMINING THE BALANCE BETWEEN OPTIMISING AN ANALYSIS FOR BEST LIMIT SETTING AND BEST DISCOVERY POTENTIAL

G.C. HILL, J. HODGES, B. HUGHEY, A. KARLE AND M. STAMATIKOS

University of Wisconsin, Madison, USA
E-mail: ghill@icecube.wisc.edu

The question of how to perform an unbiased optimisation of the final event selection (via cuts on the event observables) in a counting experiment is discussed. We wish to balance the desire to make a discovery, i.e. see a highly significant excess if a real signal is present, with the desire to have a high sensitivity, i.e. set the best upper limit if no real signal is present. In an astrophysical setting, the search usually involves a signal hypothesis where the model parameters defining the shape of the source flux spectrum are assumed known and we seek to find a confidence interval on the normalisation constant of the model. To optimise for best sensitivity in this type of analysis, the model rejection potential method[1] has been widely used in the high energy neutrino detector field (AMANDA, IceCube, ANTARES). In this approach, the final event selection region in the observable space is chosen so that the average upper limit on the source flux normalisation at a given confidence level $1 - \alpha$ is minimised. To assess the discovery potential, we suggest an analogous approach, where we choose to minimise the source flux normalisation, which if truly non-zero, would lead to the rejection of the background only hypothesis at a required significance level α with some high probability ($\equiv$ power $= 1 - \beta$). For example, we might decide to optimise the event selection region by minimising the source strength needed to see a highly significant discovery (e.g. 5 sigma) with a high probability (e.g. 90%). By comparing the resulting fluxes as a function of the event selection region, we can examine the trade off between retaining a region based on sensitivity to that based on discovery, and make a choice that balances these desires. This discovery/sensitivity optimisation has been used in the AMANDA search for neutrinos in coincidence with the gamma-ray burst GRB030329 observed in 2003 by the HETE-II satellite[2].

1. Introduction

In this paper we define ways of assessing and optimising the limit setting ("model rejection potential"[1]) and discovery ("model discovery potential"[3]) capabilities of an experiment. We examine the tradeoff between choosing cuts based on either criteria using an example of an astrophysical search for a point source of neutrinos or gamma rays and compare the results to those following a traditional approach to this problem[4].

2. Optimising for best limit setting sensitivity

The model rejection potential technique[1] (MRP) has seen widespread use in the high-energy neutrino telescope community as a method of unbiased sensitivity (best limit) optimisation of an analysis. This method optimises the event cuts to minimise the expected upper limit[5] from the experiment, assuming there is no true signal present. For a Poisson counting experiment, suppose one chose a selection cut, leaving an expected background rate of μ_b and an expected number of signal events μ_s. We are searching for a signal we believe to be described by a flux model $A\ \Phi(E)$, where E is the true energy of the incident particle. The shape of the signal flux as a function of the particle energy E is $\Phi(E)$, and the total rate of expected events is given by the constant A. We will assume the shape of $\Phi(E)$ and place a limit on the scale factor A. After averaging the flux over the detector response $\epsilon(E)$ we find the number of expected signal events $\mu_s = \int A\Phi(E)\epsilon(E)\mathrm{d}E$. The data are then unblinded and n_{obs} events are seen. The limit on the normalisation factor A is then $A_{lim} = A\ \mu(n_{obs}, \mu_b)/\mu_s$, where $\mu(n_{obs}, \mu_b)$ is the event upper limit (e.g. the upper boundary of a Feldman-Cousins[5] interval). The quantity $\mu(n_{obs}, \mu_b)/\mu_s$ is called the model rejection factor (MRF), as it shows how much the initial assumed normalisation is rescaled to reach the limit A_{lim}. Note also that the final A_{lim} is independent of the initially assumed A. The optimal choice of cut would be one that minimises the MRF and thus minimises A_{lim}, however the MRF depends on the experimentally observed number of events n_{obs}. Prior to unblinding the data, we can replace $\mu(n_{obs}, \mu_b)$ with the average[a] upper limit[5] expected from an ensemble of events with no real signal, to assess the sensitivity of

[a]The *average* upper limit is however not independent of the metric used for the parameter and should in future be replaced by the *median* upper limit[6].

the experiment in an unbiased fashion. The cut that minimises the average MRF is the one that would produce the lowest average limit over an ensemble of repeated experiments.

3. Optimising for discovery

How is a discovery defined? If we are using frequentist statistics, then we usually say we have "discovered" a real effect when the probability of the observation, or a more extreme one, under the assumption of the background only hypothesis is very small[b]. In our counting experiment from section 2, we calculate the p-value, $P(\geq n_{obs} \mid \mu_b)$ and if it is very small, e.g. $P(\geq n_{obs} \mid \mu_b) < \alpha$ where $\alpha = 5.73 \times 10^{-7}$ (area in the two-sided 5σ Gaussian tails), we might claim to have seen something interesting. For a given α, we can calculate a critical number of events, n_{crit}, where $P(\geq n_{crit} \mid \mu_b) < \alpha$. This is the minimum number of events needed to be observed in order that a p-value less than alpha is reported. If a real signal of strength μ_s is also present, then the probability (statistical power 1-β) that we would observe n_{crit} or more events is $1 - \beta = P(\geq n_{crit} \mid \mu_b + \mu_s)$. We define the "least detectable signal" μ_{lds} as the value of μ_s where this equality is satisfied for a given value of $1 - \beta$. This strength of signal would produce an observation n_{obs} leading to a p-value less than α in a fraction of $1 - \beta$ experiments. As an example, take $\mu_b = 3.0$, require $\alpha = 5.73 \times 10^{-7}$ and $1 - \beta = 0.9$. Then $n_{crit} = 16$ ($P(\geq 16 \mid 3.0) = 1.24 \times 10^{-7}$), $\mu_{lds} + \mu_b = 21.3$ ($P(\geq 16 \mid 21.3) = 0.90$), leading to $\mu_{lds} = 18.3$. We can calculate μ_{lds} as a function of μ_b for various values of α and $1 - \beta$. The results are shown in figure 1. Now, if we replace the average upper limit in the model rejection factor calculation with the least detectable signal, we can then minimise the "model discovery potential" of the experiment. Choosing the cut corresponding to the minimum MDP minimises the true signal flux required to obtain an observation at significance level α with

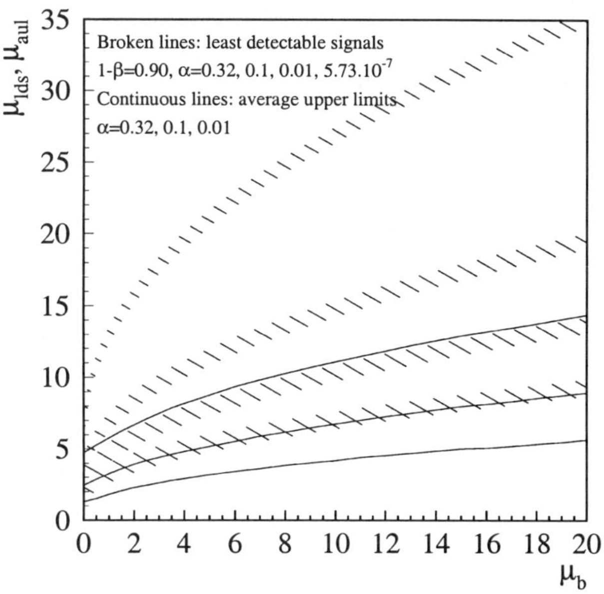

Fig. 1. Least detectable signal and average upper limits for various confidence levels, α. The LDS values are all calculated for power $(1 - \beta)$ 90%.

probability $1-\beta$. As for the model rejection potential case, the minimised true signal flux is independent of any original assumption of the signal scale. This discovery/sensitivity optimisation has been used in the AMANDA search for neutrinos in coincidence with the gamma-ray burst GRB030329 observed in 2003 by the HETE-II satellite[2]. Upon completion of this work, we learned that Giovanni Punzi had defined the same "discovery" criterion[3]. During refereeing of this paper, it was suggested that the model discovery optimisation does in fact also optimise a limit setting criterion[7]. The least detectable signal, μ_{lds}, for power $1 - \beta$ is also the standard Neyman upper limit with confidence level $1 - \beta$ that would be reported if exactly $n_{crit} - 1$ events were observed, the largest number of events for which a discovery cannot be claimed (i.e. largest number of events where the p-value is still greater than α). This turns out to be the same conclusion drawn by Punzi. Thus optimising the MRP using the Neyman upper limit at $1 - \beta$ c.l. that would be reported when just failing to reject the background-only hypothesis would lead to the same optimal cut as the MDP method with power $1 - \beta$.

[b]Of course, a small probability of the observation, or a more extreme one, under one hypothesis does not imply the truth of some alternative explanation (one needs to know priors in order to get to posterior probabilities of the hypotheses). This commonly used definition of discovery is not consistent with Bayesian decision making and this should always be kept in mind when intepreting results that use frequentist statistics.

4. Model discovery/rejection potential optimisation for an astrophysical point source search

As an example, we examine the optimisation of a search for events from a point-like source in the sky. This could correspond to a search for gamma rays or neutrinos from an astrophysical object such as an active galaxy or gamma-ray burst. The observing telescope reconstructs the arrival directions of the events to an accuracy characterised by the detector point spread function. This describes the probability of observing a reconstructed event at an angular position away from the true direction. Typically, this function is a two-dimensional Gaussian, peaked along the correct direction, and uniform in azimuth. We wish to find the optimal angular cut value using the criteria developed in sections 2 and 3. For a given angular separation, the solid angle in an annulus at an angle ψ from the source location and with width $d\psi$ goes as $\sin\psi\, d\psi$. Figure 2 shows the solid angle weighted functions, weighted with $\sin\psi$ to account for the increasing solid angle, with the azimuthal dependence projected out. Next, we integrate these functions from zero up to ψ and show the result in figure 3. The level of background has been chosen such that there is a total of one event inside a source bin of size one degree, corresponding to the width of the signal point spread function. Also shown are the 5σ, 90% power least detectable signal and 90% c.l. average upper limit as a function of ψ. The choice of best cut for MDP and MRP will come from minimising the ratios of least detectable signal and average upper limit divided by the integrated signal, as shown by the curves in figure 4. Here, we see the positions of the optimal cut for each case shown – the dotted line indicates the MDP minimum, the dashed line the MRP minimum and the solid line shows the minimum of square-root background divided by signal, a commonly used measure of optimisation. The $\sqrt{b}/s$ minimum corresponds to a cut at an angle of 1.58σ, a well known result for point source searches with Gaussian point spread functions[4]. In figure 5 we compare the optimal MRP and MDP values of the angular cut as a function of the background, where the background is again taken as the total expected events in an angular window of size equal to the signal point spread function resolution ($\sigma = 1°$). The discrete nature of the Poisson distribution accounts for the discontinuities seen in the MDP curve. For larger background values both the MRP and MDP cut approach the optimum ($\psi = 1.58\sigma$) based on the $\sqrt{b}/s$ minimisation. Next we examine the balance between opti-

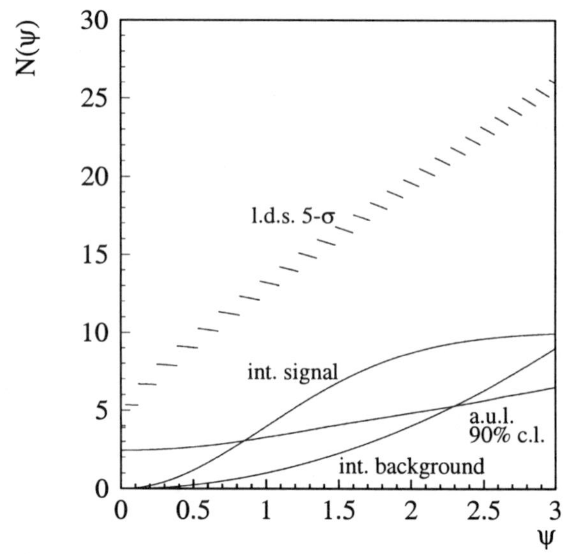

Fig. 3. Integrated signal and background functions, 90% c.l. average upper limit and 5σ, 90% power least detectable signal as a function of angular cut in a point source search.

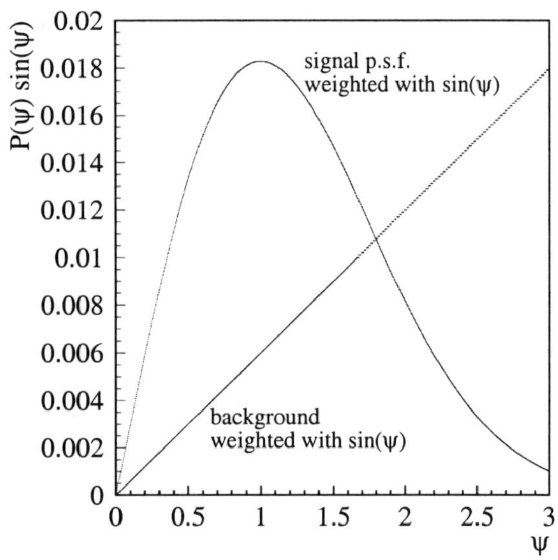

Fig. 2. Reconstruction point spread function weighted with solid angle. Ψ is in degrees.

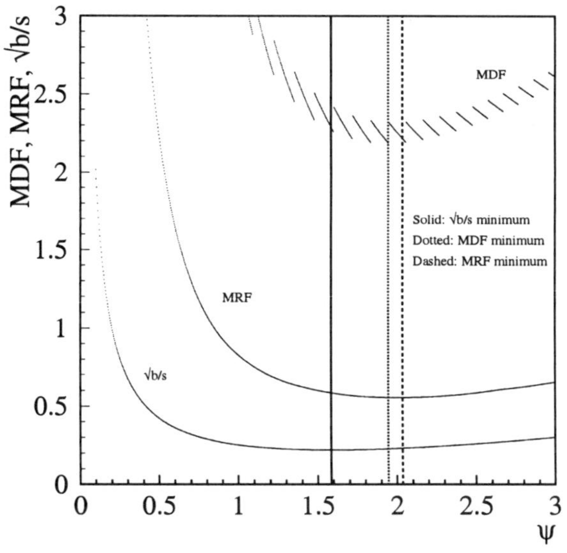

Fig. 4. MDF, MRF and $\sqrt{b}/s$ optimisation curves.

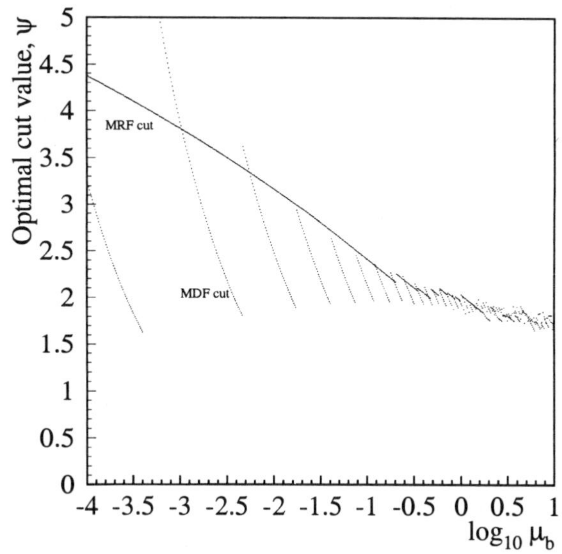

Fig. 5. Optimal MDF and MRF cut value as a function of total background contained within 1° of the source location.

misation for limit setting and discovery. We do this by looking at what would happen to the limit setting potential if we optimised for discovery, and vice versa, i.e. what the discovery potential would be if we had optimised for limit setting. For the limit case, we calculate the MRF evaluated at the best MDF cut ($MRF_{bestMDFcut}$) and take the ratio to the MRF evaluated at the best MRF cut ($MRF_{bestMRFcut}$). For the discovery case, we calculate the MDF evaluated at the best MRF cut ($MDF_{bestMRFcut}$) and take the ratio to the MDF evaluated at the best MDF cut ($MDF_{bestMDFcut}$). For the smallest background rate considered, we find that both the MRF and MDF are about 40% higher when one uses the optimised cut from the other method. As the background increases, these differences steadily decrease to about 5% at the highest background considered here. For the higher backgrounds ($\mu_b \sim 1$ or greater), neither optimisation criterion (MDF, MRF) is thus badly affected by choosing the cut based on the optimisation of the other criterion (MRF, MDF). Of course, only one case (point spread function resolution $\sigma = 1°$) has been examined in detail here. Further work would include examining the balance between MRP and MDP for different resolutions, statistical powers and significances. The relation to the similar method of Punzi[3] should also be investigated further.

Acknowledgments

GCH thanks Louis Lyons for useful comments and for pointing out that Giovanni Punzi had developed and presented a similar discovery criterion at PHYSTAT2003. Luc Demortier is thanked for very useful comments during his thorough refereeing of the work. This work was supported by the NSF under grant number OPP-0337726.

References

1. Hill, G.C. and Rawlins, K., *Astropart. Phys.* **19**, 393 (2003).
2. Stamatikos, M. (for the IceCube collaboration) *Proc. 29th International Cosmic Ray Conference* **Pune**, India (2005). (http://arxiv.org/astro-ph/0510336).
3. Just prior to attending PHYSTAT05, we learned that Giovanni Punzi had previously defined the same concept - see the proceedings of PHYSTAT2003, SLAC, Stanford (2003).
4. Alexandreas, D.E. *et al.*, *Nucl. Instrum. Methods Phys. Res.* **A328**, 570 (1993).
5. Feldman, G.J. and Cousins, R.D. *Phys. Rev. D* **57**, 3873 (1998). The "unified" method is used in the MRP optimisation; however, any upper limit formulation can be used.
6. Gary Feldman, private communication.
7. Luc Demortier, private communication.

STATISTICAL CHALLENGES FOR SEARCHES FOR NEW PHYSICS AT THE LHC

KYLE CRANMER
Brookhaven National Laboratory, Upton, NY 11973, USA
e-mail: Kyle.Cranmer@cern.ch

Because the emphasis of the LHC is on 5σ discoveries and the LHC environment induces high systematic errors, many of the common statistical procedures used in High Energy Physics are not adequate. I review the basic ingredients of LHC searches, the sources of systematics, and the performance of several methods. Finally, I indicate the methods that seem most promising for the LHC and areas that are in need of further study.

1. Introduction

The Large Hadron Collider (LHC) at CERN and the two multipurpose detectors, ATLAS and CMS, have been built in order to discover the Higgs boson, if it exists, and explore the theoretical landscape beyond the Standard Model.[1, 2] The LHC will collide protons with unprecedented center-of-mass energy ($\sqrt{s} = 14$ TeV) and luminosity (10^{34} cm^{-2}s^{-1}); the ATLAS and CMS detectors will record these interactions with $\sim 10^8$ individual electronic readouts per event. Because the emphasis of the physics program is on discovery and the experimental environment is so complex, the LHC poses new challenges to our statistical methods – challenges we must meet with the same vigor that led to the theoretical and experimental advancements of the last decade.

In the remainder of this Section, I introduce the physics goals of the LHC and most pertinent factors that complicate data analysis. I also review the formal link and the practical differences between confidence intervals and hypothesis testing.

In Sec. 2, the primary ingredients to new particle searches are discussed. Practical and toy examples are presented in Sec. 3, which will be used to assess the most common methods in Sec. 4. The remainder of this paper is devoted to discussion on the most promising methods for the LHC.

1.1. *Physics Goals of the LHC*

Currently, our best experimentally justified model for fundamental particles and their interactions is the *standard model*. In short, the physics goals of the LHC come in two types: those that improve our understanding of the standard model, and those that go beyond it.

The only particle of the standard model that has not been observed is the Higgs boson, which is key for the standard model's description of the electroweak interactions. The mass of the Higgs boson, m_H, is a free parameter in the standard model, but there exist direct experimental lower bounds and more indirect upper bounds. Once m_H is fixed, the standard model is a completely predictive theory. There are numerous particle-level Monte Carlo generators that can be interfaced with simulations of the detectors to predict the rate and distribution of all experimental observables. Because of this predictive power, searches for the Higgs boson are highly tuned and often employ multivariate discrimination methods like neural networks, boosted decision trees, support vector machines, and genetic programming.[3–5]

While the Higgs boson is key to understanding the electroweak interactions, it introduces a new problem: the *hierarchy problem*. There are several proposed solutions to the problem, one of which is to introduce a new symmetry, called supersymmetry (SUSY), between bosons and fermions. In practice, the minimal supersymmetric extension to the standard model (MSSM), with its 105 parameters, is not so much a theory as a theoretical framework.

They key difference between SUSY and Higgs searches is that, in most cases, discovering SUSY will not be the difficult part. Searches for SUSY often rely on robust signatures that will show a deviation from the standard model for most regions of the SUSY parameter space. It will be much more challenging to demonstrate that the deviation from the standard model is SUSY and to measure the fundamental parameters of the theory.[6] In order to restrict the scope of these proceedings, I shall focus on LHC Higgs searches, where the issues of hypothesis testing are more relevant.

1.2. *Challenges of LHC Environment*

The challenges of the LHC environment are manifold. The first and most obvious challenge is due to the enormous rate of uninteresting background events from QCD processes. The total interaction rate for the LHC is of order 10^9 interactions per second; the rate of Higgs production is about ten orders of magnitude smaller. Thus, to understand the background of a Higgs search, one must understand the extreme tails of the QCD processes.

Compounding the difficulties due to the extreme rate is the complexity of the detectors. The full-fledged simulation of the detectors is extremely computationally intensive, with samples of 10^7 events taking about a month to produce with computing resources distributed around the globe. This computational limitation constrains the problems that can been addressed with Monte Carlo techniques.

Theoretical uncertainties also contribute to the challenge. The background to many searches requires calculations at, or just beyond, the state-of-the-art in particle physics. The most common situation requires a final state with several well-separated high transverse momentum objects (*e.g.* $t\bar{t}jj \rightarrow bl\nu\bar{b}jj\,jj$), in which the regions of physical interest are not reliably described by leading-order perturbative calculations (due to infra-red and collinear divergences), are too complex for the requisite next-to-next-to-leading order calculations, and are not properly described by the parton-shower models alone. Enormous effort has gone into improving the situation with next-to-leading order calculations and matrix-element–parton-shower matching.[7, 8] While these new tools are a vast improvement, the residual uncertainties are still often dominant.

Uncertainties from non-perturbative effects are also important. For some processes, the relevant regions of the parton distribution functions are not well-measured (and probably will not be in the first few years of LHC running), which lead to uncertainties in rate as well as the shape of distributions. Furthermore, the various underlying-event and multiple-interaction models used to describe data from previous colliders show large deviations when extrapolated to the LHC.[9] This soft physics has a large impact on the performance of observables such as missing transverse energy.

In order to augment the simulated data chain, most searches introduce auxiliary measurements to estimate their backgrounds from the data itself. In some cases, the background estimation is a simple sideband, but in others the link between the auxiliary measurement to the quantity of interest is based on simulation. This hybrid approach is of particular importance at the LHC.

While many of the issues discussed above are not unique to the LHC, they are often more severe. At LEP, it was possible to generate Monte Carlo samples of larger size than the collected data, QCD backgrounds were more tame, and most searches were not systematics-limited. The Tevatron has much more in common with the LHC; however, at this point discovery is less likely, and most of the emphasis is on measurements and limit setting.

1.3. *Confidence Intervals & Hypothesis Tests*

The last several conferences in the vein of *PhyStat2005* have concentrated heavily on confidence intervals. In particular, 95% confidence intervals for some physics parameter in an experiment that typically has few events. More recently, there has been a large effort in understanding how to include systematic errors and nuisance parameters into these calculations.

LHC searches, in contrast, are primarily interested in 5σ discovery. The 5σ discovery criterion is somewhat vague, but usually interpreted in a frequentist sense as a hypothesis test with a rate of Type I error $\alpha = 2.85 \cdot 10^{-7}$.

There is a formal link between confidence intervals and hypothesis testing: frequentist confidence intervals from the Neyman construction are formally inverted hypothesis tests. It is this equivalence that links the Neyman-Pearson lemma* to the ordering rule used in the *unified method* of Feldman and Cousins.[10] Furthermore, this equivalence will be very useful in translating our understanding of confidence intervals to the searches at the LHC.

In some cases, this formal link can be misleading. In particular, there is not always a continuous parameter that links the fully specified null hypoth-

*The lemma states that, for a simple hypothesis test of size α between a null H_0 and an alternate H_1, the most powerful critical region in the observable x is given by a contour of the likelihood ratio $L(x|H_0)/L(x|H_1)$.

esis H_0 to the fully specified alternate H_1 in any physically interesting or justified way. Furthermore, the performance of a method for a 95% confidence interval and a 5σ discovery can be quite different.

2. The Ingredients of an LHC Search

In order to assess the statistical methods that are available and develop new ones suited for the LHC, it is necessary to be familiar with the basic ingredients of the search. In this section, the basic ingredients, terminology, and nomenclature are established.

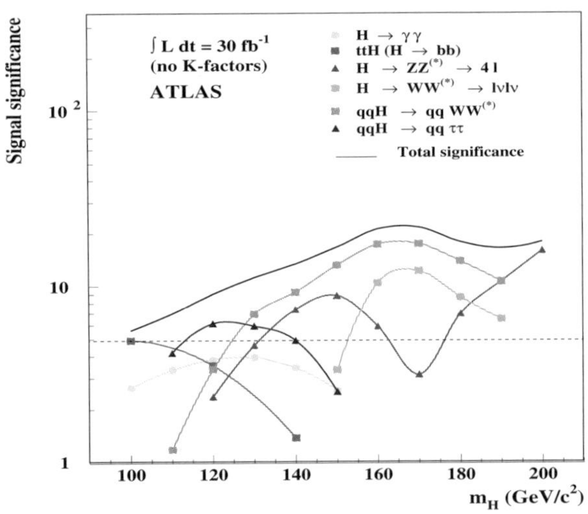

Fig. 1. Expected significance as a function of Higgs mass for the Atlas detector with 30 fb^{-1} of data.

2.1. *Multiple Channels & Processes*

Almost all new particle searches do not observe the particle directly, but through the signatures left by the decay products of the particle. For instance, the Higgs boson will decay long before it interacts with the detector, but its decay products will be detected. In many cases, the particle can be produced and decay in many different configurations, each of which is called a *search channel* (see Tab. 1). There may be multiple signal and background processes which contribute to each channel. For example, in $H \to \gamma\gamma$, the signal could come from any Higgs production mechanism and the background from either continuum $\gamma\gamma$ production or QCD backgrounds where jets fake photons. Each of these processes have their own rates, distributions for observables, and uncertainties. Furthermore, the uncertainties between processes may be correlated.

In general the theoretical model for a new particle has some free parameters. In the case of the standard model Higgs, only the mass m_H is unknown. For SUSY scenarios, the Higgs model is parametrized by two parameters: m_A and $\tan\beta$. Typically, the unknown variables are scanned and a hypothesis test is performed for each value of these parameters. The results from each of the search channels can be combined to enhance the power of the search, but one must take care of correlations among channels and ensure consistency.

The fact that one scans over the parameters and performs many hypothesis tests increases the chance that one finds at least one large fluctuation from the null-hypothesis. Some approaches incorporate the number of trials explicitly,[11] some approaches only focus on the most interesting fluctuation,[12] and some see this heightened rate of Type I error as the motivation for the stringent 5σ requirement.[13]

2.2. *Discriminating Variables & Test Statistics*

Typically, new particles are known to decay with certain characteristics that distinguish the *signal* events from those produced by *background* processes. Much of the work of a search is to identify those observables and to construct new discriminating variables (generically denoted as m). Examples include angles between particles, invariant masses, and particle identification criterion. Discriminating variables are used in two different ways: to define a signal-like region and to weight events.

The usage of discriminating variables is related to the test statistic: the real-valued quantity used to summarize the experiment. The test statistic is thought of as being ordered such that either large or small values indicate growing disagreement with the null hypothesis.

A simple "cut analysis" consists of defining a signal-like region bounded by upper- and lower-values of these discriminating variables and counting events in that region. In that case, the test statistic is simply the number of events observed in the signal like region. One expects b background events and s signal events, so the experimental sensitivity is optimized by adjusting the cut values. More sophisticated techniques use multivariate algorithms, such as neural networks, to define more complicated signal-like regions, but the test statistic remains unchanged. In these number counting analyses, the likelihood of observing n events is simply given by the Poisson model.

There are extensions to this number-counting technique. In particular, if one knows the distribution of the discriminating variable m for the background-only (null) hypothesis, $f_b(m)$, and the signal-plus-background (alternate) hypothesis, $f_{s+b}(m) = [sf_s(m) + bf_b(m)]/(s+b)$, then there is a more powerful test statistic than simply counting events. This is intuitive, a well measured 'golden event' is often more convincing than a few messy ones. Following the Neyman-Pearson lemma, the most powerful test statistic is

$$Q = \frac{L(\mathbf{m}|H_1)}{L(\mathbf{m}|H_0)} \qquad (1)$$

$$= \frac{\prod_i^{N_{chan}} Pois(n_i|s_i+b_i) \prod_j^{n_i} \frac{s_i f_s(m_{ij}) + b_i f_b(m_{ij})}{s_i+b_i}}{\prod_i^{N_{chan}} Pois(n_i|b_i) \prod_j^{n_i} f_b(m_{ij})}$$

(n_i denotes events in i^{th} channel) or equivalently

$$q = \ln Q = -s_{tot} + \sum_i^{N_{chan}} \sum_j^{n_i} \ln\left(1 + \frac{s_i f_s(m_{ij})}{b_i f_b(m_{ij})}\right). \qquad (2)$$

The test statistic in Eq. 2 was used by the LEP Higgs Working Group (LHWG) in their final results on the search for the Standard Model Higgs.[14]

At this point, there are two loose ends: how does one determine the distribution of the discriminating variables $f(m)$, and how does one go from Eq. 2 to the distribution of q for H_0 and H_1? These are the topics of the next subsections.

2.3. *Parametric & Non-Parametric Methods*

In some cases, the distribution of a discriminating variable $f(m)$ can be parametrized and this parametrization can be justified either by physics arguments or by goodness-of-fit. However, there are many cases in which $f(m)$ has a complicated shape not easily parametrized. For instance, Fig. 2 shows the distribution of a neural network output for signal events. In that case kernel estimation techniques can be used to estimate $f(m)$ in a non-parametric way from a sample of events $\{m_i\}$.[15] The technique that

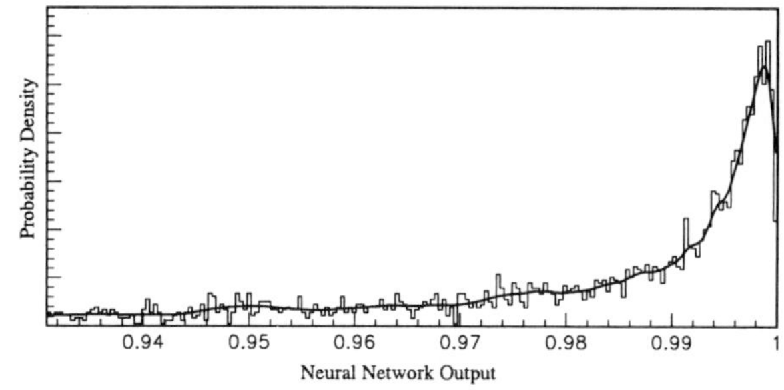

Fig. 2. The distribution of a neural network output for signal events. The histogram is shown together with $\hat{f}_1(m)$.

was used by the LHWG[14] was based on an adaptive kernel estimation given by:

$$\hat{f}_1(m) = \sum_i^n \frac{1}{nh(m_i)} K\left(\frac{m - m_i}{h(m_i)}\right), \qquad (3)$$

where

$$h(m_i) = \left(\frac{4}{3}\right)^{1/5} \sqrt{\frac{\sigma}{\hat{f}_0(m_i)}} n^{-1/5}, \qquad (4)$$

σ is the standard deviation of $\{x_i\}$, $K(x)$ is some kernel function (usually the normal distribution), and $\hat{f}_0(x)$ is the fixed kernel estimate given by the same equation but with a fixed $h(m_i)$

$$h^* = \left(\frac{4}{3}\right)^{1/5} \sigma n^{-1/5}. \qquad (5)$$

The solid line in Fig. 2 shows that the method (with modified-boundary kernels) works very well for shapes with complicated structure at many scales.

2.4. *Numerical Evaluation of Significance*

Given, $f_s(m)$ and $f_b(m)$ the distribution of $q(x)$ can be constructed. For the background-only hypothesis, $f_b(m)$ provides the probability of corresponding values of q needed to define the single-event pdf ρ_1.[†]

$$\rho_{1,b}(q_0) = \int f_b(m)\, \delta(q(m) - q_0) dm \qquad (6)$$

For multiple events, the distribution of the log-likelihood ratio must be obtained from repeated convolutions of the single event distribution. This convolution can either be performed implicitly with approximate Monte Carlo techniques,[16] or analytically

[†]The integral is necessary because the map $q(m): m \to q$ may be many-to-one.

with a Fourier transform technique.[17] In the Fourier domain, denoted with a bar, the distribution of the log-likelihood for n events is

$$\overline{\rho_n} = \overline{\rho_1}^n \tag{7}$$

Thus the expected log-likelihood distribution for background with Poisson fluctuations in the number of events takes the form

$$\rho_b(q) = \sum_{n=0}^{\infty} \frac{e^{-b} b^n}{n!} \rho_{n,b}(q) \tag{8}$$

which in the Fourier domain is simply

$$\overline{\rho_b(q)} = e^{b[\overline{\rho_{1,b}(q)} - 1]}. \tag{9}$$

For the signal-plus-background hypothesis we expect s events from the $\rho_{1,s}$ distribution and b events from the $\rho_{1,b}$ distribution, which leads to the expression for ρ_{s+b} in the Fourier domain[‡]

$$\overline{\rho_{s+b}(q)} = e^{b[\overline{\rho_{1,b}(q)} - 1] + s[\overline{\rho_{1,s}(q)} - 1]}. \tag{10}$$

This equation generalizes, in a somewhat obvious way, to include many processes and channels.

Numerically these computations are carried out with the Fast Fourier Transform (FFT). The FFT is performed on a finite and discrete array, beyond which the function is considered to be periodic. Thus the range of the ρ_1 distributions must be sufficiently large to hold the resulting ρ_b and ρ_{s+b} distributions. If they are not, the "spill over" beyond the maximum log-likelihood ratio q_{max} will "wrap around" leading to unphysical ρ distributions. Because the range of ρ_b is much larger than $\rho_{1,b}$ it requires a very large number of samples to describe both distributions simultaneously. The implementation of this method requires some approximate asymptotic techniques that describe the scaling from $\rho_{1,b}$ to ρ_b.[18]

The nature of the FFT results in a number of round-off errors and limits the numerical precision to about 10^{-16} – which limits the method to significance levels below about 8σ. Extrapolation techniques and arbitrary precision calculations can overcome these difficulties,[18] but such small p-values are of little practical interest.

[‡]Perhaps it is worth noting that $\overline{\rho(q)}$ is a complex valued function of the Fourier conjugate variable of q. Thus numerically the exponentiation in Eq. 9 requires Euler's formula $e^{i\theta} = \cos\theta + i\sin\theta$.

From the log-likelihood distribution of the two hypotheses we can calculate a number of useful quantities. Given some experiment with an observed log-likelihood ratio, q^*, we can calculate the background-only confidence level, CL_b :

$$CL_b(q^*) = \int_{q^*}^{\infty} \rho_b(q')dq' \tag{11}$$

In the absence of an observation we can calculate the expected CL_b given the signal-plus-background hypothesis is true. To do this we first must find the median of the signal-plus-background distribution $\overline{q}_{s+b}$. From these we can calculate the expected CL_b by using Eq. 11 evaluated at $q^* = \overline{q}_{s+b}$.

Finally, we can convert the expected background confidence level into an expected Gaussian significance, $Z\sigma$, by finding the value of Z which satisfies

$$CL_b(\overline{q}_{s+b}) = \frac{1 - \text{erf}(Z/\sqrt{2})}{2}. \tag{12}$$

where $\text{erf}(Z) = (2/\pi)\int_0^Z \exp(-y^2)dy$ is a function readily available in most numerical libraries. For $Z > 1.5$, the relationship can be approximated[19] as

$$Z \approx \sqrt{u - \ln u} \quad \text{with} \quad u = -2\ln(CL_b\sqrt{2\pi}) \tag{13}$$

2.5. *Systematic Errors, Nuisance Parameters & Auxiliary Measurements*

Sections 2.3 and 2.4 represent the state of the art for HEP in frequentist hypothesis testing in the absence of uncertainties on rates and shapes of distributions. In practice, the true rate of background is not known exactly, and the shapes of distributions are sensitive to experimental quantities, such as calibration coefficients and particle identification efficiencies (which are also not known exactly). What one would call a *systematic error* in HEP, usually corresponds to what a statistician would refer to as a *nuisance parameter.*

Dealing with nuisance parameters in searches is not a new problem, but perhaps it has never been as essential as it is for the LHC. In these proceedings, Cousins reviews the different approaches to nuisance parameters in HEP and the professional statistical literature.[20] Also of interest is the classification of systematic errors provided by Sinervo.[21] In Sec. 4, a few techniques for incorporating nuisance parameters are reviewed.

From an experimental point of view, the missing ingredient is some set of auxiliary measurements

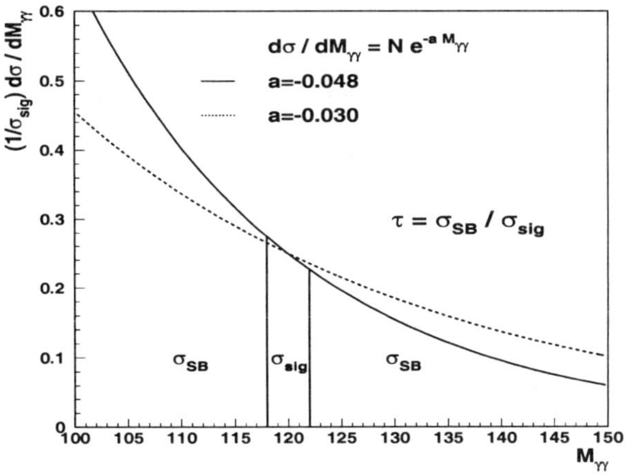

Fig. 3. The signal-like region and sideband for $H \to \gamma\gamma$ in which τ is correlated to b via the model parameter a.

that will constrain the value of the nuisance parameters. The most common example would be a sideband measurement to fix the background rate, or some control sample used to assess particle identification efficiency. Previously, I used the variable M to denote this auxiliary measurement[22]; while Linnemann,[19] Cousins,[20] and Rolke, Lopez, and Conrad[23, 24] used y. Additionally, one needs to know the likelihood function that provides the connection between the nuisance parameter(s) and the auxiliary measurements.

The most common choices for the likelihood of the auxiliary measurement are $L(y|b) = Pois(y|\tau b)$ and $L(y|b) = G(y|\tau b, \sigma_y)$, where τ is a constant that specifies the ratio of the number of events one expects in the sideband region to the number expected in the signal-like region.[§]

A constant τ is appropriate when one simply counts the number of events y in an "off-source" measurement. In a more typical case, one uses the distribution of some other variable, call it m, to estimate the number of background events inside a range of m (see Fig. 3). In special cases the ratio τ is independent of the model parameters. However, in many cases (*e.g.* $f(m) \propto e^{-am}$), the ratio τ depends on the model parameters. Moreover, sometimes the sideband is contaminated with signal events, thus the background and signal estimates can be correlated. These complications are not a problem as long as they are incorporated into the likelihood.

[§]Note that Linnemann[19] used $\alpha = 1/\tau$ instead, but in this paper α is reserved for the rate of Type I error.

The number of nuisance parameters and auxiliary measurements can grow quite large. For instance, the standard practice at Ba$\bar{\text{B}}$ar is to form very large likelihood functions that incorporate everything from the parameters of the unitarity triangle to branching fractions and detector response. These likelihoods are typically factorized into multiple pieces, which are studied independently at first and later combined to assess correlations. The factorization of the likelihood and the number of nuisance parameters included impact the difficulty of implementing the various scenarios considered below.

3. Practical and Toy Examples

In this Section, a few practical and toy examples are introduced. The toy examples are meant to provide simple scenarios where results for different methods can be easily obtained in order to expedite their comparison. The practical examples are meant to exclude methods that provide nice solutions to the toy examples, but do not generalize to the realistic situation.

3.1. *The Canonical Example*

Consider a number-counting experiment that measures x events in the signal-like region and y events in some sideband. For a given background rate b in the signal-like region, say one can expect τb events in the sideband. Additionally, let the rate of signal events in the signal-like regions – the parameter of interest – be denoted μ. The corresponding likelihood function is

$$L_P(x, y|\mu, b) = Pois(x|\mu + b) \cdot Pois(y|\tau b). \quad (14)$$

This is the same case that was considered in Refs. [20, 22–24] for $x, y = \mathcal{O}(10)$ and $\alpha = 5\%$. For LHC searches, we will be more interested in $x, y = \mathcal{O}(100)$ and $\alpha = 2.85 \cdot 10^{-7}$. Furthermore, the auxiliary measurement will rarely be a pure number counting sideband measurement, but instead the result of some fit. So let us also consider the likelihood function

$$L_G(x, y|\mu, b) = Pois(x|\mu + b) \cdot G(y|\tau b, \sqrt{\tau b}). \quad (15)$$

As a concrete example in the remaining sections, let us consider the case $b = 100$ and $\tau = 1$. Operationally, one would measure y and then find the value $x_{crit}(y)$ necessary for discovery. In the language of confidence intervals, $x_{crit}(y)$ is the value of x necessary for the $100(1-\alpha)\%$ confidence interval in μ to exclude $\mu_0 = 0$. In Sec. 4 we check the coverage (Type I error or false-discovery rate) for both L_P and L_G.

Linnemann reviewed thirteen methods and eleven published examples of this scenario.[19] Of the published examples, only three (the one from his reference 18 and the two from 19) are near the range of x,y, and α relevant for LHC searches. Linnemann's review asks an equivalent question posed in this paper, but in a different way: what is the significance (Z in Eq. 12) of a given observation x, y.

3.2. *Standard Model Higgs Searches*

The search for the standard model Higgs boson is by no means the only interesting search to be performed at the LHC, but it is one of the most studied and offers a particularly challenging set of channels to combine with a single method. Figure 1 shows the expected significance versus the Higgs mass, m_H, for several channels individually and in combination for the ATLAS experiment.[25] Two mass points are considered in more detail in Tab. 1, including results from Refs.[1, 25, 26]. Some of these channels will most likely use a discriminating variable distribution, $f(m)$, to improve the sensitivity as described in Sec. 2.3. I have indicated the channels that I suspect will use this technique. Rough estimates on the uncertainty in the background rate have also been tabulated, without regard to the classification proposed by Sinervo.

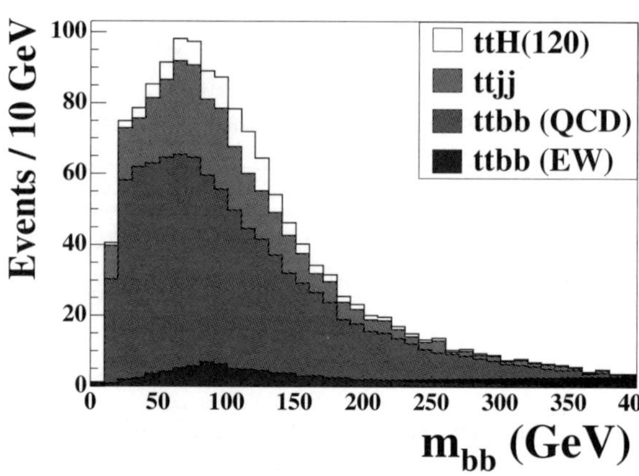

Fig. 4. The bb invariant mass spectrum for $t\bar{t}H$ signal and background processes at ATLAS.

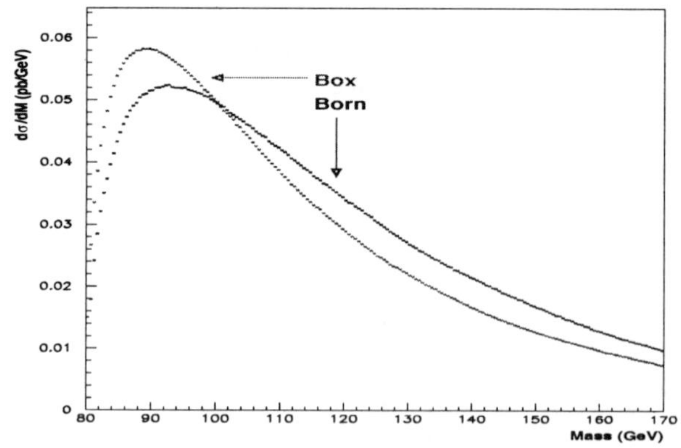

Fig. 5. Two plausible shapes for the continuum $\gamma\gamma$ mass spectrum at the LHC.

The background uncertainties for the $t\bar{t}H$ channel have been studied in some detail and separated into various sources.[26] Figure 4 shows the m_{bb} mass spectrum for this channel.¶ Clearly, the shape of the background-only distribution is quite similar to the shape of the signal-plus-background distribution. Furthermore, theoretical uncertainties and b-tagging uncertainties affect the shape of the background-only spectrum. In this case the incorporation of systematic error on the background rate most likely precludes the expected significance of this channel from ever reaching 5σ.

Similarly, the $H \to \gamma\gamma$ channel has uncertainty in the shape of the $m_{\gamma\gamma}$ spectrum from background processes. One contribution to this uncertainty comes from the electromagnetic energy scale of the calorimeter (an experimental nuisance parameter), while another contribution comes from the theoretical uncertainty in the continuum $\gamma\gamma$ production. Figure 5 shows two plausible shapes for the $m_{\gamma\gamma}$ spectrum from "Born" and "Box" predictions.

4. Review of Methods

Based on the practical example of the standard model Higgs search at the LHC and the discussion in Sec. 2, the list of admissible methods is quite short. Of the thirteen methods reviewed by Linnemann, only five are considered as reasonable or recommended. These can be divided into three classes: hybrid Bayesian-frequentist methods, methods based on the Likelihood Principle, and frequentist methods based on the Neyman construction.

¶It is not clear if this result is in agreement with the equivalent CMS result.[27]

Table 1. Number of signal and background events for representative Higgs search channels for two values of Higgs mass, m_H, with 30 fb^{-1} of data. A rough uncertainty on the background rate is denoted as $\delta b/b$, without reference to the type of systematic uncertainty. The table also indicates if the channels are expected to use a weight $f(m)$ as in Eq. 2.

channel	s	b	$\delta b/b$	dominant backgrounds	use $f(m)$	m_H (GeV)
$t\bar{t}H \to t\bar{t}bb$	42	219	~10%	$t\bar{t}jj, t\bar{t}bb$	Yes	120
$H \to \gamma\gamma$	357	11820	~ 0.1%	$\gamma\gamma, j\gamma, jj$	No	120
$qqH \to qq\tau\tau \to qqll\not{p}_T$	17	14	~10%	$Z \to \tau\tau$, $t\bar{t}$	Yes	120
$qqH \to qq\tau\tau \to qqlh\not{p}_T$	16	8	~10%	$Z \to \tau\tau$, $t\bar{t}$	Yes	120
$qqH \to qqWW^* \to qqll\not{p}_T$	28.5	47.4	~10%	$t\bar{t}, WW$	Yes	120
$qqH \to qqWW^* \to qqll\not{p}_T$	262.5	89.1	~10%	$t\bar{t}, WW$	Yes	170
$H \to ZZ \to 4l$	7.6	3.1	~ 1%	$ZZ \to 4l$	No	170
$H \to WW \to ll\not{p}_T$	337	484	~5%	$Z \to \tau\tau$, $t\bar{t}$	Yes	170

4.1. *Hybrid Bayesian-Frequentist Methods*

The class of methods frequently used in HEP and commonly referred to as the Cousins-Highland technique (or secondarily Bayes in statistical literature) are based on a Bayesian average of frequentist p-values as found in the first equation of Ref.[28]. The Bayesian average is over the nuisance parameters and weighted by the posterior $P(b|y)$. Thus the p-value of the observation (x_0, y_0) evaluated at μ is given by

$$p(x_0, y_0|\mu) = \int_0^\infty db\, p(x_0|\mu, b)P(b|y_0) \qquad (16)$$

$$= \int_{x_0}^\infty dx\, P(x|\mu, y_0) \qquad (17)$$

where

$$P(x|\mu, y_0) = \int_0^\infty db\, P(x|\mu, b)\frac{P(y_0|b)\ P(b)}{P(y_0)} \qquad (18)$$

The form in Eq. 16, an average over p-values, is similar to the form written in Cousins & Highland's article; and it is re-written in Eq. 17 to the form that is more familiar to those from LEP Higgs searches.[16, 17] Actually, the dependence on y_0 and the Bayesian prior $P(b)$ shown explicitly in Eq. 18 is often not appreciated by those that use this method.

The specific methods that Linnemann considers correspond to different choices of Bayesian priors. The most common in HEP is to ignore the prior and use a truncated Gaussian for the posterior $P(b|y_0)$, which Linnemann calls Z_N. For the case in which the likelihood $L(y|b)$ is known to be Poisson, Linneman prefers to use a flat prior, which gives rise to a Gamma-distributed posterior and Linnemann's second preferred method Z_Γ, which is identical to the ratio of Poisson means Z_{Bi} and can be written in terms of (in)complete beta functions as[19]

$$p_\Gamma = p_{Bi} = B(1/(1+\tau), x, y+1)/B(x, y+1). \quad (19)$$

The method Linnemann calls $Z_{5'}$ can be seen as an approximation of Z_N for large signals and is what ATLAS used to assess its physics potential.[1] The method was not recommended by Linnemann and was critically reviewed in Ref.[29].

$$x^{5'}_{crit}(y) = y/\tau + Z\sqrt{y/\tau(1+1/\tau)} \qquad (20)$$

4.2. *Likelihood Intervals*

As Cousins points out, the professional statistics literature seems less concerned with providing correct coverage by construction, in favor of likelihood-based and Bayesian methods. The likelihood principle states that given a measurement x all inference about μ should be based on the likelihood function $L(x|\mu)$. When nuisance parameters are included, things get considerably more complicated.

The profile likelihood function is an attempt to eliminate the nuisance parameters from the likelihood function by replacing them with their conditional maximum likelihood estimates (denoted, for example, $\hat{\hat{b}}$). The profile likelihood for L_P in Eq. 14 is given by $L(x, y|\mu_0, \hat{\hat{b}}(\mu_0))$, with

$$\hat{\hat{b}}(\mu_0) = \frac{x + y - (1+\tau)\mu_0}{2(1+\tau)} + \frac{\sqrt{(x+y-(1+\tau)\mu_0)^2 + 4(1+\tau)y\mu_0}}{2(1+\tau)}. \qquad (21)$$

The relevant likelihood ratio is then

$$\lambda_P(\mu_0|x, y) = \frac{L(x, y|\mu_0, \hat{\hat{b}}(\mu_0))}{L(x, y|\hat{\mu}, \hat{b})}, \qquad (22)$$

where $\hat{\mu}$ and $\hat{b}$ are the unconditional maximum likelihood estimates.

One of the standard results from statistics is that the distribution of $-2\ln\lambda$ converges to the χ^2 distribution with k degrees of freedom, where k is the number of parameters of interest. In our example $k = 1$, so a 5σ confidence interval is defined by the set of μ with $-2\ln\lambda(\mu|x,y) < 25$. Figure 6 shows the graph of $-2\ln\lambda(\mu|x,y)$ for $y = 100$ at the critical value of x for a 5σ discovery.

At PhyStat2003, Nancy Reid presented various adjustments and improvements to the profile likelihood which speed asymptotic convergence properties.[30] Cousins considers these methods in more detail from a physicist's perspective.[20]

Only recently was it generally appreciated that the method of MINUIT[31] commonly used in HEP corresponds to the profile likelihood intervals. The coverage of these methods is not guaranteed, but has been studied in simple cases.[23, 24] These likelihood-based techniques are quite promising for searches at the LHC, but their coverage properties must be assessed in the more complicated context of the LHC with weighted events and several channels. In particular, the distribution of q in Eq. 10 is often highly non-Gaussian.

4.3. *The Neyman Construction with Systematics*

Linnemann's preferred method, Z_{Bi}, is related to the familiar result on the ratio of Poisson means.[32] Unfortunately, the form of Z_{Bi} is tightly coupled to the form of Eq. 14, and can not be directly applied to the more complicated cases described above. However, the standard result on the ratio of Poisson means[32] and Cousins' improvement[33] are actually special cases of the Neyman construction with nuisance parameters (with and without conditioning, respectively).

Of course, the Neyman construction does generalize to the more complicated cases discussed above. Two particular types of constructions have been presented, both of which are related to the profile likelihood ratio discussed in Kendall's chapter on likelihood ratio tests & test efficiency.[34] This relationship often leads to confusion with the profile likelihood intervals discussed in Sec. 4.2.

The first method is a full Neyman construction over both the parameters of interest and the nuisance parameters, using the profile likelihood ratio as an ordering rule. Using this method, the nuisance parameter is "projected out", leaving only an interval in the parameters of interest. I presented this method at PhyStat2003 in the context of hypothesis testing,[‖] and similar work was presented by Punzi at this conference.[22, 35] This method provides coverage by construction, independent of the ordering rule used.

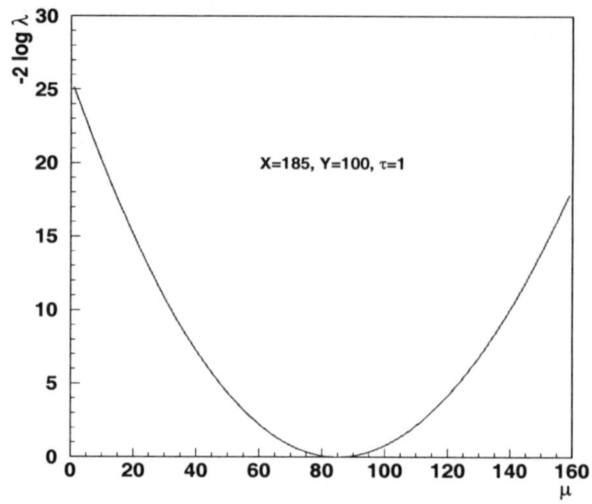

Fig. 6. The profile likelihood ratio $-2\ln\lambda$ versus the signal strength μ for $y = 100$, $\tau = 1$, and $x = x_{crit}(y) = 185$.

The motivation for using the profile likelihood ratio as a test statistic is twofold. First, it is inspired by the Neyman-Pearson lemma in the same way as the Feldman-Cousins ordering rule. Secondly, it is independent of the nuisance parameters; providing some hope of obtaining similar tests.[**] Both Punzi and myself found a need to perform some "clipping" to the acceptance regions to protect from irrelevant values of the nuisance parameters spoiling the projection. For this technique to be broadly applicable, some generalization of this clipping procedure is needed and the scalability with the number of parameters must be addressed.[††]

The second method, presented by Feldman at the Fermilab workshop in 2000, involves a Neyman construction over the parameters of interest, but

‖In simple hypothesis testing μ is not a continuous parameter, but only takes on the values $\mu_0 = 0$ or $\mu_1 = s$.

**Similar tests are those in which the critical regions of size α are independent of the nuisance parameters. Similar tests do not exist in general.

††A Monte Carlo sampling of the nuisance parameter space could be used to curb the curse of dimensionality.[22]

the nuisance parameters are fixed to the conditional maximum likelihood estimate: a method I will call the *profile construction*. The profile construction is an approximation of the full construction, that does not necessarily cover. To the extent that the use of the profile likelihood ratio as a test statistic provides similar tests, the profile construction has good coverage properties. The main motivation for the profile construction is that it scales well with the number of nuisance parameters and that the "clipping" is built in (only one value of the nuisance parameters is considered).

It appears that the CHOOZ experiment actually performed both the full construction (called "FC correct syst.") and the profile construction (called "FC profile") in order to compare with the strong confidence technique.[36]

Another perceived problem with the full construction is that bad over-coverage can result from the projection onto the parameters of interest. It should be made very clear that the coverage probability is a function of both the parameters of interest and the nuisance parameters. If the data are consistent with the null hypothesis for *any* value of the nuisance parameters, then one should probably not reject it. This argument is stronger for nuisance parameters directly related to the background hypothesis, and less strong for those that account for instrumentation effects. In fact, there is a family of methods that lie between the full construction and the profile construction. Perhaps we should pursue a hybrid approach in which the construction is formed for those parameters directly linked to the background hypothesis, the additional nuisance parameters take on their profile values, and the final interval is projected onto the parameters of interest.

5. Results with the Canonical Example

Consider the case $b_{true} = 100$, $\tau = 1$ (*i.e.* 10% systematic uncertainty). For each of the methods we find the critical boundary, $x_{crit}(y)$, which is necessary to reject the null hypothesis $\mu_0 = 0$ at 5σ when y is measured in the auxiliary measurement. Figure 7 shows the contours of L_G, from Eq. 15, and the critical boundary for several methods. The far left curve shows the simple $s/\sqrt{b}$ curve neglecting systematics. The far right curve shows a critical region with the correct coverage. With the exception of the profile

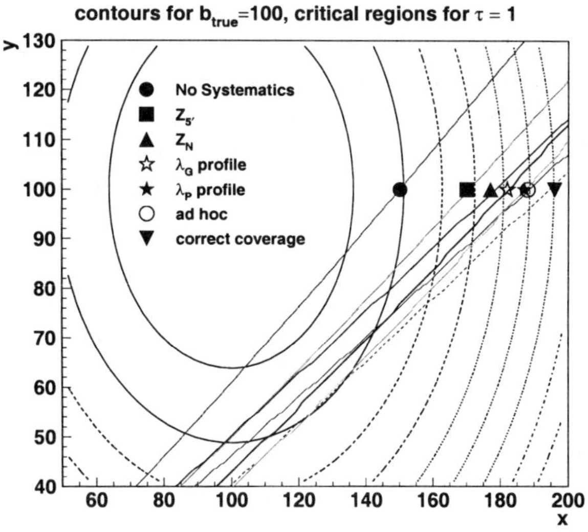

Fig. 7. A comparison of the various methods critical boundary $x_{crit}(y)$ (see text). The concentric ovals represent contours of L_G from Eq. 15.

likelihood, λ_P, all of the other methods lie between these two curves (*ie.* all of them under-cover). The rate of Type I error for these methods was evaluated for L_G and L_P and presented in Table 2.

The result of the full Neyman construction and the profile construction are not presented. The full Neyman technique covers by construction, and it was previously demonstrated for a similar case ($b = 100$, $\tau = 4$) that the profile construction gives similar results.[22] Furthermore, if the λ_P were used as an ordering rule in the full construction, the critical region for $b = 100$ would be identical to the curve labeled "λ_P profile" (since λ_P actually covers).

It should be noted that if one knows the likelihood is given by $L_G(x, y|\mu, b)$, then one should use the corresponding profile likelihood ratio, $\lambda_G(x, y|\mu)$, for the hypothesis test. However, knowledge of the correct likelihood is not always available (Sinervo's Class II systematic), so it is informative to check the coverage of tests based on both $\lambda_G(x, y|\mu)$ and $\lambda_P(x, y|\mu)$ by generating Monte Carlo according to $L_G(x, y|\mu, b)$ and $L_P(x, y|\mu, b)$. In a similar way, this decoupling of true likelihood and the assumed likelihood (used to find the critical region) can break the "guaranteed" coverage of the Neyman construction.

It is quite significant that the Z_N method under-covers, since it is so commonly used in HEP. The degree to which the method under-covers depends on the truncation of the Gaussian posterior $P(b|y)$.

Linnemann's table also shows significant under-coverage (over estimate of the significance Z). In order to obtain a critical region with the correct coverage, the author modified the region $x_{crit}(y) = x_{crit}^{Z_N}(y) + C$ and found $C = 16$ provided the correct coverage. A discrepancy of 16 events is not trivial!

Table 2. Rate of Type I error interpreted as equivalent $Z\sigma$ for various methods designed for a 5σ test. Monte Carlo events are generated via either L_G or L_P. The critical x for $y = 100$ is also listed for easy comparison.

Method	L_G ($Z\sigma$)	L_P ($Z\sigma$)	$x_{crit}(y = 100)$
No Syst	3.0	3.1	150
$Z_{5'}$	4.1	4.1	171
Z_N (Sec. 4.1)	4.2	4.2	178
ad hoc	4.6	4.7	188
$Z_\Gamma = Z_{Bi}$	4.9	5.0	185
profile λ_P	5.0	5.0	185
profile λ_G	4.7	4.7	~182

Notice that for large x, y the Bayesian-frequentist hybrid Z_N approaches $Z_{5'}$, where the the critical region is of the form $x_{crit}(y) = y/\tau + n\sqrt{y/\tau}$. Because the boundary is very nearly linear around y_0, one can find the value of n that gives the proper coverage with a little geometry. In particular, the number n needed to get a $Z\sigma$ test gives

$$x_{crit}(y) = y/\tau + Z\sqrt{1 + 1/\tau m^2}\sqrt{y/\tau} \qquad (23)$$

where

$$m = \left(1 + \frac{Z}{2\sqrt{y/\tau}}\right)^{-1} \qquad (24)$$

The m^2 factor can be seen as a correction to the $Z_{5'}$ and Z_N results. Notice that the correction is larger for higher significance tests. As an *ad hoc* method, I experimented with the Z_N method replacing τ with τm^2 in the posterior $P(b|y)$. The coverage of this *ad hoc* method is better than Z_N, but not exact because x, y are not sufficiently large.

6. Conclusions

I have presented the statistical challenges of searches at the LHC and the current state of the statistical methods commonly used in HEP. I have attempted to accurately portray the complexity of the searches, explain their key ingredients, and provide a practical example for future studies. Three classes of methods, which are able to incorporate all the ingredients, have been identified: hybrid Bayesian-frequentist methods, methods based on the Likelihood Principle, and frequentist methods based on the Neyman construction.

The Bayesian-frequentist hybrid method, Z_N, shows significant under-coverage in the toy example considered when pushed to the 5σ regime. While Bayesian might not care about coverage, significant under-coverage is undesirable in HEP. Further study is needed to determine if a more careful choice of prior distributions can remedy this situation – especially in more complex situations. The improved coverage of Z_Γ may give some guidance.

The methods based on the likelihood principle have gained a great deal of attention from HEP in recent years. While the methods appear to do well in the toy example, it requires further study to determine their properties in the more realistic situation with weighted events.

Slowly, the HEP community is coming to grips with how to incorporate nuisance parameters into the Neyman construction. Several ideas for reducing the over-coverage induced by projecting out the nuisance parameters and reducing the computational burden have been presented. A hybrid approach between the full construction and the profile construction should be investigated in more detail.

Finally, it seems that the HEP community is approaching a point where we appreciate the fundamental statistical issues, the limitations of some methods, and the benefits of others. Clearly, the philosophical debate has not ended, but there seems to be more emphasis on practical solutions to our very challenging problems.

Acknowledgments

I would like to thank the many people that helped in preparing this review. In particular, Bob Cousins, Jim Linnemann, Gary Feldman, Jan Conrad, Fredrik Tegenfeldt, Wolfgang Rolke, Nancy Reid, Gary Hill, and Stathes Paganis. I would also like to thank Louis Lyons for his continuing advice and the invitation to speak at such an enjoyable and productive conference.

This manuscript has been authored by Brookhaven Science Associates under Contract No. DE-AC02-98CH1-886 with the U.S. DOE. The U.S. Government retains, and the publisher, by accepting the article for publication, acknowledges, a world-wide license to publish or reproduce the published form of this manuscript, or allow others to do so, for the U.S. Government purposes.

References

1. ATLAS Collaboration, Detector and physics performance technical design report (volume ii) CERN-LHCC/99-15 (1999).
2. CMS Collaboration, Technical proposal CERN-LHCC/94-38 (1994).
3. H. B. Prosper Advanced Statistical Techniques in Particle Physics, Durham, England, 18-22 Mar 2002.
4. J. H. Friedman, Recent advances in predictive (machine) learning PhyStat2003 (2003).
5. K. Cranmer and R. S. Bowman, *Comp. Phys. Commun.* **167**, 165 (2005).
6. I. Hinchliffe, F. E. Paige, M. D. Shapiro, J. Soderqvist and W. Yao, *Phys. Rev.* **D55**, 5520 (1997).
7. S. Frixione and B. R. Webber, The mc@nlo event generator hep-ph/0207182, (2002).
8. A. Schalicke and F. Krauss, *JHEP* **07**, p. 018 (2005).
9. C. Buttar, D. Clements, I. Dawson and A. Moraes, *Acta Phys. Polon.* **B35**, p. 433 (2004).
10. G. J. Feldman and R. D. Cousins, *Phys. Rev.* **D57**, 3873 (1998).
11. B. Abbott *et al.*, *Phys. Rev.* **D62**, p. 092004 (2000).
12. Y. Gao, L. Lu and X. Wang, Significance calculation and a new analysis method in searching for new physics at the LHC physics/0509174, (2005).
13. G. Feldman, Concluding talk PhyStat05. These proceedings.
14. LEP Higgs Working Group, *Phys. Lett.* **B565**, 61 (2003).
15. K. Cranmer, *Comput. Phys. Commun.* **136**, 198 (2001).
16. T. Junk, *Nucl. Instrum. Meth.* **A434**, 435 (1999).
17. H. Hu and J. Nielsen, Analytic Confidence Level Calculations Using the Likelihood Ratio and Fourier Transform CERN 2000-005 physics/9906010, (2000).
18. K. Cranmer, B. Mellado, W. Quayle and Sau Lan Wu, Challenges of Moving the LEP Higgs Statistics to the LHC. PhyStat2003 physics/0312050 (2003).
19. J. Linnemann, Measures of significance in HEP and astrophysics. PhyStat2003 physics/0312059, (2003).
20. R. Cousins, Treatment of nuisance parameters in high energy physics, and possible justifications and improvements in the statistical literature. PhyStat05. These proceedings.
21. P. Sinervo, Definition and treatment of systematic uncertainties in high energy physics and astrophysics PhyStat2003, (2003).
22. K. Cranmer, Frequentist hypothesis testing with background uncertainty. PhyStat2003 physics/0310108 (2003).
23. W. A. Rolke and A. M. Lopez, *Nucl. Instrum. Meth.* **A458**, 745 (2001).
24. W. A. Rolke, A. M. Lopez and J. Conrad, *Nucl. Instrum. Meth.* **A551**, 493 (2005).
25. S. Asai et al., *Eur. Phys. J.* **C3252**, 19 (2004).
26. J. Cammin and M. Schumacher, The ATLAS discovery potential for the channel $ttH, (H \to bb)$ ATLAS Note ATL-PHYS-2003-024 (2003).
27. V. Drollinger, Th. Müller, and D. Denegri, Searching for Higgs Bosons in Association with Top Quark Pairs in the $H \to bb$ Decay Mode CMS NOTE-2001/054 (2001).
28. R. Cousins and V. Highland, *Nucl. Instrum. Meth.* **A320**, 331 (1992).
29. K. Cranmer, P. McNamara, B. Mellado, W. Quayle, and Sau Lan Wu, Confidence level calculations for $H \to W^+W^- \to l^+l^-\not{p}_T$ for $115 < M_H < 130$ GeV using vector boson fusion ATL-PHYS-2003-008 (2002).
30. N. Reid, Likelihood inference in the presence of nuisance parameters PhyStat2003, (2003).
31. F.James and M. Roos, *Comput.Phys.Commun.* **10**, 343 (1975).
32. F. James and M. Roos, *Nucl. Phys.* **B 172**, 475 (1980).
33. R. Cousins, *Nucl. Instrum. and Meth. in Phys. Res.* **A 417**, 391 (1998).
34. J. Stuart, A. Ord and S. Arnold, *Kendall's Advanced Theory of Statistics, Vol 2A (6th Ed.)* (Oxford University Press, New York, 1994).
35. G. Punzi, Ordering algorithms and confidence intervals in the presence of nuisance parameters PhyStat05 physics/0511202 (2005).
36. D. Nicolo and G. Signorelli Proceedings of Advanced Statistical Techniques in Particle Physics, Durham, England, (2002).

MACHINE LEARNING

SEPARATING SIGNAL FROM BACKGROUND USING ENSEMBLES OF RULES

JEROME H. FRIEDMAN

Department of Statistics and Stanford Linear Accelerator Center, Stanford University, Stanford, CA 94305
E-mail: jhf@stanford.edu

Machine learning has emerged as a important tool for separating signal events from associated background in high energy particle physics experiments. This paper describes a new machine learning method based on ensembles of rules. Each rule consists of a conjuction of a small number of simple statements ("cuts") concerning the values of individual input variables. These rule ensembles produce predictive accuracy comparable to the best methods. However their principal advantage lies in interpretation. Because of its simple form, each rule is easy to understand, as is its influence on the predictive model. Similarly, the degree of relevance of each of the respective input variables can be assessed. Graphical representations are presented that can be used to ascertain the dependence of the model jointly on the variables used for prediction.

1. Introduction

Predictive learning is a common application in data mining, machine learning and pattern recognition. The purpose is to predict the unknown value of an attribute y of a system under study, using the known joint values of other attributes $\mathbf{x} = (x_1, x_2, \cdots, x_n)$ associated with that system. The prediction takes the form $\hat{y} = F(\mathbf{x})$, where the function $F(\mathbf{x})$ maps a set of joint values of the "input" variables $\mathbf{x}$ to a value $\hat{y}$ for the "output" variable y. The goal is to produce an accurate mapping. Lack of accuracy is defined by the prediction "risk"

$$R(F) = E_{\mathbf{x}y} L(y, F(\mathbf{x})) \tag{1}$$

where $L(y, \hat{y})$ represents a loss or cost for predicting a value $\hat{y}$ when the actual value is y, and the expected (average) value is over the joint distribution of all variables $(\mathbf{x}, y)$ for the data to be predicted.

As an example consider the problem of separating signal from background events in a high energy particle physics experiment. Here the outcome attribute y for each event has one of two values $y \in \{signal, background\}$. The attributes $\mathbf{x}$ used for prediction are the variables measured from each event, perhaps augmented with various quantities constructed from these measurements. The prediction $\hat{y}$ also realizes one of the two values $\hat{y} \in \{signal, background\}$. A natural loss function for this two–class classification problem would be

$$L(y, \hat{y}) = \begin{cases} L_S & \text{if } \; y = signal \;\&\; \hat{y} = background \\ L_B & \text{if } \; y = background \;\&\; \hat{y} = signal \end{cases} \tag{2}$$

with $L(y, \hat{y}) = 0$ for correct predictions. Here L_S and L_B are the respective user specified costs for misclassifying signal and background events for the particular problem. The goal is to construct a mapping function $F(\mathbf{x})$ that given (2) minimizes the prediction risk (1).

Although the loss function (2) characterizes the actual goal, it cannot be directly used to construct classification functions $F(\mathbf{x})$ with most machine learning procedures. The problem is that with this loss criterion the associated risk (1) is not a continuous function of the parameters associated with the predicting function $F(\mathbf{x})$. This excludes the application of numerical optimization techniques in the search for a good solution, requiring instead far more costly combinatorial optimization methods.

In order to apply numerical optimization techniques one must approximate the discrete loss (2) with a smooth continuous one that produces the same solution, at least in the limit of infinite amount of data. For finite data sets the hope is that the solutions will be similar enough to be useful. One scores the signal events with the numerical value $y = 1$ and the background with $y = -1$. In this case the predicting function $F(\mathbf{x})$ produces a numerical score that estimates a monotone function of the probability that $y = 1$ (signal event) given the joint values of the predictor variables $\mathbf{x}$; that is, $F(\mathbf{x}) = m(\Pr[y = 1 \,|\, \mathbf{x}])$ where $m(\eta)$ is a monotonically increasing function of its argument η. Classification is accomplished by thresholding this score at an appropriate value t

$$\begin{aligned} F(\mathbf{x}) \geq t &\Rightarrow signal \\ F(\mathbf{x}) < t &\Rightarrow background. \end{aligned} \tag{3}$$

The value chosen for the threshold t is the one

that minimizes the prediction risk (1) using (2) and thereby depends on the values chosen for L_S and L_B.

Within this framework a variety of smooth surrogate loss functions have been proposed in the statistics and machine learning literatures. A commonly used criterion is squared–error loss $L(y, \hat{y}) = (y - \hat{y})^2$. In this case the predicting score function approximates $F(\mathbf{x}) = 2 \cdot \Pr[y = 1 \,|\, \mathbf{x}] - 1$. Other popular choices include $L(y, \hat{y}) = \log(1 + e^{-y \cdot \hat{y}})$ used by logistic regression in statistics, and $L(y, \hat{y}) = e^{-y \cdot \hat{y}}$ used by the AdaBoost boosting procedure (Freund and Schapire 1996) from machine learning. For these latter two loss functions the numerical score estimates the log–odds

$$F(\mathbf{x}) = \log \frac{\Pr[y = 1 \,|\, \mathbf{x}]}{1 - \Pr[y = 1 \,|\, \mathbf{x}]}.$$

Given a particular smooth surrogate $L(y, \hat{y})$, the optimal mapping ("target") function $F^*(\mathbf{x})$ is defined as the one that minimizes the prediction risk (1) over all possible functions

$$F^*(\mathbf{x}) = \arg \min_{F(\mathbf{x})} E_{\mathbf{x}y} L(y, F(\mathbf{x})). \tag{4}$$

This optimal predicting function is unknown because the distribution of the joint values of the variables $(\mathbf{x}, y)$, $p(\mathbf{x}, y)$, is unknown.

With the machine learning approach one has a data base of previously solved cases $T = \{\mathbf{x}_i, y_i, w_i\}_1^N$, called a training sample, containing known signal and background events. Here $\mathbf{x}_i$ represents the measurement variables associated with the ith event. Each signal event is assigned the value $y_i = 1$ and the background events are assigned $y_i = -1$. Each event also has a weight w_i that depends on its type; signal events are assigned weights

$$w_i = L_S \, \pi_S \, / \, N_S$$

where L_S is the cost for misclassifying a signal event (2), π_S is the fraction of signal events in *future* data to be predicted, and N_S is the total number of signal events in the training data T. Each background event receives a weight

$$w_i = L_B \, \pi_B \, / \, N_B$$

where L_B, π_B, and N_B are the corresponding quantities for the background. With this weighting the classification threshold (3) that minimizes prediction risk is $t = 0$.

These weighted training data are presumed to represent a random sample drawn from the distribution of future data to be predicted. A machine learning procedure is then applied to these training data to derive an approximation $F(\mathbf{x})$ to $F^*(\mathbf{x})$ (4). This approximation will be used to score and then classify (3) future events given only their measured variables $\mathbf{x}$. The extent to which this $F(\mathbf{x})$ so derived provides a useful approximation to $F^*(\mathbf{x})$ will depend on the nature of $F^*(\mathbf{x})$, the training sample size N, and the particular machine learning procedure employed. Different procedures are appropriate for different target functions and/or different sample sizes.

2. Ensemble learning

Learning ensembles have emerged as being among the most powerful machine learning methods (see Breiman 1996 & 2001, Freund and Schapire 1996, Friedman 2001). Their structural model takes the form

$$F(\mathbf{x}) = a_0 + \sum_{m=1}^{M} a_m f_m(\mathbf{x}) \tag{5}$$

where M is the size of the ensemble and each ensemble member ("base learner") $f_m(\mathbf{x})$ is a different function of the input variables $\mathbf{x}$ derived from the training data. Ensemble predictions $F(\mathbf{x})$ are taken to be a linear combination of the predictions of each of the ensemble members, with $\{a_m\}_0^M$ being the corresponding parameters specifying the particular linear combination. Ensemble methods differ in choice of particular base learners (function class), how they are derived from the data, and the prescription for obtaining the linear combination parameters $\{a_m\}_0^M$.

All popular ensemble methods use variants of the following generic procedure to generate the base learners used in (5). Each base learner is taken to be a simple function of the predictor variables characterized by a set of parameters $\mathbf{p} = (p_1, p_2, \cdots)$. That is,

$$f_m(\mathbf{x}) = f(\mathbf{x}; \mathbf{p}_m) \tag{6}$$

where $\mathbf{p}_m$ represents a specific set of joint parameter values indexing a specific function $f_m(\mathbf{x})$ from the parameterized class $f(\mathbf{x}; \mathbf{p})$. Particular choices for such parameterized function classes are discussed below. Given a function class the individual members of the ensemble are generated using the prescription presented in Algorithm 1.

Algorithm 1

Ensemble generation

1 $F_0(\mathbf{x}) = 0$
2 For $m = 1$ to M {
3 $\quad \mathbf{p}_m =$
$\qquad \arg\min_{\mathbf{p}} \sum_{i \in S_m(\eta)} L(y_i, F_{m-1}(\mathbf{x}_i) + f(\mathbf{x}_i; \mathbf{p}))$
4 $\quad f_m(\mathbf{x}) = f(\mathbf{x}; \mathbf{p}_m)$
5 $\quad F_m(\mathbf{x}) = F_{m-1}(\mathbf{x}) + \nu \cdot f_m(\mathbf{x})$
6 }
7 ensemble $= \{f_m(\mathbf{x})\}_1^M$

In line 3, $S_m(\eta)$ represents a different subsample of size $\eta < N$ randomly drawn without replacement from the original training data, $S_m(\eta) \subset \{\mathbf{x}_i, y_i\}_1^N$. As discussed in Friedman and Popescu 2003, smaller values of η encourage increased dispersion (less correlation) among the ensemble members $\{f_m(\mathbf{x})\}_1^M$ by training them on more diverse subsamples. Smaller values also reduce computation by a factor of N/η.

At each step m, the "memory" function

$$F_{m-1}(\mathbf{x}) = F_0(\mathbf{x}) + \nu \cdot \sum_{k=1}^{m-1} f_k(\mathbf{x})$$

contains partial information concerning the previously induced ensemble members $\{f_k(\mathbf{x})\}_1^{m-1}$ as controlled by the value of the "shrinkage" parameter $0 \leq \nu \leq 1$. At one extreme, setting $\nu = 0$ causes each base learner $f_m(\mathbf{x})$ to be generated without reference to those previously induced, whereas the other extreme $\nu = 1$ maximizes their influence. Intermediate values $0 < \nu < 1$ vary the degree to which previously chosen base learners effect the generation of each successive one in the sequence.

Several popular ensemble methods represent special cases of Algorithm 1. A "bagged" ensemble (Breiman 1996) is obtained by using squared–error loss, $L(y, \hat{y}) = (y - \hat{y})^2$, and setting $\nu = 0$, and $\eta = N/2$ or equivalently choosing S_m (line 3) to be a bootstrap sample (Friedman and Hall 1999). Random forests (Breiman 2001) introduce increased ensemble dispersion by additionally randomizing the algorithm ("arg min", line 3) used to solve for the ensemble members (large decision trees). In both cases the coefficients in (5) are set to $a_0 = \bar{y}$, $\{a_m = 1/M\}_1^M$ so that predictions are a simple average of those of the ensemble members. AdaBoost (Freund and Schapire 1996) uses exponential loss, $L(y, \hat{y}) = \exp(-y \cdot \hat{y})$ for $y \in \{-1, 1\}$, and is equivalent to setting $\nu = 1$ and $\eta = N$ in Algorithm 1. Predictions are taken to be the sign of the final memory function $F_M(\mathbf{x})$. MART (Friedman 2001) allows a variety of loss criteria $L(y, \hat{y})$ for arbitrary y, and in default mode sets $\nu = 0.1$ and $\eta = N/2$. Predictions are given by $F_M(\mathbf{x})$.

Friedman and Popescu 2003 experimented with a variety of joint (ν, η) values for generating ensembles of small decision trees, followed by a regularized regression to estimate the linear combination parameters $\{a_j\}_0^M$ (5). Given a set of base learners $\{f_m(\mathbf{x})\}_1^M$ the parameters of the linear combination are obtained by a regularized linear regression on the training data $\{\mathbf{x}_i, y_i, w_i\}_1^N$

$$\{\hat{a}_m\}_0^M = \arg \min_{\{a_m\}_0^M} \sum_{i=1}^N w_i L\left(y_i, a_0 + \sum_{m=1}^M a_m f_m(\mathbf{x}_i)\right) + \lambda \cdot \sum_{m=1}^M |\, a_m \,|. \qquad (7)$$

The first term in (7) measures the prediction risk (1) on the training sample, and the second (regularization) term penalizes large values for the coefficients of the base learners. The influence of this penalty is regulated by the value of $\lambda \geq 0$. It is well known that for this ("lasso") penalty, larger values of λ produce more overall shrinkage as well as increased dispersion among the values $\{|\, \hat{a}_m \,|\}_1^M$, often with many being set to zero (see Tibshirani 1996, Donoho *et al.* 1995). Its value is taken to be that which minimizes an estimate of future prediction risk (1) based on a separate sample not used in training, or by full (multi–fold) cross–validation. Fast algorithms for solving (7) for all values of $\lambda \geq 0$, using a variety of loss functions $L(y, \hat{y})$, are presented in Friedman and Popescu 2004. Empirical results presented in Friedman and Popescu 2003 indicated that small but nonzero values of ν ($\nu \simeq 0.01$) performed best in this context. Results were seen to be fairly insensitive to the value chosen for η provided it was small ($\eta \lesssim N/2$) and grew less rapidly than the total sample size N ($\eta \sim \sqrt{N}$) as N becomes large ($N \gtrsim 500$).

Although in principle most of these procedures can be used with other base learners, they have almost exclusively been applied with decision trees (Breiman, *et al* 1983, Quinlan 1993). This is due to the attractive properties of trees in data mining applications, and the existence of fast algorithms for inducing decision tree ensembles.

3. Rule based ensembles

The base learners considered here are simple rules. Let S_j be the set of all possible values for input variable x_j, $x_j \in S_j$, and s_{jm} be a specified subset of those values, $s_{jm} \subseteq S_j$. Then each base learner takes the form of a conjunctive rule

$$r_m(\mathbf{x}) = \prod_{j=1}^{n} I(x_j \in s_{jm}) \tag{8}$$

where $I(\delta)$ is an indicator of the truth of its logical argument; $I(\delta) = 1$ if δ is true and $I(\delta) = 0$ if δ is false. Each such base learner (8) assumes two values $r_m(\mathbf{x}) \in \{0, 1\}$. It is nonzero when all of the input variables realize values that are simultaneously within their respective subsets $\{x_j \in s_{jm}\}_1^n$. For variables that assume orderable values the subsets are taken to be contiguous intervals

$$s_{jm} = (t_{jm}, u_{jm}]$$

defined by a lower and upper limit, $t_{jm} < x_j \leq u_{jm}$. For categorical variables assuming unorderable values (names) the subsets are explicitly enumerated. Such rules (8) can be regarded as parameterized functions of $\mathbf{x}$ (6) where the parameters $\mathbf{p}_m$ are the quantities that define the respective subsets $\{s_{jm}\}$.

Note that for the case in which the subset of values s_{jm} (real or categorical) appearing in a factor of (8) is in fact the entire set $s_{jm} = S_j$, the corresponding factor can be omitted from the product. In this case the rule can be expressed in the simpler form

$$r_m(\mathbf{x}) = \prod_{s_{jm} \neq S_j} I(x_j \in s_{jm}). \tag{9}$$

The particular input variables x_j for which $s_{jm} \neq S_j$ are said to be those that "define" the rule $r_m(\mathbf{x})$. As an example, the rule

$$r_m(\mathbf{x}) = \begin{cases} I(18 \leq \text{age} < 34) \\ \cdot I(\text{marital status} \in \{\text{single}, \\ \qquad \text{living together–not married}\}) \\ \cdot I(\text{householder status} \; = \; \text{rent}) \end{cases}$$

is defined by three variables, and a nonzero value increases the odds of frequenting bars and night clubs. In high energy physics applications each rule (9) can be interpreted as an intersection of "cuts" on the variables that define the rule.

3.1. *Rule generation*

One way to attempt to generate a rule ensemble is to let the base learner $f(\mathbf{x}; \mathbf{p})$ appearing in Algorithm 1 take the form of a rule (8) and then try to solve the optimization problem on line 3 for the respective variable subsets $\{s_{jm}\}$. Such a (combinatorial) optimization is generally infeasible for more that a few predictor variables although fast approximate algorithms might be derived. The approach used here is to view a decision tree as defining a collection of rules and take advantage of existing fast algorithms for producing decision tree ensembles. That is, decision trees are used as the base learner $f(\mathbf{x}; \mathbf{p})$ in Algorithm 1. Each node (interior and terminal) of each resulting tree $f_m(\mathbf{x})$ produces a rule of the form (9).

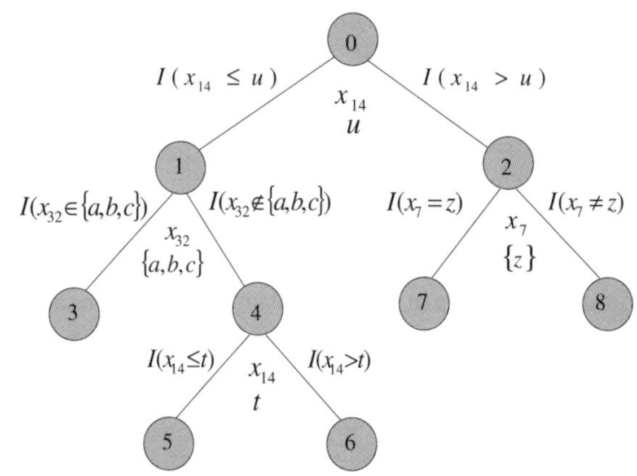

Fig. 1. A typical decision tree with five terminal nodes as described in the text.

This is illustrated in Fig. 1 which shows a typical decision tree with five terminal nodes that could result from using a decision tree algorithm in conjunction with Algorithm 1. Associated with each interior node is one of the input variables x_j. For variables that realize orderable values a particular value of that variable ("split point") is also associated with the node. For variables that assume unorderable categorical values, a specified subset of those values replaces the split point. For the tree displayed in Fig. 1

nodes 0 and 4 are associated with orderable variable x_{14} with split points u and t respectively, node 1 is associated with categorical variable variable x_{32} with subset values $\{a, b, c\}$, and node 2 is associated with categorical variable x_7 with the single value $\{z\}$.

Each edge of the tree connecting a "parent" node to one of its two "daughter" nodes represents a factor in (9) contributing to the rules corresponding to all descendent nodes of the parent. These factors are shown in Fig. 1 for each such edge. The rule corresponding to any node in the tree is given by the product of the factors associated with all of the edges on the path from the root to that node. Note that there is no rule corresponding to the root node. As examples, in Fig. 1 the rules corresponding to nodes 1, 4, 6, and 7 are respectively:

$$
\begin{aligned}
r_1(\mathbf{x}) &= I(x_{14} \leq u) \\
r_4(\mathbf{x}) &= I(x_{14} \leq u) \cdot I(x_{32} \notin \{a, b, c\}) \\
r_6(\mathbf{x}) &= I(t < x_{14} \leq u) \cdot I(x_{32} \notin \{a, b, c\}) \\
r_7(\mathbf{x}) &= I(x_{14} > u) \cdot I(x_7 = z).
\end{aligned}
$$

3.2. *Rule fitting*

The collection of all such rules derived from all of the trees $\{f_m(\mathbf{x})\}_1^M$ produced by Algorithm 1 constitute the rule ensemble $\{r_k(\mathbf{x})\}_1^K$. The total number of rules is

$$K = \sum_{m=1}^{M} 2(t_m - 1) \tag{10}$$

where t_m is the number of terminal nodes for the mth tree. The predictive model is

$$F(\mathbf{x}) = \hat{a}_0 + \sum_{k=1}^{K} \hat{a}_k r_k(\mathbf{x}) \tag{11}$$

with

$$
\begin{aligned}
\{\hat{a}_k\}_0^K = \arg\min_{\{a_k\}_0^K} &\sum_{i=1}^{N} w_i \, L\left(y_i \, , a_0 + \sum_{k=1}^{K} a_k r_k(\mathbf{x}_i)\right) \\
&+ \lambda \cdot \sum_{k=1}^{K} |\, a_k \,|.
\end{aligned} \tag{12}
$$

Fast algorithms for solving (12) for all values of $\lambda \geq 0$, and procedures for choosing a value for λ, are discussed in Friedman and Popescu 2004.

4. Rule based interpretation

The most important aspect of any predictive function $F(\mathbf{x})$ is its accuracy on future data as reflected by its prediction risk (1). Results from Friedman and Popescu 2005 suggest that rule based ensembles (11) (12) provide accuracy competitive with the best methods. However, accuracy is not the only desirable property of a predictive model. Often it is useful to be able to interpret the model to gain an understanding of how the respective input variables $\mathbf{x} = (x_1, x_2, \cdots, x_n)$ are being used to formulate predictions. This information can be used to perform "sanity checks" to see if the model is consistent with one's *a priori* domain knowledge, and to gain an *a posteriori* understanding of the system that produced the data. Such information can be used to refine the model to improve its properties.

Most ensemble as well as other machine learning methods produce "black–box" models. They are represented in an incomprehensible form making it difficult to impossible to understand how the input variables are being used for prediction. One of the primary benefits that distinguish rule based ensembles is the ability to intepret the resulting model to gain such information.

Rules of the form (9) represent easily understandable functions of the input variables $\mathbf{x}$. Although a large number of such rules participate in the initial ensemble, the fitting procedure (12) generally sets the vast majority ($\sim$ 80% to 90%) of the corresponding coefficient estimates $\{\hat{a}_k\}_1^K$ to zero and their corresponding rules are not used for prediction. As noted above, this selection property is a well known aspect of the lasso penalty in (12). The remaining rules will have varying coefficient values depending on their estimated predictive relevance. The most relevant rules can then be examined for interpretation.

A commonly used measure of relevance or importance I_k of any predictor in a linear model such as (11) is the absolute value of the coefficient of the corresponding standardized predictor. For rules this becomes

$$I_k = |\, \hat{a}_k \,| \cdot \sqrt{s_k(1 - s_k)} \tag{13}$$

where s_k is the rule support

$$s_k = \sum_{i=1}^{N} w_i \, r_k(\mathbf{x}_i) \bigg/ \sum_{i=1}^{N} w_i. \tag{14}$$

Those rules with the largest values for (13) are the most influential for prediction based on the predictive equation (11). These can then be selected and examined for interpretation.

4.1. *Input variable importance*

In predictive learning a descriptive statistic that is almost always of interest is the relative importance or relevance of the respective input variables $(x_1, x_2, \cdots, x_n)$ to the predictive model; that is, which of the variables are most influential in making predictions and which in retrospect need not have been included. For the models (11) considered here, the most relevant input variables are those that preferentially define the most influential rules appearing in the model. Input variables that frequently appear in important rules are judged to be more relevant than those that tend to appear only in less influential rules.

This concept can be captured by a measure of importance J_j of input variable x_j

$$J_j = \sum_{x_j \in r_k} I_k/m_k. \tag{15}$$

This measure sums the importances (13) of those rules (9) that contain x_j ($x_j \in r_k$) each divided by the total number of input variables m_k that define the rule. In this sense the input variables that define a rule equally share its importance, and rules with more variables do not receive exaggerated influence by virtue of appearing in multiple input variable importance measures. The distribution of $\{J_j\}_1^n$ (15) can be examined to ascertain the relative influence of each of the respective input variables on the model's predictions. Illustrations are provided in the example below.

4.2. *Partial dependence functions*

Visualization is one of the most powerful interpretational tools. Graphical renderings of the value of $F(\mathbf{x})$ as a function of its arguments provides a comprehensive summary of its dependence on the joint values of the input variables. Unfortunately, such visualization is limited to low dimensional arguments. Viewing functions of higher dimensional arguments is more difficult. It is therefore useful to be able to view the partial dependence of the approximation $F(\mathbf{x})$ on selected small subsets of the input variables. Although a collection of such plots can seldom provide a comprehensive depiction of the approximation, it can often produce helpful clues.

Let $\mathbf{z}_l$ be a chosen "target" subset, of size l, of the input variables $\mathbf{x}$

$$\mathbf{z}_l = \{z_1, \cdots, z_l\} \subset \{x_1, \cdots, x_n\},$$

and $\mathbf{z}_{\backslash l}$ be the complement subset

$$\mathbf{z}_{\backslash l} \cup \mathbf{z}_l = \mathbf{x}.$$

The approximation $F(\mathbf{x})$ in principle depends on variables in both subsets

$$F(\mathbf{x}) = F(\mathbf{z}_l\,, \mathbf{z}_{\backslash l}).$$

If one conditions on specific values for the variables in $\mathbf{z}_{\backslash l}$, then $F(\mathbf{x})$ can be considered as a function only of the variables in the chosen subset $\mathbf{z}_l$

$$F_{\mathbf{z}_{\backslash l}}(\mathbf{z}_l) = F(\mathbf{z}_l \,|\, \mathbf{z}_{\backslash l}). \tag{16}$$

In general, the functional form of $F_{\mathbf{z}_{\backslash l}}(\mathbf{z}_l)$ will depend on the particular values chosen for $\mathbf{z}_{\backslash l}$. If however, this dependence is not too strong then the averaged function

$$\bar{F}_l(\mathbf{z}_l) = E_{\mathbf{z}_{\backslash l}}[F(\mathbf{x})] = \int F(\mathbf{z}_l\,, \mathbf{z}_{\backslash l})\, p_{\backslash l}(\mathbf{z}_{\backslash l})\, d\mathbf{z}_{\backslash l} \tag{17}$$

can represent a useful summary of the "partial dependence" of $F(\mathbf{x})$ on the chosen variable subset $\mathbf{z}_l$ (Friedman 2001). Here $p_{\backslash l}(\mathbf{z}_{\backslash l})$ is the marginal probability density of $\mathbf{z}_{\backslash l}$

$$p_{\backslash l}(\mathbf{z}_{\backslash l}) = \int p(\mathbf{x})\, d\mathbf{z}_l, \tag{18}$$

where $p(\mathbf{x})$ is the joint probability density of all of the inputs $\mathbf{x}$. This complement marginal density (18) can be estimated from the training data, so that (17) becomes

$$\bar{F}_l(\mathbf{z}_l) = \sum_{i=1}^{N} w_i\, F(\mathbf{z}_l\,, \mathbf{z}_{i\backslash l}) \Big/ \sum_{i=1}^{N} w_i. \tag{19}$$

where $\mathbf{z}_{i\backslash l}$ are the data values of $\mathbf{z}_{\backslash l}$.

Partial dependence functions (19) can be used to help interpret models produced by any "black box" prediction method, such as neural networks, support vector machines, nearest–neighbors, radial basis functions, etc. They only require the value of $F(\mathbf{x})$ for specified values of $\mathbf{x}$. However, when there are a large number of predictor variables, it is very useful to have an measure of relevance (Section 4.1) to reduce the potentially large number variables, and variable combinations, to be considered.

5. Illustration

In this section we apply the RuleFit procedure to a signal/background separation problem from a high energy particle physics experiment and illustrate the various interpretational tools described in Section 4. The training data consists of 50000 Monte Carlo simulated events, half of which are signal and half are background. Details concerning the specific application and the nature of the 50 input variables are withheld at the request of the experimenters. An additional 23000 events were generated (half signal and half background) to evaluate performance. These latter ("test") events were not used to train the predictive model.

All parameters of the RuleFit procedure were set to their default values: $\nu = 0.01$ and $\eta = \min(N/2, 100 + 6\sqrt{N}) \simeq 1450$ events in Algorithm 1, four terminal nodes for each tree, and 3500 generated rules in the initial ensemble (585 trees). It is possible that performance could be improved by tuning some of these parameters for this specific application.

Applying RuleFit to the training data produced a model (11) with 410 rules having nonzero coefficients from (12). The corresponding error rate on the test data was 6.97%. Another measure of prediction quality, area under the ROC curve ("AUC"), was 0.977. Perfect prediction would have zero error rate and AUC = 1.

Figure 2 displays a graphical representation of prediction quality. The upper frame shows the distribution of the model scores $F(\mathbf{x})$ (11) for the 11500 signal events in the test sample; the lower frame shows the corresponding plot for the 11500 background events. One sees that signal events tend to have predominately higher scores than the background. Using a threshold of $t = 0$ (3) gives rise to the minimal error rate of 6.97%, with slightly more background being classified as signal than signal classified as background. Increasing the threshold value ($t > 0$) would reduce background errors leading to a purer sample of signal events at the expense of classifying more of the signal as background. Lowering the threshold ($t < 0$) would capture more of the signal at the expense of increased background contamination. In this context modifying the threshold can be viewed as changing the relative values of the misclassification costs L_S and L_B in (2).

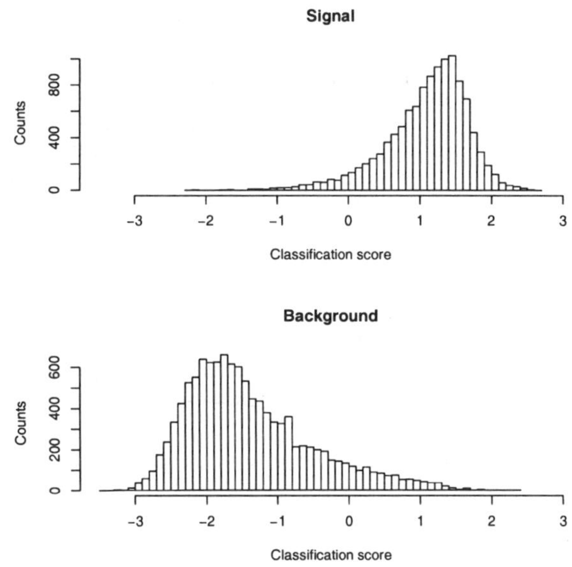

Fig. 2. Distribution of RuleFit prediction scores for signal (upper) and background (lower) test events.

This signal/background trade–off is more directly captured by the corresponding ROC curve shown in Fig. 3. Here the fraction of captured signal events (true positives) is plotted against the fraction of background contamination (false positives) as the threshold t is varied. One sees that permitting 5% background contamination allowed 90% of the signal events to be captured, whereas 10% background captures approximately 95% of the signal.

Table 1 illustrates some typical rules by displaying the five most important using (13). The first column shows the rules' relative importance normalized so that the maximum value over all rules is 100. The second column gives the coefficient $\hat{a}_k$ (11) of the corresponding rule $r_k(\mathbf{x})$. Positive coefficient values indicate that satisfying the rule ($r_k(\mathbf{x}) = 1$) increases the odds of being a signal event, where as negative values decrease the odds. The third column shows the rule's support (14). The last column shows the variables and cut values that define the corresponding rules. One sees that here all of these relatively important rules are fairly simple, typically involving two to three variables. Knowing the meaning of the variables for each of the rules could lead to insights concerning what aspects of the experiment lead to separating signal from background.

Figure 4 plots the relative importances (15) of

Table 1. The five most important rules for differentiating signal from background events.

Importance	Coefficient	Support	Rule
100	-0.16	0.45	$x_6 \leq 0.31$ & $x_{16} \leq 1117$ & $x_{32} \leq 1.31$
83	0.13	0.41	$0.025 \leq x_{14} < 0.53$ & $x_{27} < 82.4$
82	0.22	0.093	$-500 \leq x_3 < 92.6$ & $x_{21} \leq -0.022$ & $x_{39} > 1.18$
75	0.12	0.32	$x_1 \leq 5.2$ & $-500 \leq x_3 < 92.6$ & $x_{21} > -0.022$
73	-0.12	0.41	$x_1 > 4.37$ & $x_{23} \leq 160.1$ & $x_{32} \leq 1.41$

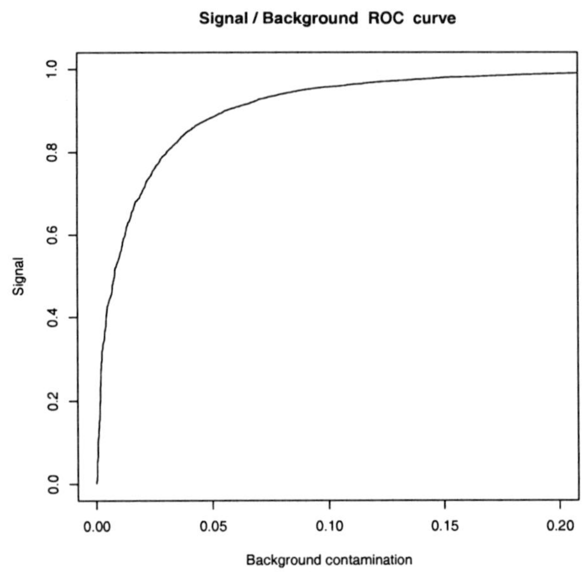

Fig. 3. ROC curve for RuleFit test predictions.

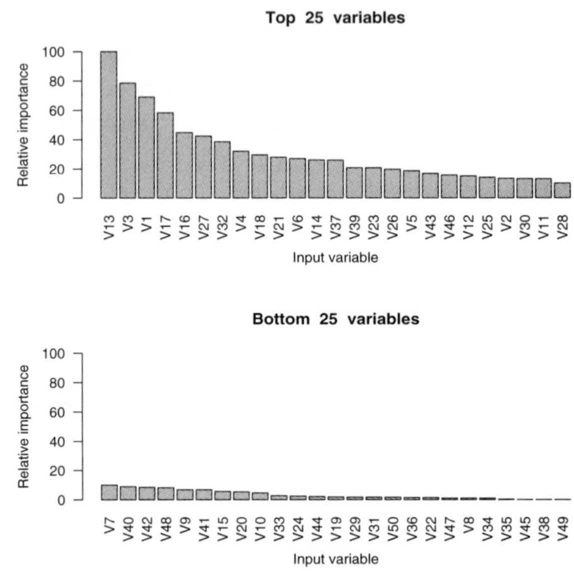

Fig. 4. Relative importances of the 50 input variables to the RuleFit predictive model.

each of the 50 input variables in (inverse) order of their importance values. Here some variables are clearly far more relevant than others to the predictive model (11). Knowing which variables are the important ones for separating signal from background can lead to insights concerning the experimental setup.

Table 2. Error rate and one minus area under the ROC curve for RuleFit models based on subsets of the most important predictor variables.

Variables	1-AUC	Error
50	0.0230	6.97
25	0.0232	7.06
20	0.0237	7.06
15	0.0264	7.60

This information can also be used to simplify the actual predictive model. This is illustrated in Table 2. Each row shows the test error rate (third column) and $1 - AUC$ (second column) for RuleFit models using subsets of the input variables. The first column shows the number of (most important – see Fig. 4) variables used out of the total of 50. One sees that training the model using only the 20 most important variables results in no significant decrease in model quality. Using only the top 15 variables degrades performance only by about 8%. Predictive models with fewer variables might be preferred if some of those variables deemed to be unimportant and thus expendable were especially difficult or expensive to measure.

Figure 5 shows plots of the single variable partial dependence of $F(\mathbf{x})$ (11) on the nine most important variables. One sees that, for example, the odds of being a signal event decrease monotonically with increasing values of the most important variable x_{13}. For the next most important variable x_3, predicted signal odds are lowest for $95 \lesssim x_3 \lesssim 170$ and

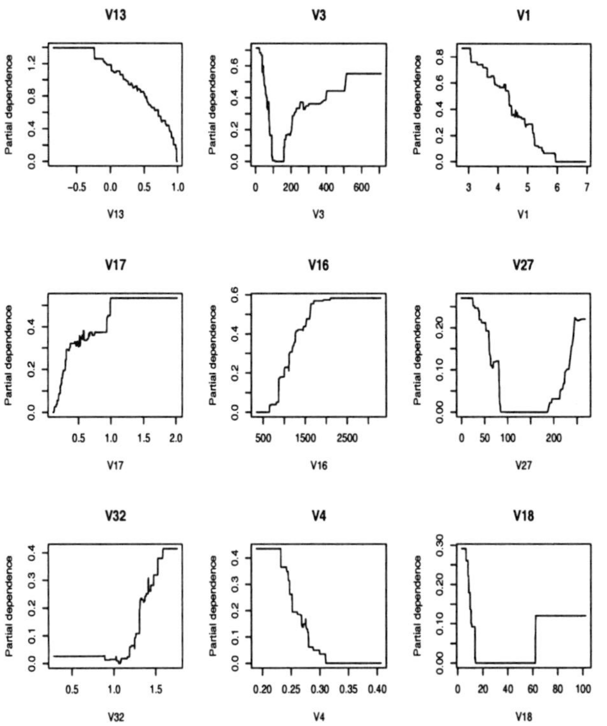

Fig. 5. Single–variable partial dependence plots of the odd of a signal event as a function of the nine most important predictor variables.

become higher for values outside this range. In general, examination of such partial dependences on the important variables provides information on how the values of the corresponding variables are being used for prediction.

More detailed information can be obtained from two–variable partial dependence plots. Figure 6 shows the partial dependence of $F(\mathbf{x})$ (11) on the joint values of selected variable pairs using several plotting formats. The upper left frame shows the partial dependence on (x_1, x_{13}) using a perspective mesh representation. One sees that signal odds increase as either of the two variables become larger. The upper right frame shows a contour plot of the partial dependence on (x_{17}, x_{13}). Here the signal odds are highest for $x_{17} \simeq 0.4$ and $x_{13} \simeq -0.4$, and decrease in all directions from that point. The lower two frames of Fig. 6 use a "heat map" to represent the respective two–variable partial dependence plots. Lowest values are shown darker while higher ones are lighter, and the highest values are white (surrounded by the lighter pixels). As an example, one sees from the lower right frame that for large values of x_1 the odds of being a signal event are low and at most depend weakly on x_3, whereas for small values of x_1 the odds strongly depend on the value of x_3. This is an example of an interaction (correlation) effect between these two variables.

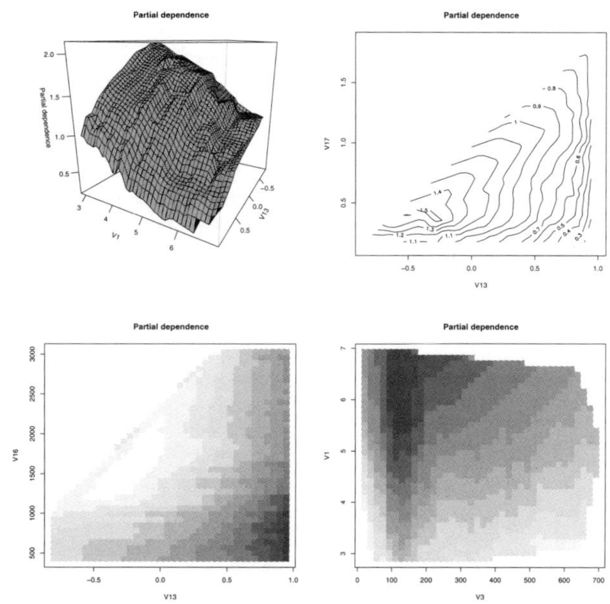

Fig. 6. Two–variable partial dependence plots of the odds of a signal event as a function of the joint values of selected variable pairs. Upper left: perspective mesh plot, upper right: contour plot, lower: heat map representation.

6. Conclusion

This paper has outlined the RuleFit technique for predictive learning and illustrated some of its features on a signal/background separation problem in high energy particle physics. A more complete description of the procedure along with its other features can be found in Friedman and Popescu 2005. A software interface to the R statistical package can be obtained from http://www-stat.stanford.edu/~jhf/RuleFit.html.

Acknowledgments

This research was partially supported by the Department of Energy under contract DE-AC02-76SF00515 and the National Science Foundation under grant DMS-97-64431.

References

1. Breiman, L. (1996). Bagging Predictors. *Machine Learning* **26**: 123-140.
2. Breiman, L. (2001). Random Forests. *Machine Learning* **45**: 5-32.
3. Breiman, L., Friedman, J. H., Olshen, R., and Stone, C. (1983). *Classification and Regression Trees.* Wadsworth.
4. Donoho, D., Johnstone, I., Kerkyacharian, G. and Picard, D. (1995). Wavelet shrinkage; asymptotia? (with discussion). *J. Royal. Statist. Soc* **57**: 201-337.
5. Freund, Y. and Schapire, R. E. (1996). Experiments with a new boosting algorithm. *Machine Learning: Proceedings of the Thirteenth International Conference*, Morgan Kauffman, San Francisco, 148-156.
6. Friedman, J. H. (2001). Greedy function approximation: a gradient boosting machine. *Annals of Statistics* **29**: 1189-1232.
7. Friedman, J. H. and Hall, P. (1999). On bagging and nonlinear estimation. *Stanford University, Department of Statistics.* Technical Report.
8. Friedman, J. H., and Popescu, B. E. (2003). Importance Sampled Learning Ensembles *Stanford University, Department of Statistics.* Technical Report.
9. Friedman, J. H., and Popescu, B. E. (2004). Gradient directed regularization for linear regression and classification. *Stanford University, Department of Statistics.* Technical report.
10. Friedman, J. H., and Popescu, B. E. (2005). Predictive learning via rule ensembles. *Stanford University. Department of Statistics.* Technical report.
11. Quinlan, R. (1993). *C4.5: Programs for Machine Learning.* Morgan Kaufmann, San Mateo.
12. Tibshirani, R. (1996). Regression shrinkage and selection via the lasso. *J. Royal. Statist. Soc. B.* **58**: 267-288.

COMMENT ON "SEPARATING SIGNAL FROM BACKGROUND USING ENSEMBLES OF RULES"

HARRISON B. PROSPER
Florida State University, Tallahassee, Florida, USA

Friedman and Popescu have introduced a novel machine learning algorithm, based on rules, that performs very well relative to other methods. We make a few general remarks about ensemble methods and comment on their particular method.

1. Introduction

It is often instructive to view things from a general perspective. This is true, in particular, of machine learning algorithms, if only because a general perspective clarifies the relationship between different algorithms, and, makes it easier to discern whether or not a "new" method is truly new. By embedding algorithms, such as AdaBoost [1], Bagging and Random Forests [2], into the framework of *ensemble learning*, Friedman and Popescu [3] have clarified the nature of these algorithms. Moreover, having understood that boosting and bagging are "merely" interesting variations on a theme, many other interesting variations spring to mind, for example, the one described by Jerry Friedman at this meeting.

2. Ensemble Learning

The idea of ensemble learning is to construct a mapping $y = F(x)$, based on some training data $\{(x_1, y_1), \cdots, (x_n, y_n)\}$, where

$$F(x) = a_0 + \sum_{m=1}^{M} a_m f_m(x),$$

and $\{f_m(x)\}$ is an *ensemble* of functions called *base learners*. The base learners are chosen from a function class, each of whose elements is labeled by the values of a set of parameters $p = (p_1, p_2, \cdots)$. Given a class of parameterized functions $f(x; p)$, an algorithmic procedure is specified to pick functions from the class, typically, by minimizing some *loss function* L. The ensemble method is very general: any function class may be used, along with any mechanism to choose from it.

As noted by Jerry Friedman, one anticipates that different procedures might be required for different data sets. The rule-based method of Friedman and Popescu, however, seems broadly applicable.

3. RuleFit

In the *RuleFit* method [3], the function class used is the class of all conjunctions

$$f_m(x) \equiv r_m(x) = A \,\&\, B \,\&\, C \cdots,$$

where A, B, etc., are simple statements with truth value 0 or 1. The function $r_m(x)$ is called a *rule*. For example, $A = E_T > 25$ GeV/c and $B = 85 < M_{ee} < 95$ GeV/c^2 might be statements typical of a signal/background classification problem in particle physics, in which one requires a jet of particles to have transverse energy exceeding 25 GeV *AND* the mass of an electron-positron pair be consistent with that of the Z boson. In practice, the class of rules is defined by the leaves of a forest of decision trees. One arrives at a leaf, appropriately, by following branches, starting at the root. Having found a set of rules, the coefficients a_m are found by minimizing the *lasso* loss function, a principal virtue of which is that, at its minimum, a large fraction of the coefficients are typically zero. Therefore, the number of rules that remain is generally far fewer than the number in the original set.

The RuleFit method has been shown to perform very well. Moreover, since its function class is based on decision trees, the method is fast. Its other advantage is that the meaning of each function $f_m(x)$, being a simple rule, is readily apparent. That being said, the function $F(x)$ is a linear sum of rules; therefore, even though each rule is easy to understand, it is less clear that $F(x)$ itself can be as readily interpreted.

An important benefit of the RuleFit method is that it provides a way to assess the importance of a variable. This is extremely useful because one can rank variables according to their importance and keep only those that are judged significantly more important than the rest. Thus can one reduce the dimensionality of the problem, and therefore the computational burden. The importance measure suggested by Friedman and Popescu is an intuitively plausible one that might be expected to work well most of the time. It is not clear, however, that it will *always* rank variables the same way as would a method in which all possible combinations of variables were tried and the subsets ranked accordingly.

4. Let a Thousand Flowers Bloom

The generality of the ensemble method invites the exploration of potentially interesting variations on that theme. One possibility, might be to use a function class defined by

$$f(x;p) = \tanh(p_0 + \sum_{i=1}^{N} p_i x_i),$$

together with any one of the standard loss functions. Given the ensemble of functions $f_m(x) = \tanh(p_{m0} + \sum_{i=1}^{N} p_{mi} x_i)$, and coefficients a_m – obtained, perhaps, by lasso regression – one would then have

$$F(x) = a_0 + \sum_{m=1}^{M} a_m \tanh(p_{m0} + \sum_{i=1}^{N} p_{mi} x_i).$$

This function is typical of the kind that appears in a neural network with N inputs, a single hidden layer of M nodes and a single output.

Acknowledgments

I would like to thank Jerry Friedman for an enlightening conversation and for providing a timely copy of his slides.

References

1. Y. Freund and R. E. Schapire, Journal of Computer and System Sciences **55**, 119 (1997).
2. Wald Lecture, Leo Breiman, http://www.stat.berkeley.edu/users/breiman/wald2002-1.pdf.
3. Jerome H. Friedman, these proceedings.

BOOSTED DECISION TREES, A POWERFUL EVENT CLASSIFIER

BYRON P. ROE[A], HAI-JUN YANG[A], AND JI ZHU[B]
[A] *Department of Physics,* [B] *Department of Statistics, University of Michigan,*
450 Church St., Ann Arbor, MI, 48109-1040
e-mail: byronroe@umich.edu

Boosted decision trees are compared with neural nets and various decision tree methods using the MiniBooNE experiment as a test bed. A discussion of methods for pruning variables and for increasing the speed of convergence are given.

1. Decision Trees and Boosting

Consider the problem of classification of events between signal and background, given a number of particle identification (PID) variables. A decision tree is a sequence of binary splits of the data. To train the tree a set of known training events is used. The results are measured using a separate set of known testing events. Consider all of the data to be on one node. The best PID variable and best place on that variable to split the data into separate signal and background is found. There are then two nodes. The process is repeated on these new nodes and is continued until a given number of final nodes (called "leaves") are obtained, or until all leaves are pure or until a node has too few events.

There are several popular criteria to determine the best PID variable and best place on which to split a node. The *gini* criterion is used here. Suppose that event i has weight W_i. The purity P of a node is defined as the weight of signal events on the node divided by the total weight of events on that node. For a given node: $gini = P(1-P)\sum_i W_i$. $gini$ is zero for $P=1$ or $P=0$. The best split is chosen as the one which minimizes $gini_{left} + gini_{right}$. The next node to split is chosen by finding that node whose splitting maximizes the change in $gini$. In this way a decision tree is built. Leaves with $P \geq 0.5$ are signal leaves and the rest are background leaves.

Decision trees are powerful, but unstable. A small change in the training data can produce a large change in the tree. This is remedied by the use of boosting. For boosting, the training events which were misclassified (a signal event fell on a background leaf or vice versa) have their weights increased (boosted), and a new tree is formed. This procedure is then repeated for the new tree. In this way many trees are built up. The score from the mth individual tree T_m is taken as $+1$ if the event falls on a signal leaf and -1 if the event falls on a background leaf. The final score is taken as a weighted sum of the scores of the individual leaves.

Two methods for boosting are considered here. The first is called AdaBoost. Define err_m = *weight misclassified/total weight for tree* m. Let $\alpha_m = \beta \log[(1-err_m)/err_m]$, where β is a constant. In the statistical literature β has been taken as one, but for the MiniBooNE experiment, $\beta = 0.5$ has been found to be the optimum value. The misclassified events have their weight multiplied by e^{α_m}. The weights are then renormalized so the sum of all of the training event weights is one. The final score is $T = \sum_{m=1}^{N_{tree}} \alpha_m T_m$.

The second method of boosting considered here is called ϵ-boost or, sometimes, shrinkage. Misclassified events have their weight multiplied by $e^{2\epsilon}$, where ϵ is a constant. For the MiniBooNE experiment, $\epsilon = 0.03$ has been optimum. (The results vary only mildly as β or ϵ are changed a bit.) The final score is $T = \sum_{m=1}^{N_{tree}} \epsilon T_m$.

ϵ-boost changes weights a little at a time, while AdaBoost can be shown to try to optimize each change in weights to minimize e^{-yT} where T is the score and y is $+1$ for a signal event and -1 for a background event. The optimum value is $T = \log prob/(1-prob)$, where $prob$ is the probability that $y=1$, given the observed PID variables. In practice, for MiniBooNE, the two boosting methods have performed almost equally well. Boosting is described as using many weak classifiers to build a strong classifier. This is seen in Figure 1. After the first few trees, the misclassification fraction for an individual tree is above 40%.

In the MiniBooNE experiment some hundreds of possible PID variables have been suggested. The most powerful of these have been selected by accept-

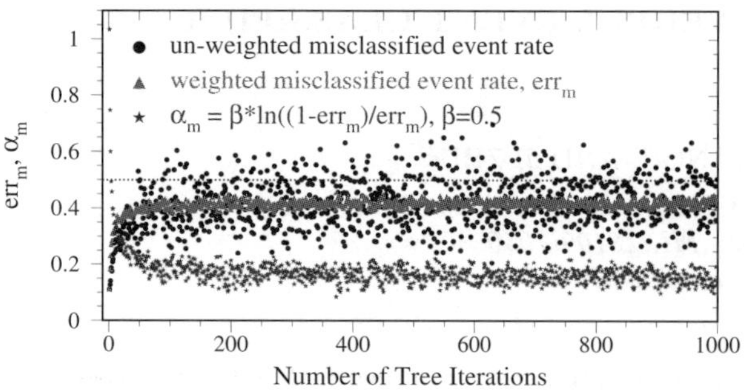

Fig. 1. The unweighted, weighted misclassified event rate (err_m), and α_m versus the number of tree iterations for AdaBoost with $\beta = 0.5$ and signal purity threshold value of 50%.

ing those which are used most often as splitting variables. Some care needs to be taken as sometimes a variable will appear unimportant for the first few trees, but then become important for later trees. Current MiniBooNE boosting trees have 80-100 PID variables. Use of more variables tends to slightly degrade the performance, probably because all of the useful information is already in the previous variables and noise without additional signal is being added. The performance has been examined varying the number of trees and the number of leaves/tree. This is shown in Figures 2. Here, relative ratio is *constant* × *fraction of background kept/fraction of signal kept* for a given signal efficiency. (Smaller is better!) Optimum results are obtained for MiniBooNE with about 1000 trees and with 45 leaves/tree. Different experiments should optimize these values for their particular data sets.

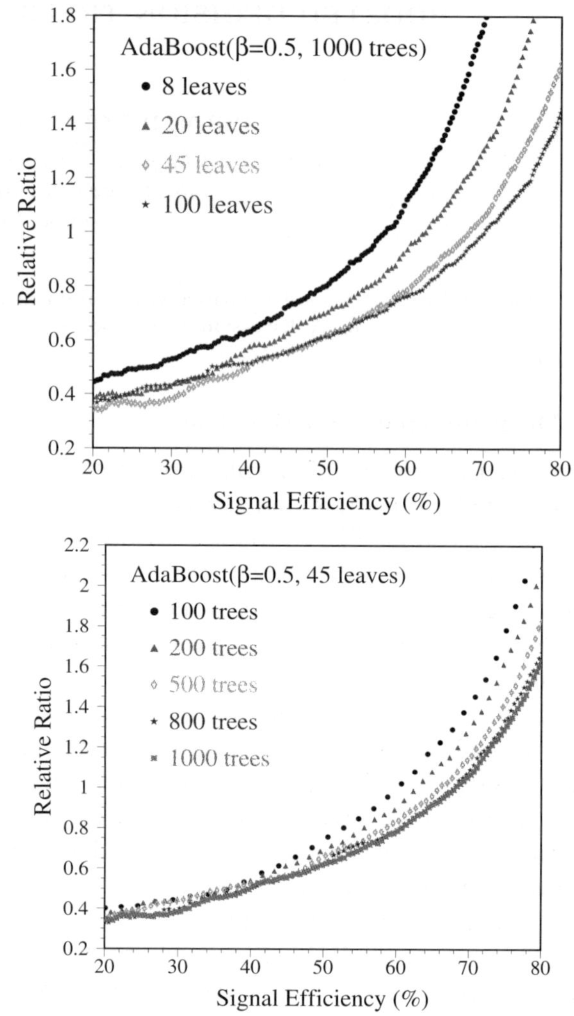

Fig. 2. Top: tuning the number of leaves when using 1000 trees. Bottom: tuning the number of trees when using 45 leaves.

2. Tests of Boosting with Other Classification Methods

Boosting was compared with artificial neural nets (ANN), which the MiniBooNE collaboration had used previously. For Figure 3 only, the relative ratio is redefined as the fraction of background kept by ANN to that for boosting for a given fraction of signal events being kept. (Larger is better for boosting!) It is seen that boosting is better than ANN by a factor of 1.2-2 for MiniBooNE data.

AdaBoost and ϵ-boost were compared with various other similar methods. Space does not permit a description of these methods; Table 1 will be of most use to those already familiar with them.

It is seen that Adaboost, ϵ-boost, ϵ-LogitBoost,

Table 1. Relative error ratio versus signal efficiency for various boosting algorithms using MiniBooNE data. Differences up to about 0.03 are largely statistical. b=0.5 means the smooth scoring function described in Section 3.

Boosting Algorithms	Parameters β,ϵ (N_{lv}, N_{tr})	Rel. ratios 50% sig. eff.
AdaBoost	0.5 (45,1000)	0.62
AdaBoost	0.8 (45,1000)	0.62
ϵ-Boost	0.03 (45,1000)	0.58
AdaBoost (b=0.5)	0.5 (45,1000)	0.60
ϵ-Boost (b=0.5)	0.03 (45,1000)	0.58
ϵ-LogitBoost	0.01 (45,1000)	0.61
ϵ-HingeBoost	0.01 (30,1000)	0.86
LogitBoost	0.1 (45,150)	0.62
Real AdaBoost	(45,1000)	0.69
Gentle AdaBoost	(45,1000)	0.67
Random Forests(RF)	(400,1000)	0.85
AdaBoosted RF	0.5 (100,1000)	0.66

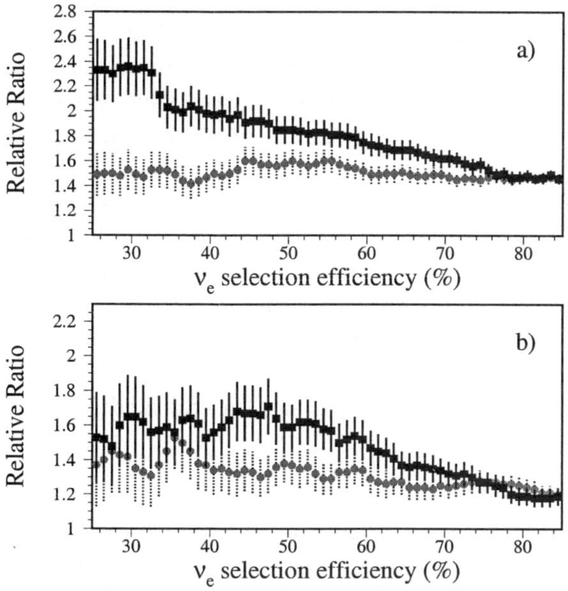

Fig. 3. Comparison of ANN and AdaBoost performance for test samples. Relative ratio (defined as the number of background events kept for ANN divided by the number of background events kept for AdaBoost) versus the intrinsic ν_e charged current quasi-elastic event selection efficiency. a) All kinds of backgrounds are combined for the training against the signal. Dots show the relative ratios for 21 training variables and boxes show them for 52 training variables. b) AdaBoost trained by signal and neutral current π^0 background. Dots show the relative ratios for 22 training variables and boxes show them for 52 training variables. All error bars shown in the figures are for Monte Carlo statistical errors only.

and LogitBoost performed similarly. The Random Forest method uses no boosting, but uses a random fraction of the training events, chosen with replacement, for each tree and a random fraction of the PID variables for each node. For the tests in Table 1, all of the PID variables were used in each node. This option is also known as "bagging". Bagging did poorly compared with AdaBoost, but had performance close to AdaBoost if boosting was added.

Post-Fitting is an attempt to reweight the trees when summing tree scores after all the trees are made. Two post-fitting attempts were made. They produced only a very modest (few percent), if any, gain.

For any of these methods, robustness, the resistance to small inaccuracies between data and training events, is important. In MiniBooNE this is being done by generating several dozen Monte Carlo event samples, each with some parameter varied by about one standard deviation. Individual PID variables which are strongly sensitive to variation are eliminated from the boosting variables. This procedure is not yet complete, but the initial results indicate that the boosting output is then quite robust.

In March 2005, a large change in the detector optical model was made requiring retuning of the reconstructions. The networks trained on the old model were tested on the new versions of the same variables. For a fixed background contamination of π^0 events, the fraction of signal kept dropped by 8.3% for boosting and by 21.4% for ANN.

ANN's tend, in practice, to be quite sensitive to a number of parameters. The temperature, hidden layer(s) size, the learning rate, feedback function, $\cdots$, must be chosen. If one multiplies one of the PID variables by two, or interchanges the order of two variables, or puts a variable in twice, the result is likely to change. For more than twenty–thirty PID variables, tuning is quite difficult and improvement in performance problematic.

For boosting many variables ($\approx$100) can be used. There are only a few parameters to optimize. The MiniBooNE experience is that once β, number of leaves, and number of trees are set, they remain about the same for all uses of boosting within the experiment. If a transformation of the PID variables x is made, $y = f(x)$, such that if $x_2 > x_1$, then $y_2 > y_1$, then the results remain identical, as they depend only on ordering. Interchanging variables or putting the same one in twice has no effect on the results.

3. Convergence Speed of Modifications to the Basic Boosting Algorithm

From Table 1, it is seen that none of the tested options for boosting proved superior to AdaBoost or ϵ-Boost for the MiniBooNE experiment. It is still possible to examine modifications to see if the computer time for convergence using the training set can be reduced. Empirically, reducing the correlations between variables has been found to speed convergence for the MiniBooNE experiment. As seen in Table 1, Random Forests with boosting does not do badly and, if optimized further, may become competitive with AdaBoost, while speeding up convergence.

In the method so far described, the score is taken as $+1$ if an event falls on a signal leaf and -1 if the event falls on a background leaf. This means that if the event falls on a leaf with purity $P = 0.51$ it gets

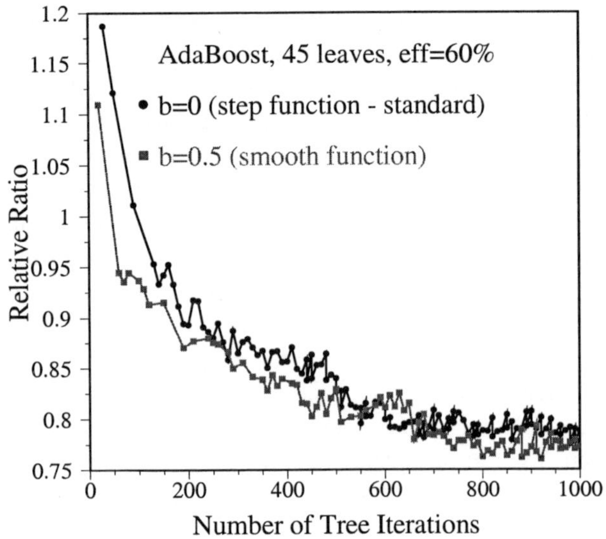

Fig. 4. Performance of AdaBoost with $b = 0$ (step function) and $b = 0.5$ (smooth square root function), $\beta = 0.5$, 45 leaves per tree, versus tree iterations.

the same score as if it fell on a leaf with purity 0.91. A "soft scoring" method can be tried using some function of the purity. Empirically it was found that if $d = 2P - 1$, then $T_m = sign(d)|d|^b$, with $b = 0.5$ worked reasonably well. The results are shown in Figure 4. It is seen that the convergence is faster for soft scoring although the end result is about the same as the standard method. From testing a number of samples it appears that, on the average, the final result is about the same for AdaBoost. There is a hint that soft scoring might be slightly better for ϵ-Boost. Since there seems no disadvantage to using soft scoring, it should be considered when one is using boosting in an analysis.

4. Conclusions

Boosting seems very robust. Given enough iterations, AdaBoost or ϵ-Boost reach an optimum level of classification which is not bettered by any variant tried. For the MiniBooNE Monte Carlo samples, boosting was better than ANN's in our tests by factors between 1.2–2. There are ways, such as smooth scoring, to increase the rate of convergence of the algorithm.

Several techniques were tried for reducing the number of variables. Selecting the variables which were most used as splitting variables seemed to work as well as any of the other methods tried.

Downloads in FORTRAN or C++ are available from:

http://www.gallatin.physics.lsa.umich.edu/~ roe/

References

1. R.E. Schapire, "The strength of weak learnability", Machine Learning 5 (2), 197-227 (1990). First suggested the boosting approach for 3 trees taking a majority vote.
2. Y. Freund, "Boosting a weak learning algorithm by majority", Information and Computation 121 (2), 256-285 (1995) Introduced using many trees.
3. Y. Freund and R.E. Schapire, "Experiments with an new boosting algorithm", Machine Learning: Proceedings of the Thirteenth International Conference, Morgan Kauffman, SanFrancisco, pp.148-156 (1996). Introduced AdaBoost.
4. J. Friedman, T. Hastie, and R. Tibshirani, "Additive logistic regression: a statistical view of boosting", Annals of Statistics 28 (2), 337-407 (2000). Showed that AdaBoost could be looked at as successive approximations to a maximum likelihood solution.
5. T. Hastie, R. Tibshirani, and J. Friedman, "The Elements of Statistical Learning", Springer (2001). Good reference for decision trees and boosting.
6. B.P. Roe, et. al., "Boosted decision trees as an alternative to artificial neural networks for particle identification", NIM A543, pp. 577-584 (2005).
7. Hai-Jun Yang, Byron P. Roe, and Ji Zhu, "Studies of Boosted Decision Trees for MiniBooNE Particle Identification", Physics/0508045, July 2005. Accepted NIM, Sept. 16, 2005.

OPTIMIZATION OF SIGNAL SIGNIFICANCE BY BAGGING DECISION TREES

I. NARSKY

356-48 California Institute of Technology, High Energy Physics, Pasadena, CA 91125, USA
E-mail: narsky@hep.caltech.edu

An algorithm for optimization of signal significance or any other classification figure of merit (FOM) suited for analysis of HEP data is described. This algorithm trains decision trees on many bootstrap replicas of training data with each tree required to optimize the signal significance or any other chosen FOM. New data are then classified by a simple majority vote of the built trees. The performance of the algorithm has been studied using a search for the radiative leptonic decay $B \to \gamma l \nu$ at *BABAR* and shown to be superior to that of all other attempted classifiers including such powerful methods as boosted decision trees. In the $B \to \gamma e \nu$ channel, the described algorithm increases the expected signal significance from 2.4σ obtained by an original method designed for the $B \to \gamma l \nu$ analysis to 3.0σ.

1. Introduction

Various pattern classification tools have been employed in analysis of HEP data to separate signal from background. One of the problems faced by HEP analysts is the indirect nature of available classifiers. In HEP analysis, one typically wants to optimize a FOM expressed as a function of signal and background, S and B, expected in the signal region. An example of such FOM is signal significance, $S/\sqrt{S+B}$, often used by physicists to express the cleanliness of the signal in the presence of statistical fluctuations of observed signal and background. None of the available popular classifiers optimizes this FOM directly. Commercial implementations of decision trees, such as CART[1], split training data into signal- and background-dominated rectangular regions using the Gini index, $Q = 2p(1-p)$, as the optimization criterion, where p is the correctly classified fraction of events in a tree node. Neural networks[2] typically minimize a quadratic classification error, $\sum_{n=1}^{N}(y_n - f(x_n))^2$, where y_n is the true class of an event, -1 for background and 1 for signal, $f(x_n)$ is the continuous value of the neural network prediction in the range $[-1,1]$, and the sum is over N events in the training data set. Similarly, AdaBoost[3] minimizes an exponential classification error, $\sum_{n=1}^{N}\exp(-y_n f(x_n))$. These optimization criteria are not necessarily optimal for maximization of the signal significance. The usual solution is to build a neural net or an AdaBoost classifier and then find an optimal cut on the continuous output of the classifier to maximize the signal significance. Alternatively, one could construct a decision tree with many terminal nodes and then combine these nodes to maximize the signal significance.

Decision trees in StatPatternRecognition[4, 5] allow the user to optimize any FOM supplied as an implementation of an abstract C++ interface included in the package. A default implementation of the decision tree includes both standard figures of merit used for conventional decision trees such as the Gini index and HEP-specific figures of merit such as the signal significance or the signal purity, $S/(S+B)$.

A decision tree, even if it directly optimizes the desired FOM, is rarely powerful enough to achieve a good separation between signal and background. The mediocre predictive power of a single decision tree can be greatly enhanced by one of the two popular methods for combining classifiers — boosting[3] and bagging[6]; the latter approach can be used in conjunction with the random forest technology[7]. This note compares predictive power of several classifiers using a search for the radiative leptonic decay $B \to \gamma l \nu$ at BABAR. It is shown that the greatest signal significance is obtained by bagging an ensemble of decision trees, with each member of the ensemble optimizing the signal significance. This study is described in more detail in two notes[4, 5] posted at the physics archive.

2. Decision Trees in StatPatternRecognition

A decision tree recursively splits training data into rectangular regions (nodes). For each node, the tree examines all possible binary splits in each dimension and selects the one with the highest FOM. This procedure is repeated until a stopping criterion, specified as the minimal number of events per tree node, is satisfied. The tree continues making new nodes until it is composed of leaves only — nodes that cannot be split without a decrease in the FOM and nodes that cannot be split because they have too few events.

As mentioned above, a conventional decision tree often uses the Gini index, $Q(p,q) = -2pq$, for split optimization, where p and $q = 1 - p$ are fractions of correctly classified and misclassified events in a given node. If a parent node with the total event weight W is split into two daughter nodes with weights W_1 and $W_2 = W - W_1$, the best decision split is chosen to maximize $Q_{\text{split}} = (W_1 Q_1 + W_2 Q_2)/W$, where Q_1 and Q_2 are figures of merit computed for the two daughter nodes. Note that a conventional decision tree treats the two categories, signal and background, symmetrically. In HEP analysis, one usually wishes to optimize an asymmetric FOM. StatPatternRecognition offers a modified splitting algorithm for this purpose. The best decision split is now chosen to maximize $Q_{\text{split}} = \max(Q_1, Q_2)$, where Q_1 and Q_2 are the asymmetric figures of merit for the daughter nodes. In case of the signal significance, the FOM is given by $Q(s,b) = s/\sqrt{s+b}$, where s and b are signal and background weights in a given node. After the tree is grown, the terminal nodes are merged to optimize the overall asymmetric FOM. The merging algorithm sorts all terminal nodes by signal purity in descending order and computes the overall FOM for the n first nodes in the sorted list with n taking consecutive values from 1 to the full length of the list. The optimal combination of the terminal nodes is given by the highest FOM computed in this manner.

This algorithm for optimization of an asymmetric FOM is nothing but an empirical solution. It is not guaranteed that this algorithm will produce a higher asymmetric FOM than the one obtained by a conventional decision tree using the Gini index or any other symmetric expression as a split criterion. It has been shown experimentally that this algorithm tends to produce higher values of the signal significance when applied to physics data sets. This note is an example of such an application.

3. Bagging Decision Trees

The predictive power of a single classifier can be enhanced by boosting[3] or bagging[6]. Both these methods work by training many classifiers, e.g., decision trees, on variants of the original training data set. A boosting algorithm enhances weights of misclassified events and reduces weights of correctly classified events and trains a new classifier on the reweighted sample. In contrast, bagging algorithms do not reweight events. Instead, they train new classifiers on bootstrap replicas of the training set. After training is completed, events are classified by the majority vote of the trained classifiers. For successful application of the bagging algorithm, the underlying classifier must be sensitive to small changes in the training data. Otherwise all trained classifiers will be similar, and the performance of the single classifier will not be improved. This condition is satisfied by a decision tree with fine terminal nodes. Because of the small node size each decision tree is significantly overtrained; if the tree were used just by itself, its predictive power on a test data set would be quite poor. However, because the final decision is made by the majority vote of all the trees, the algorithm delivers a high predictive power.

Random forest[7], typically used in conjunction with bagging, is a technique that randomly selects a subset of input variables for each decision split. This approach can make individual trees more independent of each other and increase the overall predictive power.

Boosting and bagging algorithms offer competitive predictive power. It is really hard, if possible, to predict outright which algorithm will perform better in any classification problem. For optimization of the signal significance, however, bagging is the choice favored by intuition. Reweighting events has an unclear impact on the effectiveness of the optimization routine with respect to the chosen asymmetric FOM. While it may be possible to design a reweighting algorithm efficient for optimization of a specific FOM, at present such reweighting algorithms are not known. Bagging, on the other hand, offers an obvious solution. If the base classifier directly optimizes the chosen FOM, bagging is equivalent to optimization of this FOM integrated over bootstrap replicas.

4. Separation of Signal and Background in a Search for the Radiative Leptonic Decay $B \to \gamma l \nu$ at BABAR

A search for the radiative leptonic decay $B \to \gamma l \nu$ is currently in progress at BABAR; results of this analysis will be made available to the public in the near future. The analysis focuses on measuring the B meson decay constant, f_B, which has not been previously measured.

Several samples of simulated Monte Carlo (MC) events are used to study signal and background signatures in this analysis: $B \to \gamma l \nu$ signal samples with about 1.2M events in each channel, large samples of generic B^+B^-, $B^0\bar{B}^0$, $c\bar{c}$, uds and $\tau^+\tau^-$ MC events, as well as several exclusive semileptonic modes generated separately with a typical sample size of several hundred thousand events.

Various preliminary requirements have been imposed to enhance the signal purity and at the same time reduce the MC samples to a manageable size. After these preliminary requirements have been imposed, eleven variables are included in the final optimization procedure. Distributions of these variables and more details on applied selection requirements can be found elsewhere[4].

The signal and combined background MC samples are used by various optimization algorithms to maximize the signal significance expected in 210 fb^{-1} of data. The training samples used for this optimization consist of roughly half a million signal and background MC events in both electron and muon channels, appropriately weighted according to the integrated luminosity observed in the data. The training:validation:test ratio for the sample sizes is 2:1:1. Signal MC samples are weighted assuming a branching fraction of 3×10^{-6} for each channel.

The authors of this analysis deploy an original cut optimization routine[4] for separation of signal and background. This procedure divides the available range for each variable into intervals of preselected length and finds an optimal set among all possible combinations of orthogonal cuts. Besides the original method designed by the analysts, several classifiers have been used:

- Decision tree optimizing the signal significance $S/\sqrt{S+B}$.
- Bump hunter[8] optimizing the signal significance.
- 700 boosted binary splits.
- 50 boosted decision trees with minimal node size 100 events.
- Combiner of subclassifiers trained on individual background components using boosted binary splits.
- 100 bagged decision trees with each tree optimizing the signal significance; the minimal node size has been set to 100 events.

Parameters of all classifiers have been optimized by comparing values of the statistical significance obtained for the validation samples.

Results are shown in Table 1. The output of the described bagging algorithm for the $B \to \gamma e \nu$ test data is shown in Fig. 1. The bagging algorithm provides the best value of the signal significance. It gives a 24% improvement over the original method developed by the analysts, and a 14% improvement over boosted decision trees; both numbers are quoted for the $B \to \gamma e \nu$ channel.

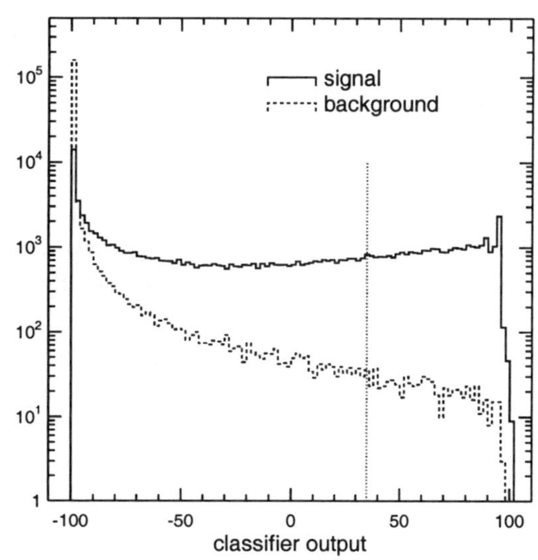

Fig. 1. Output of the bagging algorithm with 100 trained decision trees for the $B \to \gamma e \nu$ test sample. The cut maximizing the signal significance, obtained using the validation sample, is shown with a vertical line.

The bagging algorithm with decision trees optimizing the Gini index showed an 8% improvement in the $B \to \gamma e \nu$ signal significance compared to the boosted decision trees. But the signal signifi-

Table 1. Signal significances, $\mathcal{S}_{\text{train}}$, $\mathcal{S}_{\text{valid}}$, and $\mathcal{S}_{\text{test}}$, for the $B \to \gamma l \nu$ training, validation, and test samples obtained with various classification methods. The signal significance computed for the test sample should be used to judge the predictive power of the included classifiers. W_1 and W_0 represent the signal and background, respectively, expected in the signal region after the classification criteria have been applied; these two numbers have been estimated using the test samples. All numbers have been normalized to the integrated luminosity of 210 fb^{-1}. The best value of the expected signal significance is shown in boldface.

Method	$B \to \gamma e \nu$					$B \to \gamma \mu \nu$				
	$\mathcal{S}_{\text{train}}$	$\mathcal{S}_{\text{valid}}$	$\mathcal{S}_{\text{test}}$	W_1	W_0	$\mathcal{S}_{\text{train}}$	$\mathcal{S}_{\text{valid}}$	$\mathcal{S}_{\text{test}}$	W_1	W_0
Original method	2.66	-	2.42	37.5	202.2	1.75	-	1.62	25.8	227.4
Decision tree	3.28	2.72	2.16	20.3	68.1	1.74	1.63	1.54	29.0	325.9
Bump hunter with one bump	2.72	2.54	2.31	47.5	376.6	1.76	1.54	1.54	31.7	393.8
Boosted binary splits	2.53	2.65	2.25	76.4	1077.3	1.66	1.71	1.44	45.2	935.6
Boosted decision trees	13.63	2.99	2.62	58.0	432.8	11.87	1.97	1.75	41.6	523.0
Combiner of background subclassifiers	3.03	2.88	2.49	83.2	1037.2	1.84	1.90	1.66	55.2	1057.1
Bagged decision trees	9.20	3.25	**2.99**	69.1	465.8	8.09	2.07	**1.98**	49.4	571.1

cance obtained with this method was 9% worse than that obtained by the bagging algorithm with decision trees optimizing the signal significance. The 14% improvement of the proposed bagging algorithm over the boosted decision trees therefore originated from two sources: 1) using bagging instead of boosting, and 2) using the signal significance instead of the Gini index as a FOM for the decision tree optimization.

In an attempt to improve the signal significance even further, the random forest approach has been attempted with the number of randomly sampled (with replacement) input variables taking values 1, 6, and 11. No significant improvement over the bagging algorithm has been found.

This note describes a somewhat unusual application of boosted and bagged decision trees to data analysis with the ultimate goal of classification defined as maximization of the signal significance. The classifier performance in this case is driven by a small fraction of the data set included in the signal region. In a typical application of boosted decision trees, one minimizes the exponential loss averaged over the whole data set. The optimal node size for boosted decision trees is typically much larger than the optimal node size for bagged decision trees. In this analysis, the optimal node sizes for both boosted and bagged decision trees are comparable.

5. Summary

A bagging algorithm suitable for optimization of an asymmetric FOM for HEP analyses has been described. This algorithm has been shown to give a significant improvement of the signal significance in the search for the radiative leptonic decay $B \to \gamma l \nu$ at BABAR.

Acknowledgments

Thanks to Gregory Dubois-Felsmann, Byron Roe and Frank Porter for useful discussions and comments on this work. Thanks to Ed Chen for data and documentation on the $B \to \gamma l \nu$ analysis. Thanks to Harrison Prosper for presenting this work at Phystat 2005. This work is partially supported by Department of Energy under Grant DE-FG03-92-ER40701.

References

1. L. Breiman et al., *Classification and Regression Trees*, Waldsworth International, 1984.
2. S. Haykin, *Neural Networks: A Comprehensive Foundation*, Prentice Hall, 1999.
3. Y. Freund and R.E. Schapire, *A decision-theoretic generalization of on-line learning and an application to boosting*, J. of Computer and System Sciences **55**, 119-139 (1997).
4. I. Narsky, *StatPatternRecognition: A C++ Package for Statistical Analysis of High Energy Physics Data*, physics/0507143, 2005.
5. I. Narsky, *Optimization of Signal Significance by Bagging Decision Trees*, physics/0507157, 2005.
6. L. Breiman, *Bagging Predictors*, Machine Learning **26**, 123-140 (1996).
7. L. Breiman, *Random Forests*, Machine Learning **45**, 5-32 (2001).
8. J. Friedman and N. Fisher, *Bump hunting in high dimensional data*, Statistics and Computing **9**, 123-143 (1999).

NONPARAMETRIC BAYESIAN CLASSIFICATION WITH MASSIVE DATASETS: LARGE-SCALE QUASAR DISCOVERY

ALEXANDER GRAY
Georgia Institute of Technology, College of Computing
agray@cc.gatech.edu

GORDON RICHARDS
Princeton University, Department of Astrophysical Sciences
gtr@astro.princeton.edu

ROBERT NICHOL
University of Portsmouth, Institute of Cosmology and Gravitation
bob.nichol@port.ac.uk

ROBERT BRUNNER
University of Illinois, Department of Astronomy and NCSA
rb@ncsa.uiuc.edu

ANDREW MOORE
Carnegie Mellon University, School of Computer Science
awm@cs.cmu.edu

The kernel discriminant (a nonparametric Bayesian classifier) is appropriate for many scientific tasks because it is highly accurate (it approaches Bayes optimality as you get more data), distribution-free (works for arbitrary data distributions), and it is easy to inject prior domain knowledge into it and interpret what it's doing. Unfortunately, like other highly accurate classifiers, it is computationally infeasible for massive datasets. We present a fast algorithm for performing classification with the kernel discriminant exactly (i.e. without introducing any approximation error). We demonstrate its use for quasar discovery, a problem central to cosmology and astrophysics, tractably using 500K training data and 800K testing data from the Sloan Digital Sky Survey. The resulting catalog of 100K quasars significantly exceeds existing quasar catalogs in both size and quality, opening a number of new scientific possibilities, including the recent empirical confirmation of cosmic magnification which has received wide attention.

1. Introduction and Approach

Quasars are star-like objects which are not very well understood yet play a critical role in cosmology. As the most luminous (and thus the most distant) objects in the universe, they can be used as markers of the mass in the distant (early) universe. With the very recent advent of massive sky surveys such as the Sloan Digital Sky Survey (SDSS), it is now conceivable in principle to obtain a catalog of the locations of quasars which is more comprehensive than ever before, both in sky coverage and depth (distance). Such a catalog would open the door to numerous powerful analyses of the early/distant universe which were never before possible. A central challenge of this activity is the question of how to use the limited information we have in hand (a tiny set of known, nearby quasars) to extract a massive amount of more subtle information from the SDSS dataset (a large set of faint quasar candidates). In this paper we describe a method which has yielded the most comprehensive and accurate quasar catalog to date. The catalog and data methodology have been previously described [1]; here we describe the algorithm for the first time.

Nonparametric Bayesian classification. We wish to classify a set of unlabeled objects (the *test set*, or *query points*) as either stars or quasars. We first create samples of "stars" and "quasars" that will serve as *training sets* or *reference points.* The probability that an object producing data x (represented by four color measurements) is a star is proportional to the probability that x would be produced by a star, $p(x|C_1)$; this is the *likelihood* under the probability density function (pdf), which must be estimated, describing the star class C_1.

To incorporate prior or subjective information, in our case the fraction of an unseen set of objects which the user roughly expects to be stars, we use a simple application of Bayes' Rule:

$$P(C_1|x) = \frac{p(x|C_1)P(C_1)}{p(x|C_1)P(C_1) + p(x|C_2)P(C_2)}.$$

Objects with $P(C_1|x) > 0.5$ are classified as stars, which we denote by assigning class label $c(x) = C_1$, otherwise as quasars.

Kernel density estimator. For the likelihood of each class we will use the nonparametric *kernel density estimate* (KDE) [2] of the pdf, a mature statistical method which can be thought of as a generalization of the histogram, having the form

$$p(x|C) = \frac{1}{N}\sum_{i}^{N} K_h(x, x_i)$$

where the kernel is for example, the Gaussian $K_h(x, x_i) = K(h, \|x - x_i\|) = \frac{1}{C(h)} \exp \|x - x_i\|^2/h^2$ where $C(h)$ is a normalizing constant which depends on h. h, called the *bandwidth*, is the critical parameter which controls the smoothness of the estimate. In our method we use a slight generalization in which each point may be given different weights: $p(x|C) = \frac{1}{W}\sum_i^N w_i K_h(x, x_i)$ where $W = \sum_i^N w_i$, in order to possibly allow for measurement uncertainty or other prior knowledge. In the algorithm we'll work with unnormalized sums such as $\Phi(x|C) = \sum_i^N w_i K_h(x, x_i)$ so that $p(x|C) = \frac{1}{W}\Phi(x|C)$. We'll compress the notation by referring to $p(C_1|x)$ and $\Phi(x|C_1)$ as $p_1(x)$ and $\Phi_1(x)$. We refer to the Bayes classifier using KDE estimates as the *kernel discriminant* and such classification as *kernel discriminant analysis* (KDA).

Computational challenge. Training the kernel discriminant consists of finding the parameters $\{h_1, h_2\}$ which will maximize its performance at predicting the class labels of data drawn from the same distribution as the training data. We use as an estimator of this performance the leave-one-out cross-validated accuracy score. This form of classifier is highly accurate in practice. The main reason it is not commonly used is that it comes with a severe quadratic computational cost. The main problem treated in this paper is that of computing KDA classifications tractably.

2. Algorithm

We developed an algorithm which computes for each query point x_q its class label $c(x_q)$ as if the sums $\Phi_1(x)$ and $\Phi_2(x)$ had been computed *exactly*, though in many cases they need not be.

First, a space-partitioning tree data structure such as a *kd*-tree [3] is constructed on the query (testing) dataset, and another is created on the reference (training) dataset.

The idea is to maintain bounds on $p_1(x)$ and $p_2(x)$ and successively tighten them in a multiresolution manner, as nodes at increasingly finer levels of the trees are considered, until we can show that the bounds determine that one of these class probabilities must dominate the other. This is true if one is definitely greater than 0.5, definitely less than 0.5, or definitely greater than the other. Initially the class label for each query point $c(x_q)$ is recorded as "?" (unknown), and is updated to C_1 or C_2 when the bounds determine it. Efficiency over the naive algorithm is obtained to the extent that we are able to determine the label for large chunks of the query points simultaneously.

Bounds. We'll maintain various bounds during the run of the algorithm, including bounds on $\Phi_1(x)$ and $\Phi_2(x)$, *e.g.* $\Phi_1^L(x) \leq \Phi_1(x)$ and $\Phi_1^U(x) \geq \Phi_1(x)$, and bounds on $p_1(x)$ and $p_2(x)$, *e.g.* $p_1^L(x) \leq p_1(x)$ and $p_1^U(x) \geq p_1(x)$.

We'll also maintain bounds which hold for various subsets X of the points, which correspond to tree nodes, *e.g.* $\forall x \in X$: $\Phi_1^L(X) \leq \Phi_1(x)$, $\Phi_1^U(X) \geq \Phi_1(x)$, $p_1^L(X) \leq p_1(x)$, and $p_1^U(X) \geq p_1(x)$. We can utilize bounds on the class-conditional likelihoods to obtain simple bounds on the final class probability:

$$p_1^L(x) := \left(\Phi_1^L(x)\pi_1\right) / \left(\Phi_1^L(x)\pi_1 + \Phi_2^U(x)\pi_2\right)$$
$$p_1^U(x) := \left(\Phi_1^U(x)\pi_1\right) / \left(\Phi_1^U(x)\pi_1 + \Phi_2^L(x)\pi_2\right).$$

Within each node X, in an efficient bottom-up (dynamic programming) fashion, we compute and store certain properties of the class 1 points (if any) and the class 2 points (if any) which reside in the node: for each class, the bounding box of the points in that class and the sum of the weights of the points in that class, $W_1(X)$ and $W_2(X)$. Note that expressions like $W_1(X)$ are generally implemented as $X.W_1$, to use a C-like notation.

We can use these bounding boxes to compute simple lower and upper bounds on the distance between any point in a node Q and any point (of a certain class) in a node R, *e.g.*: $\forall x_q \in Q, \forall x_r \in R$ such that $c(x_r) = C_1$: $\delta_1^L(Q,R) \leq \delta_{qr}$ and $\delta_1^U(Q,R) \geq \delta_{qr}$, where $\delta_{qr} = \|x_q - x_r\|$.

Bound tightening. Let K^L and K^U be constants such that $\forall x, y$: $K^L \leq K(x,y)$ and $K^U \geq K(x,y)$ – for most kernels of interest such as the Gaussian, which are probability density functions, the lower bound is 0 and upper bound is 1. At the beginning of a run of the algorithm, the bounds are initialized using these values, *e.g.*: $\forall x_q$: $\Phi_1^L(x_q) = W_1 K^L$ and $\Phi_1^U(x_q) = W_1 K^U$. For each query x_q, the bounds $\Phi_1^L(x)$ and $\Phi_1^U(x)$ have accounted for each reference point's potential contribution to the sum in a worst-case manner.

Nodes are examined in pairs $\{Q,R\}$ – one node Q from the query tree and one node R from the reference tree. The idea is that when we see a new reference node R, we can tighten our bounds on the contribution of the reference points in R to the sum for each query point. When doing so, we must also undo the previous contribution of the points in R to each of our bounds. For example the new contribution of R to $\Phi_1^L(Q)$ is $W_1(R)K(h_1, \delta_1^U(Q,R))$ whereas the old contribution was implicitly $W_1(R)K^L$. So we update $\Phi_1^L(Q)$ by adding to it

$$\Delta\Phi_1^L(Q,R) := W_1(R)K_{h_1}(\delta_1^U(Q,R)) - W_1(R)K^L.$$

Similarly, we change $\Phi_1^U(Q)$ by adding to it

$$\Delta\Phi_1^U(Q,R) := W_1(R)K_{h_1}(\delta_1^L(Q,R)) - W_1(R)K^U.$$

Because we always move downward in the tree, these updates are always improvements to the bounds or at worst leave them unchanged.

Control flow. The order in which nodes are examined is determined by a min-priority queue which stores node-pair objects $\{Q,R\}$. Note that values such as $\Delta\Phi_1^L(\{Q,R\})$ ($\Delta\Phi_1^L$ for short) are often implemented as $\{Q,R\}.\Delta\Phi_1^L$. A node-pair object stores the change values that are computed for it, the *previous* such values (denoted by apostrophes), and its priority.

Node-pair $\{Q,R\}$ is assigned priority

$$f(Q,R) := |(\Delta\Phi_1^U - \Delta\Phi_1^{U'}) + (\Delta\Phi_2^U - \Delta\Phi_2^{U'}) \\ -(\Delta\Phi_1^L - \Delta\Phi_1^{L'}) - (\Delta\Phi_2^L - \Delta\Phi_2^{L'})|,$$

the difference in improvement (*i.e.* current values minus previous values) of the upper bounds and the lower bounds. A procedure **makePair**$(Q,R,\ldots)$ creates the node-pair structure $\{Q,R\}$ and stores the other arguments in its slots for $\Delta\Phi_1^{L'}$, $\Delta\Phi_1^{U'}$, $\Delta\Phi_2^{L'}$, $\Delta\Phi_2^{U'}$, respectively. **computeBounds**(Q,R) computes the Δ values and the priority for node-pair $\{Q,R\}$. Node-pairs are expanded further by placing every pairwise combination of their respective children on the queue. Each node-pair also stores an "undo" flag undo(Q,R) which determines whether it should be expanded. Whenever improvements are made to the bounds of a query node, they are updated in all the children of the query node with a simple recursive routine **passDown**$(Q,\ldots)$. For each node Q in the query tree we store $M(Q)$, the number of points in the tree which have known class labels (definitely C_1 or C_2). If we encounter a node for which all $N(Q)$ of the query points have known labels, we can stop recursing on it.

When both Q and R are leaf nodes, this corresponds to the base case of the recursion. In this case we compute the contribution of each point in R to each point in Q exhaustively. Because this direct type of contribution is exact and thus unchangeable, while other contributions tighten bounds which can change, it is useful to record it separately – we denote it by $\phi(x_q)$ for each query point.

In the pseudocode, for brevity, we use the convention that children of leaves point to themselves, and redundant node pairs are not placed on the priority queue. In the pseudocode, following a C-like notation for compactness, the a += b denotes $a = a + b$ and a != b denotes $\overline{(a = b)}$. The pseudocode shows a version of the algorithm which only computes and uses the bounds for one of the classes.

3. Results

Using the algorithm described, we were able tractably to estimate (find optimal parameters for) a classifer based on a large training set consisting of 500K star-labeled objects and 16K quasar-labeled objects, and predict the label for 800K faint (up to $g = 21$) query objects from 2099 deg^2 of the SDSS DR1 imaging dataset. Of these, 100K were predicted to be quasars, forming our catalog of quasar candidates. This significantly exceeds the size, faintness, and sky coverage of the largest quasar catalogs to date. Based on spectroscopic hand-validation of a subset of the candidates, we estimate that 95.0% are truly quasars, and that we have identified 94.7% of the actual quasars. These efficiency and completeness numbers far exceed those any previous catalog, making our catalog both the most comprehensive and most accurate to date. The recent empirical confirmation of cosmic magnification [4] using our catalog is an example of the scientific possibilities opened up by this work. In ongoing efforts we are exploring ways to make the method both more computationally and statistically efficient, with the goal of obtaining all 1.6M quasars we estimate are detectable in principle from the entire SDSS dataset.

References

1. Richards, G., Nichol, R., Gray, A., *et. al*, Efficient Photometric Selection of Quasars from the Sloan Digital Sky Survey: 100,000 $z > 3$ Quasars from Data Release One, *Astrophysical Journal Supplement Series* 155, 2, 257–269, 2004.
2. Silverman, B. W., Density Estimation for Statistics and Data Analysis, Chapman and Hall/CRC, 1986.
3. Preparata, F. P. and Shamos, M., *Computational Geometry*, Springer-Verlag, 1985.
4. Scranton, R., *et. al*, Detection of Cosmic Magnification with the Sloan Digital Sky Survey, *Astrophysical Journal* 633, 2, 589–602, 2005. Described in *Nature*, April 27, 2005.

kdaBase(Q, R)
forall $x_q \in Q$,
if $c(x_q) =$ "?",
forall $x_r \in R$,
if $c(x_q) = C_1$, $\phi_1(x_q) \mathrel{+}= w_r K_{h_1}(\delta_{qr})$.
if $c(x_q) = C_2$, $\phi_2(x_q) \mathrel{+}= w_r K_{h_1}(\delta_{qr})$.

$\Phi_1^L(x_q) := C(h_1)(\Phi_1^L(Q) + \phi_1(x_q) - W_1(R)K^L)$.
$\Phi_2^L(x_q) := C(h_2)(\Phi_2^L(Q) + \phi_2(x_q) - W_2(R)K^L)$.
$\Phi_1^U(x_q) := C(h_1)(\Phi_1^U(Q) + \phi_1(x_q) - W_1(R)K^U)$.
$\Phi_2^U(x_q) := C(h_2)(\Phi_2^U(Q) + \phi_2(x_q) - W_2(R)K^U)$.

$p_1^L(x_q) := \Phi_1^L(x_q)\pi_1/(\Phi_1^L(x_q)\pi_1 + \Phi_2^U(x_q)\pi_2)$.
$p_1^U(x_q) := \Phi_1^U(x_q)\pi_1/(\Phi_1^U(x_q)\pi_1 + \Phi_2^L(x_q)\pi_2)$.

if $p_1^L(x_q) \geq 0.5$, $c(x_q) := C_1$.
if $p_1^U(x_q) < 0.5$, $c(x_q) := C_2$.
if $c(x_q)$!= "?", $M(Q) \mathrel{+}= 1$.

$\phi_1^L(Q) := \min_{x_q \in Q} \phi_1(x_q)$.
$\phi_2^L(Q) := \min_{x_q \in Q} \phi_2(x_q)$.
$\phi_1^U(Q) := \max_{x_q \in Q} \phi_1(x_q)$.
$\phi_2^U(Q) := \max_{x_q \in Q} \phi_2(x_q)$.

$\Phi_1^L(Q) \mathrel{-}= W_1(R)K^L$, $\Phi_2^L(Q) \mathrel{-}= W_2(R)K^L$.
$\Phi_1^U(Q) \mathrel{-}= W_1(R)K^U$, $\Phi_2^U(Q) \mathrel{-}= W_2(R)K^U$.

kda(Q_{root}, R_{root})
$\{Q_{root}, R_{root}\} :=$ **makePair**$(Q_{root}, R_{root}, 0,0,0,0)$.
computeBounds$(\{Q_{root}, R_{root}\})$.
insertHeap$(H, \{Q_{root}, R_{root}\})$.

while H is not empty,
$\{Q, R\} :=$ **extractMin**(H).
if $M(Q) = N(Q)$, skip.
if !leaf(Q),
$\Phi_1^L(Q) := \min(\Phi_1^L(\text{ch}_1(Q)), \Phi_1^L(\text{ch}_2(Q)))$.
$\Phi_2^L(Q) := \min(\Phi_2^L(\text{ch}_1(Q)), \Phi_2^L(\text{ch}_2(Q)))$.
$\phi_1^L(Q) := \min(\phi_1^L(\text{ch}_1(Q)), \phi_1^L(\text{ch}_2(Q)))$.
$\phi_2^L(Q) := \min(\phi_2^L(\text{ch}_1(Q)), \phi_2^L(\text{ch}_2(Q)))$.
(similar for upper bounds)
$M(Q) := M(\text{ch}_1(Q)) + M(\text{ch}_2(Q))$.

$\Delta\Phi_1^L := \Delta\Phi_1^L(\{Q, R\}), \ldots$
if undo(Q, R), **passDown**$(Q, \Delta\Phi_1^L - \Delta\Phi_1^{L'}$,
$\Delta\Phi_1^U - \Delta\Phi_1^{U'}, \Delta\Phi_2^L - \Delta\Phi_2^{L'}, \Delta\Phi_2^U - \Delta\Phi_2^{U'})$.
else, **passDown**$(Q, \Delta\Phi_1^L, \Delta\Phi_1^U, \Delta\Phi_2^L, \Delta\Phi_2^U)$.

$\Phi_1^L := C(h_1)(\Phi_1^L(Q) + \phi_1^L(Q))$.
$\Phi_2^L := C(h_2)(\Phi_2^L(Q) + \phi_2^L(Q))$.
$\Phi_1^U := C(h_1)(\Phi_1^U(Q) + \phi_1^U(Q))$.
$\Phi_2^U := C(h_2)(\Phi_2^U(Q) + \phi_2^U(Q))$.
$p_1^L(Q) = \Phi_1^L\pi_1/(\Phi_1^L\pi_1 + \Phi_2^U\pi_2)$.
$p_1^U(Q) = \Phi_1^U\pi_1/(\Phi_1^U\pi_1 + \Phi_2^L\pi_2)$.

if $p_1^L(Q) \geq 0.5$, $c(Q) := C_1$.
if $p_1^U(Q) < 0.5$, $c(Q) := C_2$.
if $c(Q)$!= "?",
$M(Q) := N(Q)$. skip.

if leaf(Q) and leaf(R),
passDown$(Q, \Delta\Phi_1^L, \Delta\Phi_1^U, \Delta\Phi_2^L, \Delta\Phi_2^U)$.
kdaBase(Q, R).
else,
$\{\text{ch}_1(Q), \text{ch}_1(R)\} :=$ **makePair**$(\text{ch}_1(Q)$,
$\text{ch}_1(R), \Delta\Phi_1^L, \Delta\Phi_1^U, \Delta\Phi_2^L, \Delta\Phi_2^U)$.
$\{\text{ch}_1(Q), \text{ch}_2(R)\} :=$ **makePair**$(\text{ch}_1(Q)$,
$\text{ch}_2(R), \Delta\Phi_1^L, \Delta\Phi_1^U, \Delta\Phi_2^L, \Delta\Phi_2^U)$.
computeBounds$(\{\text{ch}_1(Q), \text{ch}_1(R)\})$.
computeBounds$(\{\text{ch}_1(Q), \text{ch}_2(R)\})$.
if $\rho(\{\text{ch}_1(Q), \text{ch}_1(R)\}) < \rho(\{\text{ch}_1(Q), \text{ch}_2(R)\})$
undo$(\{\text{ch}_1(Q), \text{ch}_1(R)\}) :=$ true.
else, undo$(\{\text{ch}_1(Q), \text{ch}_2(R)\}) :=$ true.
insertHeap$(H, \{\text{ch}_1(Q), \text{ch}_1(R)\})$.
insertHeap$(H, \{\text{ch}_1(Q), \text{ch}_2(R)\})$.

$\{\text{ch}_2(Q), \text{ch}_1(R)\} :=$ **makePair**$(\text{ch}_2(Q)$,
$\text{ch}_1(R), \Delta\Phi_1^L, \Delta\Phi_1^U, \Delta\Phi_2^L, \Delta\Phi_2^U)$.
$\{\text{ch}_2(Q), \text{ch}_2(R)\} :=$ **makePair**$(\text{ch}_2(Q)$,
$\text{ch}_2(R), \Delta\Phi_1^L, \Delta\Phi_1^U, \Delta\Phi_2^L, \Delta\Phi_2^U)$.
computeBounds$(\{\text{ch}_2(Q), \text{ch}_1(R)\})$.
computeBounds$(\{\text{ch}_2(Q), \text{ch}_2(R)\})$.
if $\rho(\{\text{ch}_2(Q), \text{ch}_1(R)\}) < \rho(\{\text{ch}_2(Q), \text{ch}_2(R)\})$
undo$(\{\text{ch}_2(Q), \text{ch}_1(R)\}) :=$ true.
else, undo$(\{\text{ch}_2(Q), \text{ch}_2(R)\}) :=$ true.
insertHeap$(H, \{\text{ch}_2(Q), \text{ch}_1(R)\})$.
insertHeap$(H, \{\text{ch}_2(Q), \text{ch}_2(R)\})$.

BAYESIAN NEURAL NETWORKS

PUSHPALATHA C. BHAT[1] AND HARRISON B. PROSPER[2]
[1]*Fermi National Accelerator Laboratory, Batavia, Illinois, USA*
[2]*Florida State University, Tallahassee, Florida, USA*

The training of neural networks can be viewed as a problem of inference, which can be addressed from a Bayesian viewpoint. This perspective leads to a method, new to the field of particle physics, called Bayesian neural networks (BNN). After a brief overview of the method we illustrate how it can be usefully deployed in particle physics research.

1. Introduction

Neural networks (NN) [1] are non-linear functions that, in principle, can model *any* (smooth) map of a set of one or more real input variables to a set of one or more outputs [2]. In this paper, we consider neural networks for binary classification. The typical application in particle physics is to separate signal from background. We also make some brief remarks on networks for regression, that is, for fitting functions. We end with an example of classification from particle physics.

2. Classification

If a network is trained with events, described by a vector of variables x, such that signal events are labeled by $t = 1$ and background events by $t = 0$, then the network output y approximates the posterior probability [3]

$$y \approx \text{Prob}(t = 1|x) = \frac{p(x|1)p(1)}{p(x|1)p(1) + p(x|0)p(0)}, \quad (1)$$

that is, the probability that an event defined by the variables x belongs to the signal class $t = 1$. $p(x|1)$ and $p(x|0)$ are the probability density functions for the signal and background classes, respectively, and $p(1)$ and $p(0)$ are the corresponding class prior probabilities. Typically, one trains with equal numbers of signal and background events, in which case the priors cancel out. The label t is referred to as the *target*.

The idea behind Bayesian neural networks (BNN) is to cast the task of training a network as a problem of inference, which is solved using Bayes' theorem. The latter is used to assign a probability density to each point w in the parameter space of the neural network. Each point w corresponds to a network defined by a specific set of parameter values. In the standard methods for training neural networks, one finds a *single* point w_0 in the parameter space, that is, a single network. In the Bayesian approach, one performs a weighted average over all points, that is, all networks. As with the standard methods, the BNN methods make use of training data $\{(t_1, x_1), \cdots, (t_N, x_N)\}$, where t_i is the *known* label associated with data x_i. The probability density assigned to point w, that is, to a network, is given by Bayes' theorem

$$\begin{aligned} p(w|t,x) &= \frac{p(t,x|w)p(w)}{p(t,x)}, \\ &= \frac{p(t|x,w)p(x|w)p(w)}{p(t|x)p(x)}, \\ &= \frac{p(t|x,w)p(w)}{p(t|x)}, \end{aligned} \quad (2)$$

where we have assumed that the data x do not depend on w, in which case $p(x|w) = p(x)$. Thus, in order to assign a probability density to a network, defined by the point w, we need the likelihood $p(t|x,w)$ and the prior density $p(w)$.

Consider a *class* of neural networks defined by the functional form

$$y(x,w) = \frac{1}{1 + \exp[-f(x,w)]}, \quad (3)$$

where

$$f(x,w) = b + \sum_{j=1}^{H} v_j \tanh(a_j + \sum_{i=1}^{P} u_{ij}\, x_i), \quad (4)$$

having P inputs, a *single* hidden layer of H hidden nodes and a single output. In the particular BNN method described here, every network has the same structure. However, as noted below, the effective number of hidden nodes could be fewer than H, if there are hidden nodes with associated weights near zero and if such networks are assigned higher probability than those with a greater number of active nodes. The parameters u_{ij} and v_j are called *weights* and a_j and b are called *biases.* Both sets of parameters are usually referred to collectively as weights, w.

Since, for a correctly trained network, the probability that $t = 1$ is $y(x, w)$, and $1 - y$ for $t = 0$, the probability of the set of targets $t = (t_1, t_2, \cdots, t_N)$, *given* the data $x = (x_1, x_2, \cdots, x_N)$, is

$$p(t|x, w) = \prod_{i=1}^{N} y^{t_i} (1-y)^{1-t_i}, \quad (5)$$

in which we have assumed the events to be independent. Given an event with data x', a reasonable estimate of the probability that it belongs to the signal class (assuming $p(0) = p(1)$) is given by the weighted average

$$\bar{y}(x'|t, x) = \int y(x', w)\, p(w|t, x)\, dw, \quad (6)$$

where the posterior density $p(w|t, x)$, given by Eq. (2), is computed using the likelihood, Eq. (5), and some prior $p(w)$, to be discussed shortly.

Currently, the only feasible way to perform the high-dimensional integral in Eq. (6) is to sample the density $p(w|t, x)$, in some appropriate way, and to approximate the integral using the average

$$\bar{y}(x'|t, x) \approx \frac{1}{K} \sum_{k=1}^{K} y(x', w_k), \quad (7)$$

where K is the number of points w sampled from $p(w|t, x)$. We note, again, that each point w corresponds to a different neural network function in the class of networks with P inputs and H hidden nodes. The average is therefore an average over *networks.*

It may happen that some of the points w correspond to networks that are tightly fit to the training data. Such networks will typically perform poorly on an independent set of events. However, if one averages over many networks, one expects to produce an estimate of the signal class probability, $y = p(1|x)$, that is less likely to be affected by "over training." Moreover, in the Bayesian approach, there is less need to limit, severely, the number of hidden nodes because a low probability density will be assigned to points w that correspond to unnecessarily large networks, in effect, pruning them away. Indeed, networks have been trained [4], successfully, that contain more weights than the number of training data! In this Bayesian approach, the network should be as large as is computationally feasible so that the class of functions defined by the network parameter space includes a subset that are good approximations to the true mapping.

3. Regression

In a regression problem, the targets t are usually sampled from a continuous set. For example, we may wish to model a function $t = f(x)$ that maps an uncorrected measurement of the transverse momentum of a jet of particles, to a corrected measurement. In this case, the target t would be the *known* "correct" value of the transverse momentum of the jet—perhaps, taken to be the transverse momentum of a Z boson recoiling against the jet, while x would be the measured jet transverse momentum, along with any other measured quantities believed to be relevant. If we wish to fit a function to these data, the form given in Eq. (5) for the likelihood of the targets is inappropriate. A better model, assuming that the noise in the targets can be modeled by a Gaussian—at least approximately, is

$$\begin{aligned} p(t|x, w) &= \prod_{i=1}^{N} \exp\,[-(t_i - f(x_i, w))^2/2\sigma^2], \\ &= \exp\,[-\sum_{i=1}^{N}(t_i - f(x_i, w))^2/2\sigma^2], \end{aligned} \quad (8)$$

with $f(x_i, w)$ given in Eq. (4). Even if the noise in the target is not Gaussian, Eq. (8) may still yield reasonable results, provided that the value of σ is chosen to match the noise level in the targets.

One advantage of modeling such mappings with neural networks is that the functional form Eq. (4) is flexible enough to model functions $t = f(x)$ in which x is multi-dimensional, and, in which one or more components of x may be statistically dependent.

4. Computing the Posterior Density

In order to compute the average in Eq. (6), it is necessary to generate a sample of points w from the posterior density, Eq. (2). Unfortunately, sampling from the posterior density is not feasible using simple numerical methods. In practice, a sample is generated using Markov Chain Monte Carlo (MCMC) methods [4]. In the MCMC method, one steps through a parameter space in such a way that points are visited with a probability proportional to the density one wishes to sample, here the posterior density $p(w|t,x)$. Points where $p(w|t,x)$ is large will be visited more often than points where $p(w|t,x)$ is small. The methods of choice for sampling complex densities, such as $p(w|t,x)$, originate in the field of computational statistical physics. The problem of moving through the network parameter space is re-cast as a problem of statistical mechanics, specifically, of a *single* particle moving through a (rather complicated) potential.

The posterior density is written as

$$p(w|t,x) = \exp\left[-V(q)\right], \tag{9}$$

where $V(q) = -\ln p(w|t,x)$ (with $q \equiv w$) is interpreted as a spatially varying "potential" through which the "particle" moves. One adds a "kinetic energy" term $T(p) = \frac{1}{2}p^2$, where p is a vector of dimensionality equal to that of the network parameter space. The "mass" of the "particle" can be taken to be unity by appropriate re-scaling. The motion of the particle is governed by its "Hamiltonian" $H = T + V$. For a Hamiltonian system, the particle will, *eventually*, visit every phase space point (q,p) arbitrarily closely in such a way that the density of points in phase space is proportional to $\exp(-H)$. By randomly (and appropriately) injecting or removing "energy" from the system, different constant energy regions of phase space $\{(p,q)\}$ can be explored. A Markov chain $q_1, q_2, \ldots q_N$ is thereby created, which converges (eventually) to a sequence of points that constitute a sample from the density $p(w|t,x)$. Since the correlation between adjacent points is very high, typically 0.9 or higher, one usually saves a point, that is, a network, after every L steps, to lessen the correlation between the saved points. It is also necessary to discard the initial part of the Markov chain because, in general, it will not be a faithful sample of the required density.

4.1. *Prior*

Every Bayesian inference requires the specification of a prior. Unfortunately, for this problem, the choice of prior is not obvious. However, experience suggests that a reasonable class to choose from is the class of Gaussian priors centered at zero, which favors smaller rather than larger weights. Smaller weights yield smoother fits to data. In the example described next, which uses the BNN package of Radford Neal[4], a Gaussian prior is specified for each weight. However, the variance for weights belonging to a given group (either *input-to-hidden weights* (u_{ij}), *hidden-biases* (a_j), *hidden-to-output weights* (v_j) or *output-bias* (b)) is chosen to be the same: σ_u^2, σ_a^2, σ_v^2, or σ_b^2, respectively. However, since we do not know, *a priori*, what these variances should be, their values are allowed to vary over a large range, while favoring small variances. This is done by assigning each variance a gamma prior

$$p(z) = \left(\frac{\alpha}{\mu}\right)^{\alpha} \frac{z^{\alpha-1} e^{-z\frac{\alpha}{\mu}}}{\Gamma(\alpha)}, \tag{10}$$

where $z = \sigma^{-2}$, and with the mean μ and shape parameter α set to some fixed plausible values. The inverse of the variance $z = \sigma^{-2}$ is sometimes referred to as the *precision*. The gamma prior is referred to as a *hyperprior* and the parameter (here the precision) for which it is a prior is called a *hyperparameter*.

5. An Example: Single Top Search

The electroweak production of single top quarks, which has not been observed so far, is an important prediction of the Standard Model. Moreover, it is potentially a sensitive probe of new physics. The observation of this process at the Fermilab Tevatron is much more challenging than the observation of top-antitop pairs [5], because of the much smaller signal to background ratio involved. We have studied the discrimination of the signal, in the channel $p\bar{p} \to tqb \to Wbqb$, from the dominant background process, $p\bar{p} \to Wb\bar{b}$, for the case in which the W boson decays into a muon (μ) and a neutrino (ν). The final state, therefore, contains a high transverse momentum muon, two b-quark jets and significant missing transverse energy due to the neutrino from the W boson.

We considered eleven kinematic variables that involve the transverse energies, spatial separation and invariant masses of the measured final state objects. All eleven variables were used as inputs to the neural networks, each with thirty hidden nodes and a single output. A Markov chain of networks was generated using the BNN software package noted above, with a training sample consisting of 1000 events each of signal and background. Five hundred iterations, of twenty MCMC steps each, were used. Neural network parameters were stored after each iteration. The results of the training are shown in Fig. 1. For both the signal and background samples, the network outputs, averaged over the last 100 networks,

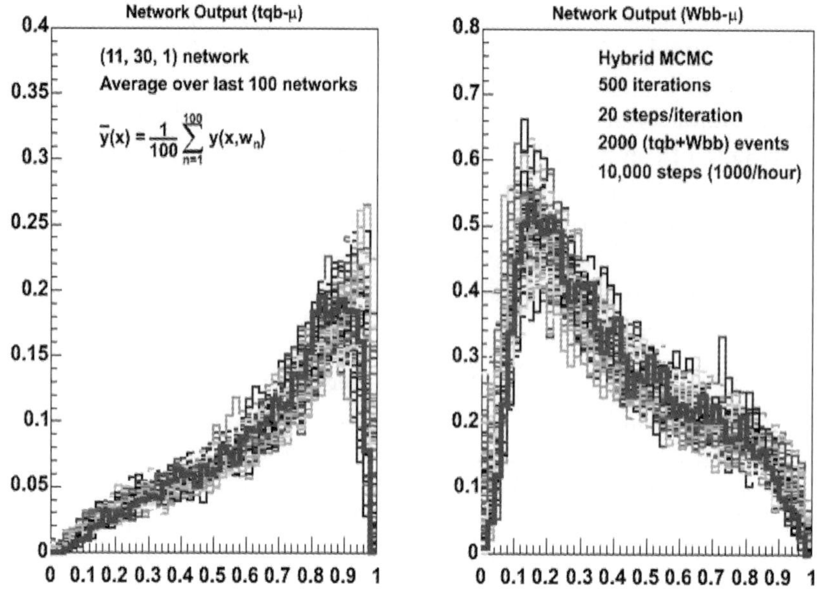

Fig. 1. The output distributions of an ensemble of neural networks, generated using a MCMC method, for Monte Carlo signal (single top – tqb channel) and background (Wbb) events. The thick histogram is the distribution of the network output averaged over the last 100 of a sequence of 500 networks, sampled from the posterior density $p(w|t,x)$.

are shown superposed on the output distributions of each of the individual networks. As one might have expected, the distributions of the 100 networks show some scatter. One expects, however, the Bayesian average to be a more robust estimate of the true signal class probability.

6. Conclusions

Bayesian learning of neural networks could take us another step closer to realizing optimal and robust results in classification problems. It also allows a fully probabilistic approach with proper treatment of uncertainties. But, of course, the key question is: does the averaging help? The answer, in principle, is yes. More to the point, we have found the answer to be yes, in practice. Figure 1 is, in effect, a comparison of 100 *single* neural networks with the Bayesian average over all of them. A study of these distributions reveals that the area under the ROC curve (the plot of the signal efficiency vs. background efficiency) is *larger* for the "Bayesian-averaged network" than for any one of the individual networks, which is an indication that the averaging helps.

The BNN method, however, is computationally demanding. In the example described above, 10 hours were required to sample 10,000 points, that is, networks, from the posterior density. A large number of points is needed so that one can abstract a subset of (several hundred) networks that are approximately statistically independent.

We have started exploring the application of the BNN method to the analysis of particle physics data. The initial results, as illustrated by the example of the single top quark search by the DØ Collaboration, at Fermilab, are very promising.

References

1. C. M. Bishop, *Neural Networks for Pattern Recognition,* (Clarendon Press, Oxford, 1998); R. Beale and T. Jackson, *Neural Computing: An Introduction,* (Adam Hilger, New York, 1991).
2. E.K. Blum and L.K. Li, "Approximation theory and feedforward networks," Neural Networks, **4**, 511-515 (1991).
3. D.W. Ruck et. al., "The multilayer perceptron as an approximation to a Bayes optimal discriminant function," IEEE Trans. Neural Networks **1 (4)**, 296-298 (1990); E.A. Wan, "Neural network classification: a Bayesian interpretation," IEEE Trans. Neural Networks **1 (4)**, 303-305 (1990).
4. R. M. Neal, *Bayesian Learning of Neural Networks,* (Springer-Verlag, New York, 1996). For a review, see: http://www.cs.utoronto.ca/~radford/ftp/review.pdf.
5. CDF Collaboration, F. Abe et. al., Phys. Rev. Lett. **74**, 2626 (1995); DØ Collaboration, S. Abachi et. al., Phys. Rev. Lett. **74**, 2632 (1995); For a review, see P.C. Bhat, H.B. Prosper and S.S. Snyder, Int. J. Mod. Phys. **13**, 5113 (1998).

SIGNAL ENHANCEMENT USING MULTIVARIATE CLASSIFICATION TECHNIQUES AND PHYSICAL CONSTRAINTS

R. VILALTA and P. SARDA

Dept. of Computer Science, University of Houston, 4800 Calhoun Rd., Houston TX 77204, USA
E-mail: {vilalta,ppsarda}@cs.uh.edu

G. MUTCHLER and B. P. PADLEY

Bonner Nuclear Lab, Rice University, 6100 Main Street, Houston, TX 77005, USA
E-mail: {mutchler,padley}@rice.edu

S. TAYLOR

Dept. of Physics and Astronomy, Ohio University, 251 Clippinger Labs, Athens, OH 45701, USA
E-mail: staylor@jlab.org

We report on an empirical comparison of several multivariate classification techniques (e.g., random forests, Bayesian classification, support vector machines) for signal identification; our experiments use K* mass as a test case. We show 1) the effect of using different cost matrices in generalization performance and 2) how information about physical constraints obtained from kinematic fitting procedures can be used to enrich the original feature representation. The latter step is done through a derivation of Λ particle parameters (e.g., momentum, energy, and mass) using kinematic fitting; the degree of fit using a χ^2 statistic is used as a new feature. Overall, our goal is to investigate how to incorporate physical constraints to improve classification performance.

1. Introduction

The purpose of this analysis is to gain insight on how to exploit multivariate techniques and physical constraints for signal classification and enhancement. Traditional techniques that exploit physical constraints use "kinematic fitting" to improve measured quantities and to provide a means to cut background. We propose an additional step where a multivariate classification technique is invoked on Monte Carlo data to generate a predictive model. The model is used to separate signal events from background events. Applying the model to real data results in a (predicted) signal distribution where evidence for the existence of a particle of interest is enhanced.

1.1. *The Physical Experiment*

We begin by describing the physical experiment. A broad band energetic photon beam (γ) hits a liquid hydrogen target, the proton (p). The photon interacts and produces a number of charged and uncharged particles. We will look for the following reaction:

$$\gamma \mathrm{p} \rightarrow \Lambda K^{*+} \tag{1}$$

$$\gamma \mathrm{p} \rightarrow \Lambda K^{+}\pi^{0} \tag{2}$$

$$\gamma \mathrm{p} \rightarrow K^{+}\mathrm{p}\pi^{-}\pi^{0} \tag{3}$$

Our data set contains information about the incident photon (γ), and three charged particles, K^+, p, and π^-. While the charged particles are detected, the uncharged ones are not seen, and must be inferred from the missing mass (e.g., π^0).

For each detected charged particle we measure the momentum p and the polar angle θ and azimuthal angle ϕ. From these quantities we can construct the three vector, $\mathbf{p} = \mathbf{i}p_x + \mathbf{j}p_y + \mathbf{k}p_z$ where $\mathbf{i}$, $\mathbf{j}$ and $\mathbf{k}$ are the unit vectors. We also measure the Time-of-Flight (TOF). From the TOF and momentum we can calculate the mass m of the particle. Finally, for each particle, we are able to construct a 4-vector, $(E, \mathbf{p})$, where $E = \sqrt{p^2 + m^2}$.

In this particular paper we focus on identifying the presence of K^{*+} after the photon-proton interaction (γp). This is in practice not of real interest, but stands as a convenient test case to assess the value behind multivariate classification techniques. Invoking these techniques is justified by the inherent difficulty in separating signal events from background events (many background reactions produce similar measured particles).

1.2. *Using Kinematic Fitting and Physical Constraints*

At first we applied the technique of kinematic fitting[1]. This technique takes advantage of constraints such as energy and momentum conservation to improve measured quantities and to provide a means to cut background. We have chosen to use the Lagrange multiplier method. First, the unknown variables are divided into a set of measured variables ($\vec{\eta}$) and a set of unmeasured variables ($\vec{\xi}$) such as the missing momentum or the 4-vector for a decay particle. For each constraint equation a new variable λ_i is introduced. These variables are the Lagrange multipliers. To find the best fit we minimize

$$\chi^2(\vec{\eta},\vec{\xi},\vec{\lambda}) = (\vec{\eta}_0 - \vec{\eta})^T V^{-1}(\vec{\eta}_0 - \vec{\eta}) + 2\vec{\lambda}^T \vec{f} \quad (4)$$

by differentiating χ^2 with respect to all the variables, linearizing the constraint equations and iterating. Here $\vec{\eta}_0$ is a vector containing the initial guesses for the measured quantities, V is the covariance matrix comprising the estimated errors on the measured quantities, and $\vec{f}$ represents the constraints such as energy and momentum conservation.

1.3. *Generating Confidence Levels*

For our purposes, we are interested in using kinematic fitting to obtain a confidence level (goodness of fit to the data). As an example, let's look into the fitting procedure as applied to the proton (p) and pi-minus (π^-) tracks with the Λ hypothesis. Explicitly, the constraint equations are as follows:

$$\vec{f} = \begin{bmatrix} E_p + E_\pi - E_\Lambda \\ \vec{p}_p + \vec{p}_\pi - \vec{p}_\Lambda \\ (y - y_\pi)p_\pi^z - (z - z_\pi)p_\pi^y \\ (x - x_\pi)p_\pi^z - (z - z_\pi)p_\pi^x \\ (y - y_p)p_p^z - (z - z_p)p_p^y \\ (x - x_p)p_p^z - (z - z_p)p_p^x \end{bmatrix} = \vec{0}. \quad (5)$$

The χ^2 distribution for this fit is the result of a fit to the histogram using the functional form of a χ^2 distribution with two degrees of freedom plus a flat background term. Explicitly,

$$f(\chi^2) = \frac{P_1}{2} e^{-P_2 \chi^2/2} + P_3. \quad (6)$$

P_2 is a measure of how close the distribution in the histogram is to an ideal χ^2 distribution, for which $P_2 = 1$. The Confidence Level (CL) is the primary measure of the goodness of fit to the data and is given by the equation

$$CL = \int_{\chi^2}^{\infty} f(z;n)dz \quad (7)$$

where $f(z:n)$ is the χ^2 probability density function with n degrees of freedom (where we have assumed normally distributed errors).

2. Using Multivariate Classification Techniques

In addition to the traditional approach of kinematic fitting, we suggest using multivariate classification techniques for signal identification and enhancement. Our approach consists of using the confidence levels (goodness of fit to the data described above) as new features into a classification problem. The resulting model implicitly uses the kinematic fitting results to further enhance the signal of interest (e.g., to enhance K^{*+}).

2.1. *The Classification Problem*

We begin by giving a brief overview of the classification problem [2, 3]. A classifier receives as input a set of training examples $T = \{(\mathbf{x}, y)\}$, where $\mathbf{x} = (a_1, a_2, \cdots, a_n)$ is a vector or point in the input space ($x \in \mathcal{X}$), and y is a point in the output space ($y \in \mathcal{Y}$). We assume T consists of independently and identically distributed (i.i.d.) examples obtained according to a fixed but unknown joint probability distribution. The outcome of the classifier is a function h (or hypothesis) mapping the input space to the output space, $h : \mathcal{X} \to \mathcal{Y}$. Function h can then be used to predict the class of previously unseen attribute vectors.

2.2. *Data for Analysis*

In our study, the output variable for each event indicates if the photon-proton interaction resulted in the production of K^{*+} (positive event) or not (negative event). Each feature vector $\mathbf{x}$ is made of 45 features. The first 4 features are confidence level numbers derived from the kinematic fits (Section 1.3). The next feature corresponds to the total energy. The last 40 features characterize 8 particles (3 of them detected and 5 inferred). Each particle is represented by energy E, momentum p, polar angle θ, azimuthal angle ϕ, and mass squared m^2.

Table 1. Columns 2-3: Mean accuracy performance (Acc.) with different misclassification costs. Numbers enclosed in parentheses represent standard deviations. Columns 4-5: Mean false positive rates (FPR) with different misclassification costs.

Analysis Technique	Acc. Equal Costs	Acc. Unequal Costs	FPR Equal Costs	FPR Unequal Costs
Naive Bayes	85.59 (0.86)	86.79* (0.78)	20.1	6.8
Support Vector Machines	87.69 (0.70)	88.29 (0.51)	18.7	1.6
Multilayer Perceptron	88.57 (0.85)	90.58 (0.73)	14.3	3.0
ADTree	88.90 (1.14)	90.81* (0.96)	11.5	3.7
Decision Tree	89.23 (0.93)	91.97* (0.87)	12.7	4.7
Random Forest	90.02 (1.12)	92.34* (0.95)	11.6	4.3

Our data set is derived using the CEBAF large angle spectrometer (CLAS). We gathered 1000 Monte Carlo signal events and 6000 Monte Carlo background events. The real data comprised about 13,500 events.

2.3. *Using Monte Carlo Data and Variable Misclassification Costs*

Our first set of experiments was limited to Monte Carlo data for which the value of the output variable of each event is known. Our study compared the performance of several classification algorithms in terms of predictive accuracy. We employed several algorithms including decision trees, support-vector machines, random forests, etc.

First we reduced the original size of the input space through a feature selection process, using information gain as the evaluation metric [3]. For each algorithm we varied the amount of misclassification costs. Table 1 shows our results. The first column describes the multivariate classification techniques used for our experiments. The second column shows accuracy estimations with equal misclassifications costs; the third column shows accuracy estimations where the cost of a false positive is 3 times more expensive than the cost of a false negative. Each result is the average of 5 trials of 10-fold cross validation each [3]. An asterisk at the top right of a number implies the difference is significant at the $p = 0.01$ level (assuming a two-tailed t-student distribution). Overall there is a significant increase in performance by adding a penalty when mislabelling background events as target events. In addition, Table 1 shows how for this particular domain, varying misclassification costs can yield a significant reduction in the false positive rate (FPR %, columns 4-5).

Our results denote a preference for the strategy behind "random forests". We have observed similar results in other experiments. Random forests have the ability to reduce the variance and bias components of error by voting over multiple decision trees using on each tree a random selection of features [4]. They exhibit robust behavior against problems with multiple systematic errors as is common to problems in particle physics.

2.4. *Signal Enhancement on Real Data*

Our next set of experiments used real data for which the value of the output variable of an event is unknown. In this case the problem is not to maximize accuracy performance (i.e., minimize a risk functional such as zero-one loss) but instead to provide enough evidence to believe that the signal event occurred multiple times during the photon-proton interaction. The goal is to find a technique able to enhance the signal distribution over the background distribution.

Our approach to deal with the signal enhancement problem is as follows. Applying a multivariate technique M on Monte Carlo data yields a predictive model h_M. One can then apply h_M on the real data to generate a histogram for the predicted signal distribution. If model h_M exhibits good performance, we expect the histogram generated through h_M to provide evidence for the occurrence of the desired signal.

To illustrate our approach Figure 1 (left) shows a histogram generated with all real data; the x-axis corresponds to the squared mass (m^2) of the signal particle (K^{*+}). Figure 1 (middle) shows a histogram generated by taking only those events predicted as signal on the real data by a classification model. Kinematic fitting variables were part of the feature vectors. We employed random forests

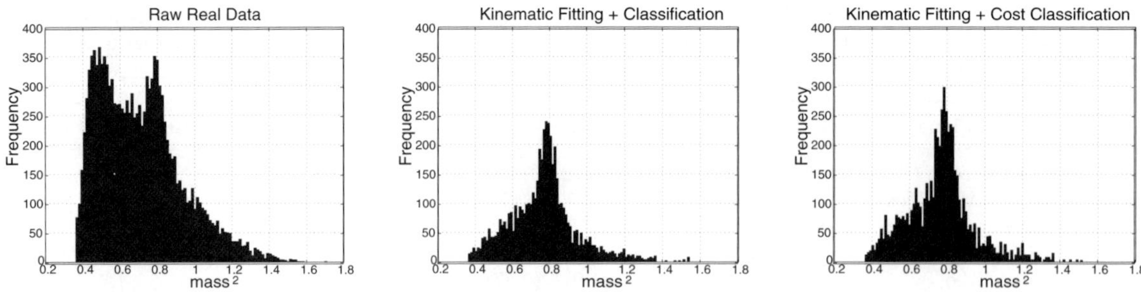

Fig. 1. Histograms using (left) real data (middle) predicted signals on real data by random forests, and (right) predicted signals on real data by random forests using cost-sensitive information. The x-axis corresponds to K^{+*} squared mass (units are in $\frac{\mathrm{GeV}^2}{c^4}$).

as the classification technique; the derived information helps isolate and enhance the signal distribution. Figure 1 (right) shows the corresponding histogram using random forests with cost sensitive classification and kinematic fitting variables. The resulting histogram shows an even larger enhancement over the signal distribution.

To quantify the difference between Figure 1 (middle) and Figure 1 (right), we computed the distance between each of these empirical distributions. We used relative entropy [5] $K(f_1||f_2)$ to compute the distance between probability distributions f_1 and f_2, where

$$K(f_1||f_2) = \sum_i f_1^i \log \frac{f_1^i}{f_2^i} \tag{8}$$

and index i varies along the values of the random variable. In our case, let f_r be the distribution for the real data (Figure 1 left), f_{kc} be the distribution for kinematic fitting and classification (Figure 1 middle), and f_{kcs} be the distribution for kinematic fitting and cost sensitive classification (Figure 1 right). We found empirically the following results: $K(f_r||f_{kc}) = 0.2798$; $K(f_r||f_{kcs}) = 0.4048$. This indicates the distribution obtained by combining kinematic fitting with cost-sensitive classification yields a new signal distribution that has a larger separation from the original real data (in terms of relative-entropy).

3. Conclusions

Our study suggests generating a predictive model over Monte Carlo data to produce a distribution over real data where a signal of interest is enhanced. Our model integrates information about physical constraints using kinematic fitting.

Our current work adds confidence levels derived from kinematic fitting as new features for classification. One unexplored area is to determine the degree to which multivariate classification techniques contribute to signal enhancement without any information derived from kinematic fitting. It is important to understand how current classification techniques can exploit information derived from physical constraints.

Acknowledgments

This material is based upon work supported by the National Science Foundation under Grants no. IIS-431130 and IIS-448542, and by the Department of Energy under Grant no. DE-FG03-93ER40772.

References

1. A. G. Frodesen (1979). *"Probability and Statistics in Particle Physics"*, Oxford University Press.
2. R. O. Duda, P. E. Hart, and D. G. Stork (2001). *"Pattern Classification"*, John Wiley Ed. 2nd Edition.
3. T. Hastie, R. Tibshirani, and J. Friedman (2001). *"The Elements of Statistical Learning, Data Mining, Inference, and Prediction"*, Springer-Verlag.
4. L. Breiman (2001). *"Random Forests"*, Machine Learning 45(1) pp. 5-32. Springer Science-Business Media.
5. T. M. Cover and J. Thomas (1991). *"Elements of Information Theory"*, Wiley-Interscience.

SOFTWARE

SOFTWARE FOR STATISTICS FOR PHYSICS

JAMES T. LINNEMANN
Michigan State University, Department of Physics & Astronomy,
E. Lansing, MI 48824, USA
E-mail: linnemann@pa.msu.edu

I discuss two workshops held in 2004 and 2005 relevant to the software environment for statistical analysis in physics and astrophysics. The first largely explored the R environment used by statisticians and the Root environment widely used in particle physics and related fields. The second was a step towards starting a repository for software useful in statistical analyses for these fields. I also discuss some of the statistical software resources on the web of relevance to physicists.

1. Introduction

The work behind this talk grew out of the PHYSTAT2003 conference, where Louis Lyons invited proposals for focused PHYSTAT workshops. After my interactions with statisticians at PHYSTAT2003, the first thing I wanted was to improve the environment for doing particle and cosmic ray physics analysis using statistical tools. I particularly wanted to enhance what I already had available in Root[1], which is our everyday working environment. I clearly suffered from R envy. The R language and environment[2] is heavily used by statisticians. The second thing I missed was a web page for physicists of pointers to implementations of statistical methods. And the third thing I felt we lacked was a web site to collect new statistical software oriented to the needs of physics. The first and the third of these were subjects of workshops, and the second I started work on myself. I'll discuss each in turn.

I must apologize if I didn't invite you personally to these workshops. Each workshop had a limited goal: to try to do *work* in a day or two (or at least to start). To me, the right way to do that is to get at least some of the right people in a room. Thus, the workshops were designed with small attendance, to concentrate on discussion rather than hearing a parade of presentations. I'll leave you to judge how successful we were.

2. R and Root

In 2004 I organized a PHYSTAT workshop[3] at MSU on statistical software, concentrating mainly, because of those able to attend, on Root and R. Two developers of major software systems attended: Luc Tierney of the R core development team (thanks to valuable contacts by Jerry Friedman, Nancy Reid), and Rene Brun, lead developer of the Root system. Astronomers also attended: Eric Feigelson, who developed the StatCodes[4] web site and who is working on the virtual observatory statistics project VOStat[5], and Tim Beers who developed the Rostat[6] robust statistics package. Physicist/developers included Harrison Prosper, Scott Snyder, Sherry Towers (TerraFerMa[7]), and three physicist R users from Fermilab: Adam Lyon, Jim Kowalkowski, and Marc Paterno.

For those not familiar with Root, I would describe its key features as follows. It provides a GUI for publication-quality graphics and for making the cuts (data sub-region selections) we physicists are so fond of. It also provides I/O which scales to petabytes data sets consisting of collections of files containing event data (with each event individually tree-structured). Root uses a histogram as its base metaphor. Its primary interface is a command prompt, which accepts C++ as a language for interpreted and compiled macros. Root is extensible, though most might not say "easily." Root contains sophisticated nonlinear fitting and reporting of multidimensional parameter errors. Its collection of statistical algorithms is small, but growing. For example, robust (to outliers) curve fitting was recently added. Anna Kreshuk's talk at this conference gives more information on recent developments in Root.

For those not familiar with R, it is an elegant

data manipulation language (R is a gnu implementation of the S language[8]), embedded in an environment rich in statistical functionality. The user sees a command prompt. Macros in R are interpreted, but heading toward byte-compilation. R is not GUI-oriented, though hooks are being built: most users are satisfied with the command line. However, S+, a commercial[9] implementation of the S language, does provide a rich GUI interface. Most S or S+ code runs happily in the R environment.

R is described by statisticians as a quick and easy interactive analysis tool, and is indeed the standard tool of professional research statisticians. So if a statistician suggests a method to you (for example bootstrapping, the lasso, bagging, boosting, cross-validation etc.), it's probably implemented in R. The R environment has as built in functions a large range of sophisticated statistical tests and graphics, many of which are not in common physics usage.

R has links to further multidimensional graphics (Ggobi), and a broad package library[10], with trivial download mechanism. R allows straightforward extensibility to new packages in R or C code. Functions and packages are often very fast if they are R-wrapped C code. R keeps data in virtual memory Data Frames, and uses vectors as its basic metaphor. R has interfaces to postgres, mysql, and other databases, and has parallel computation under development. While both Root and R are used outside their home communities, R and S documentation[2, 8] is commercially published and available at Amazon.

Susan Holmes' talk at this conference discusses data visualization largely using R tools, and Marc Paterno's talk provides further detail on R use from a physicist's perspective. Also useful is Adam Lyon's talk[11] at the MSU workshop.

There were three main results of the workshop. Eric Feigelson was confirmed in his initial inclination to use R for the basis of the VOStat project. Adam Lyon, encouraged by discussions with Luc Tierney, wrote a fairly general Root Tree reader for R. Rene Brun was perhaps further interested in R, encouraged on his existing path of adding statistical functionality to Root, and, I hope, inspired by R's elegant package mechanism[10]. Rene and I at this conference celebrated (?) a quarter century of my encouraging Rene to do even better than he has in providing an everyday environment for particle physicists.

3. Statistical Resources on the Web for Physicists

My second topic grew out of preparation for the software workshop just described. I wanted to survey what statistical resources were available on the web for physicists. Having a few lazy bones in my body, I wanted to know where I might find useful statistical software without having to write it all from scratch. In the process I developed a page of links at `http://www.pa.msu.edu/people/linnemann/stat_resources.html` .

I definitely don't want to claim there had been no effort in particle physics before mine. But to my shock, this is now the largest such page I know of. Others who had preceded me in HEP included Glen Cowan, and the CDF statistics committee. But the reason for the lack of pointer pages is, I believe, the lack of actual web statistics-oriented resources specific to physics.

Here again I suffer envy of other fields. In particular, astrophysicist Eric Feigelson has done an excellent job of surveying statistical resources at his StatCodes site[4]. Its point of view is quite general in fact – physicists should most certainly look there – though of course he is particularly interested in items relevant to astronomy and astrophysics, a few of which have found less application in particle physics. Tom Loredo[12] also has a very useful collection of links. Not surprisingly, there are many useful sites from statistics, particularly StatLib[13]. There are also quite a number of useful resources on multidimensional analysis which I included on my page. I'm sure many of you have your own favorite links to software, and I would be delighted for you to send them to me. I have avoided most references to commercial software, mainly because I have seldom seen my physicist colleagues use (i.e. pay for) commercial analysis software. Astrophysicists, however, find their productivity gains well worth the cost of the commercial IDL [14] package for analysis, interpolation and manipulation of 2 and 3D image data; it contains substantial statistical functionality as well.

4. Towards a Repository for Statistical Software for Physicists

One conclusion I drew after searching for physics-oriented statistical resources was that I was also suffering from a serious case of repository envy. Astronomy has a number of user-contributed repositories under way for analysis and statistical codes, for example those maintained by the Astronomical Software Directory Service[15] and NASA's HEASARC[16]. Even biology has *bioconductor*[17], a large collection of R software for bioinformatics. There are a few HEP repositories[18], but there is little physics-oriented analysis or statistical software on the web at present. In some ways this is surprising, as the web was invented for HEP. Assessing user interest in such a repository was the motivation for a 2005 workshop[19] Mark Fishler and I organized at Fermilab.

Behind the archive is Mark Twain's notion that if you make it sufficiently attractive for someone to write statistical code, they might actually do it for you[20]. Louis Lyons advised me that in giving this talk, I was coming to the right place to find software writers. And when I asked who in the audience had written statistical software of use to someone else, a goodly majority indeed raised their hands. I know I would find it useful to have access to the programs used to produce results for many of the talks at this conference.

The basic motivation for a software repository is sharing: don't reinvent the wheel; improve it. A repository requires some implied longevity which seems best met by having an organization rather than an individual as sponsor. Fermilab is potentially interested in such a role. Clearly a web interface is needed for upload, search, retrieval. One can envision a hierarchy of purposes, ranging from an archive for source code of software associated with physics or conference papers, through a downloadable package library (either of stand-alone packages, or packages adapted to particular frameworks), to a component library with various language or web interfaces, possibly with distribution of binaries for various platforms.

A Statistical Software Archive. The simplest repository function discussed at the workshop was an open archive (roughly analogous to arxiv.org). If you publish a statistical calculation in refereed physics papers or at statistical conferences such as this one, you could put the code in the archive, and reference it. With an archival repository available, one could hope that this becomes as much a part of the culture as submitting preprints to arxiv.org has become. Archiving offers the potential of substantial benefit for a modest effort.

The "guarantee" for users would be intentionally weak: once, the code compiled and ran on some machine and produced useful results. To allow reuse of code with credit to authors, the minimal information supplied would be the author, title, and a one-line explanation of purpose. Keywords and possibly the experiment to which it was relevant would make it easier to locate. Your grad student could start a project here, rather than from scratch, and possibly compare methods used by different experiments. Documentation would be encouraged (but not quite required). Version tracking would need to be supported by the system even at this basic level.

There are many candidates for software in such an archive: calculations of significance, limit setting programs, and goodness of fit tests come to mind, as does software for studying the behavior of statistical methods. In these areas, competing procedures exist: some are published, some not. Actual programs are very hard to find: you have to know of the method, and ask its author personally; at best, you might find some such code in your physics collaboration's CVS repository. Only a few such programs have public web interfaces (D0 or Babar have some).

A Package Download Site. A more sophisticated use of a repository would be software written explicitly for re-use (rather than archived for the historical record). Packages of this kind might be stand-alone programs, or packages for frameworks such as R or Root. Here there is a real need for well-designed conventions to support portability and simplify building and upload. Documentation now also becomes a vital issue, including of course any published references for the methods used. In this context, R's package mechanism is particularly admirable. Attaining the same level of simplicity for user and author for Root add-ons would be a real achievement. A repository sponsor can add real value by providing proper repository design to help authors reach users simply and effectively. Further value could be added by choosing packages (possibly even those originally submitted only for archiving) which are of sufficient interest to maintain for reuse at this level, and by providing assistance to authors on issues of portability, numerical techniques, base library choice, or other coding practices.

Candidates for packages of this type also spring readily to mind: multidimensional analysis packages such as Sherry Towers' TerraFerMa[7] and Ilya

Narsky's StatPatternRecognition (described at this conference). Both are currently stand-alone programs rather than framework packages.

A Linkable Toolkit. An even higher level of functionality is also conceivable, by working from the basis of such a repository. One could imagine providing a toolkit library. One might aim to support writing "toy Monte Carlo" or ensemble test writing by providing a coherent library, perhaps building on the gsl (Gnu Scientific Library), or on the *mathcore* or *mathmore* libraries envisioned within Root. This would involve interacting with users to assess existing tools and designing and supplying missing ones. One might repackage existing programs as framework packages to enhance usability. Quality control (robustness, accuracy, etc) become critical at this level of ambition. Personnel for design, coding, and maintenance would be required. One might also provide a computation service for some of the simpler algorithms, with web data entry via forms or ASCII data files.

Plans. The main outcome of this workshop was a consensus statement (included here as an Appendix). Based on this consensus, we had further negotiations with the Fermilab computing division and carried forward discussion to this conference. Marc Paterno of Fermilab attended this conference in large part to assess the interest of a wider community in the repository. At this PHYSTAT2005 conference, Marc Paterno, Louis Lyons, and I have discussed these ideas with many people. These consultations uncovered potentially interesting synergies with Root and the Cedar physics archive in the UK. We also found that considerable interest was expressed for the archive, and were strongly encouraged to get started and see how things evolve once contributions begin arriving. Thus, we are in the process of preparing a proposal to the Fermilab Computing Division to support some version of an archive. Let us know if you think its worthwhile, and pass along any advice you might have. We are currently thinking about the level of manpower required to get started, and working through computer security, copyright, and license issues associated with such a venture.

In the end, it is hardly our vision of the repository and how it might be used that matter. What counts is how the community chooses to use it, and our main motivation is to provide a forum for unleashing the creativity of that community. Whether it houses mathematica, mathcad, or matlab software for producing statistical figures for conference papers; C++ or Fortran routines used in Physical Review or Physics Letters or NIM articles; mass fitters, deconvolvers, goodness of fit or significance or limit calculators, or cunning ways of telling signal from background; whether submissions are programs or packages for R or Root; whether written in java, C, C++, perl, ruby, or python; or entirely other things, depends on what the community finds most useful.

5. The Reproducible Research Ideal

Reproducible Research[21] is an interesting concept in some ways related to the repository. The ideal is that when you write a paper, you save (in a tar archive, say) the entire environment necessary for creating the paper through scripts, and the whole paper and its figures and tables are generated by executing a single high-level script. This tar archive would of course be an excellent submission to the software repository we have discussed.

We all know the kind of problems that led to these thoughts: you ask a graduate student to pick up a project and suggest one of your papers as a starting point, but the student finds it remarkably difficult to actually reproduce the plot you suggested. To do so requires having the same data set you used several years ago, and to use the programs with all the same settings. To achieve a reasonable approximation to this ideal requires as a minimum a powerful script-oriented method of producing figures and tables (such as R or Root) and all the data used in the paper. It also implicitly implies a data set of rather modest size, and a stable set of tools. Otherwise you'd have to save the entire contents of your computer each time.

More is required, however: directory conventions, makefiles, and many other details should be conventional and stable. Arxiv.org provides a subset of such an environment: you know that you will be able to rebuild a pdf file from the latex source and eps files if you meet arxiv's requirements. This ideal is achievable for most (not all) plots shown at this conference, and for most significance and limit calculations in our physics papers. It is problematic for large HEP data sets, which are not publicly available and not necessarily permanently archived with full version control. It is also problematic for analyses which are long in duration (months to years, not hours to days). This is exacerbated when multiple analyses are combined into a single publication, as is often the case in large physics collaborations.

Still, the reproducible research ideal is well worth striving toward. Those who have created a research environment fully supporting the ideal describe it as a discipline with more benefits to authors than to readers wishing to build on the published research.

6. Conclusions

To summarize, I'd like you to take away three main points. First, R has many intrinsic attractions, and is a window to the statistics community. It should be better known in physics and astrophysics, and it is now possible to read Root trees in R. I would personally be delighted if everything in R appeared in Root, my everyday environment. Second, I started a page of web links to statistical software resources relevant to physicists and astrophysicists. If you find it useful, tell your colleagues, link to it, and more importantly, help me improve it. Third, we are trying to start a repository for statistics-oriented software of use to physicists and astrophysicists. I'd appreciate your discussing this repository within your collaboration, and encourage us (and the Fermilab Computing Division) if you think it should be pursued. And we hope you will also contribute software to the repository.

Appendix: Consensus Statement from the 2005 Fermilab Workshop

Following is a slightly abbreviated version of the consensus statement resulting from the workshop:

Currently, statistical tools are in use by individual physicists, and within collaborations. Their ultimate purpose is to make the best use of the data collected by collaborations. However, their effectiveness is limited by the lack of a straightforward mechanism for the community to share software on a wider basis, learn best practice from one another, and avoid unnecessary re-development of similar tools. Some tools are of general use (for example event classifiers, or limit calculation programs). These codes often embody standard practices within a collaboration, recent progress of understanding within our field, or implementation of important ideas developed by statisticians or within the machine learning communities. Other programs encode hard-won expertise in handing particular situations. Sharing such codes across research groups and collaborations contributes directly to the diffusion of such knowledge, and indirectly to improvement of our understanding of our data and the training of students by facilitating comparison of methods. A repository could provide, as objects of study and understanding, working codes which have been tested under realistic conditions. Such codes would also provide a point of departure for improvements, rather than having to first re-implement present ideas for lack of publicly-accessible code.

What sort of repository would support such efforts? We suggest a phased approach. The first and perhaps most important step would be a very open archival repository, where essentially anyone could upload code felt to be useful for statistical tasks in physics experiments. The repository should make it straightforward to store software used to perform calculations for a paper, and refer to those calculations in the publications: "we calculated the upper limit using a Bayesian technique assuming a flat prior in the cross section [17]", and reference [17] might read "C. Calvin & H. Hobbes, www.phystat.org/05/07/23/0013/, version 3". The repository would provide some basic expectations on what a submitting author should provide, but the absolute requirements would be purposely minimal, in order to encourage submission.

A submission should minimally include authors, an email contact address, a tar archive with code and a brief text description of what the submission does. There would be a possibility to provide keywords and an experiment of origin, but not a requirement. A read-me file would be encouraged to include documentation and the platform(s) on which the code had run. Overall, the effort required for submission should be less than or comparable to submitting a paper to arxiv.org.

Downloading code from the archive should be similarly straightforward. Search facilities from the web might start with a simple web listing of entries with a one line description, but could become more sophisticated as more entries became available. Attaching user feedback is another possible evolution path.

Fermilab would be a natural sponsor of such a repository, assuming that it could provide the desired degree of openness. The lab hosts experiments which are currently producing much innovative statistical software, and the lab intends to be a center for ongoing research in particle and astro-particle physics. This is an important activity supporting data analysis, which does not require proximity to the physical location of the experiment. And there are members of the computing division with professional interests in this area.

A longer term vision of the repository goes beyond passively archiving code. One value-added activity would be to classify the submissions to distinguish archival entries from actively maintained packages. Capture of user assessment of such packages might be particularly useful. Packages could also benefit from expertise by improving the efficiency or portability of the submitted code. Design expertise might provide standards for packages which would make them more readily usable. A particular example of interest is the elegant R package mechanism: it would be a real achievement to have design standards which would allow a similar ease of package creation and import within the Root framework. Standards might include naming conventions, package directory structure, allowed base libraries, or build tools. Other activities might include mining the submissions for likely contributions to a linkable library (for example mathmore packages), identifying and writing code for missing functionality, integrating related packages, soliciting and supporting extensions of existing code (justifiable by a broader use base than a single experiment), or actively looking for interesting software produced by the statistical software community and providing web interfaces or language translation wrappers to support use by the physics community. Another possibility is maintenance of a list of such software, perhaps building on the software link web site developed by Jim Linnemann. Such value-added activities would best evolve over time as the use of the repository grows.

We intend to submit soon a more formal request to Fermilab management, and to approach large collaborations to solicit their support for such an endeavor.

Acknowledgments

Thanks to Tom Loredo for many astronomy links, and reminding me of Reproducible Research, and to Bob Nichol for useful comments on this manuscript.

A word on references

I have omitted the initial http:// in all the web references. Many more links are available at: `http://www.pa.msu.edu/people/linnemann/stat_resources.html` .

References

1. `root.cern.ch`. This site contains links to source code, online documentation and tutorials.
2. `www.r-project.org`; Venables and Smith, An Introduction to R, Network Theory Limited (2001); Dalgaard, Introductory Statistics with R, Springer (2002); Everitt, An R and S-Plus Companion to Multivariate Analysis, Springer (2005); `zoonek2.free.fr/UNIX/48_R/all.html` (R tutorial).
3. `user.pa.msu.edu/linnemann/public/workshop`
4. `astrostatistics.psu.edu/statcodes`
5. `astrostatistics.psu.edu/vostat`
6. Beers, T.C., Flynn, K., Gebhardt, K., "Measures of Location and Scale in Clusters of Galaxies. I. A Robust Approach," 1990, Astronomical Journal, 100, 32; see also Hoaglin, Mosteller, Tukey, Understanding Robust and Exploratory Data Analysis, Wiley(2000). There is no Rostat web site.
7. `www-d0.fnal.gov/$\sim$smjt/multiv.html`
8. Becker, Chambers, and Wilks, The New S Language, Chapman and Hall (1988); Chambers and Hastie, Statistical Models in S, Chapman and Hall (1992); Venables and Ripley, S Programming, Spring (2000); Chambers, Programming with Data: A Guide to the S Language, Springer (2004).
9. `www.insightful.com`
10. `cran.us.r-project.org`
11. `user.pa.msu.edu/linnemann/public/workshop/rInHep.ppt`
12. `www.astro.cornell.edu/staff/loredo/statpy`
13. `lib.stat.cmu.edu`
14. `www.rsinc.com/idl/`
15. `asds.stsci.edu/packages.html`
16. `heasarc.gsfc.nasa.gov/docs/software.html`
17. `www.bioconductor.org`
18. `ph-sft.web.cern.ch/ph-sft`, `www.freehep.org`, `cepa.fnal.gov/CPD`, `www.cedar.ac.uk`, `www2.slac.stanford.edu/computing/top_pages/software.htm`
19. `whcdf03.fnal.gov/PHYSTATworkshop`, `user.pa.msu.edu/linnemann/public/workshop/Fermi_Program.htm`
20. Mark Twain, Adventures of Tom Sawyer (1876); see how Tom handles the chore of painting the fence around his house.
21. `www.stat.washington.edu/jaw/jaw.research.reproducible.html`

STATISTICAL COMPUTATIONS WITH ASTROGRID AND THE GRID

ROBERT NICHOL
Institute of Cosmology and Gravitation (ICG), Univ. of Portsmouth, Portsmouth, PO1 2EG, UK

GARRY SMITH
ICG Portsmouth & Institute of Astronomy, School of Physics, University of Edinburgh, UK

CHRISTOPHER MILLER
Cerro-Tololo Inter-American Observatory, NOAO, Casilla 603, LaSerena, Chile

CHRIS GENOVESE, LARRY WASSERMAN
Dept. of Statistics, Carnegie Mellon University, Pittsburgh, PA-15213, USA

BRENT BRYAN, ALEXANDER GRAY, JEFF SCHNEIDER, ANDREW MOORE
School of Computer Science, Carnegie Mellon University, Pittsburgh, PA-15213, USA

We outline our first steps towards marrying two new and emerging technologies; the Virtual Observatory (e.g, AstroGrid) and the computational grid. We discuss the construction of *VOTechBroker*, which is a modular software tool designed to abstract the tasks of submission and management of a large number of computational jobs to a distributed computer system. The broker will also interact with the AstroGrid workflow and MySpace environments. We present our planned usage of the *VOTechBroker* in computing a huge number of n–point correlation functions from the SDSS, as well as fitting over a million CMBfast models to the WMAP data.

1. Introduction

Over a petabyte of raw astronomical data is expected to be collected in the next decade (see Szalay & Gray 2001). This explosion of data also extends to the volume of parameters measured from these data including their errors, quality flags, weights and mask information. Furthermore, these massive datasets facilitate more complex analyses, e.g. nonparametric statistics, which are computationally intensive. A key question therefore is: Can existing statistical software scale-up to cope with such large datasets and massive calculations? We address this question here.

We focus here on two exciting new technologies, namely the Virtual Observatory (VO) and computational grids. However, we point the reader to Jim Linnemann's paper in these proceedings for an excellent summary of existing statistical software packages in physics and astrophysics. We also direct the reader to the recent ADASS conference proceedings and the "Mining the Sky" proceedings (www.mpa-garching.mpg.de/c̃osmo/).

2. N–point Correlation Functions

As a case study of the types of massive calculations planned for the next generation of astronomical surveys and analyses, we discuss here the galaxy n-point correlation functions. These have a long history in cosmology and are used to statistically quantify the degree of spatial clustering of a set of data points (e.g. galaxies). There is a hierarchy of correlation functions, starting with the 2-point correlation function, which measures the joint probability of a data pair, as a function of their separation r, compared to a Poisson distribution, i.e., $dP_{12} = N^2 dV_1\, dV_2(1 + \xi(r))$, where dP_{12} is the joint probability of an object being located in both search volumes dV_1 & dV_2, and N is the space density of objects. $\xi(r)$ is the 2-point correlation function and is zero for a Poisson distribution. If $\xi(r)$ is positive, then the objects are more clustered on scales of r than expected, and vice versa for negative values.

The next in the series is the 3-point correlation function, which is defined as $dP_{123} = N^3 dV_1\, dV_2\, dV_3(1 + \xi_{12}(r_{12}) + \xi_{23}(r_{23}) + \xi_{13}(r_{13}) +$

$\xi_{123}(r_{12}, r_{23}, r_{13})$), where $\xi_{12}, \xi_{12}, \xi_{12}$ are the 2-point functions for the three sides (r_{12}, r_{23}, r_{13}) of the triangle and ξ_{123} is the 3–point function. Likewise, one can define a 4-point, 5-point etc., correlation function. The reader is referred to Peebles (1980) for a full discussion of these n-point correlation functions including their importance to cosmology (see also the recent lecture notes of Szapudi 2005). We also refer the reader to Landy & Szalay (1993) and Szapudi & Szalay (1998) for a discussion of the practical details of computing the N–point functions.

Naively, the computation of the n–point correlation functions scale as $O(R^n)$, where R is the number of data–points in the sample. As one can see, even with existing galaxy surveys from the Sloan Digital Sky Survey (SDSS), where $R \sim 10^6$–10^7, such correlation functions quickly become untractable to compute. In recent years, there has been a number of more efficient algorithms developed to beat this naive scaling. For example, the International Computational Astrostatistics (inCA; www.incagroup.org) group has developed a new algorithm based on the use of the multi–resolutional KD-tree data structure (mrKDtrees). This software, known as *npt*, is publicly available (www.autonlab.org/autonweb/software/10378.html), and has been discussed previously in Gray et al. (2003), Nichol et al. (2001) and Moore et al. (2000). Briefly, mrKDtrees represent a condensed data structure in memory, which is used to efficiently answer as much of any data query as possible, i.e., pruning the tree in memory. The key advance of our *npt* algorithm is the use of "n" trees in memory together to compute an n–point function. See also Alex Gray's contribution in this volume.

3. Computing Correlation Functions

Even with an efficient algorithm, the computation of higher–order correlation functions is intensive. In detail, the n–point correlation functions require a large number of sequential calls to the *npt* code. These include computing the cross–correlation between the real data (called D) and a random dataset (called R), which is used to mimic the edge effects in the real data. As outlined in Szapudi & Szalay (1998), each estimation of a 3–point correlation functions, for a given bin of triangular shape (i.e., $r_{12} \pm \Delta_{r_{12}}$, $r_{23} \pm \Delta_{r_{23}}$, $r_{13} \pm \Delta_{r_{13}}$, requires seven separate source counts over the whole dataset, namely $DDD, DDR, DRR, RRR, DD, RR, DR$. Therefore, if one wished to probe $\sim 10^2$ triangle configuration, then $\sim 10^3$ sequential *npt* jobs are required. This can rise rapidly if one wishes to estimate errors on the n–point functions using either jack-knife resampling (i.e., removing subregions of the data and then re-computing the correlation functions), or a large ensemble of mock catalogs (derived from simulations). Such computations are well-suited to large clusters or grid of computers.

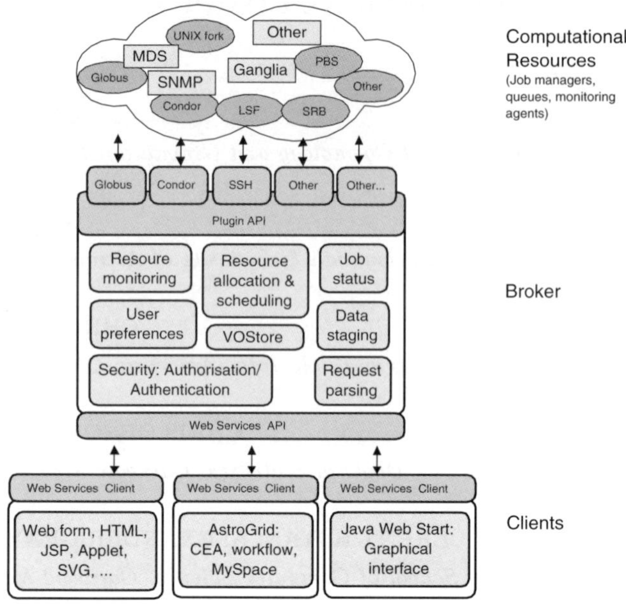

Fig. 1. The architecture of the VOTechBroker and how it interacts with the Grid, VO and our statistical algorithms. The *npt* algorithm is a "Client" (at the bottom) and interacts with the "Broker" via a web–form (HTML) to define the basic parameters needed to run the algorithm and define the resources needed. Eventually we plan to interact with the "Broker" via the AstroGrid workflow environment, allowing the submission of jobs as well as the storage of the input data and results in MySpace. There can be multiple "Clients" to the "Broker".

In recent years, we have used computational resources like TeraGrid (www.teragrid.org) and COSMOS (www.damtp.cam.ac.uk/cosmos/) to perform the computation of the n–point correlation functions for the SDSS main galaxy sample and the SDSS Luminous Red Galaxy (LRG) sample. Our experience shows that the management and scheduling of such a large number of jobs on these massive machines is laborious and tedious. To ease this problem, we are working on *VOTechBroker*, which is a tool that joins two new and emerging technologies; the VO and computational grids.

4. VOTechBroker

AstroGrid (www.astrogrid.org) is a PPARC-funded project to create a working Virtual Observatory for UK and international astronomers. AstroGrid works closely with other VO initiatives around the world (via the International Virtual Observatory Alliance; IVOA) and is part of the Euro–VO initiative in Europe. In particular, the work outlined here has been performed as part of the EU–funded VOTech project, which aims to complete the technical preparation work for the construction of a European Virtual Observatory. Specifically, VOTech is undertaking R&D into data–mining and visualization tools, which can be integrated into the emerging VO and computational grid infrastructure. Therefore, VOTech will build upon existing or emerging standards and infrastructure (e.g. IVOA standards and AstroGrid middleware), as well as looking at standards from W3C and GGF.

As part of the VOTech research, we are engaged in developing the *VOTechBroker*. The key design goals of the broker are to: *i)* Remove the execution and management of a large number of jobs (like *npt*) from the user in a transparent and reusable way; *ii)* Accommodate different grid infrastructures (e.g. condor, globus etc.); *iii)* Locate suitable resources on the grid and optimize the submission of jobs; *iv)* Monitor the status and success of jobs; *v)* Combine with AstroGrid MySpace and workflow environments to allow easy management of job submission and final results (as well as utilizing other algorithms within the VO). In Figure 1, we show the schematic design of the broker architecture which illustrates the modular and "plug-in" design philosophy we have adopted. This is required as one of the key requirements of *VOTechBroker* is that it should be straightforward to add new algorithms, resources and middleware (e.g. a different job submission tool or protocol).

We have implemented the core functionality of *VOTechBroker* and are presently testing it by submitting $\sim 10^4$ *npt* jobs on both the UK National Grid Service (www.ngs.ac.uk), COSMOS supercomputer and a local condor pool of machines. The key ingredients of the present *VOTechBroker* include GridSAM (an open-source job submission and monitoring web service from the London e-Science Centre), the UK e-Science X.509 certificates, MyProxy (a repository for X.509 Public Key Infrastructure security credentials), and the Job Submission Description Language (JSDL; a standard description of job execution requirements to a range of resource managers from the Global Grid Forum). At present, the *VOTechBroker* provides a web-form interface to just the *npt* algorithm discussed above but is modular in design so other algorithms can be easily added via other web–forms. Results from the *VOTechBroker* will soon be placed in a user's AstroGrid MySpace. In the near future, we will interface the broker with other computational resources, e.g., TeraGrid (see below), and the AstroGrid workflow.

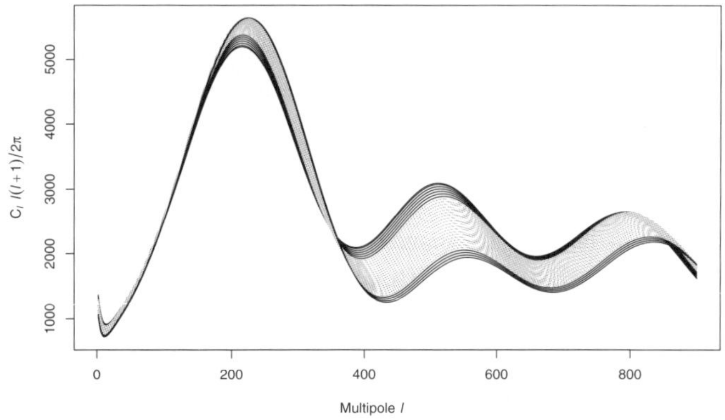

Fig. 2. Using CMBfast, we have varied Ω_b (baryon fraction) and determined which models lie within the 95% confidence ball around $f(X_i)$. For this illustration, we have kept all other parameters in these CMBfast models fixed at their fiducial values. The gray models are within the confidence ball, while the others are outside the ball indicating they are "bad fits" to the data (at the 95% confidence). We get an allowed range of $0.0169 < \Omega_b < 0.0287$.

5. Nonparametric Statistics

In addition to the need for new statistical software that scales-up to petabyte datasets, we also require new algorithms and computational resources that exploit the emerging power of nonparametric statistics. As discussed in Wasserman et al. (2001), such nonparametric methods are statistical techniques that make as few assumptions as possible about the process that generated the data. Such methods are more flexible than more traditional parametric methods that impose rigid and often unrealistic assumptions. With large sample sizes, nonparametric methods make it possible to find subtle effects which might otherwise be obscured by the assumptions built into parametric methods.

In Genovese et al. (2004), we discuss the application of nonparametric techniques to the analysis of the power spectrum of anisotropies in the Cosmic Microwave Background (CMB). For example, one can ask the simple question: How many peaks are detected in the WMAP CMB power spectrum? This question is hard to answer using parametric models for the CMB (e.g. CMBfast models) as these models possess multiple peaks and troughs, which could potentially be fit to noise rather than real peaks in the data. To solve this, we have performed a non-parametric analysis of the WMAP power spectrum (Miller et al. 2003), which involves explaining the observed data (Y_i) as $Y_i = f(X_i) + c_i$ where $f(X_i)$ is an orthogonal function (expanded as a cosine basis $\beta_i \cos(i\pi X_i)$) and c_i is the covariance matrix. The challenge is to "shrink" $f(X_i)$ to keep the number of coefficients (β_i) to a minimum. We achieve this using the method of Beran (2000), where the number of coefficients kept is equal to the number of data points. This is optimal for all smooth functions and provides valid confidence intervals. We also use monotonic shrinkage of β_i, specifically the nested subset selection (NSS). The main advantage of this methodology is that it provides a "confidence ball" (in N dimensions) around $f(X_i)$, allowing non-parametric interferences like: Is the second peak in the WMAP power spectrum detected? In addition, we can test parametric models against the "confidence ball" thus quickly assessing the validity of such models in N dimensions. This is illustrated in Figure 2.

6. Massive Model Testing

We are embarked on a major effort to jointly search the 7–dimensional cosmological parameter–space of $\Omega_m, \Omega_{DE}, \Omega_b, \tau$, neutrino fraction, spectral index and H_0 using parametric models created by CMBfast and thus determine which of these models fit within the confidence ball around our $f(X_i)$ at the 95% confidence limit. Traditionally, this is done by marginalising over the other parameters to gain confidence intervals on each parameter separately. This is a problem in high-dimensions where the likelihood function can be degenerate, ill-defined and under-identified. Unfortunately, the nonparametric approach is computationally intense as millions of models need to searched, each of which takes $\simeq 3$ minute to run.

To mitigate this problem, we have developed an intelligent method for searching for the surface of the confidence ball in high-dimensions based on Kriging. Briefly, kriging is a method of interpolation which predicts unknown values from data observed at known locations (also known as Gaussian process regression, which is a form of Bayesian inference in Statistics). There are many different metrics for evaluating the kriging success; we use here the "Straddle" method which picks new test points based both on the overall distance from previous searched points, as well as being predicted to be close to the boundary of the confidence ball. We have also developed a heuristic algorithm for searching for "missed peaks" in the likelihood space by searching models along the path joining previously detected peaks. We find no "missed peaks", which illustrates our kriging algorithm is effective in finding the surface of the confidence ball in this high dimensional space.

We have distributed the CMBfast model computations over a local condor pool of computers. In Figure 3, we show preliminary results from this high-dimension search for the surface of the confidence ball and present **joint** 2D confidence limits on pairs of the aforementioned cosmological parameters. These calculations represent 6.8 years of CPU time to calculate over one million CMBfast models. In the near future, we will move this analysis to TeraGrid, using *VOTechBroker*, and plan 10 million models to fully map the surface of the confidence ball. We will also make available a Java–based web service for accessing these models, and the WMAP confidence ball, thus allowing other users to rapidly combine their data with our WMAP constraints e.g., doing a joint constraint from LSS and CMB data. We are also working on possible convergence tests, and visualization tools within VOTech, to access this high-dimensional data.

7. Summary

The two examples given here – massive model testing of the WMAP data using nonparametric statistics and higher–order correlation functions of SDSS galaxies – represent a growing trend in astrophysics and cosmology for massive statistical computations. Our plan is to develop the *VOTechBroker* to provide a power framework within which such massive astronomical analyses can be performed. As discussed,

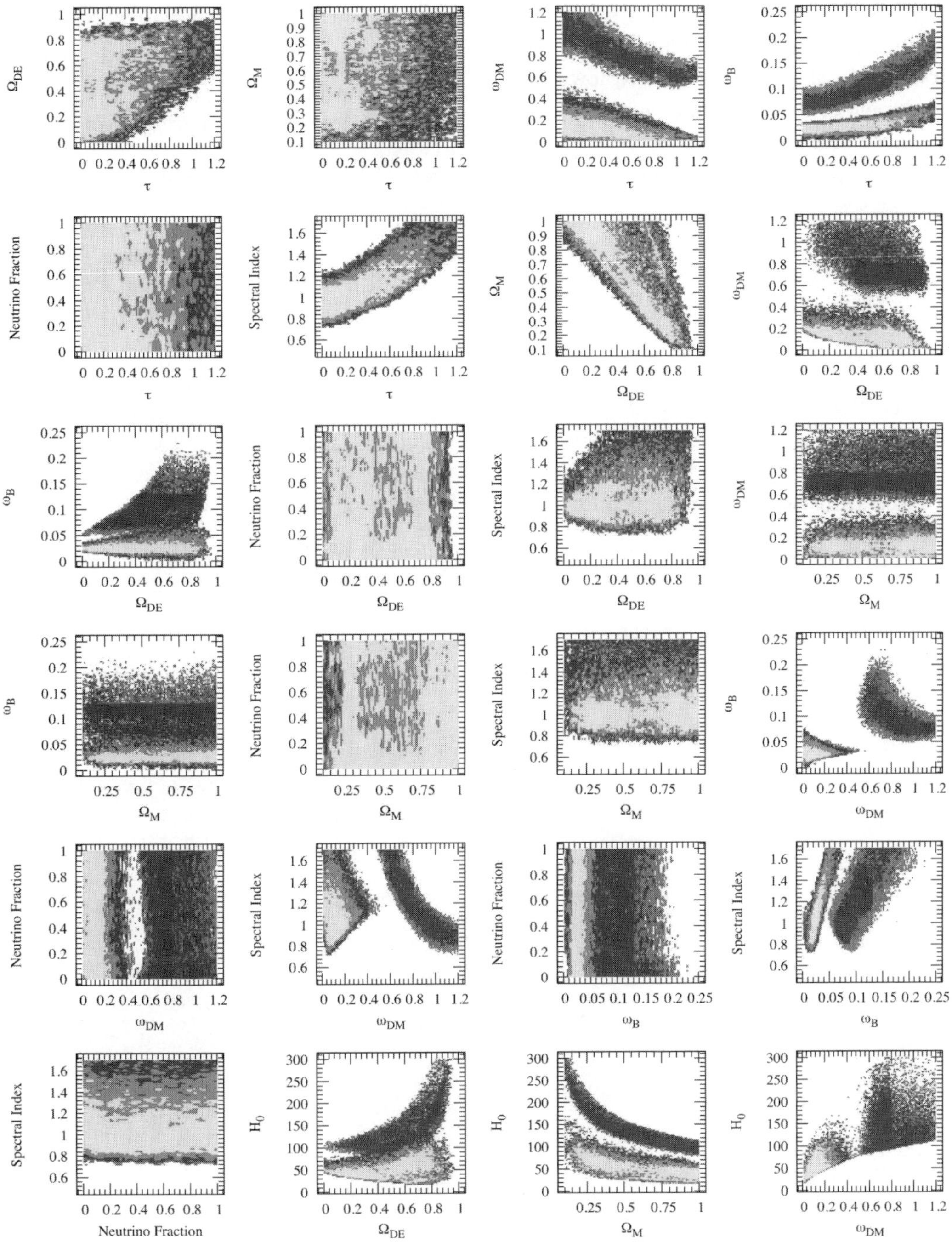

Fig. 3. The results of our 7–dimensional parameter search using 1.2 million models from CMBfast. The lightest shading are models excluded at the 34% level. The mid-scale shading are models excluded by the 68% ball and the darkest is the 95% confidence ball.

the main goals of the *VOTechBroker* are to abstract from the user (either a person or another program) the complexities of job submission and management on computational grids, as well as being a modular "plug–in" design so other algorithms and software can be easily added. Finally, we plan to integrate *VOTechBroker* into the AstroGrid workflow and MySpace environments, so it becomes a natural repository for a host of advanced statistical algorithms that scale-up in preparation for petabyte-scale datasets and analyses.

Acknowledgments

We thank our collaborators and colleagues in inCA, VOTech, AstroGrid, SDSS and VO projects. The work presented here was partly funded by NSF ITR Grant 0121671 and through the EU VOTech and Marie Curie programs. RCN thanks the organisers of the Phystat05 meeting for their invitation. GS thanks the VOTech and University of Edinburgh for his funding (see eurovotech.org for details).

References

1. Beran, R., *J. Amer. Stat. Assoc.* **95**, 155 (2000).
2. Genovese, C., *et al.*, *Statistical Science*, 2004, *astro-ph/0410140* .
3. Gray, A., *et al. Conference Proceeding for ADASS XIII*, 2003, *astro-ph/0401121* .
4. Landy, S.D. & Szalay, A., *Astrophysical Journal* **412**, 64 (1993).
5. Miller, C.J., *et al. Astrophysical Journal* **565**, 67 (2002).
6. Moore, A.W., *et al. Conference Proceeding for "Mining the Sky"*, 2000, *astro-ph/0012333*
7. Nichol, R.C., *et al. Conference Proceedings for "Statistical Challenges in Modern Astronomy III"*, (2001), *astro-ph/0110230* .
8. Peebles, P.J.E., *Large-scale structure in the Universe*, Princeton University Press, 1980.
9. Szalay, A. & Gray J., *Science* **293**, 2037, (2001).
10. Szapudi, I., (2005), *astro-ph/0505391* .
11. Szapudi, I. & Szalay, A., *Astrophysical Journal* **494**, 41 (1998).
12. Wasserman, L., *et al. Conference Proceedings for "Statistical Challenges in Modern Astronomy III"*, 2001, *astro-ph/0112050* .

${}_s\mathcal{P}lot$: A QUICK INTRODUCTION

M. PIVK
CERN, 1211 Geneva 23, Switzerland
E-mail: muriel.pivk@cern.ch

The paper advocates the use of a statistical tool dedicated to the exploration of data samples populated by several sources of events. This new technique, called ${}_s\mathcal{P}lot$, is able to unfold the contributions of the different sources to the distribution of a data sample in a given variable. The ${}_s\mathcal{P}lot$ tool applies in the context of a Likelihood fit which is performed on the data sample to determine the yields of the various sources.

1 Introduction

This paper describes a new technique to explore a data sample when the latter consists of several sources of events merged into a single sample of events. The events are assumed to be characterized by a set of variables which can be split into two components. The first component is a set of variables for which the distributions of all the sources of events are known: below, these variables are referred to as the *discriminating* variable. The second component is a set of variables for which the distributions of some sources of events are either truly unknown or considered as such: below, these variables are referred to as the *control* variables.

The new technique, termed ${}_s\mathcal{P}lot$ [a], allows one to reconstruct the distributions for the control variable, independently for each of the various sources of events, without making use of any *a priori* knowledge on this variable. The aim is thus to use the knowledge available for the discriminating variables to be able to infer the behavior of the individual sources of events with respect to the control variable. An essential assumption for the ${}_s\mathcal{P}lot$ technique to apply is that the control variable is uncorrelated with the discriminating variables.

The ${}_s\mathcal{P}lot$ technique is developed in the context of a maximum Likelihood method making use of the discriminating variables. Section 2 is dedicated to the definition of fundamental objects necessary for the following. Section 3 presents an intermediate technique, simpler but inadequate, which is a first step towards the ${}_s\mathcal{P}lot$ technique. The ${}_s\mathcal{P}lot$ formalism is then developed in Section 4 and its properties explained in Section 5. An example of ${}_s\mathcal{P}lot$ at work is provided in Section 6 and some applications are described in Section 7. Finally, the case where the control variable is correlated with the discriminating ones is discussed in Section 8.

2 Basics and definitions

One considers an unbinned extended maximum Likelihood analysis of a data sample in which are merged several species (signal and background) of events. The log-Likelihood is expressed as:

$$\mathcal{L} = \sum_{e=1}^{N} \ln \Big\{ \sum_{i=1}^{\mathrm{N_s}} N_i \mathrm{f}_i(y_e) \Big\} - \sum_{i=1}^{\mathrm{N_s}} N_i \ , \qquad (1)$$

where

- N is the total number of events considered,
- $\mathrm{N_s}$ is the number of species of events populating the data sample,
- N_i is the (non-integral) number of events expected on the average for the ith species,
- y represents the set of discriminating variables, which can be correlated with each other,
- $\mathrm{f}_i(y_e)$ is the value of the Probability Density Function (pdf) of y for the ith species and for event e.

The log-Likelihood $\mathcal{L}$ is a function of the $\mathrm{N_s}$ yields N_i and, possibly, of implicit free parameters designed to tune the pdfs on the data sample. These parameters as well as the yields N_i are determined by maximizing the above log-Likelihood.

The crucial point for the reliability of such an analysis is to use an exhaustive list of sources of events combined with an accurate description of all the pdfs f_i. If the distributions of the control variables are known (resp. unknown) for a particular

[a] The ${}_s\mathcal{P}lot$ technique is the subject of a publication [1] where details of the calculations and more examples can be found.

source of events, one would like to compare the expected distribution for this source to the one extracted from the data sample (resp. determine the distribution for this source) [b].

The control variable x which, by definition, does not explicitly appear in the expression of $\mathcal{L}$, can be:

1. totally correlated with the discriminating variables y (x belongs to the set y for example). This is the case treated in Section 3.
2. uncorrelated with y. This is the subject of Section 4.
3. partly correlated with y. This case is discussed Section 8.

In an attempt to have access to the distributions of control variables, a common method consists of applying cuts which are designed to enhance the contributions to the data sample of particular sources of events. However, the result is frequently unsatisfactory: firstly because it can be used only if the signal has prominent features to be distinguished from the background, and secondly because of the cuts applied, a sizeable fraction of signal events can be lost, while a large fraction of background events may remain.

The aim of the ${}_s\mathcal{P}lot$ formalism developed in this paper is to unfold the true distribution (denoted in boldface $\mathbf{M}_{\rm n}(x)$) of a control variable x for events of the $\rm n^{th}$ species (any one of the $\rm N_s$ species), from the sole knowledge of the pdfs of the discriminating variables $\rm f_\mathit{i}$, the first step being to proceed to the maximum Likelihood fit to extract the yields N_i. The statistical technique ${}_s\mathcal{P}lot$ allows to build histograms in x keeping all signal events while getting rid of all background events, and keeping track of the statistical uncertainties per bin in x.

3 First step towards ${}_s\mathcal{P}lot$: ${}_{\rm in}\mathcal{P}lot$

In this Section, as a means of introduction, one considers a variable x assumed to be totally correlated with y: x is a function of y. A fit having been performed to determine the yields N_i for all species, one can define naively, for all events, the weight

$$\mathcal{P}_{\rm n}(y_e) = \frac{N_{\rm n}{\rm f}_{\rm n}(y_e)}{\sum_{k=1}^{\rm N_s} N_k{\rm f}_k(y_e)} , \quad (2)$$

which can be used to build an estimate, denoted $\tilde{\rm M}_{\rm n}$, of the x-distribution of the species labelled n (signal or background):

$$N_{\rm n}\tilde{\rm M}_{\rm n}(\bar{x})\delta x \equiv \sum_{e\subset\delta x} \mathcal{P}_{\rm n}(y_e) , \quad (3)$$

where the sum runs over the events for which the x value lies in the bin centered on $\bar{x}$ and of total width δx.

In other words, $N_{\rm n}\tilde{\rm M}_{\rm n}(\bar{x})\delta x$ is the x-distribution obtained by histogramming events, using the weight of Eq. (2). To obtain the expectation value of $\tilde{\rm M}_{\rm n}$, one should replace the sum in Eq. (3) by the integral

$$\left\langle \sum_{e\subset\delta x} \right\rangle \longrightarrow \int {\rm d}y \sum_{j=1}^{\rm N_s} N_j{\rm f}_j(y)\delta(x(y)-\bar{x})\delta x . \quad (4)$$

Similarly, identifying the number of events N_i as determined by the fit to the expected number of events, one readily obtains:

$$\left\langle N_{\rm n}\tilde{\rm M}_{\rm n}(\bar{x}) \right\rangle \equiv N_{\rm n}\mathbf{M}_{\rm n}(\bar{x}) . \quad (5)$$

Therefore, the sum over events of the naive weight $\mathcal{P}_{\rm n}$ reproduces, on average, the true distribution $\mathbf{M}_{\rm n}(x)$. Plots obtained that way are referred to as ${}_{\rm in}\mathcal{P}lots$: they provide a correct means to reconstruct $\mathbf{M}_{\rm n}(x)$ only insofar as the variable considered is **in** the set of discriminating variables y. These ${}_{\rm in}\mathcal{P}lots$ suffer from a major drawback: x being fully correlated to y, the pdfs of x enter implicitly in the definition of the naive weight, and as a result, the $\tilde{\rm M}_{\rm n}$ distributions cannot be used easily to assess the quality of the fit, because these distributions are biased in a way difficult to grasp, when the pdfs $\rm f_\mathit{i}(\mathit{y})$ are not accurate. For example, let us consider a situation where, in the data sample, some events from the $\rm n^{th}$ species show up far in the tail of the $\rm M_n(\mathit{x})$ distribution which is implicitly used in the fit. The presence of such events implies that the true distribution $\mathbf{M}_{\rm n}(x)$ must exhibit a tail which is not accounted for by $\rm M_n(\mathit{x})$. These events would enter in the reconstructed ${}_{\rm in}\mathcal{P}lot$ $\tilde{\rm M}_{\rm n}$ with a very small weight, and they would thus escape detection by the above procedure: $\tilde{\rm M}_{\rm n}$ would be close to $\rm M_n$, the distribution assumed for x. Only a mismatch in the core of the x-distribution can be revealed with ${}_{\rm in}\mathcal{P}lots$. Stated differently, the error bars which can be attached to each individual bin of

[b] Removing one of the discriminating variables from the set y before performing again the maximum Likelihood fit, one can consider the removed variable as a control variable x, provided it is uncorrelated with the others.

$\tilde{\mathrm{M}}_{\mathrm{n}}$ cannot account for the systematical bias inherent to the ${}_{\mathrm{in}}\mathcal{P}lots$.

4 The ${}_s\mathcal{P}lot$ formalism

In this Section one considers the more interesting case where the two sets of variables x and y are uncorrelated. Hence, the total pdfs $\mathrm{f}_i(x,y)$ all factorize into products $\mathbf{M}_i(x)\mathrm{f}_i(y)$. While performing the fit, which relies only on y, no *a priori* knowledge of the x-distributions is used.

One may still consider the above distribution $\tilde{\mathrm{M}}_{\mathrm{n}}$ (Eq. (3)), using the naive weight of Eq. (2). However in that case, the expectation value of $\tilde{\mathrm{M}}_{\mathrm{n}}$ is a biased estimator of $\mathbf{M}_{\mathrm{n}}$:

$$\begin{aligned}\left\langle N_{\mathrm{n}}\tilde{\mathrm{M}}_{\mathrm{n}}(\bar{x})\right\rangle &= \int \mathrm{d}y\mathrm{d}x \sum_{j=1}^{\mathrm{N_s}} N_j\mathbf{M}_j(x)\mathrm{f}_j(y)\delta(x-\bar{x})\mathcal{P}_{\mathrm{n}} \\ &= N_{\mathrm{n}}\sum_{j=1}^{\mathrm{N_s}}\mathbf{M}_j(\bar{x})N_j\int \mathrm{d}y\frac{\mathrm{f_n}(y)\mathrm{f}_j(y)}{\sum_{k=1}^{\mathrm{N_s}}N_k\mathrm{f}_k(y)} \qquad (6)\\ &\neq N_{\mathrm{n}}\ \mathbf{M}_{\mathrm{n}}(\bar{x})\ .\end{aligned}$$

Here, the naive weight is no longer satisfactory because, when summing over the events, the x-pdfs $\mathbf{M}_j(x)$ appear now on the right hand side of Eq. (4), while they are absent in the weight. However, one observes that the correction term in the right hand side of Eq. (6) is related to the inverse of the covariance matrix, given by the second derivatives of $-\mathcal{L}$:

$$\mathbf{V}^{-1}_{\mathrm{n}j} = \frac{\partial^2(-\mathcal{L})}{\partial N_{\mathrm{n}}\partial N_j} = \sum_{e=1}^{N}\frac{\mathrm{f_n}(y_e)\mathrm{f}_j(y_e)}{(\sum_{k=1}^{\mathrm{N_s}}N_k\mathrm{f}_k(y_e))^2}\ . \qquad (7)$$

On average, one gets:

$$\langle\mathbf{V}^{-1}_{\mathrm{n}j}\rangle = \int \mathrm{d}y\frac{\mathrm{f_n}(y)\mathrm{f}_j(y)}{\sum_{k=1}^{\mathrm{N_s}}N_k\mathrm{f}_k(y)}\ . \qquad (8)$$

Therefore, Eq. (6) can be rewritten:

$$\left\langle\tilde{\mathrm{M}}_{\mathrm{n}}(\bar{x})\right\rangle = \sum_{j=1}^{\mathrm{N_s}}\mathbf{M}_j(\bar{x})N_j\left\langle\mathbf{V}^{-1}_{\mathrm{n}j}\right\rangle\ . \qquad (9)$$

Inverting this matrix equation, one recovers the distribution of interest:

$$N_{\mathrm{n}}\mathbf{M}_{\mathrm{n}}(\bar{x}) = \sum_{j=1}^{\mathrm{N_s}}\langle\mathbf{V}_{\mathrm{n}j}\rangle\left\langle\tilde{\mathrm{M}}_j(\bar{x})\right\rangle\ . \qquad (10)$$

Hence, when x is uncorrelated with the set y, the appropriate weight is not given by Eq. (2), but is the covariance-weighted quantity (thereafter called sWeight) defined by:

$$\boxed{{}_s\mathcal{P}_{\mathrm{n}}(y_e) = \frac{\sum_{j=1}^{\mathrm{N_s}}\mathbf{V}_{\mathrm{n}j}\mathrm{f}_j(y_e)}{\sum_{k=1}^{\mathrm{N_s}}N_k\mathrm{f}_k(y_e)}}\ . \qquad (11)$$

With this sWeight, the distribution of the control variable x can be obtained from the ${}_s\mathcal{P}lot$ histogram:

$$N_{\mathrm{n}}\ {}_s\tilde{\mathrm{M}}_{\mathrm{n}}(\bar{x})\delta x \equiv \sum_{e\subset\delta x}{}_s\mathcal{P}_{\mathrm{n}}(y_e)\ , \qquad (12)$$

which reproduces, on average, the true binned distribution:

$$\left\langle N_{\mathrm{n}}\ {}_s\tilde{\mathrm{M}}_{\mathrm{n}}(x)\right\rangle = N_{\mathrm{n}}\mathbf{M}_{\mathrm{n}}(x)\ . \qquad (13)$$

The fact that the covariance matrix $\mathbf{V}_{ij}$ enters in the definition of the sWeights is enlightening: in particular, the sWeight can be positive or negative, and the estimators of the true pdfs are not constrained to be strictly positive.

5 ${}_s\mathcal{P}lot$ properties

Beside satisfying the essential asymptotic property Eq. (13), ${}_s\mathcal{P}lots$ bear properties which hold for finite statistics.

The distribution ${}_s\tilde{\mathrm{M}}_{\mathrm{n}}$ defined by Eq. (12) is guaranteed to be normalized to unity and the sum over the species of the ${}_s\mathcal{P}lots$ reproduces the data sample distribution of the control variable. These properties rely on maximizing the Likelihood:

- Each x-distribution is properly normalized. The sum over the x-bins of $N_{\mathrm{n}}\ {}_s\tilde{\mathrm{M}}_{\mathrm{n}}\delta x$ is equal to N_{n}:

$$\sum_{e=1}^{N}{}_s\mathcal{P}_{\mathrm{n}}(y_e) = N_{\mathrm{n}}\ . \qquad (14)$$

- In each bin, the sum over all species of the expected numbers of events equals to the number of events actually observed. In effect, for any event:

$$\sum_{l=1}^{\mathrm{N_s}}{}_s\mathcal{P}_l(y_e) = 1\ . \qquad (15)$$

Therefore, an ${}_s\mathcal{P}lot$ provides a consistent representation of how all events from the various species are distributed in the control variable x. Summing up the $\mathrm{N_s}$ ${}_s\mathcal{P}lots$, one recovers the data sample distribution in x, and summing up the number of events entering in a ${}_s\mathcal{P}lot$ for a given species, one recovers the yield of the species, as it is provided by the fit. For instance, if one observes an excess of events for a particular n^{th} species, in a given x-bin, this excess is effectively accounted for in the number of events N_{n} resulting from the fit. To remove these events implies a corresponding decrease in N_{n}. It remains to gauge how significant is an anomaly in the x-distribution of the nth species.

The statistical uncertainty on $N_{\mathrm{n}}\ {}_s\tilde{\mathrm{M}}_{\mathrm{n}}(x)\delta x$ can be defined in each bin by

$$\sigma[N_{\mathrm{n}}\ {}_s\tilde{\mathrm{M}}_{\mathrm{n}}(x)\delta x] \;=\; \sqrt{\sum_{e\subset\delta x}({}_s\mathcal{P}_{\mathrm{n}})^2}\ . \qquad (16)$$

The above properties Eqs. (13)-(15) are completed by the fact that the sum in quadrature of the uncertainties Eq. (16) reproduces the statistical uncertainty on the yield N_{n}, as it is provided by the fit. In effect, the sum over the x-bins reads:

$$\sum_{[\delta x]}\sigma^2[N_{\mathrm{n}}\ {}_s\tilde{\mathrm{M}}_{\mathrm{n}}\delta x] \;=\; \mathbf{V}_{\mathrm{nn}}\ . \qquad (17)$$

Therefore, for the expected number of events per x-bin indicated by the ${}_s\mathcal{P}lots$, the statistical uncertainties are straightforward to compute using Eq. (16). The latter expression is asymptotically correct, and it provides a consistent representation of how the overall uncertainty on N_{n} is distributed in x among the events of the nth species. Because of Eq. (17), and since the determination of the yields is optimal when obtained using a Likelihood fit, one can conclude that the ${}_s\mathcal{P}lot$ technique is itself an optimal method to reconstruct distributions of control variables.

6 Illustrations

An example of ${}_s\mathcal{P}lot$ at work is taken from the analysis where the method was first used [2,3]. One deals with a data sample in which three species are present: $B^0\to\pi^+\pi^-$ and $B^0\to K^+\pi^-$ are signals and the main background comes from $e^+e^-\to q\bar{q}$. The variable which is not incorporated in the fit is called ΔE and is used here as the control variable x. The detailed description of the variables can be found in Refs. [2,3].

The left plot of Fig. 1 shows the distribution of ΔE after applying a cut on the Likelihood ratio. Therefore, the resulting data distribution concerns a reduced subsample for which statistical fluctuations cannot be attributed unambiguously to signal or to background. For example, the excess of events appearing on the left of the peak is likely to be attributed to a harmless background fluctuation.

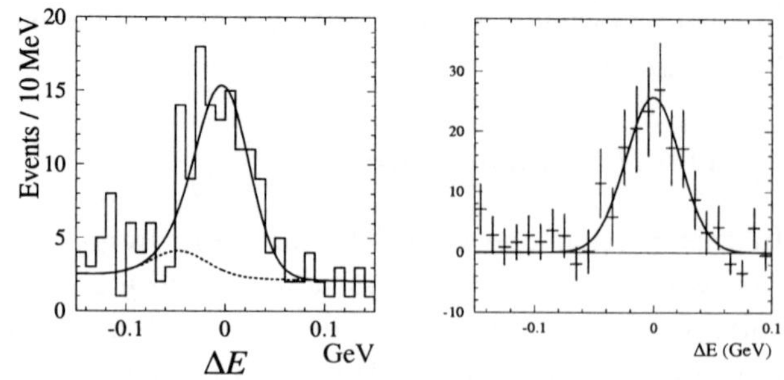

Figure 1. Signal distribution of the ΔE variable. The left figure is obtained applying a cut on the Likelihood ratio to enrich the data sample in signal events (about 60% of signal is kept). The right figure shows the ${}_s\mathcal{P}lot$ for signal (all events are kept).

Looking at the right plot of Fig. 1, which is a signal ${}_s\mathcal{P}lot$, one can see that these events are signal events, not background events. The pdf of ΔE which is used in the conventional fit for the whole analysis is superimposed on the ${}_s\mathcal{P}lot$. When this pdf is used, the events in excess are interpreted as background events while performing the fit. Further studies have shown [2] that these events are in fact radiative events, *i.e.* $B^0\to\pi^+\pi^-\gamma$. When ignored in the analysis they lead to underestimates of the branching ratios by about 10%. The updated results [4] for the $B^0\to\pi^+\pi^-$, $K^+\pi^-$ analysis, now taking into account the contribution of radiative events, show agreement with the estimate made in Ref. [2].

7 Applications

Beside providing a convenient and optimal tool to cross-check the analysis by allowing distributions of control variables to be reconstructed and then compared with expectations, the ${}_s\mathcal{P}lot$ formalism can be applied also to extract physics results, which would otherwise be difficult to obtain. For example, one may be willing to explore some unknown physics involved in the distribution of a variable x. Or, one may be interested to correct a particular yield provided by the Likelihood fit from a selection efficiency which is known to depend on a variable x, for which the pdf is unknown. Provided one can demonstrate (*e.g.* through Monte-Carlo simulations) that the variable x exhibits weak correlation with the discriminating variables y.

To be specific, one can take the example of a three body decay analysis of a species, the signal, polluted by background. The signal pdf inside the two-dimensional Dalitz plot is assumed to be not known, because of unknown contributions of resonances, continuum and of interference pattern. Since the x-dependence of the selection efficiency $\epsilon(x)$ can be computed without *a priori* knowledge of the x-distributions, one can build the efficiency corrected two-dimensional ${}_s\mathcal{P}lots$ (cf. Eq. (12)):

$$\frac{1}{\epsilon(\bar{x})} N_{\rm n}\ {}_s\tilde{\rm M}_{\rm n}(\bar{x})\delta x = \sum_{e\subset\delta x} \frac{1}{\epsilon(x_e)} {}_s\mathcal{P}_{\rm n}(y_e) \ , \qquad (18)$$

and compute the efficiency corrected yields:

$$N_{\rm n}^{\epsilon} = \sum_{e=1}^{N} \frac{{}_s\mathcal{P}_{\rm n}(y_e)}{\epsilon(x_e)} \ . \qquad (19)$$

Analyses can then use the ${}_s\mathcal{P}lot$ formalism for validation purposes, but also, using Eq. (18) and Eq. (19), to probe for resonance structures and to measure branching ratios [5].

8 Correlation between variables

Correlations between variables, if not trivial, are usually assessed by Monte-Carlo simulations. In case significant correlations are observed, one may still use the ${}_s\mathcal{P}lot$ weight of Eq. (11), but then there is a caveat. The distribution obtained with ${}_s\mathcal{P}lot$ cannot be compared directly with the marginal distribution of x. In that case, one must rely on Monte-Carlo simulation, and apply the ${}_s\mathcal{P}lot$ technique to the simulated events, in order to obtain Monte-Carlo ${}_s\mathcal{P}lots$. It is these Monte-Carlo ${}_s\mathcal{P}lots$ which are to be compared to the ${}_s\mathcal{P}lot$ obtained with the real data. Stated differently, the ${}_s\mathcal{P}lot$ can still be applied to compare the behaviour of the data with the Monte-Carlo expected behavior, but it loses its simplicity.

9 Conclusion

The technique presented in this paper applies when

- one examines a data sample originating from different sources of events,
- a Likelihood fit is performed on the data sample to determine the yields of the sources,
- this Likelihood uses a set y of discriminating variables,
- keeping aside a control variable x which is statistically uncorrelated to the set y.

By building ${}_s\mathcal{P}lots$, one can reconstruct the distributions of the control variable x, separately for each source present in the data sample. Although no cut is applied (hence, the ${}_s\mathcal{P}lot$ of a given species represents the whole statistics of this species) the distributions obtained are pure in a statistical sense: they are free from the potential background arising from the other species. The more discriminating the variables y, the clearer the ${}_s\mathcal{P}lot$ is. The technique is straightforward to implement; it is available in the ROOT framework under the class TSPlot[6]. It features several nice properties: both the normalizations and the statistical uncertainties of the ${}_s\mathcal{P}lots$ reflect the fit ouputs.

References

1. M. Pivk and F.R. Le Diberder, *Nucl. Inst. Meth.* A 555, 356-369, 2005 (`physics/0402083`).
2. M. Pivk, Thèse de l'Université Paris VII, BABAR-THESIS-03/012 (2003), available (in French) at http://tel.ccsd.cnrs.fr (ID 00002991).
3. The BABAR Collaboration, *Phys. Rev. Lett.* **89**, 281802 (2002).
4. The BABAR Collaboration, `hep-ex/0508046`.
5. The BABAR Collaboration, *Phys. Rev. Lett* **93**, 181805 (2004).
6. http://root.cern.ch/root/htmldoc/TSPlot.html

EASY DATA ANALYSIS USING R

M. PATERNO

Fermi National Accelerator Laboratory, P.O. Box 500, Batavia, IL 60510, USA
E-mail: paterno@fnal.gov

The **R** system is a mature, freely available, and widely used "language and environment for statistical computing and graphics", based upon the award-winning S language developed by John Chambers. It combines modern graphical displays of data, a convenient and easy to learn programming language, and an extremely wide variety of statistical tools, to provide a tool that is both easy to use and powerful. Unfortunately, **R** is relatively little-known in the physics community. In this paper, I will present some of the features of **R** that have proved useful in the everyday work I have been involved with at Fermilab.

1. Introduction

The exploration of data is an intensive intellectual task. It is the job of data analysis software to support the process of learning from data, by providing tools to visualize and analyze data. To be successful, a software system should be both powerful and easy to use.

The award-winning[1] **S** system[2] and its free-software implementation **R**[3] are perhaps the most widely-used data analysis systems available. As described on the **R** Project's web page, "**R** is a language and environment for statistical computing and graphics." **R** has been successful because it provides excellent graphical tools, a convenient and powerful language for data manipulation, and many modern data analysis techniques.

In this paper, I describe some of the features of **R** that have made it the preferred analysis environment for many. In each of the following sections, I concentrate on one of the advantages provided by **R**.

2. High Quality Graphics

R provides plots we commonly use: *e.g.*, histograms and (x, y) plots. In addition to the data themselves, such plots can include fits to the data (performed with a wide variety of fitting methods), error bars, textual and mathematic annotations, and color, over all of which the user has fine-grained programmatic control. The default values of the graphical parameters of **R** have been carefully chosen to match the strengths and weaknesses of human perception (see the books by Cleveland[4, 5] and references therein).

R also provides a variety of useful plot types which are *not* widely known to the physics community. These include (among others): *dot plots*, *splom* (scatter plot matrix), *box-and-whisker plots*, and *quantile* and *QQ plots*. **R** provides additional special-purpose plots. Many statistical tools come with dedicated plot styles, *e.g.* clustering techniques with associated dendrogram plots.

2.1. *The dot plot*

Previously-mentioned studies indicate that human perception is poor at interpreting the *pie chart*, because the eye is not good at comparing relative areas. The *dot plot* allows much clearer presentation of such data. Figure 1, showing the leptonic branching fractions of the Z boson, compares a pie chart and a dot plot. Unlike a pie chart, the dot plot can show both numeric values and uncertainties.

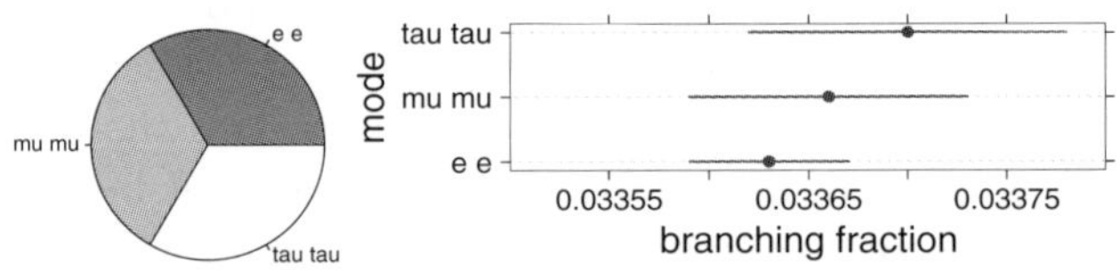

Fig. 1. A comparison between a pie chart (left) and a dot plot (right), showing the measured leptonic branching fractions of the Z boson (PDG 2004).

2.2. *The scatter plot matrix*

The *scatter plot matrix*, or *splom*, (figure 2) is a useful tool for quickly identifying pairs of quantities with interesting relationships. It shows all pairwise associations between observations of a set of measured quantities. Each measured quantity appears in one column, and also in one row, of the matrix of scatter plots. Interesting correlations are easily visible. It is important to note that the scatter plots are unbinned—so that no features in the correlations are lost, due to unfortunate binning.

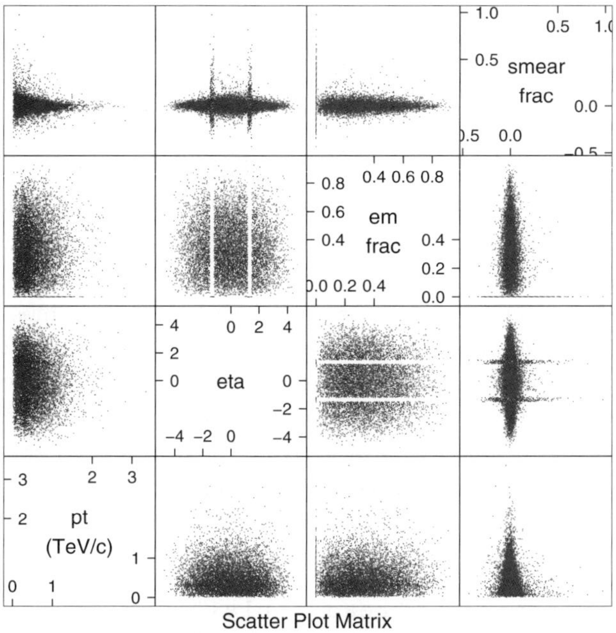

Fig. 2. A sample splom, showing correlations between "measured" features of simulated jets in a toy simulation.

2.3. *The box-and-whisker plot*

The *box-and-whisker plot*, or *box plot*, (figure 3) provides a concise summary of many of the interesting features of a 1-d distribution. For a distribution with long tails, or a distribution that is asymmetric, the mean and standard deviation are sometimes misleading. Unless the symmetry of a distribution is known, more "robust" statistics (*e.g.* median, quartiles) may be more informative. The box plot summarizes these statistics, and others.

The box plot shows the median, first and third quartiles, and "hinges" of the data, as well as "outliers". Let r be the interquartile range. The upper hinge is defined by the value of the data closest to, but not within, a distance of $1.5r$ from the upper quartile. The lower hinge is defined similarly, with respect to the first quartile. Outliers are those data still further from the median than the hinges; each is shown individually on the box plot.

In high energy physics, the "profile histogram" is often used to summarize a 2-d distribution. The "profile histogram" displays, for each bin in x, the

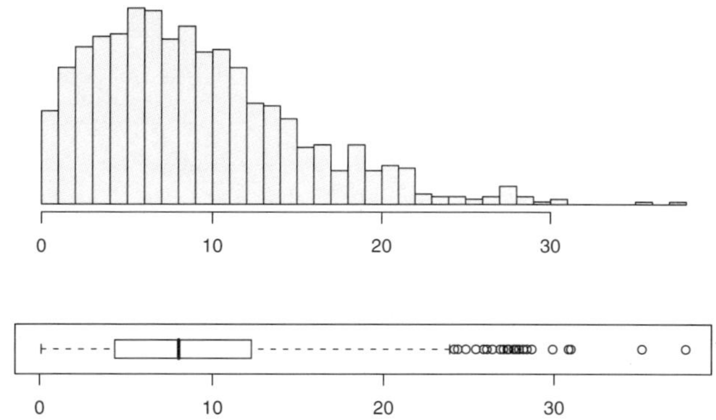

Fig. 3. A sample histogram and a box plot showing the same data.

mean and standard deviation of y. But if the distribution in one or more bins is not symmetric, or has outliers, a multi-way box plot (which presents a box plot of y for each bin in x) can be much more informative; figure 4 shows an example.

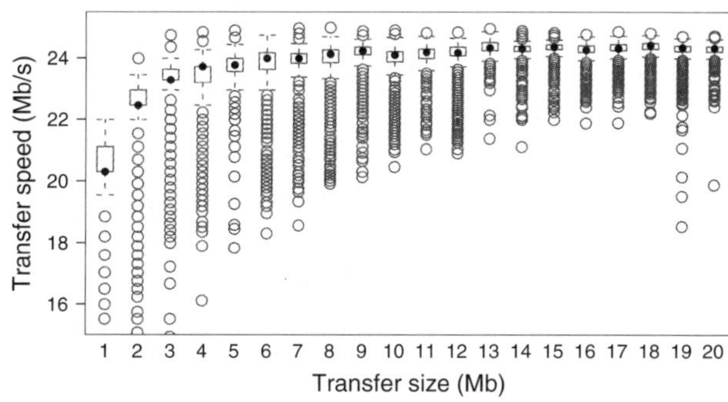

Fig. 4. A sample multi-way box plot, showing the variation of data transfer speed with the size of the data block transferred for a prototype database server. The bulk of each distribution is contained in a very small box; the wide range of the outliers would have yielded a misleading profile histogram.

2.4. *The quantile and QQ plots*

The previously-mentioned studies show human perception is poor at comparing similar histograms. Quantile plots (cumulative distributions) are somewhat easier to distinguish, and can be drawn unbinned, thus alleviating the troubles caused by binning. Quantile-quantile (QQ) plots are still easier to compare, and allow identification of even small differences between similar distributions. The QQ plot (figure 5) shows a comparison of the distributions of x and y formed by plotting the quantiles of x against the quantiles of y.

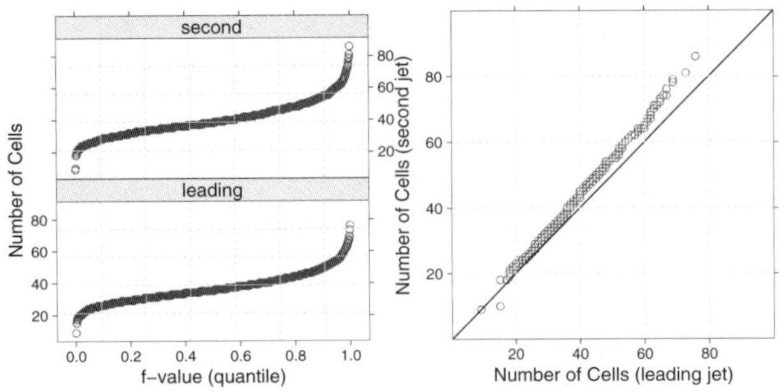

Fig. 5. Left: quantile plots of the number of cells comprising the leading and second jets in a simulated event sample from the CMS experiment. Right: a QQ plot comparing the same quantities.

3. Convenient Environment

Convenience and ease-of-use are crucial aspects of a data exploration system. This is especially important for those of us who analyze data only sporadically, or who must come back to a particular analysis after a period engaged in other pursuits. An environment which serves as a platform for learning from data should not itself be a barrier. **R** provides a number of conveniences.

An **R** session can be saved to disk, in a platform neutral format, allowing the application state to be recovered at a later time. Use of a different directory to save each different "analysis" (session files, data files, and **R** scripts) provides a clear and convenient organization capable of supporting many ongoing analyses.

R can read data from many sources: tabular data files (either local or remote, *via* HTTP), common spreadsheet formats (*e.g.* Excel), and a variety of databases (Oracle, MySQL, SQLite, PostgreSQL, or any ODBC database). It is not very hard to extend this ability to read additional file formats. For example, local development at Fermilab[6] has provided the ability for **R** to read ROOT[7] trees, such as those of the CMS experiment's reconstruction framework.

The **R** core software comes with a significant amount of functionality. Additional functionality is available through *packages*. Distributed package management is integrated into the system, with an ease of use similar to other successful products, such as Perl's CPAN and Linux's *yum* utility. A uniform documentation model is observed, in part helped by enforcement by the package building system. Users have all the tools to create (and even distribute) their own packages, and to contribute them to the **R** user community. Discovery and installation of new packages is extremely easy; one can visit `http://cran.r-project.org/` to see what is available, or use the `install.packages` or `update.packages` functions in an **R** session.

4. An Abbreviated Example of Use

In working on the data acquisition system for the (late) BTeV experiment[8], we needed to analyze a simulation of the pixel detector, to determine how the expected data rate varied with beam luminosity. The proposed BTeV pixel detector contained 32 independent detector panels, called "stations". The simulated data consisted of a record of the number clusters ("triplets") of activated pixels in each station for each simulated event. Simulation output was converted with a simple Python program to a text file, and read with **R**:

```
> stations = read.table("btev.dat")
```

This creates a *data frame*, here named "stations", which behaves in many ways like a table. We can discover the number of rows, and the names of the columns, in the data frame "stations", and print a few rows:

```
> nrow(stations)
[1] 554218
> stations[1:3,]
  nint idx station ntrip
1    1   1       0     0
2    1   1       1     0
3    1   1       2    24
```

Plots can be created from a data frame using a wide variety of functions. To view set of histograms, one uses:

```
> histogram(~ntrip|station,
   data=stations,
   subset=(station %in% 2:10 &
           nint==6))
```

The above command produces a set of histograms, each showing the distribution of `ntrip` (the number of triplets); this creates one histogram for each station, using data from the data frame "stations", but showing only those stations numbered 2 through 10, and furthermore only showing those data for which the simulated number of interactions ("nint") is 6. This example shows the conciseness and power of **R**'s data manipulation language. Once a few basic constructions are learned, they can be applied consistently in many other contexts. For example, to create a multi-way box plot, showing the distribution of the number of triplets for all stations, still limited to those data with 6 simulated interactions, one would use:

```
> bwplot(station~ntrip,
    data=stations, subset=(nint==6))
```

Next we wanted to group data: to sum `ntrip` over all stations for each event, *i.e.* for rows with equal `idx` and `nint`. This can be done with the function `aggregate`:

```
> events=aggregate(stations.ntrip,
    by=list(idx=station$idx,
            nint=station$nint), sum)
```

`aggregate` is one of many high-level data manipulation functions provided by **R**; the rich supply of such functions provides much of the expressiveness of the **R** language.

And after a little fixing of names, we can print some results:

```
> str(events)
'data.frame':   17878 obs. of  3 variables:
 $ idx: num 1 2 3 4 5 6 7 8 9 11 ...
 $ nint: Ord.factor w/ 11 levels  ...
 $ ntrip: int 26 86 15 8 70 11 3 64 7 17 ...
```

`str` shows the structure of its argument, and is useful for a short summary.

Finally, we looked at the distributions of total number of triplets in each event. We tested, using QQ plots, our suspicion that these could be described by the Weibull distribution. The plots show that only at the extreme (1-2%) high tail do the data differ from the Weibull fit. Figure 6 shows two of these plots, for those events with 5 and 10 interactions.

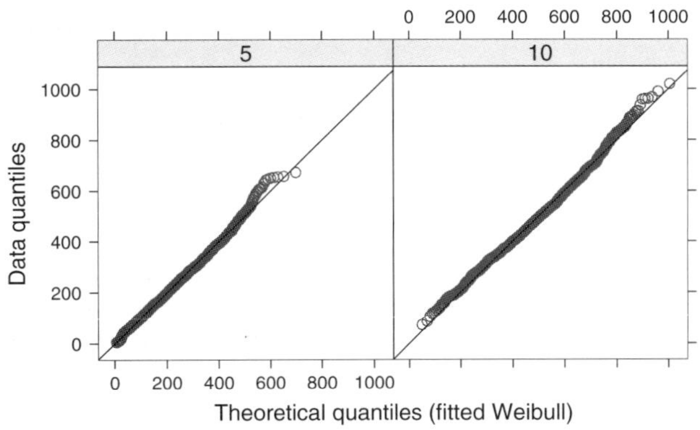

Fig. 6. Sample QQ plots, comparing the distribution of number of triplets to the fitted Weibull distribution.

5. Many Useful Add-on Packages

R makes available an enormous variety of analysis tools, including neural networks, decision trees, many types of curve fitting, bootstrapping, clustering, Markov chain Monte Carlo, and genetic algorithms. The **S** language (and so also **R**) is, more than any other language, the "common tongue" of statistical research. It is used for reference implementations of many analysis techniques, and often provides the earliest (or only) implementation of new statistical techniques.

As of the time of this writing, there were 590 packages and bundles available in the main **R** repository, and 122 more at the next largest site. Many of these packages present not just one tool, but a large *family* of tools.

6. Conclusion

The ease with which one can explore and understand data is important. My colleagues and I have found **R** to provide excellent graphical tools, an easy to learn, powerful, and convenient language for data manipulation, and a host of modern data analysis techniques. **R** allows us to concentrate on our data, not on our tools.

References

1. "ACM: Software System Award", `http:/www.acm.org/announcements/ss99.html`.
2. J.M. Chambers, *Programming with Data*, (Springer, 2004).
3. "The R Project for Statistical Computing", `http://www.r-project.org/`.
4. W.S. Cleveland, *Visualizing Data*, (Hobart Press, Summit, New Jersey, 1993).
5. W.S. Cleveland, *The Elements of Graphing Data*, (Hobart Press, Summit, New Jersey, 1994).
6. A.L. Lyon, private communication.
7. "The ROOT System Home Page", `http://root.cern.ch`.
8. "BTeV Home Page", `http://www-btev.fnal.gov`.

STATISTICS IN ROOT

R. BRUN, A. KRESHUK, L. MONETA
PH/SFT, CERN, Geneva, Switzerland

The ROOT [1] system is an Object Oriented framework for large scale data handling applications. Advanced statistical analysis tools constitute an important part of the system. This paper describes ROOT's mathematical and statistical libraries. A general overview of the system is given, with special attention payed to recently added methods.

1. Organization of mathematical and statistical libraries in ROOT

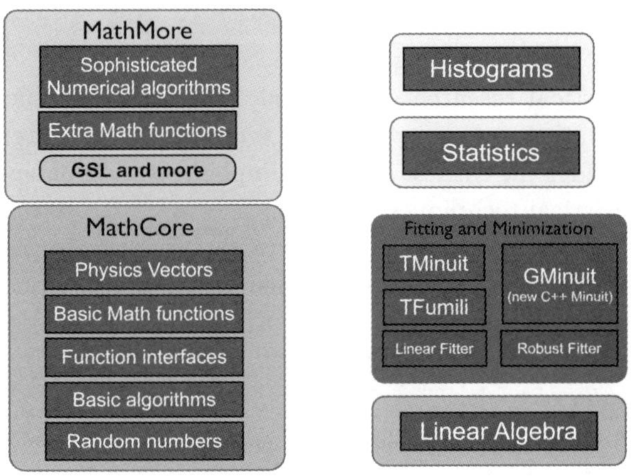

Fig. 1. New ROOT Math Components

The ROOT project includes a MATH work package that provides and supports a coherent set of mathematical and statistical libraries. The Existing Mathematical library provided by ROOT and by the SEAL[2] project are being reorganized into new libraries with the aim to avoid duplication and to facilitate support in the long term. The new structure, shown in figure 1, consists of these main components:

- **MathCore**: self-consistent minimal set of mathematical functions and C++ classes for the basic needs of HEP numerical computing. It is released as an independent library.
- **MathMore**: package incorporating functionality which might be needed for an advanced user (as opposed to MathCore which addresses the primary needs of users).
- **Linear Algebra**: library containing classes describing vector and matrix operations in arbitrary dimensions and of various types, such as symmetric or sparse matrix, and completed with linear algebra algorithms.
- **Fitting and minimization**: classes implementing various types of fitting methods, including the newly added linear and robust fitters and a set of libraries for different function minimization algorithms like MINUIT [5] and FUMILI [6], which can be loaded at run time by using the plug-in manager system.
- **Histogram library**: library containing the classes for one, two and three dimensional histograms and profiles.
- **statistical library**: package grouping the various statistical algorithms of ROOT like neural networks for multivariate analysis or classes for computing confidence levels. The algorithms are presently spread out in various ROOT libraries, but we expect in the future to group them together in a single package.

The detailed description of the above components is given in the following sections.

2. MathCore

MathCore provides the basic and most used mathematical functionality. It consists up to now of:

- commonly used special functions like the Gamma, Beta and Error function
- mathematical functions used in statistics such as probability density functions for the major distributions (normal, Poisson, binomial, Breit-Wigner, etc..)
- the physics and geometry vector package containing classes for specialized vectors in 3D and 4D and their operations.

The special functions in MathCore (and MathMore) are implemented following the same naming scheme as proposed in the next C++ Standard Library (see C++ extension proposal [3]). Extensive tests of these newly introduced mathematical functions have been performed by comparing the numerical results obtained with the functions from other packages like Mathematica or NagC. Often an accuracy at the level of 10^{-16} (double numerical accuracy) is reached for functions such as the Gamma and the Error function, improved with respect to the functions previously present in ROOT TMath. In future we expect to include in MathCore a random number generator package, which combined with the statistical functions, will provide functionality to generate random numbers according to common used statistical distributions. MathCore is a self-consistent component which can be released as an independent library and used outside the ROOT framework.

3. MathMore

This package incorporates more advanced mathematical functionality to extend MathCore. The need for separating the functionality is twofold. In order to keep the size of the core of ROOT reasonable, only the most used mathematical functionality is included in it. Secondly, there are licensing issues concerning some of the more advanced functionality which uses the GNU Scientific Library (GSL) [4], as the GSL is distributed under the GPL license, while ROOT is under LGPL. One of the design goals is to hide the implementation. Presently the mathematical functionality from GSL is used underneath. It would be very easy to shift to use another numerical package and be completely transparent to the user and straightforward for the developer. At present MathMore is composed of the following parts:

- special functions like *Bessel* functions of various types and fractional order, elliptic integrals, *Laguerre* and *Legendre* polynomials, hypergeometric functions
- cumulative distribution functions and their inverse for *Chi-Squared*, *Gamma*, *F* and *Student's* distributions and their inverses. There are also the inverses of the CDF's of the *Breit-Wigner*, *Exponential*, *Gaussian*, *Lognormal* and *Uniform* distributions.
- classes for numerical algorithms like differentiation, various types of adaptive and non-adaptive numerical integration, interpolation and root finding algorithms for one dimensional functions

4. Minimization and fitting

ROOT contains two general purpose minimization packages - Minuit[5] and Fumili[6] and a smaller class `TLinearFitter` specific for fitting functions linear in parameters. Unlike Minuit, Fumili is a specialized method for minimizing least squares and log-likelihood functions, so it makes some approximations to the second derivatives, which are not valid for a general minimization method. This allows it to converge faster. When the packages are used directly, the user has to provide the objective function to minimize. Otherwise, the Fit(...) interface methods are provided, which perform following operations:

- For histograms (classes `TH1, TH2, TH3`):
  - $\chi 2$ method
  - Log likelihood method
- For graphs
  - Unweighted (class `TGraph`)
  - Weighted (class `TGraphErrors`)
  - Taking into account asymmetry of errors (class `TGraphAsymmErrors`)
- For `TTrees`
  - Same as for histograms
  - Unbinned log likelihood method

4.1. *Linear fitting*

In ROOT v4.03/04, a special class `TLinearFitter` was introduced to separate fitting of linear and non-linear models. Now linear fitting requires only one pass over the data, and this data is not copied anywhere, which makes it very convenient for large datasets. Also, the computation time decreased substantially and the user doesn't have to set initial parameter values any more.

4.2. *Robust fitting*

The classical least squares fitting procedures are known to be very sensitive to bad observations. Even one very bad outlier can make it produce results arbitrarily far from the true parameter values. To fit such "contaminated" datasets, an extension for robust fitting was added to the `TLinearFitter` class.

Least Trimmed Squares (LTS) regression was introduced by *Rousseeuw* and *Leroy* in [7], and the approximate Fast-LTS algorithm for large data sets was proposed in [8]. This algorithm tries to find a subset of h points (out of n) that have the smallest sum of squared residuals. The parameter h - the number of good points in the dataset is set by the user and should lie between $n/2$ and n; the default value is around $n/2$. The algorithm is highly robust, with breakdown point $(n-h)/n$. In ROOT, the robust fitter can be called either by specifying option "rob" in the Fit(...) interface function of histograms, graphs and trees, or by using the extended linear fitter directly.

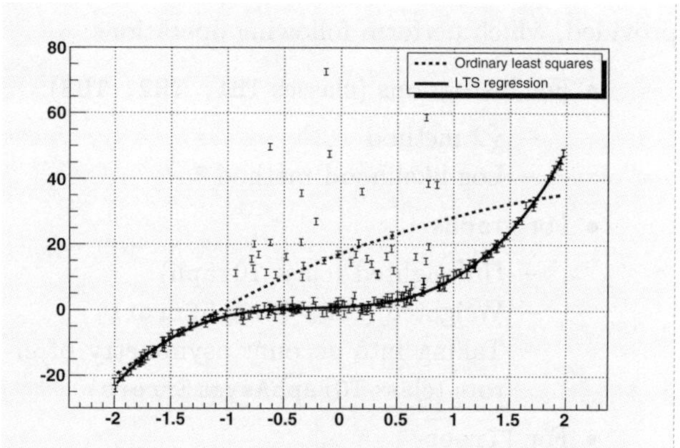

Fig. 2. LTS fit compared to ordinary least squares

4.3. *New version of Minuit*

Other recent developments include the new version of MINUIT which has been re-designed and re-implemented in the C++ language. The new package enhances all the functionality of the original Fortran version. The improvements from basing on an object oriented design are an increased flexibility, easy maintainability in the long term and opening to extensions such as integration of new algorithms, new functionality, or changes in user interfaces. Various extensive tests have been performed to study and validate the numerical quality, convergence power and computational performances of the new version. We are expecting to integrate the new version inside ROOT in one of the next development releases. In this process we also plan to redesign the current interface in ROOT to integrate in a coherent way all the new developments related to fitting and minimization.

4.4. *RooFit*

Starting from version 5.02/00 (June 2005), the RooFit package[9], developed by W. Verkerke and D. Kirkby, is distributed together with ROOT. This toolkit contains a collection of "standard" probability distribution functions and allows easy construction of new complex models. It also provides automatic normalization of PDFs.

5. Smoothing and peak finding

5.1. *TSpectrum*

Class `TSpectrum`, written by M. Morhac [10] is used for 1-dimensional background estimation, deconvolution, smoothing, orthogonal transforms, peak search and fitting. Extensions to 2 and 3 dimensions are being developed now.

5.2. *Smoothing*

Various methods for smoothing graphs are available in the class `TGraphSmooth`. It includes Friedman's "super smoother", kernel smoother and lowess.

6. Multidimensional methods

6.1. *MCD*

When data contains outliers, classical estimates of location and covariance are no longer reliable. In the multivariate case, outliers cannot be detected visually, and, when many outliers are present, Mahalanobis distance-based approaches are also not sufficient. In this case, it's better to use more resistant techniques. The Minimum Covariance Determinant estimator is a highly robust estimator of multivariate location and scatter [11]. In ROOT it is implemented in the class `TRobustEstimator`. Like the LTS regression algorithm, it tries to find a subset of h points (out of n) which has the covariance matrix with the lowest determinant. Then the mean of this subset is taken as the MCD location estimate and the subset classical covariance matrix as the MCD scatter estimate. As with LTS, the breakdown point of this estimator is $(n-h)/n$, where h is the number of good points in the dataset and should lie between $n/2$ and n.

6.2. *Principal components analysis*

Principal components analysis is a classical statistical technique for reducing dimensionality of data while still keeping as much information as possible. It is implemented in ROOT's class `TPrincipal`.

6.3. *Multidimensional fit*

Class `TMultiDimFit` approximates a multidimensional function with monomials, Chebyshev or Legendre polynomials.

6.4. *Neural networks*

Class `TMultiLayerPerceptron` describes a multilayer perceptron neural network. Six learning methods are available: stochastic minimization, steepest descent, steepest descent with fixed step size, conjugate gradients with Polak-Ribiere or Fletcher-Reeves updating formula and BFGS method[12].

7. Confidence intervals

7.1. *TLimit*

Class `TLimit` computes 95% C.L. limits using the Likelihood ratio semi-Bayesian method[13]

7.2. *TRolke*

Class `TRolke` computes confidence intervals for the rate of Poisson in the presence of background and efficiency uncertainties. The method seems to have satisfactory frequentist properties[14].

7.3. *TFeldmanCousins*

Class `TFeldmanCousins` computes the C.L. upper limit using the Feldman-Cousins method[15].

8. Future plans

Our short to medium term plans include a new random number package, improvement of the fitting interface, statistical plots (quantile-quantile plot, boxplot and spiderplot), sPlot[16], Fast Fourier Transforms and the Loess local polynomial regression fitting algorithm.

References

1. http://root.cern.ch.
2. R. Chytracek *et al.*, *Nuclear Instruments And Methods* **A534**, 115 (2004).
   See also http://www.cern.ch/seal.
3. W. Brown and M. Paterno, "A proposal to Add Mathematical Special Functions to the C++ Standard Library", WG21/N1422 = J16/03-0004.
4. M. Galassi et al, The GNU Scientific Library Reference Manual - Second Edition, ISBN = 0954161734 (paperback).
5. F. James, "MINUIT Reference Manual", CERN Program Library Writeup D506.
6. S. Yashchenko, "New method for minimizing regular functions with constraints on parameter region", Proceedings of CHEP'97 (1997).
7. P.J. Rousseeuw and A.M. Leroy, "Robust Regression and Outlier Detection", 1987 (New York, Wiley).
8. P.J. Rousseeuw and K.Van Driessen, "Computing LTS Regression for Large Datasets", *Estadistica* **54**, 163 (2002).
9. http://roofit.sourceforge.net.
10. M. Morhac et al., "Background elimination methods for multidimensional coincidence gamma-ray spectra", *Nuclear Instruments and Methods* **A401**, 113-132 (1997).
11. P.J. Rousseeuw and K.Van Driessen, "A fast algorithm for the minimum covariance determinant estimator", *Technometrics* **41**, 212 (1999).
12. http://schwind.home.cern.ch/schwind/MLP/doc/.
13. T. Junk, *Nuclear Instruments and Methods* **A434**, 435-443 (1999).
14. W. Rolke, A. Lopez, J. Conrad, *Nuclear Instruments and Methods* **A551**, 493-503 (2005).
15. G.J. Feldman and R.D. Cousins, "Unified approach to the classical statistical analysis of small signals", *Phys.Rev.* **D57**, 3873 (1998).
16. M. Pivk, F. Le Diberder, physics/0402083, to be published in *Nuclear Instruments and Methods.*

THE ROOFIT TOOLKIT FOR DATA MODELING

W. VERKERKE

University of California Santa Barbara, Santa Barbara, CA 93106, USA

D. KIRKBY

University of California Irvine, Irvine CA 92697, USA

RooFit is a library of C++ classes that facilitate data modeling in the ROOT environment. Mathematical concepts such as variables, (probability density) functions and integrals are represented as C++ objects. The package provides a flexible framework for building complex fit models through classes that mimic math operators, and is straightforward to extend. For all constructed models RooFit provides a concise yet powerful interface for fitting (binned and unbinned likelihood, χ^2), plotting and toy Monte Carlo generation as well as sophisticated tools to manage large scale projects. RooFit has matured into an industrial strength tool capable of running the BABAR experiment's most complicated fits and is now available to all users on SourceForge[1].

1. Introduction

One of the central challenges in performing a physics analysis is to accurately model the distributions of observable quantities $\vec{x}$ in terms of the physical parameters of interest $\vec{p}$ as well as other parameters $\vec{q}$ needed to describe detector effects such as resolution and efficiency. The resulting model consists of a "probability density function" (PDF) $F(\vec{x}\,;\vec{p},\vec{q})$ that is normalized over the allowed range of the observables $\vec{x}$ with respect to the parameters $\vec{p}$ and $\vec{q}$.

Experience in the BaBar experiment has demonstrated that the development of a suitable model, together with the tools needed to exploit it, is a frequent bottleneck of a physics analysis. For example, some analyses initially used binned fits to small samples to avoid the cost of developing an unbinned fit from scratch. To address this problem, a general-purpose toolkit for physics analysis modeling was started in 1999. This project fills a gap in the particle physicists' tool kit that had not previously been addressed.

A common observation is that once physicists are freed from the constraints of developing their model from scratch, they often use many observables simultaneously and introduce large numbers of parameters in order to optimally use the available data and control samples.

2. Overview

The final stages of most particle physics analysis are performed in an interactive data analysis framework such as PAW[2] or ROOT[3]. These applications provide an interactive environment that is programmable via interpreted macros and have access to a graphical toolkit designed for visualization of particle physics data. The RooFit toolkit extends the ROOT analysis environment by providing, in addition to basics visualization and data processing tools, a language to describe data models. The core features of RooFit are:

- A *natural and self-documenting vocabulary* to build a model in terms of its building blocks (e.g., exponential decay, Argus function, Gaussian resolution) and how they are assembled (e.g., addition, composition, convolution). A template is provided for users to add new PDFs specific to their problem domain.
- A *data description language* to specify the observable quantities being modeled using descriptive titles, units, and any cut ranges. Various data types are supported including real valued and discrete valued (e.g. decay mode). Data can be read from ASCII files or ROOT ntuples.
- *Generic support for fitting* any model to a dataset using a (weighted) unbinned or binned maximum likelihood, or χ^2 approach
- *Tools for plotting data with correctly calculated errors*, Poisson or binomial, and superimposing correctly normalized projections of a multidimensional model, or its components.
- *Tools for creating event samples from any model with Monte Carlo techniques*, with some variables possibly taken from a prototype dataset, e.g. to more accurately model the statistical fluctuations in a particular sample.

• *Computational efficiency.* Models coded in RooFit should be as fast or faster than hand coded models. An array of automated optimization techniques is applied to any model without explicit need for user support.
• *Bookkeeping tools for configuration management,* automated PDF creation and automation of routine tasks such as goodness-of-fit studies.

3. Object-Oriented Mathematics

To keep the distance between a physicists' mathematical description of a data model and its implementation as small as possible, the RooFit interface is styled after the language of mathematics. The object-oriented ROOT environment is ideally suited for this approach: each mathematical object is represented by a C++ software object. Table 1 illustrates the correspondence between some basic mathematical concepts and RooFit classes.

Table 1. Correspondence between mathematical concepts and RooFit classes.

Concept	Math Symbol	RooFit class name
Variable	x, p	RooRealVar
Function	$f(\vec{x})$	RooAbsReal*
PDF	$F(\vec{x};\vec{p},\vec{q})$	RooAbsPdf*
Space point	$\vec{x}$	RooArgSet
Integral	$\int_{\vec{x}_{min}}^{\vec{x}_{max}} f(\vec{x})d\vec{x}$	RooRealIntegal
List of points	$\vec{x}_k$	RooAbsData*

* Abstract base classes

Composite objects are built by creating all their components first. For example, a Gaussian probability density function with its variables is created as follows:

```
RooRealVar x("x","x",-10,10) ;
RooRealVar m("m","mean",0) ;
RooRealVar s("s","sigma",3) ;
RooGaussian g("g","gauss(x,m,s)",x,m,s) ;
```

Each object has a name, the first argument, and a title, the second argument. The name serves as unique identifier of each object, the title can hold a more elaborate description of each object and only serves documentation purposes.

Function objects are linked to their ingredients: the function object g *always* reflects the values of its input variables x,m, and s. The absence of any explicit invocation of calculation methods allows for true symbolic manipulation in mathematical style.

RooFit implements its data models in terms of probability density functions. The normalization of probability density functions, traditionally one of the most difficult aspects to implement, is handled internally by RooFit: all PDF objects are automatically normalized to unity. If a specific PDF class doesn't provide its normalization internally, a variety of numerical techniques are used to calculate the normalization.

Composition of complex models from elementary PDFs is straightforward: a sum of two PDFs is a PDF, the product of two PDFs is a PDF. The RooFit toolkit provides a set of 'operator' PDF classes that represent the sum of any number of PDFs, the product of any number of PDFs and the convolution of two PDFs.

Existing PDF building blocks can be tailored using standard mathematical techniques by substituting a variable with a formula expression. Free-form interpreted C++ function and PDF objects are available to glue together larger building blocks. The universally applicable composition operators and free-style interpreted functions make it possible to write probability density functions of arbitrary complexity in a straightforward mathematical form.

4. Composing and Using Data Models

We illustrate the process of building a model and its various uses with a simple one-dimensional yield fit example.

The RooFit models library provides more than 20 basic probability density functions that are commonly used in high energy physics applications, including basic PDFs such Gaussian, exponential and polynomial shapes, physics inspired PDFs, e.g. decay functions, Breit-Wigner, Voigtian, Argus shape, Crystal Ball shape, and non-parametric PDFs (histogram and KEYS[4]).

In the example below we use two such PDFs: a Gaussian and an ARGUS background function:

```
// Observable
RooRealVar mes("mes","mass_ES",-10,10) ;

// Signal model and parameters
RooRealVar mB("mB","m(B0)",0) ;
RooRealVar w("w","Width of m(B0)",3) ;
RooGaussian G("G","G(meas,mB,width)",mes,mB,w) ;
```

```
// Background model and parameters
RooRealVar m0("m0","Beam energy / 2",-10,10) ;
RooRealVar k("k","ARGUS slope parameter",3) ;
RooArgusBG A("A","A(mes,m0,k)",mes,m0,k) ;

// Composite model and parameter
RooRealVar f("f","signal fraction",0,1) ;
RooAddPdf M("M","G+A",RooArgList(G,A),f) ;
```

The RooAddPdf operator class M combines the signal and background component PDFs with two parameters each into a composite PDF with five parameters:

$$M(m_{ES}; m_B, w, m_0, k, f) = f \cdot G(m_{ES}; w, g) + (1 - f) \cdot A(m_{ES}; m_0, k).$$

Once the model M is constructed, a maximum likelihood fit can be performed with a single function call:

```
M.fitTo(*data) ;
```

Fits performed this way can be unbinned, binned and/or weighted, depending on the type of dataset provided. The result of the fit, the new parameter values and their errors, are immediately reflected in the RooRealVar objects that represent the parameters of the PDF, mB,w,m0,k and f. Parameters can be fixed in a fit or bounded by modifying attributes of the parameter objects prior to the fit:

```
m0.setConstant(kTRUE) ;
f.setRange(0.5,0.9) ;
```

Visualization of the fit result is equally straightforward:

```
RooPlot* frame = mes.frame() ;
data->plotOn(frame) ;
M.plotOn(frame) ;
M.plotOn(frame,Components("A"),
                LineStyle(kDashed)) ;
frame->Draw()
```

A RooPlot object represents a one-dimensional view of a given observable. Attributes of the RooRealVar object mes provide default values for the properties of this view (range, binning, axis labels). Figure 1 shows the result of the frame->Draw() operation in the above code fragment.

The default error bars drawn for a dataset are asymmetric and correspond to a Poisson confidence interval equivalent to 1σ for each bin content. The curve of the PDF is automatically normalized to the number of events of the dataset last plotted in the same frame. The points of the curve are chosen by an adaptive resolution-based technique: the deviation between the function value and the curve will not exceed a given tolerance regardless of the binning of the plotted dataset.

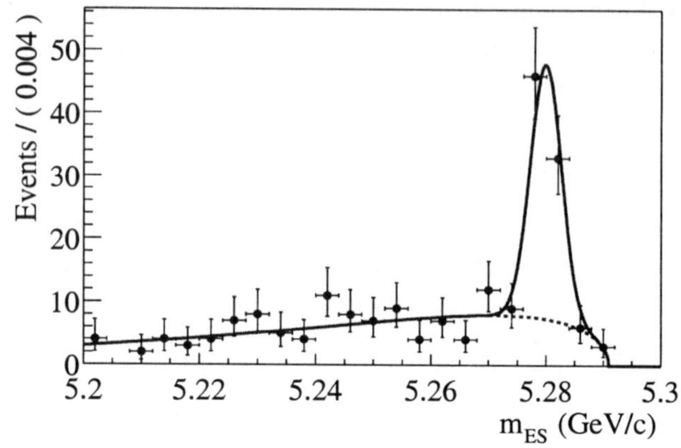

Fig. 1. One dimensional plot with histogram of a dataset, overlaid by a projection of the PDF M. The histogram errors are asymmetric, reflecting the Poisson confidence interval corresponding to a 1σ deviation. The PDF projection curve is automatically scaled to the size of the plotted dataset.

The plotOn() methods of datasets and functions accept optional arguments that modify the style and contents of what is drawn. The second M.plotOn() call in the preceding example illustrates some of the possibilities for functions: only the A component of the composite model M is drawn and the line style is changed to a dashed style. Similarly, the presentation of datasets can be changed, for example a sum-of-weights error ($\sqrt{\Sigma_i w_i^2}$) can optionally be selected for use with weighted datasets.

5. Efficiency and Optimal Function Calculation

As the complexity of fits increases, efficient use of computing resources becomes increasingly important. To speed up the evaluation of probability density functions, optimization techniques such as value caching and factorized calculations can be used.

Traditionally such optimizations require a substantial programming effort due to the large amount of bookkeeping involved, and often result in incomplete use of available optimization techniques due to lack of time or expertise. Ultimately such optimizations represent a compromise between development cost, speed and flexibility.

RooFit radically changes this equation as the object-oriented structure of its PDFs allows centrally provided algorithms to analyze any PDFs structure and to apply generic optimization techniques to it. Examples of the various optimization techniques are:

• *Precalculation of constant terms.* In a fit, parts of a PDF may depend exclusively on constant parameters. These components can be precalculated once and used throughout the fit session.

• *Caching and lazy evaluation.* Functions are only recalculated if any of their input has changed. The actual calculation is deferred to the moment that the function value is requested.

• *Factorization.* Objects representing a sum, product or convolution of other PDFs can often be factorized from a single N-dimensional problem to a product of N easier-to-solve 1-dimensional problems.

• *Parallelization.* Calculation of likelihoods and other goodness-of-fit quantities can, due to their repetitive nature, easily be partitioned into set of partial results that can be combined a posteriori. RooFit automates this process and can calculate partial results in separate processes, exploiting all available CPU power on multi-CPU hosts.

Optimizations are performed automatically and tailored to each potentially CPU intensive operation. This realizes the maximum available optimization potential for every operation at no cost for the user.

6. Data and Project Management Tools

As analysis projects grow in complexity, users are often confronted with an increasing number of logistical issues and bookkeeping tasks that may ultimately limit the complexity of their analysis. RooFit provides a variety of tools to ease the creation and management of large numbers of datasets and probability density functions such as:

• *Discrete variables.* A discrete variable in RooFit is a variable with a finite set of named states. The naming of states, instead of enumerating them, facilitates symbolic notation and manipulation.

• *Automated PDF building.* A common analysis technique is to classify the events of a dataset D into subsets D_i, and simultaneously fit a set of PDFs $P_i(\vec{x}, \vec{p_i})$ to these subsets D_i. In cases where individually adjusted PDFs $P_i(\vec{x}, \vec{p_i})$ can describe the data better than a single global PDF $P(\vec{x}, \vec{p})$, a better statistical sensitivity can be obtained in the fit. Often, such PDFs do not differ in structure, just in the value of their parameters. RooFit offers a utility class to automate the creation of the PDFs $P_i(\vec{x}, \vec{p_i})$: given a prototype PDF $P(\vec{x}, \vec{p})$ and a set of rules that explain how the prototype should be altered for use in each subset (e.g. "Each subset should have its own copy of parameter foo") this utility builds the entire set of PDFs $P_i(\vec{x}, \vec{p_i})$.

• *Project configuration management.* Advanced data analysis projects often need to store and retrieve the projection configuration, such as initial parameters values, names of input files and other parameters that control the flow of execution. RooFit provides tools to store such information in a standardized way in easy-to-read ASCII files. The use of standardized project management tools promotes structural similarity between analyses and increases a user's ability to understand other RooFit projects and to exchange ideas and code.

7. Development Status

RooFit was initially released as RooFitTools in 1999 in the BaBar collaboration and has over the years been adopted by virtually all BaBar physics analyses. Analysis topics include searches for rare B decays, measurements of B branching fractions and CP-violating rate asymmetries, time-dependent analyses of B and D decays to measure lifetime, mixing, and symmetry properties, and Dalitz analyses of B decays to determine form factors. Since October 2002 RooFit is available to the entire HEP community: the code and documentation repository has been moved from BaBar to SourceForge, an OpenSource development platform, which provides easy and equal access to all HEP users. (http://roofit.sourceforge.net). Since July 2005 RooFit is also bundled with ROOT releases, starting with ROOT version 5.02-00.

References

1. http://roofit.sourceforge.net
2. R. Brun et al., *Physics Analysis Workstation*, CERN Long Writeup Q121
3. R. Brun and F Rademakers, *ROOT - An Object Oriented Data Analysis Framework*, Proceedings AIHENP'96 Workshop, Lausanne, Sep. 1996, Nucl. Inst. & Meth. in Phys. Res. A 389 (1997) 81-86. See also http://root.cern.ch
4. K. Cranmer, *Kernel Estimation in High-Energy Physics*, Comp. Phys. Comm **136**, 198-207 (2001).

AN UPDATE ON THE GOODNESS-OF-FIT STATISTICAL TOOLKIT

B. MASCIALINO, M. G. PIA

INFN - Sezione di Genova, Via Dodecaneso 33, 16146 Genova, Italy
E-mail: Barbara.Mascialino@ge.infn.it, MariaGrazia.Pia@ge.infn.it

A. PFEIFFER, A. RIBON

CERN, CH 1211, Geneve 23, Switzerland
E-mail: Andreas.Pfeiffer@cern.ch, Alberto.Ribon@cern.ch

P. VIARENGO

IST - National Cancer Research Institute, Largo Rosanna Benzi 10, 16132 Genova, Italy
E-mail: paolo.viarengo@istge.it

The present project aims to develop an open-source and object-oriented software Toolkit for statistical data analysis. Its statistical testing component (the Goodness-of-Fit Statistical Toolkit) contains a variety of one dimensional Goodness-of-Fit tests, from Chi-squared to Kolmogorov-Smirnov, to less known, but generally much more powerful tests such as Anderson-Darling, Cramèr-von Mises, Kuiper, Watson, ... The GoF Statistical Toolkit is open-source and downloadable from the web, with its user and software documentation.
The component-based design allowed an extension of the GoF Statistical Toolkit: less known, but generally more powerful GoF tests based on EDF-statistics have been recently added to the toolkit. A much more complete variety of GoF inferences is now offered the user, and "standard" GoF tests have been complemented by more "exotic" ones. The weighted formulations of some GoF tests (Kolmogorov-Smirnov and Cramèr-von Mises) have been implemented. Approximations of the distribution of some of the existing GoF tests to the Chi-squared one (Kolmogorov-Smirnov, Cramèr-von Mises, and Watson approximations) are now available in the GoF Statistical Toolkit.
Moreover, a layer for user input from ROOT objects has been easily added recently, thanks to the component-based architecture.
We present the recent improvements and extensions of the GoF Statistical Toolkit, describing the new statistics methods implemented, and an outlook towards future developments.

1. Introduction

Data comparison is an essential part of all physics experiments. Classical statistical inference techniques are based on fairly specific assumptions regarding the nature of the underlying distribution. Usually both its form and some parameter values must be explicitly stated in the null hypothesis, and this requires a certain level of knowledge about what is going to be compared. When we have no information on the distribution of the data, a variety of statistical techniques is available in literature: distribution-free or non-parametric procedures. Non-parametric testing, in fact, allows the formulation of a hypothesis which is not a statement about parameter values. Non-parametric statistics include Goodness-of-Fit (GoF) testing. These tests measure the compatibility of a random sample with a theoretical probability distribution function (the *one-sample problem*) or between the empirical distributions of two different populations perhaps coming from the same theoretical distribution (the *two-sample problem*).

A project is in progress, aiming at the development of an object-oriented software toolkit for statistical data analysis. The GoF Statistical Comparison component of the toolkit provides algorithms to solve the two-sample problem in a variety of use cases typical of physics experiments. The GoF Statistical Toolkit[1] is an easy to use, up-to-date and versatile tool for data distribution comparison in physics analysis. The GoF Statistical Toolkit has been released, it is downloadable from the web together with user and software process documentation[2].

Thanks to its flexible design the GoF Statistical Toolkit has recently been extended, providing to the user additional "less known" GoF tests among which he/she can choose.

We present the recent improvements and extensions of the GoF Statistical Toolkit, describing the architecture of the extended system, the new statistical methods implemented, and an outlook towards future developments.

2. The GoF Statistical Toolkit

2.1. *GoF tests*

The GoF Statistical Toolkit offers a variety of GoF tests to solve the two-sample problem, and attention will be focused on this topic. We will deal with one-dimensional tests only.

Many inferences are available to solve the two-sample problem, testing the null hypothesis that the distributions F, G underlying the two samples are equal

$$H_0 : \mathrm{F} = \mathrm{G}$$

without specifying the common distribution function. For this reason, these tests are distribution-free under the null hypothesis when the observations are independent and identically distributed. In most application of GoF techniques, the alternative hypothesis is composite (i.e. it depends on unspecified parameters), as it gives little or no information on the distribution of the data, and simply states that the null hypothesis is false:

$$H_1 : \mathrm{F} \neq \mathrm{G}.$$

Non-parametric statistics provides a variety of GoF tests to solve the two-sample problem. These tests can be roughly divided into two major groups:

(1) tests of Chi-squared type: the test statistic computation is based on the weighted squared difference in the number of categorical observations[3],
(2) tests based on empirical distribution function (EDF) statistics: the test statistic measures the distance between the two EDFs F and G[3].

The first class of tests is used to assess the fit of models for binned data. The Chi-squared test can be useful also in case of unbinned data; in this case the researcher is compelled to group events into classes, converting the distribution from unbinned to binned, and sacrificing in this way a good deal of the information conveyed by the distribution itself.

The second class can be applied to both binned and unbinned data; it evaluates the difference between F and G, and the value of the corresponding test statistic is computed as a mathematical function of the vertical differences between the two EDFs. These tests can be classified according to the mathematical form used to evaluate the difference between the two EDFs:

- tests based on uniform distance, in which the test statistic involves the maximum difference between the two EDFs at one point (the Kolmogorov family of tests),
- tests based on quadratic distance, in which the test statistic measures the sum of the weighted squared distance between F and G (the Cramèr-von Mises family of tests).

A wide set of algorithms aimed at testing the compatibility of the distributions of two variables have been gathered together in the GoF Statistical Toolkit[1]. The first release of the GoF Statistical Toolkit provided the user with a variety of two-sample GoF tests, from Chi-squared test to tests based on the maximum distance between the two EDFs (Kolmogorov-Smirnov, Kuiper, Goodman), to tests based on the weighted quadratic distance between the two EDFs (Cramèr-von Mises, Anderson-Darling).

Thanks to its flexible design, the GoF Statistical Toolkit has recently been extended, implementing other less known GoF tests. The component-based design allowed new GoF tests to be added, without the need to change the code of the already existing tests. The user is therefore allowed to choose among a wider set of tests, including:

- modifications of Kolmogorov-Smirnov and Cramèr-von Mises tests, introducing appropriate non-negative weight functions (documented in the statistical literature) in order to give various weights to the differences $|F - G|$ and $(F - G)^2$ respectively;
- modifications of Kolmogorov-Smirnov, Cramèr-von Mises, and Watson tests, approximating the test statistics to a χ^2 (i.e. the modified test statistics follow a Chi-square distribution).

Therefore, the up-dated GoF Statistical Toolkit juxtaposes the well known Chi-squared test with a wide variety of GoF tests based on EDF statistics, covering all the tests we could find in the statistical literature to solve the two-sample problem by means of EDF statistics.

2.2. *User layer*

The component-based design uses object-oriented techniques together with generic programming. The adoption of AIDA[4] for the user layer decouples the usage of the GoF Toolkit from any concrete analysis system the user may have adopted in his/her analysis.

The user is shielded from the complexity of both the core component design of the GoF Statistical Toolkit and the computational aspects of the mathematical algorithms implemented. All the user has to do is to provide the two distributions he/she wants to compare, to choose the most appropriate GoF test (in practice writing one line of code) and to run the comparison. The comparison returns the user a statistical comparison result object, giving access to the computed value of the test statistics, the number of degrees of freedom and the quality of the copmparison (p-value).

A layer for user input from ROOT[5] objects has recently been added, thanks to the component-based architecture.

3. Power of GoF Tests

A test is considered powerful if the probability of accepting the null hypothesis when it is wrong is low. It must be stressed that with a set of non-parametric tests, power evaluation can be quantified only specifying the alternatives in detail.

A quantitative comparative study to evaluate the power of the GoF tests contained in the GoF Statistical Toolkit is in progress. The aim is to provide the users of the GoF Statistical Toolkit a guideline for the practical choice of the most suitable two-sample GoF test under general non-parametric conditions.

4. Conclusions

The GoF Statistical Toolkit represents an up-to-date and versatile tool for data comparison in physics analysis.

Nowadays the GoF Statistical Toolkit represents one of the most complete system available to face the two-sample EDF GoF hypothesis testing, both in the physics and in the statistical data analysis domains. It is the first statistical software system providing such a variety of sophisticated and powerful Goodness-of-Fit algorithms in high energy physics.

References

1. G. A. P. Cirrone *et al.*, *IEEE Trans. Nucl. Sci.* **51**, 2056 (2004).
2. www.ge.infn.it/geant4/analysis/HEPstatistics/
3. R. B. D'Agostino and M. A. Stephens, *Goodness-of-Fit techniques*, Dekker, 1986.
4. G. Barrand *et al.*, *Proc. of CHEP* (2001).
5. R. Brun *et al.*, *Nucl. Meth. Instr. A* **389**, 81 (1997).

CEDAR: COMBINED E-SCIENCE DATA ANALYSIS RESOURCE

ANDY BUCKLEY

Institute for Particle Physics Phenomenology, Dept. of Physics, Science Laboratories, South Road, Durham, England, DH7 7QQ
E-mail: Andy.Buckley@durham.ac.uk

The CEDAR collaboration is developing a set of tools for tuning and validating models of high-energy physics processes by comparing the predictions of Monte Carlo event generators with data from experiments. The core CEDAR program is to interface the Durham HepData database of experimental measurements and the UCL JetWeb system for automated event generator validation: eventually JetWeb will use HepData records for its comparisons with Monte Carlo predictions. Other aspects of the project include defining XML data formats, building a software development environment for high-energy physics projects and providing an archive of HEP computation software.

1. Introduction

Although the Standard Model is extraordinarily successful in describing a wide range of phenomena, processes involving non-perturbative QCD cannot be calculated explicitly and require some phenomenological modelling. This is particularly true of hadronic collisions, since the final state is influenced by the parton distribution functions (PDFs) of the colliding beams, by multiple soft interactions between partons (the "underlying event") and by the hadronisation of the out-going partons. Accurate modelling of such hadronic processes is crucial for robust interpretation of data from the LHC.

The non-perturbative aspects of high-energy processes are typically simulated by Monte Carlo event generators, which typically introduce several free or nearly free parameters. These can only be constrained by fitting the model predictions to the experimental data. However, this is far from a trivial task as the experimental conditions vary widely, involving different beam particles, different regions of phase space and complicated observables. The variables may be highly correlated, and tuning to a limited set of data may result in non-physical predictions in an un-fitted region of the parameter space.

The CEDAR project[1, 2] exists to provide a standard, robust and simple system for performing simultaneous data vs. model comparisons. Its main focus is the integration of the HepData[3] and JetWeb[4] services, improving JetWeb's ability to constrain Monte Carlo simulation parameters. The rest of this article will describe the projects comprising CEDAR and how this goal can be achieved.

2. HepData Upgrade

The existing HepData system is based on a hierarchical database, which is accessed via legacy FORTRAN routines and suffers from a lack of maintainability or support. To make HepData more flexible and suitable for remote access by JetWeb, HepData is being migrated to a relational database system with a re-designed data model. Rather than query the database directly, Web users will query a Java-based front-end which will present the data records in a choice of formats. These are foreseen to include HTML-formatted data tables, plain text, HepML records (see Section 5) and AIDA XML[12].

3. JetWeb Upgrade

The JetWeb system is also undergoing a substantial update as part of the CEDAR programme. In its present incarnation, JetWeb consists of a MySQL database of data from experiments and Monte Carlo simulations, and a Java servlet engine which performs comparisons between these. A typical use of JetWeb is to specify a number of generator parameters and a number of events via a Web interface; JetWeb then determines if sufficient Monte Carlo data is already available and distributes simulation jobs if not. The comparisons between data and Monte Carlo predictions are obtained using the FORTRAN "HZTool" library[5].

HZTool is maintained by CEDAR with subroutines for specific measurements being contributed by various authors both inside and outside the CEDAR group. Under CEDAR, a new version of HZTool (v4.0) has been released: in this the generator "steering" components required for JetWeb integration have been extracted into an external package called HZSteer[6].

Work has begun in CEDAR to develop an object-oriented replacement for HZTool, to be called "Robust Independent Validation of Experiment and Theory" (Rivet)[7]. Guiding design principles of Rivet include the implementation in OO C++ and compatibility with existing standard data formats such as HepMC[11] and AIDA[12]. Rivet aims to minimise dependence on generator-specific features: this will make it easier to incorporate new Monte Carlo generators than is currently the case with HZTool.

4. Connecting JetWeb and HepData

The next version of JetWeb will communicate with HepData to get the experimental data used for its comparisons to MC data, rather than maintaining its own separate database. The details of the data exchange are still to be decided, but it is likely that the data models used by HepData and JetWeb will be partially unified. This allows the HepData objects to be "serialised", sent to JetWeb and used directly, rather than involving an intermediate format.

5. HepML

Since interoperability is a major concern for CEDAR, a family of XML-based data formats is being developed to describe HepData records and generator configurations. The formats are defined by a set of XML schemas, and the family as a whole is referred to as HepML[8]. These can be easily translated into other representations such as AIDA XML, plain text, XHTML or many others, which makes flexible output modes for HepData possible.

CEDAR's XML schemas are separate from the HepML developed within the MCDB project[9], which is primarily a format for event records. Eventually, the two formats may be incorporated into a more general family of HEP schemas.

6. HepForge and HepCode

CEDAR provides a centralised development environment for high energy physics software projects, called HepForge[10]. This currently hosts the core CEDAR projects (HZTool, HZSteer, JetWeb & HepML) and also Herwig++[14], ThePEG[15], LHAPDF[16], Jimmy[17], fastNLO[18], KtJet[19] and RunMC[20]. Other HEP projects are encouraged to use HepForge.

The facilities currently offered by HepForge include the Subversion and CVS code management systems with Web-based viewers, a bug tracker, a wiki for documentation and communication between developers and mailing lists for developer contact, project announcements and discussion.

HepForge will eventually be used to provide a final portion of CEDAR, named HepCode[13]. This is a project to provide access to well-defined versions of Monte Carlo generator programs, parton distribution functions and other high-energy physics calculation codes. In its current state, HepCode is simply a list of programs with links to where they can be downloaded and some information on which processes they calculate, implementation language, and so-on. The aim is for HepForge to be used to maintain this list and to store each released version of the code: this will be most readily achieved if projects who wish their code to be in HepCode register a HepForge project and archive their releases there.

Acknowledgements

The CEDAR team would like to thank the UK Particle Physics & Astronomy Research Council (PPARC) for their generous support of CEDAR.

References

1. J. M. Butterworth *et al.*, hep-ph/0412139, presented at CHEP'04, Interlaken, September 2004
2. `http://www.cedar.ac.uk`
3. `http://durpdg.dur.ac.uk/hepdata/`
4. J. M. Butterworth and S. Butterworth, *Comput. Phys. Commun.* **153**, 164 (2003); `http://jetweb.cedar.ac.uk/`
5. HZTool package, manual and tutorial: `http://hepforge.cedar.ac.uk/hztool/`
6. `http://hepforge.cedar.ac.uk/hzsteer/`
7. `http://www.cedar.ac.uk/rivet/`
8. `http://hepforge.cedar.ac.uk/hepml/`
9. P. Bartalini *et al.*, hep-ph/0404241
10. `http://hepforge.cedar.ac.uk`
11. `http://mdobbs.home.cern.ch/mdobbs/HepMC/`
12. `http://aida.freehep.org/index.thtml`
13. `http://www.cedar.ac.uk/hepcode/`
14. `http://hepforge.cedar.ac.uk/herwig/`
15. `http://hepforge.cedar.ac.uk/thepeg/`
16. `http://hepforge.cedar.ac.uk/lhapdf/`
17. `http://hepforge.cedar.ac.uk/jimmy/`
18. `http://hepforge.cedar.ac.uk/fastnlo/`
19. `http://hepforge.cedar.ac.uk/ktjet/`
20. `http://hepforge.cedar.ac.uk/runmc/`

VISUALISATION

VISUALISING DATA

SUSAN HOLMES
Statistics Department, Stanford University, USA
Biometry-INRA- Montpellier, France

Keeping as much information about data as possible allows for exploration of possible structure without apriori knowledge. We will show examples of this Exploratory Data Analysis approach championed by Tukey and developed at AT&T and Bell labs. The labs and follow-up developers such as R. Gentleman and R. Ihaka [1] have made high quality software available (S,R,`xgobi,ggobi`) that makes even visualisation of high dimensional data highly effective. We finish the presentation with a small example of how visualisation has helped us improve a method for estimating the bias in coin tossing.

1. Statistics is not just statistics

Often outsiders think of statistics as the boiling down of large data sets to one or two number summaries. Such compression rarely works unless the data are distributed according to a simple parametric family such as a normal, gamma or Poisson. In real experiments the data come in the form of large matrices. Often we need to look at the data to understand what information the matrix contains and how to compress it. We will show examples of such data-mapping setups. This presentation can only be considered a brief introduction into a well developed field that has filled entire books (Wilkinson[2], Chambers et al.[3], Cleveland[4, 5]).

2. One dimensional data

As stated previously, if the data come from a nice parametric family, a few numbers suffice to capture all the information in the vector, but evaluating how well the data fit the parametric model is a preliminary hurdle. As was shown in Lauritzen[6] there are good summary statistics for such evaluation. However visual inspection of the fit conveys more details on the data.

2.1. *Distribution evaluation*

Histograms are not very useful as they do not provide us with a good visual evaluation of a distribution. It is hard to differentiate between a bell shaped curve and a symmetric heavy tailed distribution. We will show in an example how more sophisticated plots enable immediate recognition of departures from the distributional forms under study.

2.2. *Random matrix data*

We chose a simple random matrix type of data for this illustration. Take the QR decompostion of a matrix filled with uniform, independent, identically distributed entries, the decomposition is performed through a simple Grahm-Schmidt algorithm. We fill a 1000×1000 data matrix with random numbers from a Uniform(-1,1). We then find the QR decomposition of the matrix and multiply Q by $\sqrt{1000}$, as we follow the columns the data become more and more normal. Histograms rarely show normality although they can be used for detecting multimodalities. A more sophisticated plot is a `qqplot` that plots the quantiles of the observed distribution versus the theoretical quantiles of a target distribution. In our comparisons below, we use the Normal as our reference distribution. If the data are normal as in the bottom of Figure 1, we can see that the points follow the diagonal line perfectly. In general considering residuals from the theoretical or expected values allows for more precise visual evaluations. These pictures were generated with the `qqnorm` command in R, which is the best software for visualising data. R has a large Exploratory Data Analysis (EDA) component provided by the AT&T labs under the guidance of the master of data visualization, John Tukey, who invented the term EDA, stem and leaf plots, boxplots, projection pursuit, and many more brilliant visualisation wonders[7].

3. Bivariate data

3.1. *Two continuous variables: scatterplots*

As soon as the size of the data sets exceed a hundred or so points, the overlay of points hides the actual

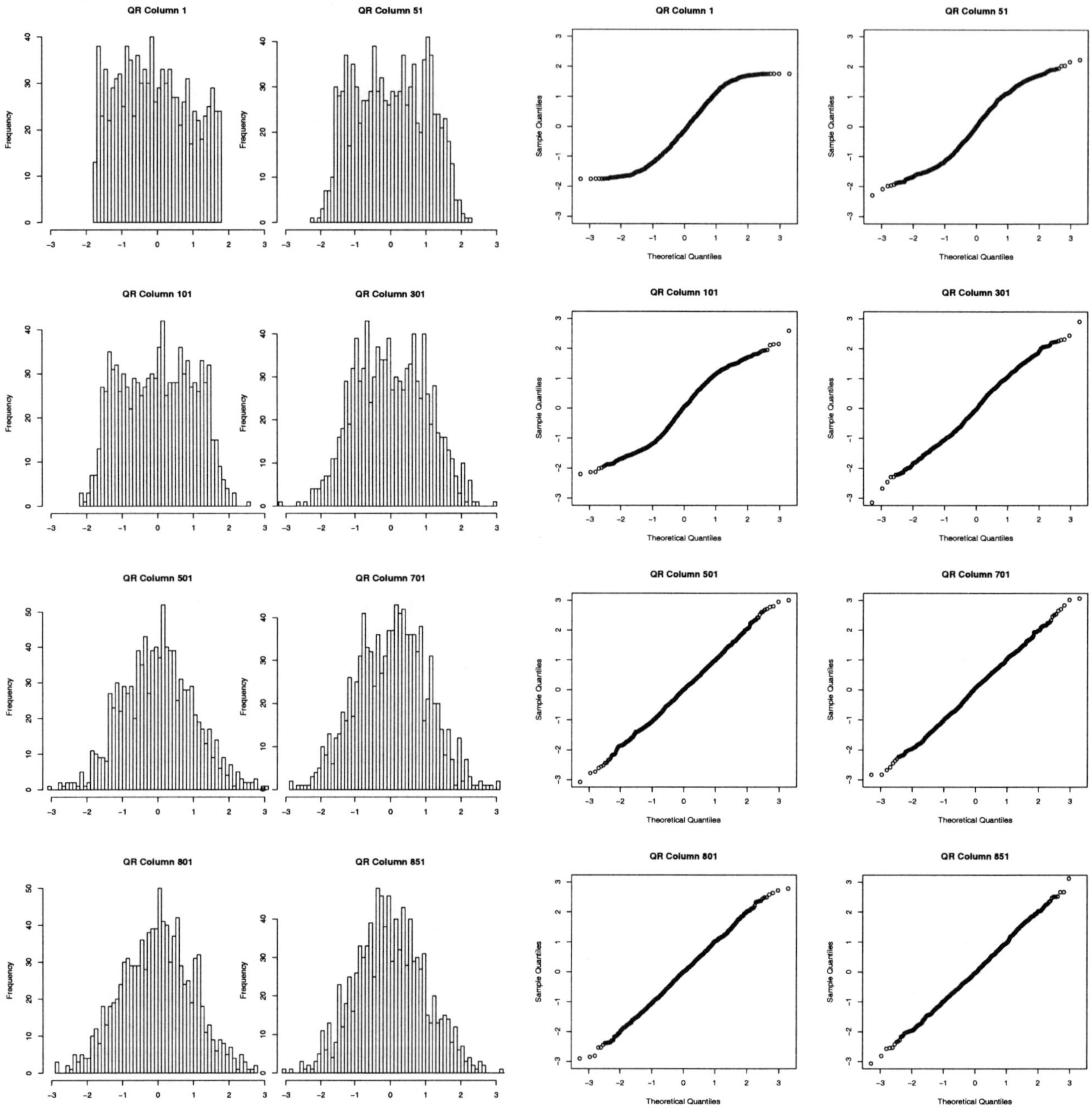

Figure 1: Histograms, qqplots of random matrix data.

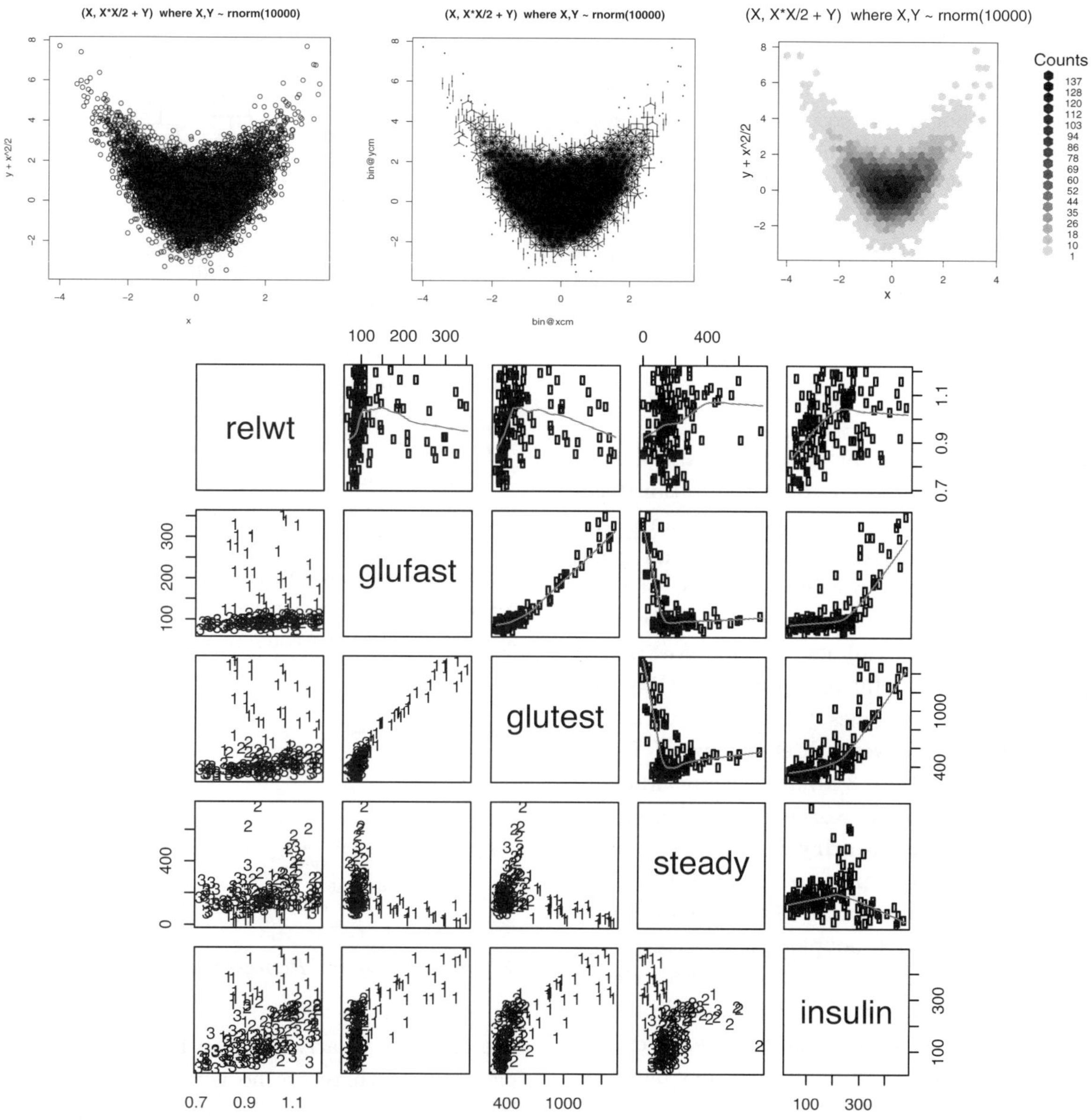

Figure 2: A variety of scatterplots of data.

distributional density of a bivariate distribution, the upper part of Figure 2 shows on the left the overlay that occurs with an ordinary scatterplot and two ways of avoiding it, the first using a sunflower plot which adds petals to the points according to how many overlaid instances occur(`sunflowerplot`). In the case of this data the improvement only occurs on the outer layer of the cloud, a better plot is provided by hexagonal binning (`hexbin`) which provides a two dimensional equivalent to a histogram with hexagonal bins. Usually we are interested in the relation between the two variables that are plotted. The lower part of Figure 2 shows multivariate scatterplots of several variables measured on three groups of diabetes patients and healthy subjects. This matrix of plots was generated by the `pairs` command in `R`, with the upper right hand panels showing a smooth curve fit to the data as well, and the lower left hand plots showing the symmetric plots with the points labeled according to their diabetic phenotype. The variables are named along the relevant diagonal, thus the third plot in the second row shows a scatterplot with **steady** on the horizontal axis and **glufast**(glucose after fasting) on the vertical axis. We see the smooth curve shows us that after the initial drop **glufast** stays constant. The symmetric plot shows us that the patients with low **glufast** are all in group 1 (overt diabetic).

3.2. *Boxplots*

When comparing a continuous variable and a categorical variable, the most useful visualisation technique is the boxplot. Here we will show an examples where we summarize measurements on genes that are differentially expressed in T-cells of three different kinds; see the details of the study done using microarrays in Holmes et al.[8] We ranked the genes by their adjusted p-value. We find about 160 interesting genes (continuous variables) which differentiate between three groups of T-cells (categorical variable with 3 levels: Effector, Memory and Naive). For biologists working with microarrays this is **The dreaded laundry list**: we need to visualise the information contained in this table of 160 genes. Here are boxplots of three such genes, but as we cannot look at all 160 boxplots, it is better to use the angles between the medians in the three classes and plot them around a circle. Here we didn't standardise the data by reducing all the measurements to have variance 1 as we would have if they had been measured on very different scales.

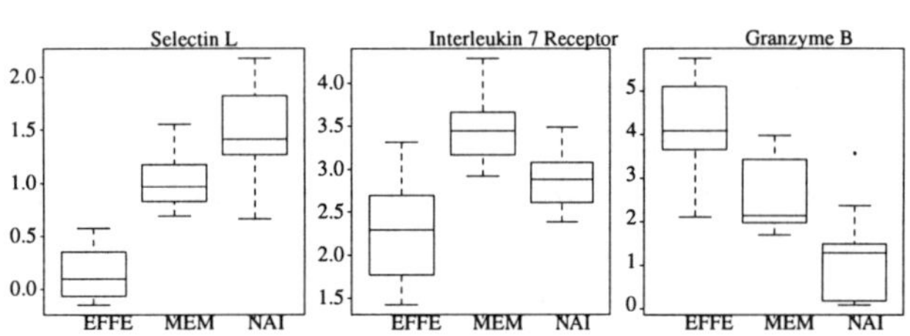

Figure 3: Boxplots of three genes for 3 types of T cells.

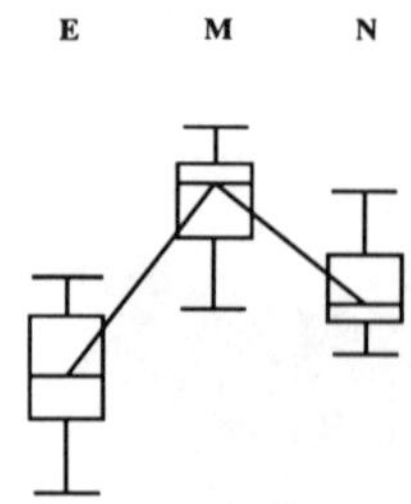

Figure 4: Building an angular summary of a boxplot.

1) Consider the medians of the gene expression patterns for effector, memory, and naive cells, in that order.
2) Take x = median(memory) - median(effector) and y = median(naive) - median(memory) ('effective slopes').
3) Combine into the coordinate (x,y), and normalize to length 1, i.e. multiply by a constant c so that $(cx)^2 + (cy)^2 = 1$
4) Map the point onto the unit circle.
5) An angle is attached to each point with (1,0) being angle 0 and rotating counter-clockwise.

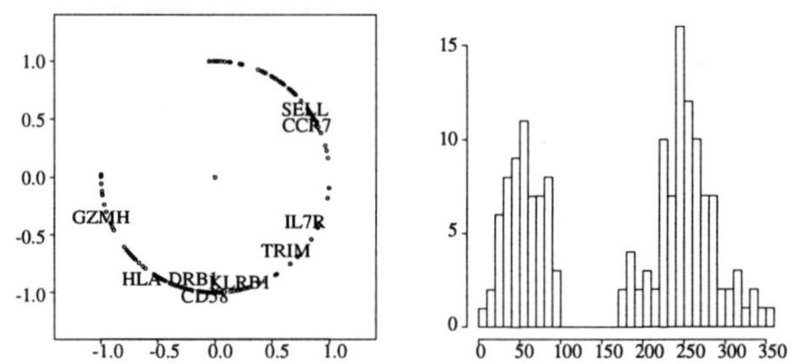

Figure 5: Angular Representation and its histogram.

We see very clearly the gap in the angles showing clearly that there are no genes that are downregulated in the Memory cell types compared to the other two.

4. More than two dimensions

When a scatterplot is not possible because the data have too many variables, we have to choose clever ways of creating 'maps' of the data that will show up the features that we are most interested in. Sometimes these features are known ahead of time as in the case of the diabetes data where there is a natural grouping of the patients. We want to find a mapping that separates out the three groups of patients. In the second study, we have categorical data we want to seriate, and we need to discover an underlying hidden gradient.

4.1. *Choosing axes that explain a categorical variable*

One can take linear combinations of the original variables such as those in the diabetes data set shown above that maximize an intergroup variance, when the groups are given – this is called **Linear Discriminant Analysis**. The first and second discriminant axes are linear combinations of the the original variables with coefficients:

	relwt	glufast	glutest	steady	insulin
axis 1:	-0.17	2.03	-3.87	0.00	-0.41
axis 2:	-0.49	2.29	-2.11	-0.72	0.06

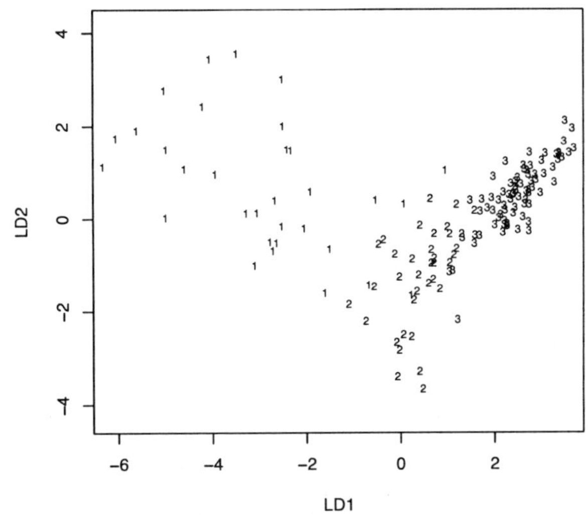

Figure 6: Linear Discriminant Variables.
(1='Overt Diabet.' 2='Chem. Diabet.' 3='Normal')

Figure 6 shows the display for the diabetes data, we can see that the groups are well separated in this planar map.

4.2. *Counts and categorical variables*

Contingency table data containing cross-tabulated counts of categorical variables can be visualised by using the chisquare distances between rows or columns and a generalized singular value decomposition called correspondence analysis[9] which has proved invaluable for finding hidden rankings (called ordinations) or clusters of multicategory count data. As an example we take data analysed by Cox and Brandwood [10] who wanted to seriate Plato's works using the proportion of sentence endings in a given book, with a given stress pattern. We propose the use of correspondence analysis on the table of frequencies of sentence endings (for a detailed analysis see Charnomordic and Holmes[11]).

The first 10 profiles (as percentages) look as follows:

	Rep	Laws	Crit	Phil	Pol	Soph	Tim
UUUUU	1.1	2.4	3.3	2.5	1.7	2.8	2.4
-UUUU	1.6	3.8	2.0	2.8	2.5	3.6	3.9
U-UUU	1.7	1.9	2.0	2.1	3.1	3.4	6.0
UU-UU	1.9	2.6	1.3	2.6	2.6	2.6	1.8
UUU-U	2.1	3.0	6.7	4.0	3.3	2.4	3.4
UUUU-	2.0	3.8	4.0	4.8	2.9	2.5	3.5
--UUU	2.1	2.7	3.3	4.3	3.3	3.3	3.4
-U-UU	2.2	1.8	2.0	1.5	2.3	4.0	3.4
-UU-U	2.8	0.6	1.3	0.7	0.4	2.1	1.7
-UUU-	4.6	8.8	6.0	6.5	4.0	2.3	3.3

.......etc (there are 32 rows in all)

The eigenvalue decomposition (called the scree plot) of the chisquare distance matrix (see [11]) shows that two axes out of a possible 6 (the matrix is of rank 6) will provide a summary of 85% of the departure from independence. This justifies the use of a planar representation to provide a good visual summary of the data.

	Eigenvalue	inertia %	cumulative %
1	0.09170	68.96	68.96
2	0.02120	15.94	84.90
3	0.00911	6.86	91.76
4	0.00603	4.53	96.29
5	0.00276	2.07	98.36
6	0.00217	1.64	100.00

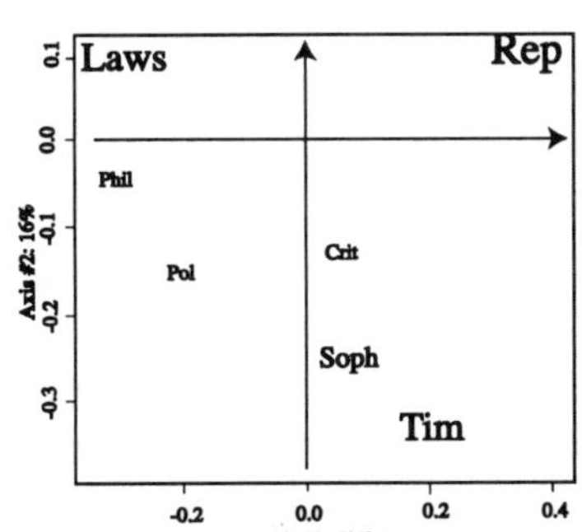

Figure 7: Correspondence Analysis of Plato's Works.

We can see from the plot that there is a seriation that in most cases follows a parabola or arch[12] from Laws on one extreme being the latest work and Republica being the earliest among those studied.

4.3. *Multidimensional scaling*

A common situation is outlined in the mapping of observations for which we have a natural distance. We need to create a planar or three dimensional representation that represents these distances with the least distortion. This is well fulfilled by what is called multidimensional scaling.

Multidimensional Scaling [13,14,15] proposes to make the best low dimensional map of the observations given by a distance matrix.

	Marseille	Milan	Munich	Paris	Rome	Stockh.	Vienna
Athens	2865	2282	2179	3000	817	3927	1991
Barcelona	521	1014	1365	1033	1460	2868	1802
Brussels	1011	925	747	285	1511	1616	1175
Calais	1059	1077	977	280	1662	1786	1381
Cherbourg	1101	1209	1160	340	1794	2196	1588
Cologne	1035	911	583	465	1497	1403	937
Copenhagen	1778	1537	1104	1176	2050	650	1455
Geneva	425	328	591	513	995	2068	1019
Gibralta	1693	2185	2565	1971	2631	3886	2974
Hamburg	1479	1238	805	877	1751	949	1155

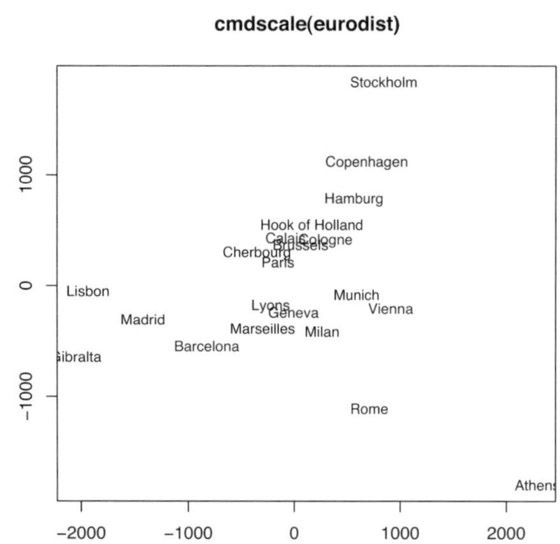

Figure 8: Multidimensional Scaling Plane of Distances.

Some consider correspondence analysis to be a special case of MDS where the distances approximated are the Chisquare distances.

4.4. *Data projection methods*

We have already seen three projection methods. The reason for trying to find low dimensional projections of the data that contain most of the relevant information comes from what is commonly known as the *curse of dimensionality* which makes nonparametric multivariate density estimation[16] impossible. Each small multivariate window will contain too few points to be useful in estimating the density. This is why dimension reduction techniques are so essential. Projecting data to detect structure can also be done by choosing an axis that maximizes the variance along a direction[17]; this is Principal Components Analysis (PCA).

An innovative extension to this idea was to change the optimization criteria to capture directions in which the data appear far from Normal. This was introduced to solve visualisation problems in particle physics by Friedman, Tukey and Kruskal who invented Projection Pursuit[18, 19]. `Xgobi`[20] and `ggobi`[21] provide implementations of these methods, where instead of rotating the data manually one follows the computer's suggestion of projections in the *'most interesting'* direction, sometimes exhibiting clusters in the data. These methods find linear combinations of the original variables onto which the data either project into clumps or multimodal distributions or have skewed or long tailed distributions. A very high quality but complicated multivariate R package for linear multivariate visualisations is `ade4`[22] with which some of the presented graphics were made.

However what if the data show non linear structure that no linear combination of the original variables will capture? Then we have to use more recent generalizations of these methods that look for lower dimensional manifolds to which the data lie abnormally close. These new nonparametric methods include Local linear Embedding LLE [23] which use local distances to find structure, and ISOMAP [24]: which 'unfolds the data' if it follows a manifold.

4.5. *Detecting hidden categorical variables: clustering*

When trying to detect hidden hierarchical structure as between genes for instance, a hierarchical representation can be quite useful. Figure 9 shows a heatmap of the gene expression intensities with hierarchical clustering of the rows and columns. See Gordon[25] for more details as to the construction of clustering trees. Each measurement is coded on a grayscale and both the genes and the patients are clustered using the `heatmap` command in R.

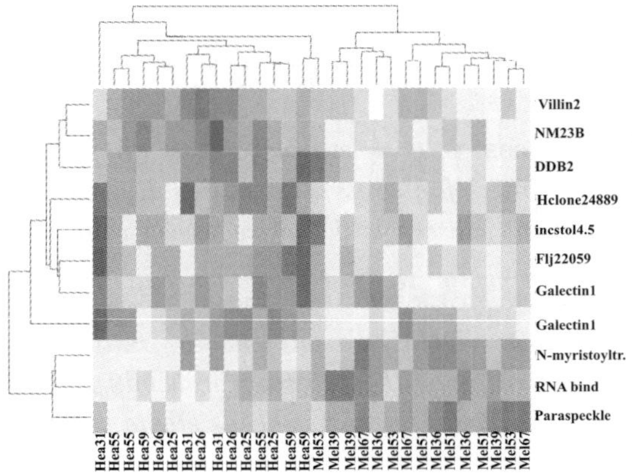

Figure 9: Hierarchical Clustering Heatmap.

Clustering can be particularly useful in the visual evaluation of a model. By clustering both the observed data and the simulated data we can try to detect their differences. If the observed data appear on the fringe of the data simulated according to the model, the chances are the model is wrong.

4.6. *Visualisation of tree stability*

We can often use these visualisation techniques in a confirmatory context. For instance, the bootstrap[26] is a popular nonparametric resampling scheme that provides useful approximations to sampling distributions of estimators, even multivariate ones. In the clustering study shown in Figure 9, we estimated the hierarchical clustering of the patients given a set of relevant genes. We can resample the genes many times, say 100 times in this example, and compare the bootstrap trees to the original tree using a distance between trees[27]. We have 101 by 101 matrix of distances between the bootstrapped tree and the original one that can tell us what small perturbations of the data would do to the estimated trees.

As in the case of European cities above, we use the matrix of distances and search for a low dimensional space in which to embed the objects in such a way as to reproduce as faithfully as possible the distances between objects using MDS[28]. In order to appreciate the quality of the approximation, we consider the plot of eigenvalues of the distance matrix. This is called a screeplot [17] and is always a necessary preliminary precaution as the main reason of instability in all these techniques based on eigendecomposition comes from choosing to split two very close eigenvalues, retaining one and rejecting the other.

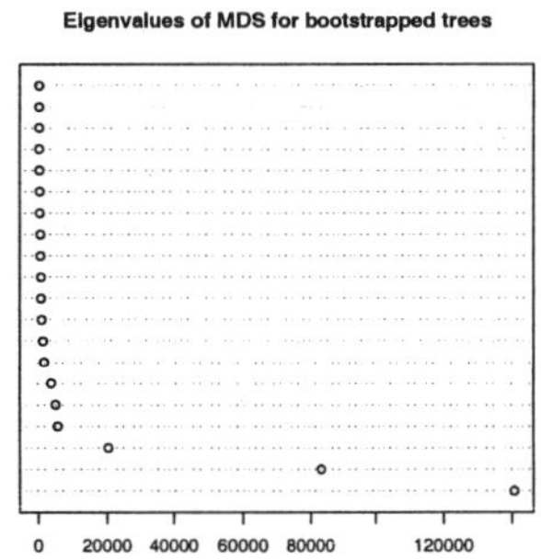

Figure 10: Screeplot of Eigenvalues.

In this case we can see that retaining either 2 or 3 axes would be reasonable, cutting off at 4 dimensions would not. Here is the planar representation obtained with the first two eigenvectors representing 85% of the information in the distance matrix.

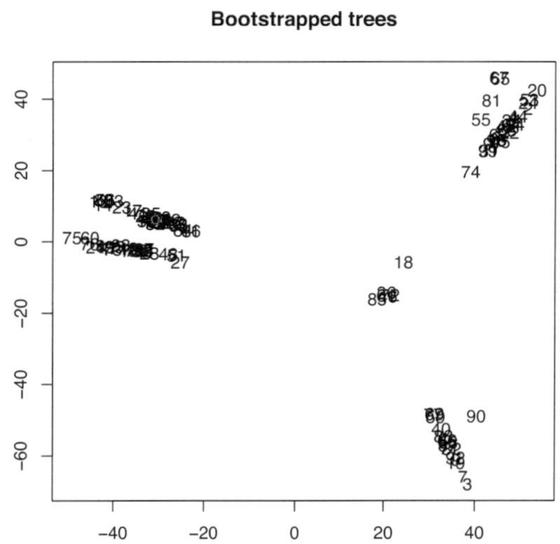

Figure 11: Structured Groups of Resampled Trees.

This is the first plane provided by the MDS algorithm. A histogram of these distances would have just shown a bimodality in the distances, those that are near to the original tree (that is in the upper left-hand group) and those that are far. Here we see the structure of the groupings of the trees into a distinctive pattern. We followed up on this map of the data by characterising the groups by which genes were absent in these resamples.

5. Dynamical bias in the coin toss: visualisation gives the answer

Here we will give an example of an analysis of a physical experiment where visualisation proved the key to improving the data analysis and estimation of the fundamental parameter of interest. Coin-tossing is a basic example of a random phenomenon. Naturally tossed coins obey the laws of mechanics (we neglect air resistance) and their flight is determined by their initial conditions.

Joe Keller carried out a study [29] of the physics assuming that the coin spins about an axis through its plane. Then, the initial upward velocity and the rate of spin determine the final outcome. Keller showed that in the limit of large initial velocity and large rate of spin, a vigorous flip, caught in the hand without bouncing, lands heads half the time.

Diaconis, Holmes and Montgomery (2004) [30] (DHM) take precession into account: the fact that real flips often precess a fair amount changes the conclusions of Keller's work.

Consider first a coin starting heads up and hit exactly in the center so it goes up without turning like a pizza. We call such a flip a "total cheat coin", because it always comes up the way it started.

For such a toss, the angular momentum vector $\vec{M}$ lies along the normal to the coin.

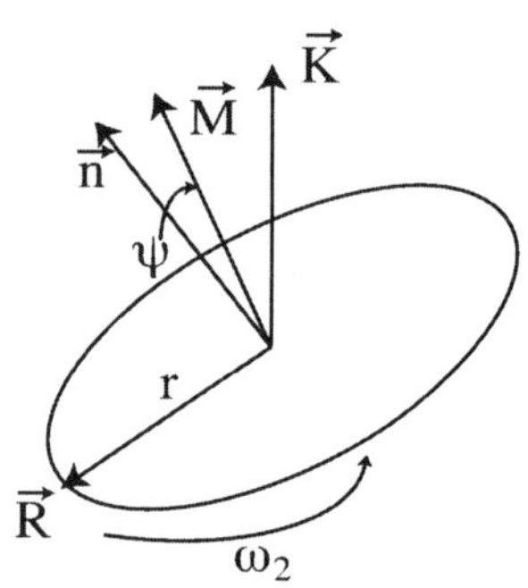

Figure 12: Coordinates of Precessing Coin.

DHM prove that the angle ψ between $\vec{M}$ and the normal to the coin stays constant. If this angle is less than 45°, the coin never turns over. It wobbles around and always comes up the way it started. Magicians and gamblers can carry out such controlled flips which appear visually indistinguishable from normal flips. For Keller's analysis, $\vec{M}$ is assumed to lie in the plane of the coin making angle 90° with the normal to the coin.

Theorem 1. *(DHM) For a coin tossed starting heads up at time 0, the cosine of the angle between the normal to the coin at time t and the up direction is*

$$f(t) = A + B\ \cos(\omega_N t) \tag{1}$$

with $A = \cos^2\psi, B = \sin^2\psi,\ \omega_N = \|\vec{M}\|/I_1, I_1 = \frac{1}{4}mR^2 + \frac{1}{3}mh^2$ *for coins with radius R, thickness h and mass m. Here ψ is the angle between the angular momentum vector $\vec{M}$ and the normal at time $t = 0$.*

Theorem 2. *(DHM) For all smooth, compactly supported densities g, the limiting probability of heads $p(\psi)$ with ψ fixed, given that heads starts up, is given by*

$$p(\psi) = \begin{cases} \frac{1}{2} + \frac{1}{\pi}\sin^{-1}(\cot^2(\psi)) \text{ if } \frac{\pi}{4} < \psi < 3\pi/4 \\ 1 \text{ if } 0 < \psi < \pi/4 \quad \text{ or } \frac{3\pi}{4} < \psi < \pi \end{cases}$$

We wanted to use Theorems 1 and 2 to find out what is the empirical distribution of ψ when real people toss coins. Thus we could be able to decide whether coin tossing is fair. Our empirical study used a high-speed slow motion camera. The projection of a circle onto the plane of the camera is an ellipse. Using image analysis techniques we fit the ellipses to the images of the tossed coin. A simple function of the lengths of the major and minor axes gives the normal to the coin in three-space. As explained, these normals spin in a circle about the angular momentum vector which stays fixed during the coin's flight. This gives an estimate of ψ.

Slow Motion Photography

We used a high-speed slow motion camera to record fifty coin flips. We found it best to film at about 600 frames per second. In contrast, the slow motion feature on standard camcorders shoot at about 60 frames per second. This is much too slow to give any useful data. A circular disc projected onto a plane results in an ellipse. From each ellipse the major and minor axis were determined. From these, as described below, the normal to the coin in three dimensions can be estimated. The sequence of (a) coin images (b) coins with fitted ellipses can be viewed by the reader at `http://www-stat.stanford.edu/~susan/coins/` .

At this stage, for each flip, we have a sequence of fitted normal vectors in three dimensions, centered at the coin's center of gravity. According to the theory, these normals lie on a circle centered at the fixed angular momentum vector. The radius of this circle thus gives an estimate of the angle ψ associated to the flip. Of course, the circles can be fit from just a few points. We used about 20 points/flip and again checked visually to see if these looked as if they lay on a circle.

The plane of the camera is fixed throughout. In spatial coordinates (X_1, X_2, X_3), the $(X_1, X_2, 0)$ plane will be identified with the camera plane and the line $(0, 0, X_3)$ is the orthogonal to the camera plane. At a fixed time, the coin is in a fixed position in 3-space. We observe, and can accurately estimate, the major and minor axes of the elliptical projection of the coin on the plane. Without loss of generality we assume the coin has radius 1. Let $\vec{A} = (A_1, A_2, 0)$ be a unit vector in the plane of the camera centered at the ellipse center along the major axis. Let $\vec{B} = (B_1, B_2, 0)$ be an orthogonal vector along the minor axis. Thus $|\vec{A}| = 1$ and $|\vec{B}| = \cos\theta$ for some angle θ, $0 \leq \theta \leq \pi/2$. This description of $\vec{A}, \vec{B}$ involves a choice of $\pm$ sign which we will deal with in a moment. Throughout, we assume that the coin has been parallel translated so that its center lines up with the center of the ellipse.

Let $\vec{U}, \vec{V}$ be the unit vectors on the coin, which project to $\vec{A}, \vec{B}$ respectively so that in fact $\vec{A} = \vec{U}$. Let $\vec{K} = (0, 0, 1)$ be the direction orthogonal to the camera plane.

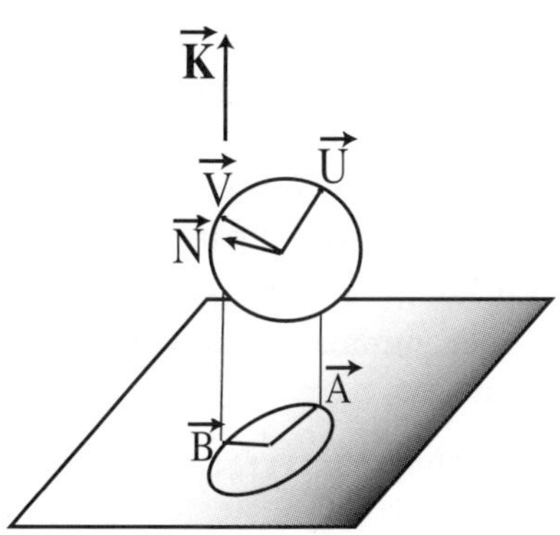

Figure 13: The disk shaped coin is projected onto an ellipse in the camera plane.

Lemma With notation as above the normal $\vec{N}$ to the coin is

$$\vec{N} = (\epsilon_1 A_2 \sqrt{1 - (B_1^2 + B_2^2)}, \quad \epsilon_2 A_1 \sqrt{1 - (B_1^2 + B_2^2)},\ \epsilon_3 (A_1 B_2 - A_2 B_1))$$

for some choice of signs $\epsilon_i = \pm 1$. Once the normals are obtained we can visualize them in three dimensions using `ggobi`[21].

```
library(Rggobi)
load('n27s.save')
ggobi(n27s)
load('n27u.save')
ggobi(n27u)
```

When the normals were represented in three dimensional space with ggobi, we obtained a very disappointing picture. The indeterminacy in the signs introduced into the normal coordinates leads to data that does not lie nicely on a circle as it should. We used the interactive feature of `ggobi` to explore the projections but they did not lie in a plane, let alone lie on a circle.

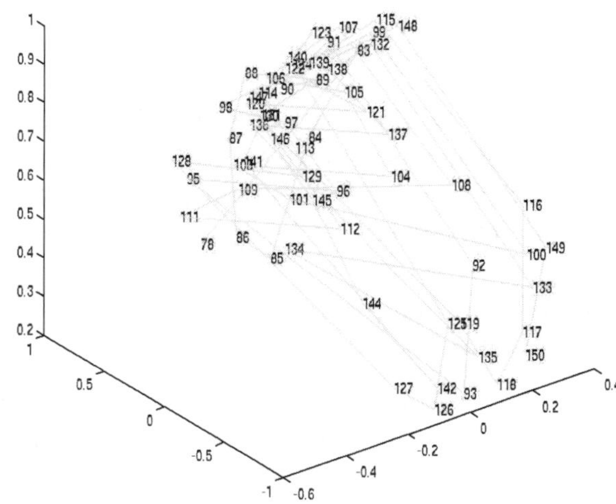

Figure 14: The normals originally have a scrambled sign pattern.

The visualisation of the data enabled us to test various fixes to the sign indeterminacy which scrambled the coordinates. The solution was found once we had the idea of incorporating the time ordering in the labelling of the points. Then we found the choice of signs by continuity, choosing (at each time frame) the choice of eight sign patterns that makes the inner product of the current normal and previous normal as close as possible to a constant, while keeping the curvature continuous.

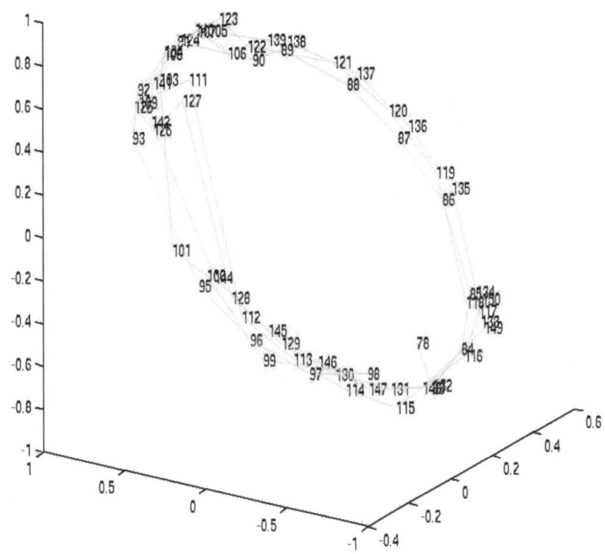

Figure 15: How the normal vectors sit around a circle in 3-D.

Unscrambled by appropriate choice of signs, all lie clearly around the circle. Our theory implies that these points lie in a plane in three dimensional space. We fit the plane using least squares. The distance between the plane and the origin gives us d from which we can find $\psi = \cos^{-1}(d)$.

Visualisation suggests a better method of estimation

However we also noted that once unscrambled, the data gave us a revelation as to another method of measuring angular momentum. We note that there are several visible triplets of points, (88,121,137) and (86,119,135) for instance, which actually correspond to images that are exactly a flip apart and thus are in Berry phase[31]. We will explain how this provides another estimation method for the angular momentum. We can see that points are coupled together after they have completed a rotation. By comparing these images we will be able to measure the precession even better.

Theorem 3. Each time the normal vector completes one full cycle around the angular momentum vector, the coin has precessed by the angle:

$$\Delta A = -\frac{\omega_{pr}}{\omega_N} 2\pi \sim -\pi \cos(\psi) \quad \text{as} \quad h \downarrow 0.$$

Remark: When $\psi \simeq 0$ so that $\vec{M}$ is nearly aligned with the vertical, we have $\Delta A \simeq \pi$. In other words, every time the normal vector precesses around once, the coin rotates approximately 180°.

We now have two methods for estimating ψ from photographs: using this 'Berry phase' from Theorem 3, or reconstructing the normal's time evolution and figuring out the radius of the resulting circle on the sphere. We have used four such tosses to check the two methods. For instance for toss No 27, we have an estimate of $\hat{\psi} = 1.48$ with the Berry phase method and for Tosses No 30: $\hat{\psi} = 1.47$ and for Toss No 32: $\hat{\psi} = 1.40$, and for Toss No 33: $\hat{\psi} = 1.36$.

Figure 16: Berry Phase. These images are separated by exactly one coin flip.

5.1. *The results*

Of our 50 flips, 27 gave useful final results. From the measured values of ψ, the probability $p(\psi)$ was calculated from Theorem 2. The estimated probabilitites range from 0.500 to 0.545. The 27 probabilities are displayed in a stem and leaf plot. The first row of this plot shows the values 0.500,0.500,0.501,... indicating occurrences of flips for which $p(\psi)$ took on these value. The next-to-last row shows no occurrences between 0.540 and 0.545. The last row shows the single outlying value 0.545. Following this are the five number summary, the mean and the standard deviation.

```
50 | 001111111222333334
50 | 555
51 | 3
51 |
52 | 3
52 | 9
53 | 34
53 |
54 |
54 | 5
```

Figure 17: Stem and leaf plot of estimates for $p(\psi)$.

The stem and leaf visualisation is the one used by all statisticians when deciding on how to grade students

(A+, A, A-, B+,...) because it shows clearly any natural groupings.

```
Five number  summary of probabiblities:
   Min. 1st Qu.  Median    3rd Qu.    Max.
 0.5001  0.5011  0.5027   0.5052    0.5448
 Mean= 0.5083         sd      = 0.0125
```

The mean of the probabilities is 0.508. We have rounded this up to the 0.51 quoted in DHM.

Acknowledgements

I would like to thank R Core team of developers who have made all the R tools I have shown here, and to the Xgobi and Ggobi creators and developers: Andreas Buja, Di Cook and Debbie Swayne in particular. Thanks to Persi Diaconis for discussions on the random matrix problem and Projection Pursuit and for our joint work with Richard Montgomery on the coin toss problem, to Peter Lee and Michael He for our joint work on the T-cell gene expression study, and to Louis Lyons and Eric Feigelson for their careful reading of the original article. This work was supported by the NSF-DMS 0241246.

References

1. R. Ihaka and R. Gentleman, *Journal of Computational and Graphical Statistics* **5**, 299 (1996).
2. L. Wilkinson, D. Wills, D. Rope, A. Norton and R. Dubbs, *The Grammar of Graphics (Statistics and Computing)* (Springer, NY, 2005).
3. J. M. Chambers, W. S. Cleveland and P. A. Tukey, *Graphical methods for data analysis.* (Wadsworth, CA, 1983).
4. W. S. Cleveland, *The Elements of Graphing Data* (Hobart Press, Summit,NJ, 1994).
5. W. S. Cleveland, *Visualizing Data* (Hobart Press, Summit, NJ, 1993).
6. S. Lauritzen, *Goodness of Fit*, in *Phystat Proceedings*, eds. L. Lyons and M. K. Unel (World Scientific, 2005).
7. W. S. Cleveland, *The Collected Works of John W. Tukey: Graphics 1965-1985, Volume V, Graphics : 1965-1985* (Wadsworth, Pacific Grove,CA, 1988).
8. S. Holmes, M. He, T. Xu and P. P. Lee, *Proc. National Academy Sciences* **102**, 5519 (2005).
9. M. Greenacre, *Theory and Applications of Correspondence Analysis* (Academic Press, NY., 1984).
10. D. R. Cox and L. Brandwood, *J. Roy. Statist. Soc. Ser. B* **21**, 195 (1959).
11. B. Charnomordic and S. Holmes, *Statistical Graphics and Computing Newlsetter* **12** (2001).
12. M. Hill and H. Gauch, *Vegetatio* **42**, 47 (1980).
13. J. C. Gower and D. Hand, *Biplots* (Chapman and Hall, London, 1996).
14. J. B. Kruskal, *Psychometrika* **29**, 115 (1964).
15. R. N. Shepard, *Science* **210**, 390 (1980).
16. D. W. Scott, *Multivariate Density Estimation : Theory, Practice, and Visualization* (Wiley, NY, 1992).
17. K. Mardia, J. Kent and J. Bibby, *Multiariate Analysis* (Academic Press, NY., 1979).
18. J. H. Friedman and J. W. Tukey, *IEEE Transactions on Computers* **C-23**, 881 (1974).
19. M. A. Fisherkeller, J. H. Friedman and J. W. Tukey, PRIM-9: An interactive multidimensional data display and analysis system, in *Dynamic Graphics for Statistics*, 1988.
20. D. F. Swayne, D. Cook and A. Buja, *JCGS* **7**, p. 1 (1998).
21. D. F. Swayne, D. T. Lang, A. Buja and D. Cook, *Comput. Stat. Data Anal.* **43**, 423 (2003).
22. D. Chessel, A. B. Dufour and J. Thioulouse., *R News* **4**, 5 (2004).
23. S. T. Roweis and L. K. Saul, *Science* **290**, 2323 (2000).
24. J. B. Tenenbaum, V. d. Silva and J. C. Langford, *Science* **290**, 2319 (2000).
25. A. D. Gordon, *Classification* (Chapman and Hall Ltd, 1999).
26. B. Efron and R. Tibshirani, *An Introduction to the Bootstrap* (Chapman and Hall, London, 1993).
27. L. Billera, S. Holmes and K. Vogtmann, *Adv. Appl. Maths* **27**, 771 (2001).
28. J. B. Kruskal, Multidimensional scaling and other methods for discovering structure, in *Statistical Methods for Digital Computers*, eds. K. Enslein, A. Ralston and H. S. Wilf (Wiley, 1977) pp. 296–339.
29. J. B. Keller, *American Mathematical Monthly* , 191 (1986).
30. P. Diaconis, S. Holmes and R. Montgomery, *SIAM* (2005).
31. R. Montgomery, *Am. J. Physics* **59**, 394 (1991).

ASTROPHYSICS

ASTROPHYSICS WITH TERABYTES OF DATA

ALEXANDER S. SZALAY
Department of Physics and Astronomy, The Johns Hopkins University
Baltimore. MD 21218, USA
szalay@jhu.edu

Unprecedented data sizes in astronomy are creating a new challenge for statistical analysis. In these large datasets statistical errors are typically much smaller that the systematic errors. Due to the exponential growth of data volume, optimal algorithms with poor scaling behavior are becoming untenable. New approaches with suboptimal but fast algorithms are required.

1. Introduction

1.1. *Evolving Science*

Computational Science is an emerging new branch of most scientific disciplines. A thousand years ago, science was primarily *empirical*. Over the last 500 years each discipline has grown a *theoretical* component. Theoretical models often motivate experiments and generalize our understanding. Today most disciplines have both empirical and theoretical branches. In the last 50 years, most disciplines have grown a third, *computational* branch (e.g. empirical, theoretical and computational ecology, or physics, or linguistics).

Computational Science has meant simulation. It grew out of our inability to find closed form solutions for complex mathematical models. Computers can simulate these complex models.

Computational Science has been evolving to include information management. Scientists are faced with mountains of data that stem from four converging trends: (1) the flood of data from new scientific instruments driven by Moore's Law – doubling their data output every year or so; (2) the flood of data from simulations; (3) the ability to economically store petabytes of data online; and (4) the Internet and computing Grid that makes all these archives accessible to anyone anywhere.

Acquisition, organization, query, and visualization tasks scale almost linearly with data volumes. By using parallelism, these problems can be solved within fixed times (minutes or hours).

1.2. *Emerging New Paradigms*

As a result of this data explosion, there are new emerging paradigms not only in the way how we collect our data, but also in how we publish and analyze them. The traditional method of science consisted of the data collection as its first step, followed by the analysis, then publication.

With today's terabyte size data sets, collected by large collaborative teams, the data need to be first properly organized, usually into on-line databases, before their analysis can even begin. Since these large, data intensive projects take at least 5-6 years, most of their data will only migrate from the project databases to a central repository at the end of the project, i.e. most of the data in central archives will be at least 3 years old. When data sizes are doubling each year, this means that centralized data sets will never exceed 12% of all data available for science.

Since these large projects are scattered over the world, most of the world's scientific data can only be accessed by successfully federating these diverse resources.

1.3. *Living in an Exponential World*

Detector sizes of our astronomical survey instruments are improving exponentially, since they are based on the same technology as computer CPUs. Consequently, astronomy data volumes are approximately doubling every year. This even exceeds the rate of Moore's law, describing the speedup of CPUs and growth of storage. This trend results from the emergence of large-scale surveys, like 2MASS, SDSS or 2dFGRS. Soon there will be almost all-sky data in more than ten wavebands. These large-scale surveys have another important characteristic: they are each obtained by a single group, with sound statistical plans and well-controlled systematics.

As a result, the data are becoming increasingly more homogeneous, and approach a fair sample of the Universe. This trend has brought a lot of advances in the analysis of the large-scale galaxy distribution. Our goal today is to reach an unheard-of level of accuracy in measuring both the global cosmological parameters and the shape of the power spectrum of primordial fluctuations. The emerging huge data sets from wide field sky surveys pose interesting issues, both statistical and computational. One needs to reconsider the notion of optimal statistics.

These large, homogenous datasets are also changing the way we approach their analysis. Traditionally, statistics in cosmology has primarily dealt with how to extract the most information from the small samples of galaxies we had. This is no longer the case: there are redshift surveys of 300,000 objects today; soon there will be a million measured galaxy redshifts. Angular catalogs today have samples in excess of 50 million galaxies; soon they will have 10 billion (LSST). In the observations of the CMB, COBE had a few thousand pixels on the sky, MAP will have a million, PLANCK will have more than 10 million. Thus, shot noise and sample size is no longer an issue. The limiting factors in these data sets are the systematic uncertainties, like photometric zero points, effects of seeing, uniformity of filters, etc.

The statistical issues are changing accordingly: it is increasingly important to find techniques that can be desensitized to systematic uncertainties. Many traditional statistical techniques in astronomy focused on `optimal' techniques. It was generally understood, that these minimized the statistical noise in the result, but they are quite sensitive to various systematics. Also, they assumed infinite computational resources. This was not an issue when sample sizes were in the thousands. But, many of these techniques involve matrix diagonalizations or inversions and so the computational cost scales as the 3rd power of matrix size. Samples a thousand times larger have computational costs a billion times higher. Even if the speedup of our computers keeps up with the growth of our data, it cannot keep pace with such powers. We need to find algorithms that scale more gently. In the near future we hypothesize that only algorithms with *NlogN* scaling will remain feasible.

As the statistical noise decreases with larger samples, another effect emerges: *cosmic variance*. This error term reflects the fact that our observing position is fixed at the Earth, and at any time we can only study a fixed – albeit ever increasing – region of the Universe. This provides an ultimate bound on the accuracy of any astronomical measurement. We should carefully keep this effect in mind when designing new experiments.

2. Astrophysical Motivation

2.1. *Precision Cosmology*

We are entering the era of precision cosmology. The large new surveys with their well-defined systematics are key to this transition. There are many different measurements we can make that each constrain combinations of the cosmological parameters. For example, the fluctuations in the cosmic Microwave Background (CMB) around the multipole l of a few hundred are very sensitive to the overall curvature of the Universe, determined by both dark matter and dark energy (deBernardis et al 2000, Netterfield et al 2002).

Due to the expansion of the Universe, we can use redshifts to measure distances of galaxies. Since galaxies are not at rest in the frame of the expanding Universe, their motions cause an additional distortion in the line-of-sight coordinate. This property can be used to study the dynamics of galaxies, inferring the underlying mass density. Local redshift surveys can measure the amount of gravitating dark matter, but they are insensitive to the dark energy. Combining these different measurements (CMB + redshift surveys), each with their own degeneracy can yield considerably tighter constraints than either of them independently. We know most cosmological parameters to an accuracy of about 10% or somewhat better today. Soon we will be able to reach the regime of 2-5% relative errors, through both better data but also better statistical techniques.

The relevant parameters include the age of the Universe, t_0, the expansion rate of the Universe, also called Hubble's constant H_0, the deceleration parameter q_0, the density parameter Ω, and its components, the dark energy, or cosmological constant Ω_Λ, the baryonic+dark matter Ω_m, the baryon fraction f_B, and the curvature Ω_k. These are not independent from one another, of course. Together, they determine the dynamic evolution of the Universe, assumed to be homogeneous and isotropic, described by a single scale factor $a(t)$. For a Euclidian (flat) Universe $\Omega_\Lambda+\Omega_m=1$.

One can use both the dynamics, luminosities and angular sizes to constrain the cosmological parameters. Distant supernovae have been used as standard candles to get the first hints about a large cosmological constant. The angular size of the Doppler-peaks in the CMB fluctuations gave the first conclusive evidence for a flat universe, using the angular diameter-distance relation. The gravitational infall manifested in redshift-space distortions of galaxy surveys has been used to constrain the amount of dark matter.

These add up to a remarkably consistent picture today: a flat Universe, with Ω_Λ=0.65±0.05, Ω_m=0.35±0.05. It would be nice to have several independent measurements for the above quantities. Recently, new possibilities have arisen about the nature of the cosmological constant – it appears that there are many possibilities, like quintessence, that can be the dark energy. Now we are facing the challenge of coming up with measurements and statistical techniques to distinguish among these alternative models.

There are several parameters used to specify the shape of the fluctuation spectrum. These include the amplitude σ_8, the root-mean-square value of the density fluctuations in a sphere of 8 Mpc radius, the shape parameter Γ, the redshift-distortion parameter β, the bias parameter b, and the baryon fraction $f_B=\Omega_B/\Omega_m$. Other quantities, like the neutrino mass also affect the shape of the fluctuation spectrum, although in more subtle ways than the ones above (Seljak and Zaldarriega 1996).

The shape of the fluctuation spectrum is another sensitive measure of the Big Bang at early times. Galaxy surveys have traditionally measured the fluctuations over much smaller scales (below 100 Mpc), where the fluctuations are nonlinear, and even the shape of the spectrum has been altered by gravitational infall and the dynamics of the Universe. The expected spectrum on very large spatial scales (over 200 Mpc) was shown by COBE to be scale-invariant, reflecting the primordial initial conditions, remarkably close to the predicted Zeldovich-Harrison shape.

There are several interesting physical effects that will leave an imprint on the fluctuations: the scale of the horizon at recombination, the horizon at matter-radiation equality, and the sound-horizon—all between 100-200 Mpc (Eisenstein and Hu 1998). These scales have been rather difficult to measure: they used to be too small for CMB, too large for redshift surveys. This is rapidly changing. New, higher resolution CMB experiments are now covering sub-degree scales, corresponding to less than 100 Mpc comoving, and redshift surveys like 2dF and SDSS are reaching scales well above 300 Mpc.

We have yet to measure the overall contribution of baryons to the mass content of the Universe. We expect to find the counterparts of the CMB Doppler bumps in galaxy surveys as well, since these are the remnants of horizon scale fluctuations in the baryons at the time of recombination. The Universe behaved like a resonant cavity at the time. Due to the dominance of the dark matter over baryons the amplitude of these fluctuations is suppressed, but with high precision measurements they should be detectable.

A small neutrino mass of a few electron volts is well within the realm of possibilities. Due to the very large cosmic abundance of relic neutrinos, even such a small mass would have an observable effect on the shape of the power spectrum of fluctuations. It is likely that the sensitivity of current redshift surveys will enable us to make a meaningful test of such a hypothesis. One can also use large angular catalogs, projections of a 3-dimensional random field to the sphere of the sky, to measure the projected power spectrum. This technique has the advantage that dynamical distortions due to the peculiar motions of the galaxies do not affect the projected distribution. The first such analyses show promise.

2.2. *Large Surveys*

As mentioned in the introduction, some of the issues related to the statistical analysis of large redshift surveys, like 2dF (Percival et al 2002), or SDSS (York et al 2000, Pope et al. 2005) with nearly a billion objects are quite different from their predecessors with only a few thousand galaxies. The foremost difference is that shot-noise, the usual hurdle of the past, is irrelevant. Astronomy is different from laboratory science because we cannot change the position of the observer at will. Our experiments in studying the Universe will never approach an ensemble average; there will always be an unavoidable *cosmic variance* in our analysis. By studying a larger region of the Universe (going deeper and/or wider) can decrease this term, but it will always be present in our statistics.

Systematic errors are the dominant source of uncertainties in large redshift surveys today. For example photometric calibrations, or various instrumental and natural foregrounds and backgrounds contribute bias to the observations. Sample selection is also becoming increasingly important. Multicolor surveys enable us to invert the observations into physical quantities, like redshift, luminosity and spectral type. Using these broadly defined '*photometric redshifts*', we can select statistical subsamples based upon approximately rest-frame quantities, for the first time allowing meaningful comparisons between samples at low and high redshifts.

Various effects, like small-scale nonlinearities, or redshift space distortions, will turn an otherwise homogeneous and isotropic random process into a non-isotropic one. As a result, it is increasingly important to find statistical techniques, which can reject or incorporate some of these effects into the analysis. Some of these cannot be modeled analytically; we need to perform Monte-Carlo simulations to understand the impact of these systematic effects on the final results. The simulations themselves are also best performed using databases.

Data are organized into databases, instead of the flat files of the past. These databases contain several well-designed and efficient indices that make searches and counting much faster than brute-force methods. No matter which statistical analyses we seek to perform, much of the analysis consists of data filtering and counting. Up to now most of this has been performed off-line. Given the large samples in today's sky surveys, offline analysis is becoming increasingly inefficient – scientists want to be able to interact with the data. Here we would like to describe our first efforts to integrate large-scale statistical analyses with the database. Our analysis would have been very much harder, if not entirely infeasible, to perform on flat files.

3. Statistical Techniques

3.1. *The Two-point Correlation Function*

The most frequent techniques used in analyzing data about spatial clustering are the two-point correlation functions and various power spectrum estimators. There is an extensive literature about the relative merits of each of the techniques. For an infinitely large data set in principle both techniques are equivalent. In practice, however, there are subtle differences: finite sample size affects the two estimators somewhat differently, edge effects show up in a slightly different fashion and there are also practical issues about computability and hypothesis testing, which are different for the two techniques.

The two point correlations are most often computed via the LS estimator (Landy and Szalay 1992)

$$\xi(r) = \frac{DD - 2DR + RR}{RR}$$

which has a minimal variance for a Poisson process. DD, DR and RR describe the respective normalized pair counts in a given distance range. For this estimator and for correlation functions in general, hypothesis testing is somewhat cumbersome. If the correlation function is evaluated over a set of differential distance bins, these values are not independent, and their correlation matrix also depends on the three and four-point correlation functions, less familiar than the two-point function itself. The brute-force technique involves the computation of all pairs and binning them up, so it scales as $O(N^2)$. In terms of modeling systematic effects, it is very easy to compute the two-point correlation function between two points.

Another popular second order statistic is the power spectrum P(k), usually measured by using the FKP estimator (Feldman et al 1994). This is the Fourier-space equivalent of the LS estimator for correlation functions. It has both advantages and disadvantages over correlation functions. Hypothesis testing is much easier, since in Fourier space the power spectrum at two different wavenumbers are correlated, but the correlation among modes is localised. It is determined by the *window-function*, the Fourier transform of the sample volume, usually very well understood. For most realistic surveys the window function is rather anisotropic, making angular averaging of the three-dimensional power spectrum

Figure 1. The layout of a single 'stripe' of galaxy data in the SDSS angular catalog, with the mask overlay. The stave-like shape of the stripe is due to stripe layout over the sphere. The vertical direction is stretched considerably. The narrow white boxes represent areas around bright stars that need to be 'masked' out from the survey. This illustrates the complex geometry and the edge effects we have to consider.

estimator somewhat complicated. During hypothesis testing one is using the estimated values of P(k), either directly in 3D Fourier space, or compressed into quadratic sums binned by bands. Again, the 3rd and 4th order terms appear in the correlation matrix. The effects of systematic errors are much harder to estimate.

3.2. *Hypothesis Testing*

Hypothesis testing is usually performed in a parametric fashion, with the assumption that the underlying random process is Gaussian. We evaluate the log likelihood as

$$\ln L(\Pi) = -\frac{1}{2}x^T C^{-1} x - \frac{1}{2}\ln |C|$$

where x is the data vector, and C is its correlation matrix, dependent on the parameter vector Π. There is a fundamental lower bound on the statistical error, given by the Fisher matrix, easily computed. This is a common tool used these days to evaluate the sensitivity of a given experiment to measure various cosmological parameters. For more detailed comparisons of these techniques see Tegmark et al (1998). This algorithm requires the inversion of C, usually an N^3 operation, where N is the dimension of the matrix.

What would an ideal method be? It would be useful to retain much of the advantages of the 2-point correlations so that the systematics are easy to model, and those of the power spectra so that the modes are only weakly correlated. We would like to have a hypothesis testing correlation matrix without 3rd and 4th order quantities. Interestingly, there is such a method, given by the Karhunen-Loeve transform. In the following subsection we describe the method, and show why it is a useful framework for the analysis of the galaxy distribution. Then we discuss some of the detailed issues we had to deal with over the years to turn this into a practical tool.

One can also argue about parametric and non-parametric techniques, like using bandpowers to characterize the shape of the fluctuation spectrum. We postulate, that for the specific case of redshift surveys it is not possible to have a purely non-parametric analysis. While the shape of the power spectrum itself can be described in a non-parametric way, the distortions along the redshift direction are dependent on a physical model (gravitational infall). Thus, without an explicit parameterization or ignoring this effect no analysis is possible.

3.3. *Karhunen-Loeve Analysis of Redshift Surveys*

The Karhunen-Loeve (KL) eigenfunctions (Karhunen 1947, Loeve 1948) provide a basis set in which the distribution of galaxies can be expanded. These eigenfunctions are computed for a given survey geometry and fiducial model of the power spectrum. For a Gaussian galaxy distribution, the KL eigenfunctions provide optimal estimates of model parameters, i.e. the resulting error bars are given by the inverse of the Fisher matrix for the parameters (Vogeley & Szalay 1996). This is achieved by finding the orthonormal set of eigenfunctions that optimally balance the ideal of Fourier modes with the finite and peculiar geometry and selection function of a real survey. The KL method has been applied to the Las Campanas redshift survey by Matsubara, Szalay & Landy (2000) and to the PSCz survey by Hamilton, Tegmark & Padmanabhan (2001).

The KL transform is often called *optimal subspace filtering* (Therrien 1992), describing the fact that during the analysis some of the modes are discarded. This offers distinct advantages. If the measurement is composed of a signal that we want to measure (gravitational clustering) superposed on various backgrounds (shot-noise, selection effects, photometric errors, etc) which have slightly different statistical properties, the diagonalization of the correlation matrix can potentially segregate these different types of processes into their own subspaces. If we select our subspace carefully, we can actually improve on the signal to noise of our analysis.

The biggest advantage is that hypothesis testing is very easy and elegant. First of all, all KL modes are orthogonal to one another, even if the survey geometry is extremely anisotropic. Of course, none of the KL modes can be narrower than the survey window, and their shape is clearly affected by the survey geometry. The orthogonality of the modes represents repulsion between the modes, they cannot get too close; otherwise they could not be orthogonal. As a result the KL modes are dense-packed into Fourier-space, thus optimally representing the information enabled by the survey geometry.

Secondly, the KL transform is a linear transformation. If we do our likelihood testing over the KL-transform of the data, the likelihood correlation matrix contains only second order quantities. This avoids problems with 3 and 4-point functions. All these advantages became very apparent when we applied the KL method to real data.

4. Working with a Database

4.1. *Why use a Database?*

The sheer size of the data involved makes it necessary to store the data in a database – there are just too many objects to organize into directories of files. We originally started to use databases solely for this reason. Astronomers in general use databases only to store their data, but when they do science, they generate a flat file, usually in a simple table. Then they use their own code for the scientific analysis. Mostly the databases are remote. One has to enter queries into a web form, and retrieve the data as either an ASCII or binary table.

We have been working on creating the Science Archive for the Sloan Digital Sky Survey. We are now using a relational database, Microsoft's SQL Server, as the back-end database.

The database contains much more than just the basic photometric or spectroscopic properties of the individual objects. We have computed, as an add-on, the inversion of the photometric observations into physical, rest-frame parameters, like luminosity, spectral type, and of course a photometric redshift. Much of the training for this technique was obtained from the spectroscopic observations. These are stored in the database as an ancillary table. Information about the geometry of the survey, how it is organized into stripes, strips, runs, camera columns and fields, is also stored in the database. The value of seeing (the blur of images caused by the atmosphere) is monitored and saved in the Field table. The 'blind' pixels of the survey, caused by a bright star, satellite, meteorite or a ghost in the camera are also saved in the database as an extended object, with their convex hull, a bounding circle and a bounding rectangle.

We are using this database for subsequent scientific analysis. We have a local copy of the data at Johns Hopkins, stored in a relatively high performance, yet inexpensive database server. While building our applications to study the correlation properties of galaxies, we have discovered that many of the patterns in our statistical analysis involve tasks that are much better performed inside the database than outside, on flat files. The database gives high-speed sequential search of complex predicates using multiple CPUs, multiple disks, and large main memories. It also has sophisticated indexing and data combination algorithms that compete favorably with hand-written programs against flat files. Indeed, we see cases where multi-day batch file runs are replaced with database queries that run in minutes.

4.2. *Going Spatial*

In order to efficiently perform queries that involve spatial boundaries, we have developed a class library based upon a hierarchical triangulation of the sky (Kunszt et al 2001) to handle searches over spherical polygons. We added the class library as an extension to SQL Server, so its functions can be called directly inside the database.

In order to generate meaningful results for the clustering, we need to create a well-defined, statistically fair sample of galaxies. We have to censor objects that are in areas of decreased sensitivity or bad seeing. We also have to be aware of the geometry of the censored areas. We created these 'masks' using plain database queries for fields of bad seeing, rectangles around bright stars and other trouble spots. In the current release of the database these regions are derived by processing images that contain flag information about every pixel on the sky.

We have also implemented a library of database procedures that perform the necessary computational geometry operations inside the database, and they also perform logarithmic-time search procedures over the whole database (Szalay et al 2005), fully integrated with our previous approach, based on the Hierarchical Triangular Mesh.

4.3. *Building Statistical Samples*

We analyzed a large sample of galaxies from the photometric observations of the Sloan Digital Sky Survey. The data extend over an area of about 3000 square degrees, organized in long, 2.5 degree wide 'stripes'. The stripes are organized into 12 'camcols', corresponding to the detector camera columns, and those are broken up into 'fields' that are pipeline processing units. We downloaded about 50 million galaxies from the project database at Fermilab, and created a custom database of this downloaded data, using Microsoft SQL Server. Each galaxy in the database had a five-color photometry, and an angular position, plus a description of which stripe, camcol, and field it belongs to. Next, we computed the photometric redshifts, absolute luminosities, rest-frame types, and their covariances for each of the 50 million galaxies. These derived data were also inserted into the database. Using this database, we can write SQL queries that generate a list of objects that satisfy a given selection criterion and that are in the angular statistical sample of galaxies. We can also write SQL to get a list of masks, with their rectangular boundaries.

The selection criteria for the redshift survey were much more complex: they involve observing objects selected in 2.5 degree stripes, then observed spectroscopically with a plug-plate of 3 degrees diameter. The centers of the plates were selected to overlap in higher density areas, since optical fibers cannot be placed closer than 55" in a given observation. This complex pattern of intersections and our sampling efficiency in each of the resulting 'sectors' was calculated using our spherical polygon library.

Subsequently we created a separate database for the redshift-space analysis that has both our statistical sample and a Monte-Carlo simulation of randomly distributed objects that were generated per our angular sampling rate. The size of this latter data set is about 100 million points.

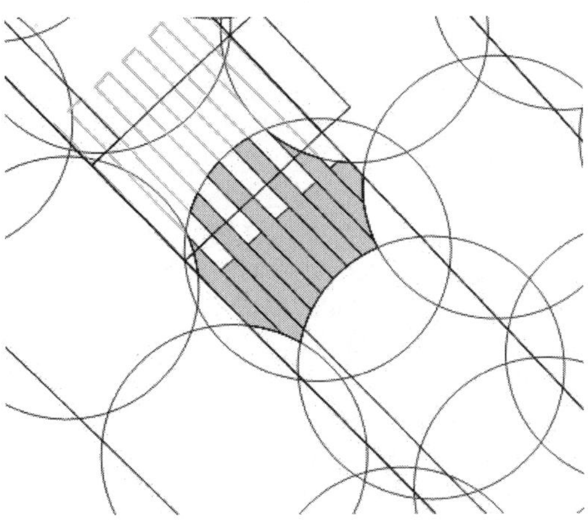

Figure 2. An example of the complex spatial geometry required to describe the statistical completeness of the Sloan Digital Sky Survey data. The regions of uniform sampling rate are formed by the homogeneous intersections of the circular '*tiles*', and the elongated '*camcols*'.

5. Next Generation Data Analysis

5.1. *Scaling of Optimal Algorithms*

Exponentially growing astronomy data volumes pose serious new problems. Most statistical techniques labeled as 'optimal' are based on several assumptions that were correct in the past, but are no longer valid. Namely, the dominant contribution to the variance is no longer statistical – it is systematics, computational resources cannot handle N^2 and N^3 algorithms -- N has grown from 10^3 to 10^9, and cosmic variance can no longer be ignored.

5.2. *Advanced Data Structures*

What are the possibilities? We believe one answer lies in clever data structures, borrowed from computer science to pre-organize our data into a tree-hierarchy, and having the computational cost dominated by the cost of sorting, an *NlogN* process. This is the approach taken by A. Moore and collaborators in their tree-code (Moore et al. 2001).

5.3. *Approximate but Fast Heuristic Algorithms*

Another approach is to use approximate statistics, as advocated by Szapudi et al (2001). In the presence of a cosmic variance, an algorithm that spends an enormous amount of CPU time to minimize the statistical variance to a level substantially below the cosmic variance can be very wasteful. One can define a cost function that includes all terms in the variance and a computational cost, as a function of the accuracy of the estimator. Minimizing this cost-function will give the best possible results, given the nature of the data and our finite computational resources. We expect to see more and more of these algorithms emerging. One nice example of these ideas is the fast CMB analysis developed by Szapudi et al (2002), which reduces the computations for a survey of the size of Planck from 10 million years to approximately 1 day!

5.4. *Challenges*

We have discussed several of the issues arising in spatial statistics. These include the need of fast (*N*log*N*) algorithms for correlation and power spectrum analyses. These also need to be extended to cross-correlations among different surveys, like galaxies and CMB to look for the Integrated Sachs-Wolfe (ISW) effect. These require an efficient sky pixelization and fast harmonic transforms of all relevant data sets. Higher order clustering methods are becoming increasingly more relevant and of discriminating value. Their scaling properties are increasingly worse.

Time-domain astronomy is coming of age. With new surveys like PanStarrs and LSST we will have billions of objects with multiple epoch observations, sampled over a large dynamic range of time intervals. Their real-time classification into transients, periodic variable stars, moving asteroids will be a formidable challenge.

With large all-sky surveys reaching billions of objects, each having hundreds of attributes in the database, creating fast outlier detection with discovery value will be one of the most interesting applications, that will connect the large surveys to the new large telescopes that can do the individual object follow-up.

We need to develop techniques which are robust with respect to the systematic errors. Hypothesis testing will soon involve millions of different models in very high dimensional spaces. Visualization of complex models is going to be a major part of the statistical comparisons.

Summary

Several important new trends are apparent in modern cosmology and astrophysics: data volume is doubling every year, the data is well understood, and much of the low level processing is already done by the time the data is published. This makes it much easier to perform additional statistical analyses. At the same time many of the outstanding problems in cosmology are inherently statistical, either studying the distributions of typical objects (in parametric or non-parametric fashion) or finding the atypical objects: extremes and/or outliers. Many of traditional statistical algorithms are infeasible because they scale as polynomials of the size of the data. Today, we find that more and more statistical tools use advanced data structures and/or approximate techniques to achieve fast computability.

In both applications we presented, the databases and the computations performed inside were an essential component of the analysis and enabled us to deal with much larger datasets. We also integrated some of our tools with the database itself: like generating plots of galaxy surface densities or the whole angular correlation code itself.

In the not too distant future, when our data sets grow another order of magnitude, only *NlogN* algorithms will remain feasible—the cost of computation will become a very important ingredient of an optimal algorithm. Such an evolution in our approach to astrostatistics can only be accomplished with an active and intense collaboration of astronomers, statisticians and computer scientists.

Acknowledgments

AS would like to acknowledge support from the National Science Foundation AST-0407308, and grants from the W.M. Keck Foundation and the Gordon and Betty Moore Fundation.

References

1. de Bernardis, P. et al. 2000, Nature, 404, 955
2. Eisenstein, D.J. & Hu, W. 1998, ApJ, 496, 605
3. Feldman, H.A., Kaiser, N. & Peacock, J.A. 1994, ApJ, 426, 23
4. Hamilton, A.J.S., Tegmark, M., Padmanabhan, N., 2000, MNRAS, 317, L23
5. Karhunen, H. 1947, Ann. Acad. Science Finn. Ser. A.I. 37
6. Kunszt,P.Z. et al.2001, *Mining the Sky: Proc. of the MPA/ESO/MPE workshop, Garching, A.J.Banday, S. Zaroubi, M. Bartelmann (ed.)*, (Springer-Verlag Berlin Heidelberg), pp. 631-637
7. Landy, S.D. and Szalay, A.S. 1992, ApJ, 394, 25
8. Loeve, M. 1948, Processes Stochastiques et Mouvement Brownien, (Hermann, Paris, France)
9. Matsubara, T., Szalay, A. S., Landy, S. D., 2000, ApJ, 535, L1
10. Moore,A.W. et al, 2001, *Mining the Sky: Proc. of the MPA/ESO/MPE workshop, Garching, A.J.Banday, S. Zaroubi, M. Bartelmann (ed.)*, (Springer-Verlag Berlin Heidelberg), 71-82.
11. Netterfield, C.B. et al. 2002, ApJ, 571, 604.
12. Percival, W.J., et al. 2002, MNRAS, 337, 1068.
13. Pope, A. et al. 2004, Ap.J. 607, 655.
14. Seljak, U. & Zaldarriaga, M. 1996, ApJ, 469, 437
15. Szalay, A.S. et al 2005, Microsoft Technical report MSR-TR-2005-123.
16. Szapudi,I., Prunet,S., Pogosyan,D., Szalay,A.S. and Bond,J.R. 2001, *Astrophys. J. Lett.*, **548**, 115.
17. Szapudi, I., Prunet,S. & Colombi, S. 2001, Ap.J.Lett 561, 11
18. Tegmark, M. et al 1998, ApJ, 499, 555
19. Therrien, C. W. 1992, Discrete Random Signals and Statistical Signal Processing, (New Jersey: Prentice-Hall).
20. Vogeley, M.S., Szalay, A.S., 1996, ApJ, 465, 34
21. York, D. G. et al. 2000, AJ, 120, 1579

MULTISCALE GEOMETRIC ANALYSIS OF THE 2DF DATA

JEAN-LUC STARCK
DAPNIA/SEDI-SAP, Service d'Astrophysique, CEA-Saclay, 91191 Gif-sur-Yvette, France
E-mail: jstarck@cea.fr
Laboratoire APC, Collège de France
11 place Marcelin Berthelot 75231 Paris Cedex 05

VICENT J. MARTÍNEZ
Observatori Astronòmic, Universitat de València, Apartat de Correus 22085, E-46071 València, Spain
E-mail:martinez@uv.es

ENN SAAR
Tartu Observatoorium, Tõravere, 61602, Estonia
E-mail:saar@aai.ee

In order to investigate the recent discovery of a discrepancy between the 2DF data and the Λ CDM simulations, we have applied a Multiscale Geometric Analysis (MGA) on the 2DF data. We report in this paper the results of this study.

1 Introduction

The distribution of galaxies seen in the available galaxy redshift catalogues shows complex structures such as voids, filaments, walls, or clusters. In order to compare the data with the simulations resulting from the cosmological models, we need to extract statistical or morphological information from the data. The two-point correlation (2CF), extensively used by Peebles [1], is certainly the most popular indicator to describe the spatial clustering of the galaxy distribution. Many different 2CF estimators have been proposed in the past [2,3,4]. A detailed description of these estimators may be found in refs. [5,6] and they are compared in refs. [7,8]. The two-point correlation function can been generalized to the N-point correlation function [9,10]. Other statistical measures to characterize the spatial distribution of points have also been developed, such as the void probability function [11], the multifractal approach [12], the Minkowski functionals [13,14], the J function [15,16], the minimal spanning tree [17,18,19], or the wavelet [20,21,22,23,24]. The Sloan Digital Sky Survey (Early Data Release) has recently been analyzed using a 3D Genus Statistics [25] and results were consistent with that predicted by simulations of a Λ-dominated spatially-flat cold dark matter model. The Genus is calculated by (i) convolving the data by a kernel, generally a Gaussian, (ii) setting to zero all values under a threshold ν in the obtained distribution, and (iii) taking the difference D between the number of holes and the number of isolated regions. The Genus curve $G(\nu)$ is obtained by varying the threshold level ν. The first step of the algorithm, the convolution by a Gaussian, may be dramatic for the description of filaments, which are spread out along all directions [26]. It has been shown that replacing the Gaussian smoothing by a wavelet denoising leads to much more reliable results [26]. The wavelet-Genus method has been applied to both the 2DF data and a set of 22 Λ CDM simulations and the 2DF genus curve is clearly not compatible with the simulations [26]. Figure 1 shows the wavelet genus function of the 2DF data. The solid line is the genus for the 2DF data and the crosses are the mean genus for 22 realizations of the λ-CDM simulations with the 3σ error bars.

Question: How to explain the discrepancy ? In [26], the discrepancy was attributed to the presence of a super cluster in the data, which was not in the simulation. Therefore, even if there is a discrepancy, the λ-CDM model is still considered as a good model for representing the 2DF data. In order to better investigate this difference between the 2DF and the λ-CDM simulations, we have achieved a Multiscale Geometric Analysis (MGA) [27] of the 2DF data. Section 2 presents the MGA approach and the

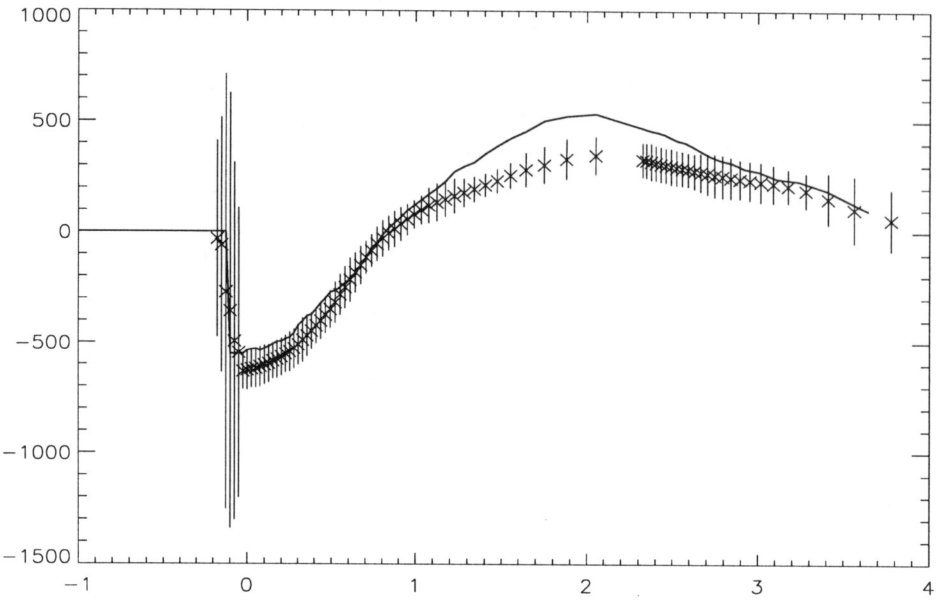

Figure 1. Wavelet Denoising+Genus: 2DF and Lambda-CDM Sim. The solid line is the genus for the wavelet denoised 2DF data and the crosses are the mean genus for 22 wavelet denoised realizations of the λ-CDM simulations with the 3σ error bars.

data (simulations and 2DF data) are described in section 3. Results are given in section 4.

2 Multiscale Geometric Analysis

As the data contain clusters, filaments and sheets, it has recently been proposed to analyze the data with three multiscale transforms, each of them being well adapted for representing only one kind of feature [27]. Wavelets represent well isotropic features (i.e. clusters in 3D), while more recent geometric multiscale methods such the beamlet and the ridgelet represent well data containing respectively filaments and sheets.

For each $a > 0$, $b_1, b_2, b_3 \in \mathbf{R}^3$, the *wavelet* is defined by

$$\psi_{a,b_1,b_2,b_3} : \mathbf{R}^3 \rightarrow \mathbf{R}$$
$$\psi_{a,b_1,b_2,b_3}(x_1,x_2,x_3) = a^{-3/2} \cdot \psi(\frac{x_1-b_1}{a}, \frac{x_2-b_2}{a}, \frac{x_3-b_3}{a})$$

The *ridgelet* function is defined by:

$$\psi_{a,b,\theta_1,\theta_2} : \mathbf{R}^3 \rightarrow \mathbf{R}$$
$$\psi_{a,b,\theta_1,\theta_2}(x_1,x_2,x_3) = a^{-1/2} \cdot \psi((x_1 \cos\theta_1 \cos\theta_2 + x_2 \sin\theta_1 \cos\theta_2 + x_3 \sin\theta_2 - b)/a)$$

and the *beamlet* function is defined by:

$$\psi_{a,b_1,b_2,\theta_1,\theta_2} : \mathbf{R}^3 \rightarrow \mathbf{R}$$
$$\psi_{a,b_1,b_2,\theta_1,\theta_2}(x_1,x_2,x_3) = a^{-1/2} \cdot \psi((-x_1 \sin\theta_1 + x_2 \cos\theta_1 + b_1)/a, (x_1 \cos\theta_1 \cos\theta_2 + x_2 \sin\theta_1 \cos\theta_2 - x_3 \sin\theta_2 + b_2)/a)$$

Figure 2 shows an example of a 3D wavelet function and Figure 3 shows respectively examples of a ridgelet function (left) and a beamlet function (right). The ridgelet function is a wavelet function in the direction defined by the line (θ_1, θ_2), and it is constant along the orthogonal plane to this line. The beamlet function is constant along lines of direction (θ_1, θ_2), and a 2D wavelet function along the plane orthogonal to this direction. More details about the implementation of these 3D multiscale transforms can be found in ref [27].

Local 3D Ridgelet and Beamlet Transform

The ridgelet (resp. beamlet) transform is optimal to find sheets (resp. filaments) of the size of the cube. To detect smaller sheets (resp. filaments), a partitioning must be introduced [28]. The cube c is decomposed into blocks of lower side-length b so that for a $N \times N \times N$ cube, we count N/b blocks in each direction. After the block partitioning, the transform is

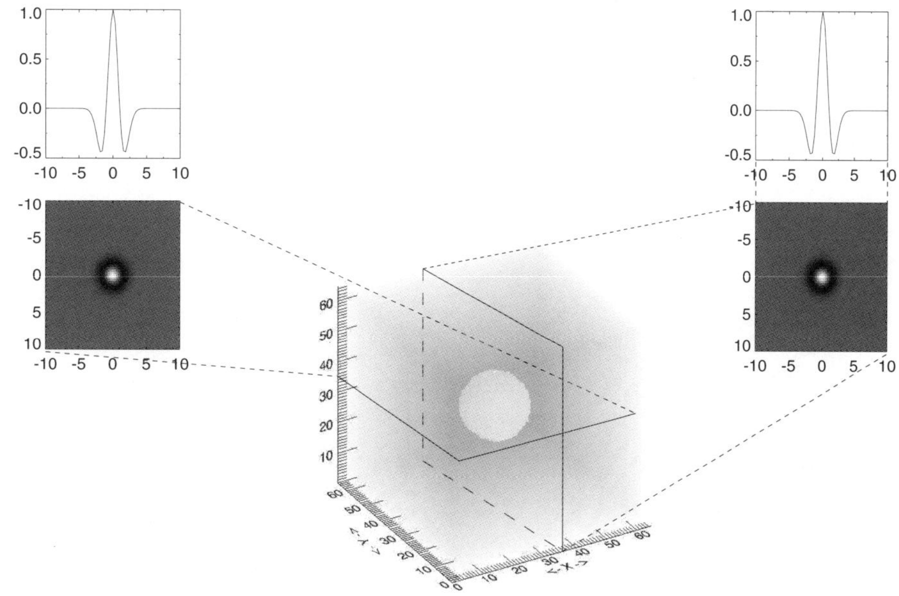

Figure 2. Example of wavelet function.

tuned for sheets (resp. filaments) of size $b \times b$ (resp. b) and of thickness a_j, a_j corresponding to the different dyadic scales used in the transformation.

MGA statistic

Hence, we have three different multiscale transforms, and for the beamlet and the ridgelet transform, we can also use several block sizes when analyzing the data. In the following, we will use the following seven decompositions:

1. 3D Isotropic Wavelet Transform with 4 dyadic scales.
2. 3D Ridgelet Transform using a block size of 8 Mpc and two scales. Here the scale is related to the width of the ridgelet function, its length being fixed by the block size.
3. 3D Ridgelet Transform using a block size of 16 Mpc and three scales.
4. 3D Ridgelet Transform using a block size of 32 Mpc and three scales.
5. 3D Beamlet Transform using a block size of 8 Mpc and two scales. Here the scale is related to the width of the beamlet function, its length being fixed by the block size.
6. 3D Beamlet Transform using a block size of 16 Mpc and three scales.
7. 3D Beamlet Transform using a block size of 32 Mpc and three scales.

For each scale of each transform, we calculate the Kurtosis value (i.e. $K = \frac{1}{N\sigma^4}\sum_k (x_k - \bar{x})^4 - 3 = \frac{1}{\sigma^4}(\bar{x^4} - 4\bar{x}\bar{x^3} + 6\bar{x^2}\bar{x}^2 - 3\bar{x}^4) - 3$). Positive K implies a higher peak and larger wings than the Gaussian distribution with the same mean and variance. Negative K means a wider peak and shorter wings.

3 2DF Data

The best available redshift catalog to study morphology of the galaxy distribution at present is the 2dF Galaxy Redshift Survey (2dFGRS) [29]. It fills large compact volume(s) in space and includes more than a quarter of million of galaxies. This is a flux-limited catalog and therefore the density of galaxies decreases with distance. For statistical analysis of such of surveys, a weighting scheme that compensates for the missing galaxies at large distances has

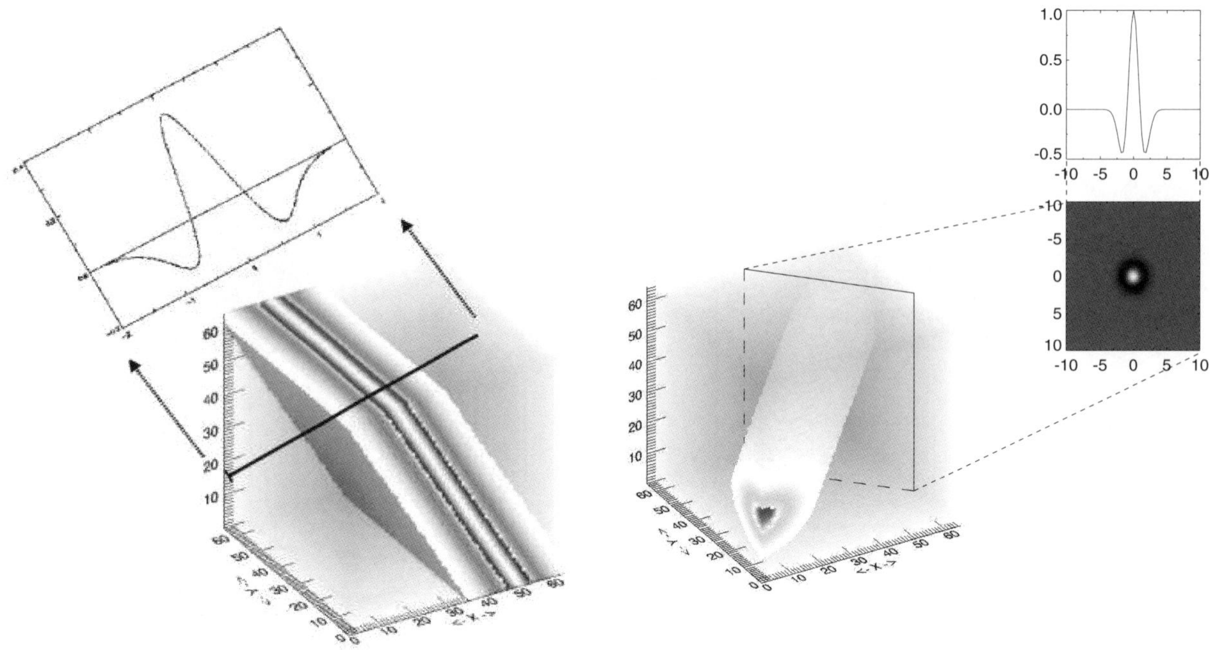

Figure 3. Examples of ridgelet function (left) and beamlet function (right).

to be used. Usually, each galaxy is weighted by the inverse of the selection function [6]. However, the resulting densities will have different resolution at different locations, and will not be suitable for morphological studies.

At the cost of discarding many surveyed galaxies, one can alternatively use volume-limited samples. In this case, the variation in density at different locations depends only on the fluctuations of the galaxy distribution itself. We have used the volume-limited samples prepared by the 2dF team for scaling studies [30,31], and kindly sent to us by Darren Croton. As our basic sample, we chose the catalog with absolute luminosities in the range $-19 > M_{B_J} - 5\log_{10} h > -20$ (the type dependent $k + e$ correction [32] has been applied to the magnitudes). This sample contains galaxies with luminosity around L_*. This catalog is the largest of the 2dF volume-limited catalogs, and as [33] point out, it provides optimal balance between the surveyed volume and the number density of galaxies. Although the catalog does not suffer from luminosity incompleteness, it is slightly spectroscopically incomplete, mainly due to missing galaxies because of fiber collisions. The incompleteness parameter has been determined by every galaxy by the 2dF team; when calculating densities, each galaxy can be weighted by the inverse of this parameter.

We split the volume-limited sample into the Northern and Southern subsamples, and cut off the numerous whiskers in the plane of the sky to obtain compact volumes. The geometry of the Northern sample is similar to a flat slice, while the Southern sample is enclosed between two cones of opening angles of 64.5° and 55.5°. When we tried to cut cuboidal volumes (bricks) from the Southern sample cone, we ended up with small brick volumes. We report in this paper only the analysis of the Northern subsamples.

In order to obtain a compact volume, we choose the angular limits for the Northern sample as $-4.5° \leq \delta \leq 2.5°$ and $149.0° \leq \alpha \leq 209.0°$. The slice lies between two cones defined by the δ limits. The right ascension limits cut the cones by planes from both sides, and there are two additional cuts by two spheres. The radii of the spheres are fixed by the original data, and depend only on the chosen absolute magnitude limits (and on the cosmological model). For our sample they are: $R_1 = 61.1\ h^{-1}$ Mpc, $R_2 = 375.6\ h^{-1}$ Mpc.

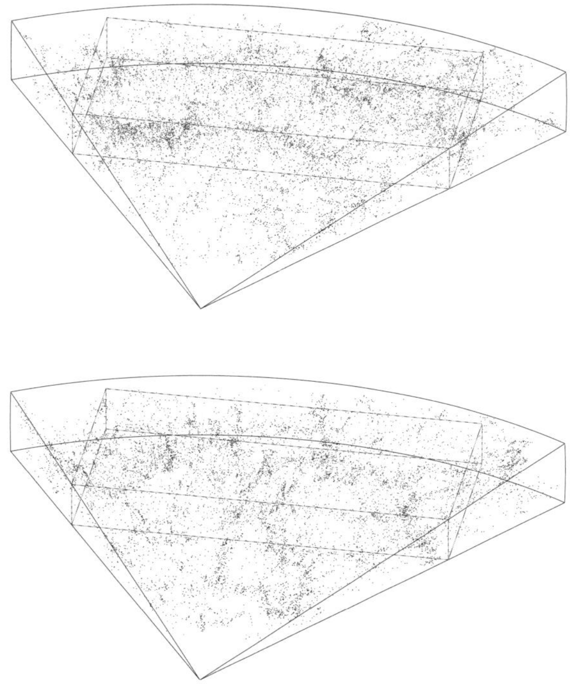

Figure 4. The volume-limited cuboidal sample analyzed in this paper drawn from the Northern slice of the 2dFGRS (top) and from a mock realization.

As this sample is pretty flat, we cut from it a maximal volume cuboidal window, a "brick" with dimensions of $254.0 \times 133.1 \times 31.1\ h^{-1}$ Mpc, with 8487 galaxies (see Fig. 4). This gives for the per-particle-volume size $d = 5.0\, h^{-1}$ Mpc.

3.1 Mock catalogs

In order to estimate sample errors of the Minkowski functionals, we use mock catalogs, provided by the 2dF team. [32] created 22 mock catalogs for the 2dF-GRS that have been used by the 2dfGRS team to measure the influence of cosmic variance of different statistics, such as correlation functions, counts-in-cells, the void probability function, clustering of groups, etc. [30,31,33,34]. The mock catalogs were extracted from the Virgo Consortium ΛCDM Hubble volume simulation, and a biasing scheme described in [35] was used to populate the dark matter distribution with galaxies. The catalogs were created by placing observers in the Hubble volume, applying the radial and angular selection functions of the 2dFGRS, and translating the positions and velocities of galaxies into redshift space. No luminosity clustering dependence is present in the mock catalogs.

The mock catalogs represent typical volumes of space. The real 2dF catalog, however, includes two superclusters, one in the Northern, another in the Southern subsample (see a thorough discussion in [31]). The Northern supercluster is especially prominent in our $M \in [-19, -20]$ catalog; all mock samples for this catalog have less galaxies than the 2dF sample, as the mocks were normalized by the total number of galaxies in both subsamples. We cut mock bricks from the mock samples, too, as we did for the real 2dF data; the mean number of galaxies in the mock bricks is 1.36 times smaller than in the 2dF brick.

4 MGA and the 2DF

We have applied the seven decompositions to the 2DF data and the 22 mock catalogs.

Figure 5 shows the Kurtosis for each scale of the wavelet transform. Crosses with the 2σ (continuous line) and 3σ (dotted line) error bars represent

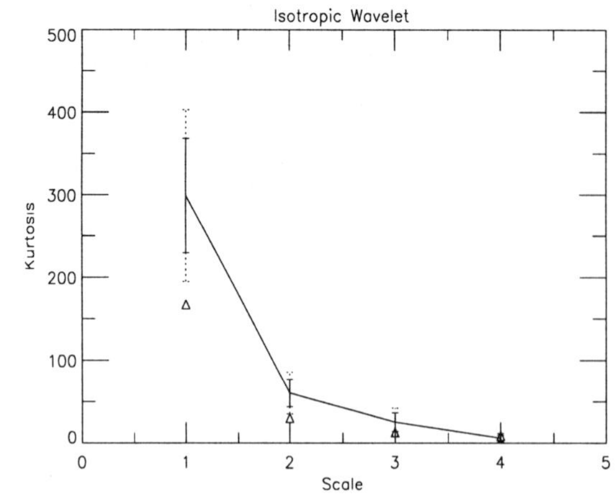

Figure 5. Isotropic WT Kurtosis versus scale. Crosses with the 2σ (continuous line) and 3σ (dotted line) error bars represent the mean kurtosis for the 22 simulation and the triangle represents the kurtosis relative to the wavelet scale of the 2DF data. The x-axis is the wavelet scale, corresponds to the analysis of isotropic structures of size 2^x Mpc.

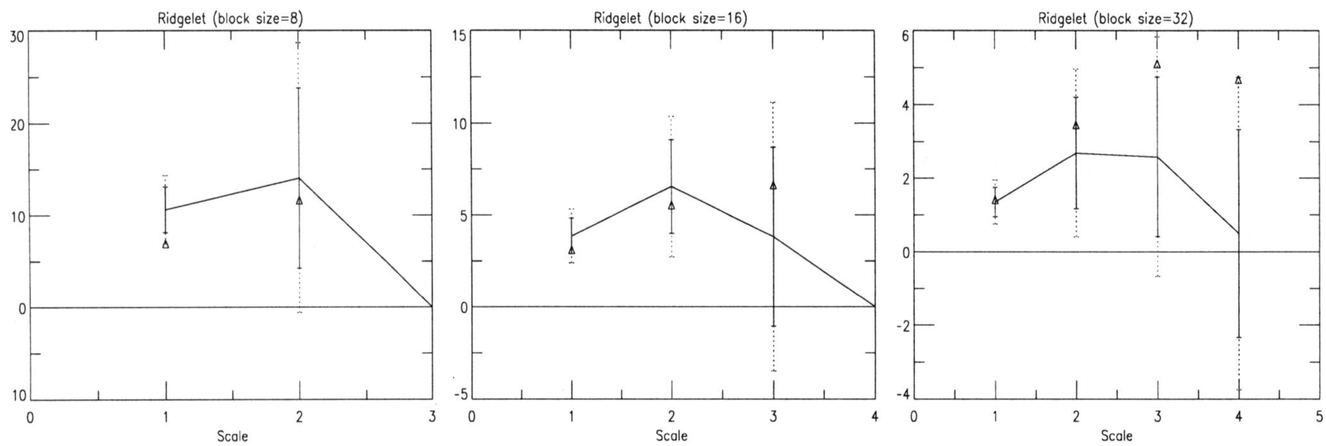

Figure 6. Ridgelet Kurtosis. The x-axis is the ridgelet scale, and corresponds to the analysis of structures of size $8 \times 8 \times 2^x$ Mpc (left), $16 \times 16 \times 2^x$ Mpc (middle) and $32 \times 32 \times 2^x$ Mpc (right).

the mean kurtosis for the 22 simulation and the triangle represents the kurtosis relative to the wavelet scale of the 2DF data. Figure 7 shows the kurtosis for the three beamlet transforms and Figure 6 shows the kurtosis for the three ridgelet transforms.

We can see that a strong discrepancy between the data and the simulations appears at the finest scale of the wavelet transform. It is also visible at the finest scale of the beamlet transform (block size 8). The last scale of the ridgelet transform (block size =32) shows also a difference, however not at a 3σ level. The main difference between the data and the simulations is clearly related to the smallest scales. As it is the wavelets which detect this difference, it is certainly the distribution of clusters (and not the distribution of filaments and walls) which is different.

A kurtosis exess in the simulated data can be due to a larger number of clusters and/or a few clusters with a larger amplitude. The first possibility would be in contradiction with the genus curve (see Figure 1) which indicates that the real high-density haloes (galaxy groups) are more concentrated than the mock galaxy groups. The second hypothesis, also supported by a visual inspection of the first and second wavelet scales, seems more adequate. If the real data contains more faint clusters, the simulations contain a few more prominent clusters which create a kurtosis exess.

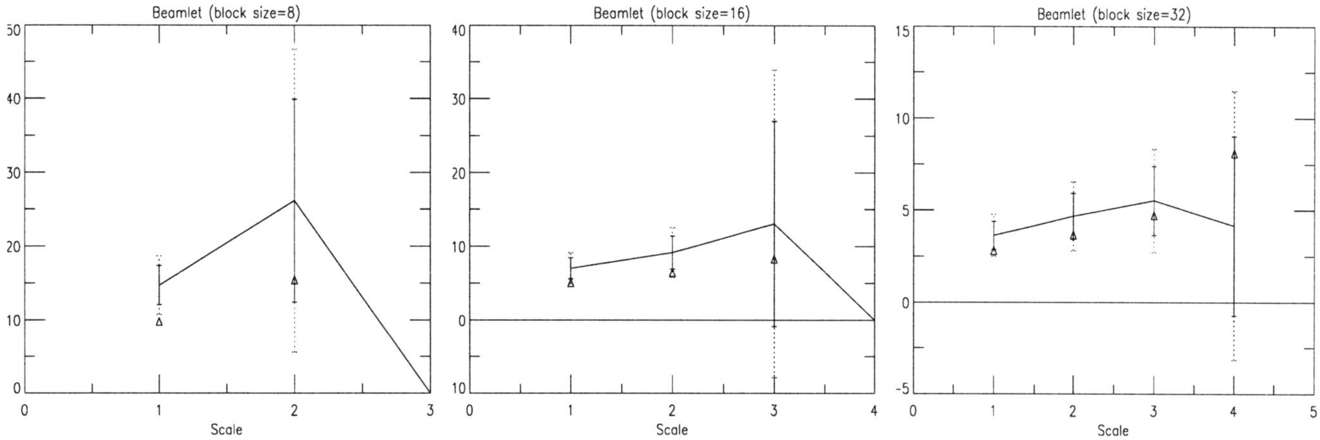

Figure 7. Beamlet Kurtosis. The x-axis is the beamlet scale, and corresponds to the analysis of structures of size $8 \times 2^x \times 2^x$ Mpc (left), $16 \times 2^x \times 2^x$ Mpc (middle) and $32 \times 2^x \times 2^x$ Mpc (right).

The supercluster has been suspected to be responsible of the genus curve difference between the simulation and the data in [26], but the MGA analysis leads to the conclusion that the problem is more likely related to the non-linear regime in the simulation which does not reflect the observations. It is probably due to individual simulated dark-matter haloes badly populated with galaxies, but it could also be due to some departure from the simplest models of primordial fluctuations of dark energy.

4.1 MGA:EARLY-LATE Type Galaxies

In this section, we have separated the 2DF data set into two parts, the early type galaxies (ETG) (3826 galaxies) and the late type galaxies (3913 galaxies) (LTG). It is well known that ETG are more clustered than LTG and this has also already be seen in the 2DF data [31]. However, we can wonder if the ELT-LTG clustering property is dependent on the type of structure. For instance, is this ELT-LTG property more important in cluster than in filaments or walls ?

In order to answer this question, we have performed a separate MGA analysis of the two catalogs, i.e. we have applied the seven transformations to both catalogs and calculated the kurtosis in the different scales. In order to normalize all values, we have also computed the kurtosis for 100 simulated cubes which contains each around 3870 galaxies randomly distributed (Poisson noise). A mean kurtosis and a standard deviation has been calculated for each transform and for each scales. The kurtosis derived from the two catalogs have been normalized using the mean values and the standard deviation values.

Figures 8, 9 and 10 shows the kurtosis for the two catalogs and the different transforms. The solid lines correspond to the ETG kurtosis and the dashed line to the LTG kurtosis. It is clear that for all transforms and all scales, the ETG presents a higher kurtosis than the LTG. This suggests that in all types of structures (filaments, clusters, walls), the ETG are more concentrated than the LTG. Both the ridgelet kurtosis and the beamlet kurtosis increase when the block size increases. This suggest that filaments and walls of the size of at least 32 Mpc exists in the data. Filaments with a width between 2 and 4 Mpc dominate (second scale of the beamlet transform) while walls seem to be thinner (1 or 2 Mpc) as the first scale of the ridgelet transform always presents a higher kurtosis.

In order to evaluate if a difference in the respective concentration exits between the three kind of features, we keep for each of the seven transformations only the higher normalized kurtosis along the scales. Hence, we built the variable $K^{(ETG)}_{max}(t)$, t being the transform number (i.e. $t = 1,..7$ for respectively the wavelet transform, the ridgelet transform for a block size equals to 8,16,32, and the beamlet transform for a block size equals to 8,16,32). Figure 11 shows the ratio $R_K(t) = \frac{K^{(ETG)}_{max}(t)}{K^{(LTG)}_{max}(t)}$. For all the transforms, the ratio $R_K(t)$ is between two and

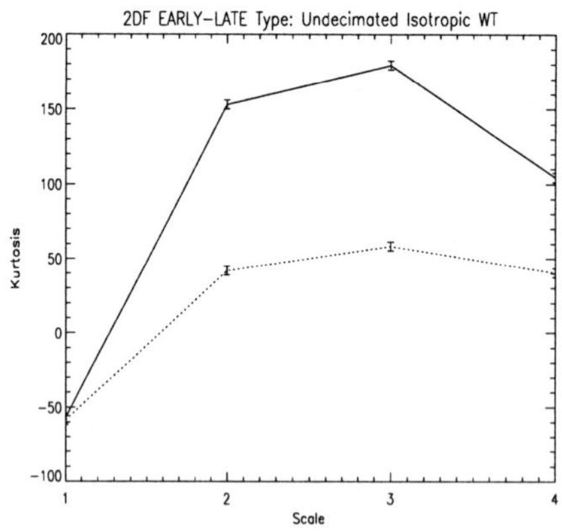

Figure 8. Isotropic WT Kurtosis of both the ETG (solid line) and the LTG (dotted line).

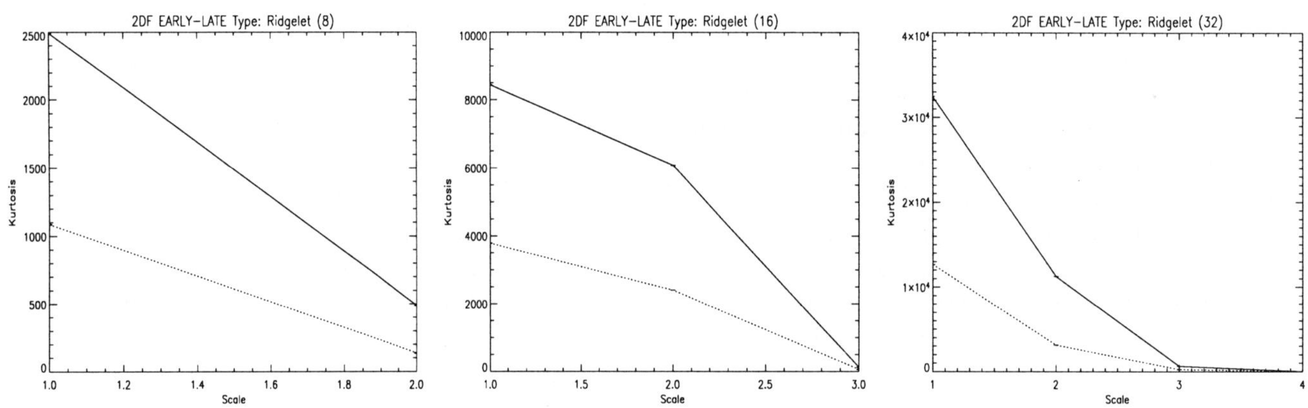

Figure 9. Ridgelet Kurtosis of both the ETG (solid line) and the LTG (dotted line).

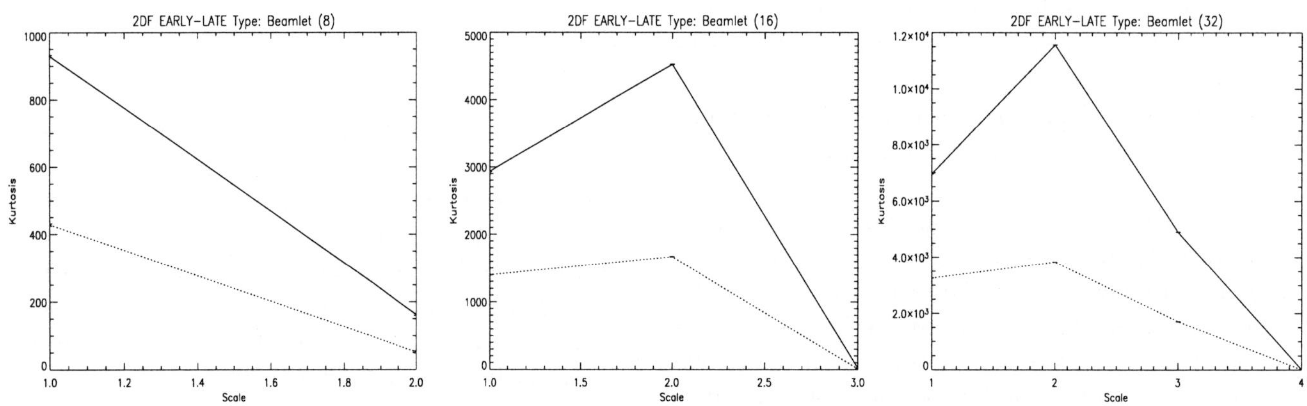

Figure 10. Beamlet Kurtosis of both the ETG (solid line) and the LTG (dotted line).

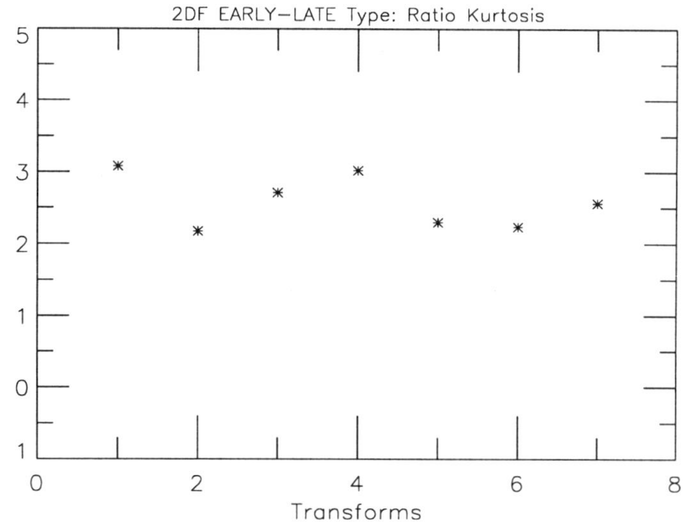

Figure 11. Kurtosis ratio $R_K = K_{max}^{(ETG)}/K_{max}^{(LTG)}$ between ETG and LTG versus the transformation. x-axis corresponds respectively for $x = 1$ to 7 to the wavelet transform, the ridgelet transform with block size equals 8, 16, 32 and the beamlet transform with block size equals 8, 16, 32.

three, which shows a remarkable stability in the clustering properties. It seems that the rates of ETG and LTG are relatively the same in all kinds of structures (i.e. filaments, walls and clusters).

Acknowledgments

We thank Darren Croton for providing us with the 2dF volume-limited data and explanations, David Donoho and Joe Silk for discussions. This work has been supported by the University of Valencia through a visiting professorship for Enn Saar, by the Spanish MCyT project AYA2003-08739-C02-01 (including FEDER), by the Generalitat Valenciana project GRUPOS03/170, and by the National Science Foundation grant DMS-01-40587 (FRG), and by the Estonian Science Foundation grant 4695.

References

1. P. Peebles, *The Large-Scale Structure of the Universe* (Princeton University Press, 1980).
2. M. Davis and P. Peebles, *Astrophysical Journal* **267**, 465(April 1983).
3. A. Hamilton, *Astrophysical Journal* **417**, p. 19(November 1993).
4. S. Landy and A. Szalay, *Astrophysical Journal* **412**, 64(July 1993).
5. M. Kerscher, *Astronomy and Astrophysics* **343**, 333(March 1999).
6. V. J. Martínez and E. Saar, *Statistics of the Galaxy Distribution* (Chapman and Hall/CRC press, Boca Raton, 2002).
7. M. Pons-Bordería, V. Martínez, D. Stoyan, H. Stoyan and E. Saar, *Astrophysical Journal* **523**, 480 (1999).
8. M. Kerscher, I. Szapudi and A. S. Szalay, *Astrophysical Journal Letters* **535**, L13(May 2000).
9. S. Szapudi and A. S. Szalay, *Astrophysical Journal Letters* **494**, p. L41(February 1998).
10. P. Peebles, The galaxy and mass N-point correlation functions: a blast from the past, in *Historical Development of Modern Cosmology*, eds. V. Martínez, V. Trimble and M. Pons-Bordería (ASP Conference Series, Astronomical Society of the Pacific, 2001).
11. S. Maurogordato and M. Lachieze-Rey, *Astrophysical Journal* **320**, 13 (1987).
12. V. J. Martínez, B. J. T. Jones, R. Domínguez-Tenreiro and R. van de Weygaert, *Astrophysical Journal* **357**, 50 (1990).
13. K. R. Mecke, T. Buchert and H. Wagner, *Astronomy and Astrophysics* **288**, 697(August 1994).
14. M. Kerscher, Statistical analysis of large-scale structure in the Universe, in *Statistical Physics and Spatial Statistics: The Art of Analyzing and*

Modeling Spatial Structures and Pattern Formation, eds. K. Mecke and D. Stoyan (Lecture Notes in Physics 554, 2000).
15. M. V. Lieshout and A. Baddeley, *Statistica Neerlandica* **50**, 344 (1996).
16. M. Kerscher, M. J. Pons-Bordería, J. Schmalzing, R. Trasarti-Battistoni, T. Buchert, V. J. Martínez and R. Valdarnini, *Astrophysical Journal* **513**, 543 (1999).
17. S. P. Bhavsar and R. J. Splinter, *Monthly Notices of the Royal Astronomical Society* **282**, 1461(October 1996).
18. L. G. Krzewina and W. C. Saslaw, *Monthly Notices of the Royal Astronomical Society* **278**, 869(February 1996).
19. A. G. Doroshkevich, D. L. Tucker, R. Fong, V. Turchaninov and H. Lin, *Monthly Notices of the Royal Astronomical Society* **322**, 369(April 2001).
20. E. Escalera, E. Slezak and A. Mazure, *Astronomy and Astrophysics* **264**, 379(October 1992).
21. E. Slezak, V. de Lapparent and A. Bijaoui, *Astrophysical Journal* **409**, 517 (1993).
22. V. J. Martínez, S. Paredes and E. Saar, *Monthly Notices of the Royal Astronomical Society* **260**, 365 (1993).
23. A. Pagliaro, V. Antonuccio-Delogu, U. Becciani and M. Gambera, *Monthly Notices of the Royal Astronomical Society* **310**, 835(December 1999).
24. T. Kurokawa, M. Morikawa and H. Mouri, *Astronomy and Astrophysics* **370**, 358(May 2001).
25. C. Hikage, Y. Suto, I. Kayo, A. Taruya, T. Matsubara, M. S. Vogeley, F. Hoyle, J. R. I. Gott and J. Brinkmann, *Publications of the Astronomical Society of the Japan* **54**, 707(October 2002).
26. V. J. Martínez, J.-L. Starck, E. Saar, D. L. Donoho, S. C. Reynolds, P. de la Cruz and S. Paredes, *Astrophysical Journal* **634**, 744(December 2005).
27. J.-L. Starck, V. Martinez, D. Donoho, O. Levi, P. Querre and E. Saar, *Eurasip Journal* **15**, 2455 (2005).
28. E. J. Candès, *Applied and Computational Harmonic Analysis* **6**, 197 (1999).
29. M. Colless *et al.*, *The 2dF Galaxy Redshift Survey: Final Data Release*, tech. rep., astro-ph/0306581 (2003).
30. D. J. Croton *et al.*, *Monthly Notices of the Royal Astronomical Society* **352**, 1232(August 2004).
31. D. J. Croton *et al.*, *Monthly Notices of the Royal Astronomical Society* **352**, 828(August 2004).
32. P. Norberg, S. Cole, C. M. Baugh, C. S. Frenk, I. Baldry, J. Bland-Hawthorn, T. Bridges, R. Cannon, M. Colless, C. Collins, W. Couch, N. J. G. Cross, G. Dalton, R. De Propris, S. P. Driver, G. Efstathiou, R. S. Ellis, K. Glazebrook, C. Jackson, O. Lahav, I. Lewis, S. Lumsden, S. Maddox, D. Madgwick, J. A. Peacock, B. A. Peterson, W. Sutherland and K. Taylor, *Monthly Notices of the Royal Astronomical Society* **336**, 907(November 2002).
33. C. M. Baugh, D. J. Croton, E. Gaztañaga, P. Norberg, M. Colless, I. K. Baldry, J. Bland-Hawthorn, T. Bridges, R. Cannon, S. Cole, C. Collins, W. Couch, G. Dalton, R. De Propris, S. P. Driver, G. Efstathiou, R. S. Ellis, C. S. Frenk, K. Glazebrook, C. Jackson, O. Lahav, I. Lewis, S. Lumsden, S. Maddox, D. Madgwick, J. A. Peacock, B. A. Peterson, W. Sutherland and K. Taylor, *Monthly Notices of the Royal Astronomical Society* **351**, L44(June 2004).
34. N. D. Padilla, C. M. Baugh, V. R. Eke, P. Norberg, S. Cole, C. S. Frenk, D. J. Croton, I. K. Baldry, J. Bland-Hawthorn, T. Bridges, R. Cannon, M. Colless, C. Collins, W. Couch, G. Dalton, R. De Propris, S. P. Driver, G. Efstathiou, R. S. Ellis, K. Glazebrook, C. Jackson, O. Lahav, I. Lewis, S. Lumsden, S. Maddox, D. Madgwick, J. A. Peacock, B. A. Peterson, W. Sutherland and K. Taylor, *Monthly Notices of the Royal Astronomical Society* **352**, 211(July 2004).
35. S. Cole, S. Hatton, D. H. Weinberg and C. S. Frenk, *Monthly Notices of the Royal Astronomical Society* **300**, 945(November 1998).

APPLICATION OF A MULTIDIMENSIONAL WAVELET DENOISING ALGORITHM FOR THE DETECTION AND CHARACTERIZATION OF ASTROPHYSICAL SOURCES OF GAMMA RAYS

S. W. DIGEL

Stanford Linear Accelerator Center, Stanford University, Stanford, CA 94309, USA
E-mail: digel@stanford.edu

B. ZHANG

Quantitative Image Analysis Unit URA CNRS 2582, Institut Pasteur, 25-28, Rue du Docteur Roux, 75724 Paris Cedex 15, France
E-mail: bzhang@pasteur.fr

J. CHIANG

Joint Center for Astrophysics/Physics Department, University of Maryland, Baltimore County, 1000 Hilltop Circle, Baltimore, MD21250, USA & SLAC
E-mail: jchiang@slac.stanford.edu

J. M. FADILI

GREYC CNRS UMR 6072, Image Processing Group, 6, Bd Marechal Juin,14050 Caen Cedex, France
E-mail: Jalal.Fadili@greyc.ensicaen.fr

J.-L. STARCK

Service dÁstrophysique CEA/Saclay, Orme des Merisiers, Bat 709, 91191 Gif-sur-Yvette Cedex, France & Department of Statistics, Stanford University
E-mail: jstarck@cea.fr

Zhang, Fadili, & Starck have recently developed a denoising procedure for Poisson data that offers advantages over other methods of intensity estimation in multiple dimensions. Their procedure, which is nonparametric, is based on thresholding wavelet coefficients. The restoration algorithm applied after thresholding provides good conservation of source flux. We present an investigation of the procedure of Zhang et al. for the detection and characterization of astrophysical sources of high-energy gamma rays, using realistic simulated observations with the Large Area Telescope (LAT). The LAT is to be launched in late 2007 on the Gamma-ray Large Area Space Telescope mission. Source detection in the LAT data is complicated by the low fluxes of point sources relative to the diffuse celestial background, the limited angular resolution, and the tremendous variation of that resolution with energy (from tens of degrees at $\sim$30 MeV to 0.1° at 10 GeV). The algorithm is very fast relative to traditional likelihood model fitting, and permits immediate estimation of spectral properties. Astrophysical sources of gamma rays, especially active galaxies, are typically quite variable, and our current work may lead to a reliable method to quickly characterize the flaring properties of newly-detected sources.

1. Introduction

The high-energy gamma-ray sky will be studied with unprecedented sensitivity by the Large Area Telescope (LAT) to be launched by NASA on the GLAST mission in late 2007. The catalog of gamma-ray sources from the previous mission in this energy range, EGRET on the Compton Gamma-Ray Observatory, has approximately 270 sources [1]. For the LAT, several thousand gamma-ray sources are expected to be detected, with much more accurately determined locations, spectra, and light curves.

We would like to reliably detect as many celestial sources of gamma rays as possible, and to recognize when a known source is varying or when a faint, previously unknown source flares up to a detectable level. The time scales of flares can be minutes to weeks. In general we will not know in advance where on the sky the sources are that we will detect; projections are that the LAT will detect several times more blazars (a class of active galaxy) than are currently known from observations at other wavelengths.

The fluxes of celestial gamma rays are low, requiring long exposure times for the $\sim$1 m^2 effective area of the LAT (by far the largest effective collecting area ever in the GeV range). An additional complicating factor is that diffuse, celestial background from the Milky Way itself (which originates in cosmic-ray interactions with interstellar gas and radiation) makes a relatively intense, structured background emission, on which the point sources of interest are superposed. Point sources are so called because they are spatially unresolved. The few very brightest gamma-ray sources will provide approximately 1 detected gamma ray per minute when they are in the field of view of the LAT. The celestial background of the Milky Way will provide about 2 gamma rays per second, distributed over the $\sim$2 sr field of view.

For previous high-energy gamma-ray missions, the standard method of source detection has been model fitting — maximizing the likelihood function while moving trial point sources around in the region of the sky being analyzed. This approach has been driven by the limited photon counts and the relatively limited resolution of gamma-ray telescopes. However, at the sensitivity of the LAT even a relatively 'quiet' part of the sky may have 10 or more point sources close enough together to need to be modeled simultaneously when maximizing the (computationally expensive) likelihood function. For this reason, and because we would like to be able to detect spatially resolved sources that do not necessarily have simple shapes (such as a supernova remnant interacting with a gas cloud), non-parametric algorithms for detecting sources are being investigated.

The new wavelet denoising procedure by Zhang, Fadili, & Starck[2] and its application to simulated LAT data are described in the sections that follow.

2. Characteristics of the Data

The LAT (Fig. 1) is a photon-counting detector, converting gamma rays into positron-electron pairs for detection. The trajectories of the pair are tracked and their energies measured in order to reconstruct the direction and energy of the gamma ray.

The energy range of the LAT is very broad, approximately 20 MeV – 300 GeV. At energies below a few hundred MeV, the reconstruction and tracking efficiencies are lower, and the angular resolution is poorer, than at higher energies. The PSF width varies from about 3.5° at 100 MeV to better than 0.1° (68% containment) at 10 GeV and above. Owing to large-angle multiple scattering in the tracker, the PSF has broad tails; the 95%/68% containment ratio may be as large as 3.

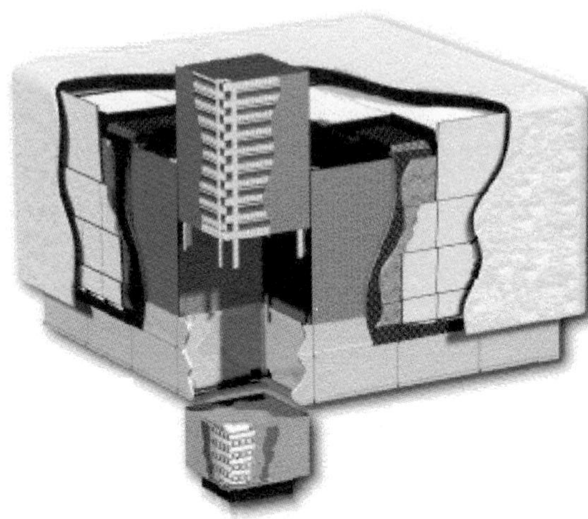

Fig. 1. Cutaway view of the LAT. The LAT is modular; one of the 16 towers is shown with its tracking planes revealed. High-energy gamma rays convert to electron-positron pairs on tungsten foils in the tracking layers. Their trajectories are measured very precisely using silicon strip tracking layers and the energies are determined with the CsI calorimeter at the bottom. The array of plastic scintillators that cover the towers provides an anticoincidence signal for cosmic rays. The outermost layers are a thermal blanket and micrometeoroid shield. The overall dimensions are 1.8 × 1.8 × 0.75 m.

3. Wavelet Poisson Intensity Estimation: Overview

Wavelet denoising of LAT data has application as part of an algorithm for quickly detecting celestial sources of gamma rays. The fundamental inputs to high-level analysis of LAT data will be energies, directions, and times of the detected gamma rays. (Pointing history and instrument live times are also inputs for exposure calculations.) For the analysis presented here, we consider the LAT data for some range of time to have been binned into 'cubes' $v(x, y, E)$ of spatial coordinates and energy, because, as we shall see, the wavelet denoising can be applied in multiple dimensions, and so permits estimation of counts spectra. The motivations for filtering data with Poisson noise in the wavelet domain are well known — sources of small angular size are localized in wavelet space.

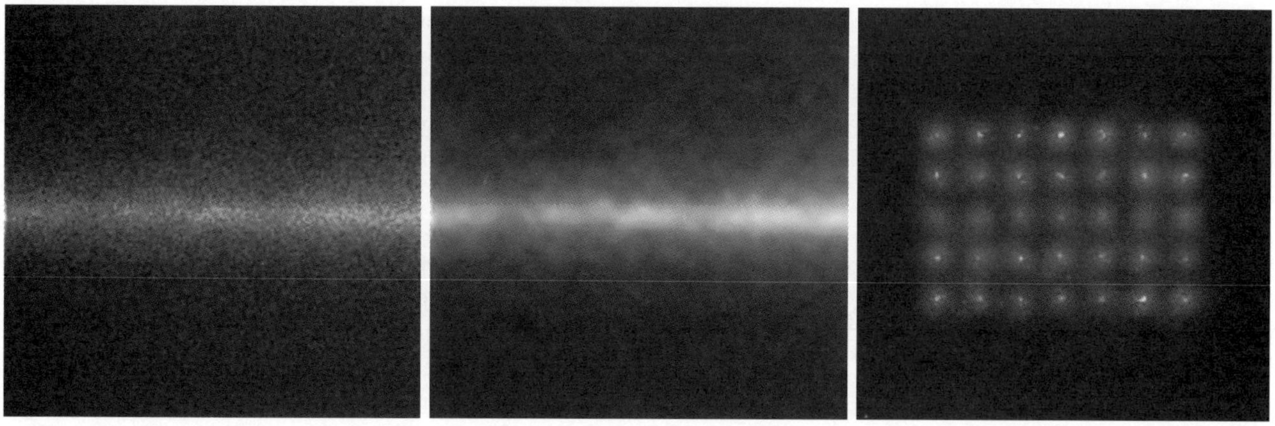

Fig. 2. Example of application of the wavelet denoising algorithm for detecting sources in simulated LAT data. An array of point sources, spaced by 4° was superimposed against the bright celestial background of the inner Milky Way and the exposure from a one-month sky was survey simulated. The point sources are identical, with flux 5×10^{-8} cm^{-2} s^{-1} (>100 MeV) and photon spectral index -2. *(left)* Before denoising; counts summed over energy >100 MeV. *(center)* After application of the denoising algorithm without providing a model for the celestial background. In this case, the algorithm estimates the background intensity from the data. *(right)* After application of the denoising algorithm, which was given a model of the background.

Many wavelet filtering methods have been developed based, e.g., on transformations (including variance stabilizing), direct filtering (like Wiener filtering in the wavelet domain), and Bayesian approaches; see Zhang et al.[2] for a review. The recent results of Zhang et al. on wavelet filtering of Poisson data extend the work of Kolaczyk[3] on filtering via hypothesis testing. The appeal of hypothesis testing methods is that they allow quantitative control of significance.

4. Denoising via Wavelet Domain Hypothesis Testing

4.1. *Formulation of the Method*

The background intensity, if not known *a priori*, is assumed to be constant over the region of the sky being analyzed. The background can be estimated from the approximation coefficients at a coarser scale.

For each wavelet coefficient w, the null hypothesis is H_0: w is consistent with the background; and the alternative hypothesis is H_1: w is inconsistent with the background. In the filtering, the H_0 coefficients, which correspond to consistency with the background within statistical fluctuations, are zeroed. The H_1 coefficients, representing regions of wavelet space with significant change from the background, are retained. This is a controlled (via the user-specified p-value) hard thresholding scheme. The coefficients are tested separately. The probability of false detection, i.e., false passing of H_1, is upper bounded by p

$$E(\frac{N_{H_0}^{reject}}{N_{H_0}}) \leq p, \tag{1}$$

where $N_{H_0}^{reject}$ is the number of coefficients satisfying H_0 but rejected by the corresponding hypothesis test and N_{H_0} is the total number of coefficients satisfying H_0.

For computational tractability, Zhang et al. use the Haar wavelet, because the pdf of the wavelet coefficient conditioning on H_0 of a Poisson process is known in closed form, a non-central Chi-Square distribution. Zhang et al. also show that the same distributions and thresholds apply to the coefficients of biorthogonal Haar (BH) wavelets, which relatively speaking provide good preservation of regularity.

Using the Fisher normal approximation (see Zhang et al.), a threshold can be derived for every Gaussian significance level. This thresholding method has a greater detection power than Kolaczyk's method[3] and Zhang et al.[2] have extended it to non-constant, or unknown *a priori*, backgrounds.

4.2. *Extension to the Energy Dimension*

The hypothesis testing approach can be extended straightforwardly to a third dimension. A 2-

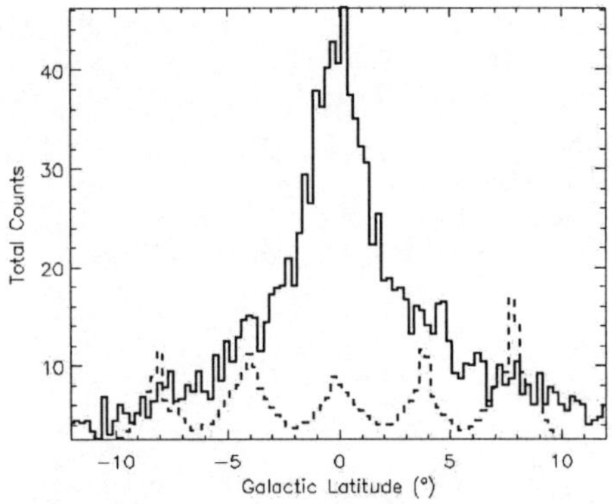

Fig. 3. Solid curve: Latitude profile, averaged over the central 1° of longitude, for the left-hand panel of Fig. 2, i.e., the simulated point sources and celestial background before denoising. Dashed curve: The same profile showing the point-like sources remaining after denoising with a model for the background, corresponding to the right-hand panel of Fig. 2. *This profile has been scaled up by a factor of 10 for clarity.*

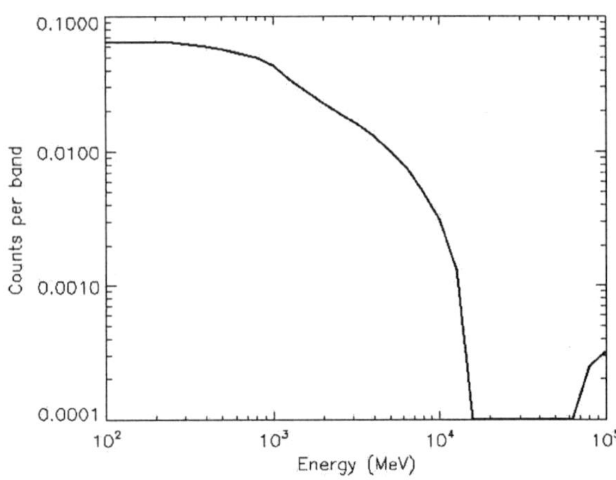

Fig. 4. Example of denoised spectrum from a source in Fig. 2. The distribution of photon counts has approximately the E^{-1} slope expected in counts per band for a differential E^{-2} spectrum.

dimensional BH transform is applied to each image (one per energy band). To each BH spatial coefficient, a 1-dimensional BH transform is applied along the energy axis, and the hypothesis testing estimator is applied. Then the inverse transformations are applied along the energy axis and spatially to obtain the denoised multi-spectral data.

5. Application to Simulated LAT Data

Figure 2 illustrates the application of the wavelet denoising algorithm for source detection against the bright, structured celestial emission of the Milky Way. This is literally just an illustration with an artificial arrangement of point sources, but the intensity of the celestial background and the flux of the sources are realistic. The region used is in fact more or less a worst case, having the brightest and most strongly varying background of any place on the sky. The center panel of Fig. 2 is the result of requiring the algorithm to estimate the background itself and the right-hand panel shows the result when a background model is provided. The profiles in Fig. 3 show that the bright, structured emission of the Milky Way can totally mask the faint sources; this is why the denoising without a prior background model failed to find the sources. However, the result is quite promising when the denoising is made with a good model for the celestial background gamma-ray emission.

Figure 4 shows a denoised counts spectrum for one of these sources. The spectral characteristics of the denoising have not been studied in detail, but the figure illustrates that the spectral slope is approximately what would be expected for the power-law spectra (photon index -2) of the input sources. Some roll-over is expected owing to the decline of the effective area of the LAT below ~300 MeV.

6. Summary

We have described the motivations for identifying a reliable nonparametric source detection algorithm to apply to GLAST LAT data. For the relatively short time ranges over which we will want to study sources, the data will be squarely in the low counts regime with widely varying response functions and significant celestial backgrounds. We are exploring the hypothesis-testing wavelet denoising algorithm of Zhang et al. It can use a model for the background for increased sensitivity. The algorithm also can be applied in the energy dimension and so allows spectra to be recovered.

Acknowledgments

This work was supported in part by U. S. Department of Energy contract DE-AC02-76SF00515

References

1. R. C. Hartman et al., *Astrophys. J. Supp.* **123**, 79 (1999).
2. B. Zhang, B., J. Fadili, & J-L. Starck, Tech. Rep. CEA-Saclay (2005).
3. E. D. Kolaczyk, *J. Amer. Stat. Assoc.* **94**, 920 (1999).

HIGHER CRITICISM STATISTIC: THEORY AND APPLICATIONS IN NON-GAUSSIAN DETECTION

J. JIN

Statistics Department, Purdue University, 150 N. University Street, West Lafayette, IN 47907, USA
E-mail: jinj@stat.purdue.edu

Higher Criticism is a statistic recently proposed by Donoho and Jin[5]. It has been shown to be effective in resolving a very subtle testing problem: whether n normal means are all zero versus a small fraction is nonzero. Higher Criticism is also useful for non-Gaussian detection in Cosmic Microwave Background (CMB) data. In this report, we review the theory developed in Donoho and Jin[5] and discuss the use of Higher Criticism for two settings: detecting the non-Gaussian component in a superposed image of CMB and cosmic strings (CS), and detecting non-Gaussianity in the WMAP first year data.

1. Introduction

The Cosmic Microwave Background (CMB) is the relic radiation emitted when the universe was about 380,000 years old. It is an almost perfect black body at a temperature of ≈ 2.726 Kelvin. The Standard Inflation model predicts that temperature anisotropies of the CMB (i.e. small angular fluctuations of the temperature) are the imprint of the initial density perturbations which gave rise to the large scale galaxies we see today. The study of the CMB is expected to improve our understanding of the very early universe, and it is of great interest to cosmologists.

The standard Inflationary model predicts that temperature anisotropies in the CMB have a Gaussian distribution. However, many other models (e.g. multi-field inflation[2], super string and topological defects[6, 7, 10]) as well as secondary effects (inverse Compton scattering etc.) predict deviations from a Gaussian distribution. The goal of non-Gaussian detection is to disentangle different non-Gaussian sources from one another.

The wavelet transformation is a powerful approach for non-Gaussian detection, and many wavelet-based methods have been investigated (see page 3 in Jin[8] for references to these works). Particularly, it was shown in Aghanim *et al.*[1] and P. Vielva *et al.*[13] that the excess kurtosis of the wavelet coefficients outperformed all other methods.

However, the effectiveness of a detection tool depends highly on the underlying non-Gaussianities: a detection tool can be sensitive to some types of non-Gaussianities, but totally immune to other types. It is thus of interest to introduce more statistical tools to this field, and to compare their strengths as well as weaknesses. Higher Criticism is one of these new tools.

2. Higher Criticism

Higher Criticism (HC) was first proposed in Donoho and Jin[5] for a multiple comparison setting, where it was shown to be effective in resolving a very subtle testing problem: whether n normal means are all zero versus a small fraction of them being nonzero. Higher Criticism can also be viewed as a goodness-of-fit measure, and a tool for non-Gaussian detection.

Consider a setting in which we have n observations $\{X_i\}_{i=1}^n$. The problem of non-Gaussian detection is to test the following hypothesis: H_0 : $X_i \overset{iid}{\sim} N(0,1)$, where for simplicity we assumed that the data are standardized. To implement Higher Criticism[5, 8, 3], we first obtain individual p-values: $p_i = P\{N(0,1) \geq X_i\}$, we then sort them in ascending order $p_{(1)} < p_{(2)} < \ldots < p_{(n)}$, and calculate the normalized z-scores:

$$HC_{n,i} = \sqrt{n} \cdot [|i/n - p_{(i)}|]/[\sqrt{p_{(i)}(1-p_{(i)})}].$$

The Higher Criticism statistic is then defined as $HC_n^* = \max_{\{1 \leq i \leq n\}} HC_{n,i}$.

The rationale behind the normalization is that, when the hypothesis H_0 is indeed true, then for almost all i (except when i is close to 1 or n), $HC_{n,i} \approx N(0,1)$, and moreover $HC_n^* \approx \sqrt{2 \log \log n}$. Thus a large HC_n^* value implies non-Gaussianity.

2.1. *Sensitive to Unusually Large Amount of Moderate Significances*

In a data set, the extreme value refers to the data point which is largest in absolute value. It is a well-known result in statistics that out of n samples from the standard Gaussian the extreme value $\approx \sqrt{2\log n}$. In contrast, moderate significances refer to the tiny portion of the data points that are slightly smaller (in absolute values) than the extreme value, e.g. data points $\approx \sqrt{\log n}$. The proportion of moderate significances is very small, e.g. $P\{N(0,1) \geq \sqrt{\log n}\}$, the proportion of samples $\geq \sqrt{\log n}$, approximately equals to $n^{-1/2}$.

In Donoho and Jin[5], the authors have considered a sparse normal mean problem: we have n observations from $X_i \sim N(\mu_i, 1)$, with all $\mu_i = 0$ except a possible tiny fraction ϵ_n of them satisfying $\mu_i = \mu_n$, where ϵ_n and μ_n depend on n but not on i. The goal is to test whether the sparse mean effect is present or not, or equivalently to test whether $\epsilon_n = 0$ or $\epsilon_n > 0$. They considered a range of (ϵ_n, μ_n) which concerns the situation of "very sparse signal with moderate significant amplitude": on one hand, ϵ_n is too small so that the sparse mean effect can't be detected by statistics based on moments (cumulants, kurtosis, etc.); on the other hand, as the signals are only moderately significant, the sparse mean effect can't be detected by merely looking at the extreme values.

It was proved in Donoho and Jin[5] that the Higher Criticism statistic is optimally adaptive in detecting the sparse normal mean effect. Roughly put, for fixed ϵ_n, whenever μ_n is large enough so that it is possible to reliably tell that $\epsilon_n > 0$, the Higher Criticism statistic is able to do so.

We now take a heuristic approach for understanding the mechanism of Higher Criticism. The sparse mean effect can be thought of as the situation in which one has n samples from the standard Gaussian, and now you want to sneak in a bunch of μ_n by the following two steps: (a) randomly select a tiny portion of the samples, leave others untouched, and (b) add μ_n to each selected samples. The problem is then to tell whether such a process has occured or not. Higher Criticism works by picking a sequence of significance levels and asking whether there are too many samples found above each significance level. If the answers are all "no", the Higher Criticism claims Gaussian and nothing is found, but claims non-Gaussian otherwise. Higher Criticism uses the normalized z-score for deciding whether there are too many samples found above each significance level or not: $HC_{n,\alpha} = \sqrt{n}[\{\text{Fraction at Level } \alpha\} - \alpha]/\sqrt{\alpha(1-\alpha)}$; when all samples are truly from the standard Gaussian, $HC_{n,\alpha} \approx N(0,1)$ and should be relatively small, so a large $HC_{n,\alpha}$ implies non-Gaussianity. Thus Higher Criticism works across the full range of significance levels, looking for evidence against possible types of "sneak-in" we mentioned above.

We now come back to the sparse normal mean problem. The strongest evidence for the presence of the sparse mean effect is that when you look at the portion of data points of moderate significance, there are too many moderate significances than there would be if the null hypothesis is true (i.e. all samples are truly from the standard Gaussian). Higher Criticism immediately reports a very large normalized z-score and rejects the null. This property of Higher Criticism is sensitive to an unusually large amount of moderate significances.

2.2. *Useful for Locating NonGaussianity*

The previous section pointed out that Higher Criticism is useful for locating the non-Gaussianity. To illustrate this point, suppose Higher Criticism picks all levels from 0 to 1 with 1% increment. Suppose the answers at levels $\alpha = 10\%$ and $\alpha = 9\%$ are "yes", while those at other levels are "no". Then on the one hand we are told that too many samples are observed at Level 10%. On the other hand we are told that *not* too many samples are observed at Level 8%. We then conclude that there are too many samples that fall between Level 8% and Level 10%, and this *slice* of data is suspected of non-Gaussianity. Notice here that the extreme values don't have to be more "non-Gaussian".

3. Detecting Cosmic Strings

We have considered using the Higher Criticism statistic for detecting cosmic strings (CS). Refer to Jin[8] for detailed discussion of the following results.

We consider a setting in which we have a superposed image of a simulated map of CMB and CS: $Y = \sqrt{1-\lambda}CMB + \sqrt{\lambda}CS$, and we are interested in testing whether $\lambda = 0$ or not. The simulated map of

CS was kindly provided by F. Bouchet. Though the real map of the CMB exists (from WMAP), we used simulated CMB maps instead in order to ensure that there is no non-Gaussianity in the maps.

Since the pixel values of the simulated CMB are correlated, working in the frequency domain is more convenient than in the space domain. Let $\{X_i\}_{i=1}^n$ be the wavelet coefficients of Y, then $X_i = \sqrt{1-\lambda}z_i + \sqrt{\lambda}w_i$, where $z_i \stackrel{iid}{\sim} N(0,1)$ are the transform coefficients of CMB and $w_i \stackrel{iid}{\sim} W$ are the coefficients of the CS map. The distribution of W is unknown, but is symmetrical and has a heavy-tail. Without loss of generality, both $\{z_i\}_{i=1}^n$ and $\{w_i\}_{i=1}^n$ have been standardized with standard deviations equal to 1. The testing problem is then equivalent to testing a null hypothesis H_0 under which $X_i \stackrel{iid}{\sim} N(0,1)$ versus an alternative hypothesis $H_1^{(n)}$ under which $X_i = \sqrt{1-\lambda}z_i + \sqrt{\lambda}w_i$. We are interested in which pair (λ, W) do the two hypotheses asymptotically merge together so that no test can separate them, versus which pair of (λ, W) the two hypotheses asymptotically separate from each other. By saying asymptotically, we mean n tends to ∞.

Clearly, if we fix $\lambda > 0$, then when n gets larger and larger, the difference between the two hypotheses becomes increasingly large, and eventually it is trivial to tell one from another. Thus the interesting range for λ is that it tends to 0 as n tends to ∞, so we set $\lambda = \lambda_n = n^{-r}$, $0 < r < 1$. At the same time, motivated by the heavy-tailed behavior of W, we assume that the tail probability of W decays algebraically:

$$\lim_{x\to\infty} x^\alpha P\{|W| \geq x\} = C_\alpha,$$

where C_α is a constant.

Intuitively, as λ_n is algebraically small, we expect that the majority of relatively smaller samples from W will not have much influence on testing. Instead, a tiny fraction of very large samples from W would play the decisive role. This turns out to be true, and there is a *threshold effect* for the testing problem. We call the curve $r = \rho^*(\alpha)$ in the α-r plane the *detection boundary*: if (r, α) falls below the detection boundary, then the null and the alternative hypothesis separate asymptotically; if (r, α) falls above the detection boundary, the null and the alternative merge asymptotically. It turns out that $\rho^*(\alpha) = 2/\alpha$ when $\alpha < 8$ and $1/4$ otherwise.

We now compare the asymptotical performance of the excess kurtosis and Higher Criticism. If $\alpha > 8$ or the 8-th moment of W exists, then the excess kurtosis is better than Higher Criticism. When (r, α) falls into the region that $\{(r,\alpha) : \alpha > 8, \frac{2}{\alpha} < r < 1/4\}$, then asymptotically the excess kurtosis has full power for detection, while the power of Higher Criticism tends to 0. If on the other hand $\alpha < 8$, then Higher Criticism is better than the excess kurtosis. When (r, α) falls into the region $\{(r,\alpha) : \alpha < 8, 1/4 < r < \frac{2}{\alpha}\}$, then asymptotically Higher Criticism will have full power, while the power of the excess kurtosis tends to 0.

The phenomenon can be explained as follows. Take $\alpha = 5$ for example. When you look at the data, before you notice any difference in the excess kurtosis, the largest sample from W is quite apparent, so detectors concentrated on the tail are more sensitive. However, when α ranges between 5 and ∞, the tail is gradually thinned out, and at some point, it will not tell you anything by merely looking at the data tail. You need to shift your attention to relatively smaller samples, or the bulk of the data, for which the excess kurtosis is more sensitive. It is interesting to study the α parameter corresponding to the tail behavior of W. Our study[8] supports the assumption that W has a power law tail: implementing the Hill estimator[9] gives $\alpha \approx 6.1$, where the standard error of this estimate approximately equals to 0.9.

Finally, the above result is highly asymptotical. It would be interesting to investigate the performances for moderately large n. Reports in this direction are included in Jin[8].

4. WMAP First Year Data

We have implemented Higher Criticism to analyze the WMAP first year data. The detailed study is in Cayon *et al.*[3]. We work with the WMAP data from the LAMBDA website (*lambda.gsfc.nasa.gov*). We construct a weighted combination of released foreground cleaned Intensity Maps at bands Q, V, and W (refer to Cayon *et al*[3] for details). We then generated 5,000 Gaussian simulations of CMB maps (including observational constraints imposed by noise and beam profiles), and take the wavelet transforms for each simulated map as well as the WMAP map. Finally, we carry out the statistical analysis on the wavelet coefficients.

The wavelet basis we used is the Spherical Mexican Hat. The wavelet coefficients we obtained do not fit very well with *iid* Gaussian samples. One reason is that the wavelet basis is not orthogonal. Despite this, Higher Criticism can still be used as a criterion for non-Gaussianity: it can be thought that the larger the Higher Criticism value, the larger the deviation from Gaussianity. We thus take the approach in which we compare the Higher Criticism values of the 5,000 simulated maps with that of the WMAP data, and claim non-Gaussianity if 99% of the simulated CMB maps have a smaller Higher Criticism value than that corresponding to the WMAP. It would be interesting to try the analysis with some orthogonal basis; we leave this for future study.

In addition to Higher Criticism, we have also implemented the excess kurtosis to the above setting. The Higher Criticism reports non-Gaussian detection at 99.46%. In comparison, the excess kurtosis is slightly better by reporting non-Gaussian detection at 99.7%.

However, Higher Criticism has more to offer. We pointed out earlier in the report that Higher Criticism can be used to automatically identify a tiny fraction of data as suspected of non-Gaussianity. We isolated 490 wavelet coefficients, at the scale of 5 degrees of the WMAP data. In detail, we set a threshold t_0 as the 1%-upper percentile of the 5,000 Higher Criticism values (HC_n^*) based on simulated CMB maps. Then out of all wavelet coefficients of the WMAP, we select those with an associated normalized z-score ($HC_{n,i}$) larger than t_0.

Last, we map these 490 wavelet coefficients back to pixels in the WMAP. There are two ways to do the mapping. In the first approach, we map each coefficient back to all pixels involving the coefficients, i.e. all pixels convoluted with the wavelet basis when calculating this coefficient. Notice that each of the coefficients naturally maps back to a cluster of pixels. It is interesting to note that the 490 pixels, and those correlated with them by the wavelet convolution, are at the cold spot found by Vielva et al.[13] and Cruz et al.[4]. In the second approach, we map each coefficient back to only one pixel: the one at the center of the pixel-cluster mentioned above. This way of mapping has the advantage of visualization. By the second approach, the selected 490 coefficients map back to a ring on the outer part of the cold spot. Interestingly, the "coldest" wavelet coefficient (i.e. largest in absolute value but is negative) maps back to a pixel in the center part of the cold spot, which is not in the ring. We clarify here that, both in this report and in Cayon *et al*[3], our result doesn't attempt to conclude that there is a ring structure in the WMAP map. Instead, the ring visualizes the position of pixels corresponding to the 490 moderately significant wavelet coefficients we extracted.

5. Conclusions

We introduced the Higher Criticism statistic for non-Gaussian detection. We have studied the application of Higher Criticism to the detection of cosmic strings and to the WMAP first year data. Higher Criticism is useful in applications by adding discussions to the field of non-Gaussian detection.

Acknowledgments

We would like to thank Louis Lyons, the anonymous referee, Laura Cayon, Jean-Luc Starck, and Anna Treaster for valuable discussions. We acknowledge the use of the Legacy Archive for Microwave Background Data Analysis (LAMBDA). Support for LAMBDA is provided by the NASA Office of Space Science. This work was partially supported by NSF Grant DMS-0505423.

References

1. N. Aghanim *et al.*, *Astronomy and Astrophysics* **406**, 797 (2003)
2. F. Benardeau and J. Uzan, *Phys. Rev. D* **66**, 103506 (2002).
3. L. Cayon *et al.*, *Mon. Not. Roy. Astron. Soc.* **362**, 826 (2005).
4. M. Cruz *et al.*, *Mon. Not. Roy. Astron. Soc.* **356**, 29 (2005).
5. D. Donoho and J. Jin, *Ann. Statist.* **32,3**, 962 (2004).
6. A. Gaugui *et al.*, *Astrophysical Journal* **430**, 447 (1994).
7. A. H. Jaffe, *Phys. Rev. D* **49**, 3893 (1994).
8. J. Jin *et al.*, *to appear in EURASIP Journal on Applied Signal Processing*, (2004).
9. B. M. Hill, *Ann. Statist.* **3**, 1163 (1975).
10. X. Luo, *Astrophysical Journal Letter* **427**, L71 (1994).
11. G. R. Shorack and J. A. Wellner, *Empirical Processes with Applications to Statistics, John Wiley & Sons*, (1986).
12. J. Starck *et al.*, *Astronomy and Astrophysics* **416**, 9 (2004).
13. P. Viela *et al.*, *Astrophysical Journal* **609**, 22 (2004).

EXPECTED PRINCIPAL COMPONENT ANALYSIS OF COSMIC MICROWAVE BACKGROUND ANISOTROPIES

SAMUEL LEACH

Astrophysics Sector, SISSA–ISAS, via Beirut 4, Trieste 34014, Italy

E-mail: leach@sissa.it

We implement and test an approach for measuring the primordial power spectrum of density perturbations given observations of the cosmic microwave background anisotropy spectrum. The method depends on exploiting the fact that the linear response of the CMB anisotropy spectrum with respect to the primordial power spectrum model parameters is well understood, as well as the noise properties of the CMB detectors. This puts us in the luxurious position of being able to precompute an accurate and useful representation of a Fisher matrix, from which a set of orthonormal power spectrum modes can be obtained. The full power spectrum mode plus nuisance parameter space can be integrated out using Markov chain Monte Carlo, and all the information concerning the primordial power spectrum is compressed onto a series of mode amplitudes which can then be easily compared with theoretical models.

1 Introduction

High signal to noise, high resolution, multifrequency observations of the Cosmic Microwave Background (CMB) are providing us with a fascinating opportunity to probe many diverse sectors of astrophysics and of our cosmological model. In the near term future an ensemble of ground-based, balloon-borne and satellite observations of CMB temperature and polarization anisotropies will provide us with a window on the basic model of linear perturbations to a photon–baryon fluid coupled to dark matter potentials via gravity, on the reionization epoch, on re-scattering of CMB photons by hot cluster gas (the Sunyaev–Zel'dovich effect), and on gravitational lensing of the CMB by the intervening dark matter distribution. Each of these phenomena poses interesting challenges for data analysis, the most fundamental of which is the fact that CMB data is *correlated*, and hence a global analysis of the entire data set must be attempted in order to fully exploit the science—see the monologue by Dodelson[1] for a recent treatment of the physics and data analysis of the CMB.

Here our focus is on measuring the *primordial power spectrum* which seeds both the oscillations in the photon–baryon fluid and gravitational instability in the dark matter sector, leading to structure formation. The basic hope is that the details of the primordial power spectrum (its shape, its Gaussianity or otherwise) will shed light on whatever mechanism in the early universe is responsible for actually generating the primordial power spectrum itself. At present the dominant early universe paradigm that emerged back in the late 1970's and early 1980's, and is by now not without many observational successes, is the celebrated *inflation* model.

2 The basic problem

The desired primordial power spectrum $\mathcal{P}(k)$ is related to the observed anisotropy spectrum of the surface of last scattering C_ℓ via

$$C_\ell = \frac{2\pi}{\ell(\ell+1)} \int d\ln k\, \mathcal{P}(k)\, T_\ell^2(k; \{\omega_i\}) + N_\ell, \quad (1)$$

where the dependence of the numerically calculable CMB transfer functions $T_\ell(k)$ on a set of cosmological parameters $\{\omega_i\}$ has been written in explicitly as well a Gaussian isotropic noise term, N_ℓ. Amongst the main science goals of all recent CMB observations has been the determination of these cosmological parameters, which is made possible by assuming some reasonable form for $\mathcal{P}(k)$ such as a smooth power-law. What kind of approaches are possible if we drop these model-motivated assumptions?

There is in fact a satisfactory solution to this rather generic data analysis problem given by Hu and Okamoto[2] in which a *Fisher matrix principal component analysis* (PCA) approach can be taken. We have implemented this method[3], a reference which also contains more of the details as well as an entry point to the literature for other approaches to the same reconstruction problem that have been investigated.

The essence of the method at hand is to con-

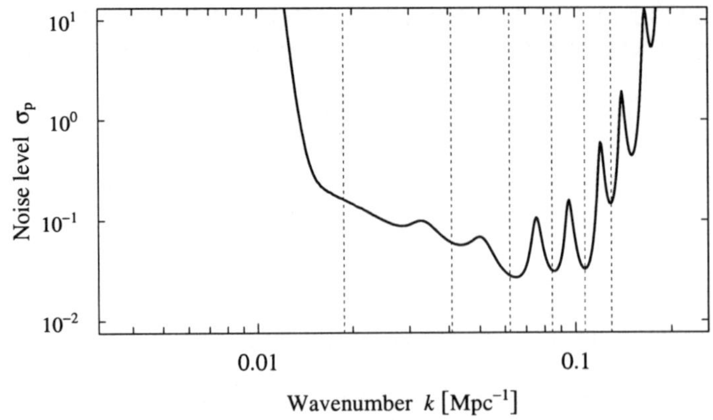

Figure 1. Illustrating the window of sensitivity to the primordial power spectrum for a *Planck*-like instrument. Here σ_P gives the approximate 1σ error on measurements of the primordial power spectrum using bandpowers with $\delta \ln k \sim 0.05$. The vertical lines indicate the position of the temperature acoustic peaks. The cosmological parameters have been fixed, so some degrading of the sensitivity is expected.

struct an *orthonormal power spectrum model*

$$\frac{\mathcal{P}(k)}{\mathcal{P}_0} = m_0 + \sum_{a=1}^{a_{\max}} m_a \mathcal{S}_a(k), \tag{2}$$

which is designed to satisfy the expectation $\langle m_a m_b \rangle = \sigma_a^2 \delta_{ab}$. Which orthonormal basis should we choose? Clearly the variation in the power spectrum modes $\mathcal{S}_a(k)$ should reflect our expectations of where observations are at their most sensitive, and hence there is a link with the Fisher information matrix, which is often associated with forecasting the expected sensitivity of a given instrumental specification.

Before sketching the details of the method however, we state the broader working assumptions that we rely on, which may be relevant when trying to implement this method in other contexts outside the realm of CMB anisotropies:

1. The initial perturbations are pure Gaussian adiabatic modes entering in Eq. (1) via a single physical component $\mathcal{P}(k)$.
2. The transfer functions $T_\ell(k)$ can be accurately calculated and are fast to evaluate. We make use of the CMB anisotropy code CAMB[4].
3. The noise model N_ℓ is known, and hence the Fisher matrix $\mathbf{F}_{ij}$ (to be described below) for a given instrument can be calculated. Tegmark, Taylor and Heavens[5] give an exposition of the Fisher matrix formalism in the context of cosmology.
4. A method for exploring a 20–50 dimensional posterior parameter space is available. Here we make use of the Markov Chain Monte Carlo method, as implemented in the state-of-the-art CosmoMC[a] code by Lewis and Bridle[6].
5. The main science driver behind the PCA analysis, however, is the prospect of the large data set being gathered over the next five years or so.

[a] publicly available at: http://cosmologist.info/cosmomc

Going more into the details, there is a basic pre-processing step which involves constructing the Fisher matrix

$$\mathbf{F}_{ij} = \sum_{\ell=2}^{\ell_{\max}} \frac{2\ell+1}{2} \mathrm{Tr}[\mathbf{D}_{\ell i} C_\ell^{-1} \mathbf{D}_{j\ell} C_\ell^{-1}], \tag{3}$$

where

$$\mathbf{D}_{\ell i} = \frac{\partial C_\ell}{\partial p_i}\Big|_{\mathrm{fid}} = \frac{2\pi}{\ell(\ell+1)} \int d\ln k\, \mathcal{P}_0 T_\ell^2(k)\, W_i(\ln k), \tag{4}$$

which is evaluated for some fixed fiducial values for the cosmological parameters. We can take our power spectrum test function W_i to be the triangle window

$$W_i(\ln k) = \max\left[1 - \left|\frac{\ln k - \ln k_i}{\Delta \ln k}\right|, 0\right]. \tag{5}$$

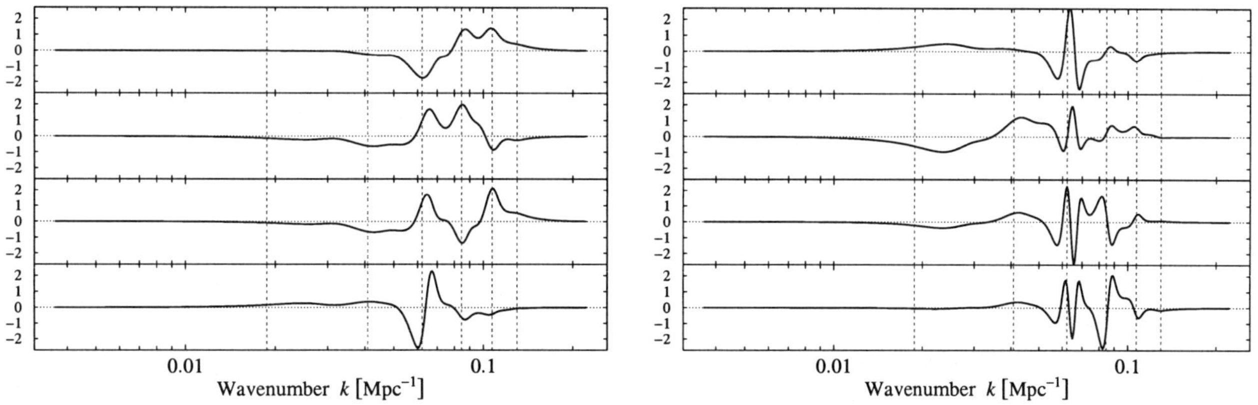

Figure 2. Illustrating PCA modes, $\mathcal{S}_a(k)$, 1–8 which have been generated assuming a *Planck*-like noise model. The vertical lines indicate the position of the temperature acoustic peaks.

The Fisher matrix Eq. (3) thus encodes the transfer of power from k-space to ℓ-space via the functions $\mathbf{D}_{\ell i}$, and the signal plus noise model via $C_\ell^{-1} = (S_\ell + N_\ell)^{-1}$. As usual the Cramer–Rao bound, given by the diagonal component of the inverse Fisher matrix, gives a useful handle on the best-case scenario for the sensitivity to the observables, and in this context reveals the possible observable range of scales, which we display in Figure 1.

The desired PCA modes $\mathcal{S}_a(k)$ are simply the (suitably normalised) orthonormal eigenvectors of the inverse Fisher matrix $\mathbf{F}_{ij}^{-1}$, and in Figure 2 we show the first eight PCA modes which resemble Fourier modes localised in the acoustic peak region $0.01 < k < 0.2\,\mathrm{Mpc}^{-1}$, displaying rapid oscillations at the acoustic peak scales. As a brief aside, the Fisher matrix pre-processing step was implemented using the R environment[7], which allows for matrix manipulations using the LAPACK linear algebra library.

The PCA mode amplitudes m_a can then be appended to the usual list of cosmological parameters to be integrated out using the now fairly standard and accessible MCMC technique. By construction the posterior distribution will be close to an uncorrelated Gaussian, which can then be used as a new likelihood function with respect to model-motivated power spectrum parametrisations. The model predictions for the PCA mode amplitudes are simply a convolution of the theoretical power spectrum over the PCA modes

$$m_a = \int d\ln k\, \mathcal{S}_a(k) \frac{\mathcal{P}}{\mathcal{P}_0}(k). \qquad (6)$$

Incidentally, the likelihood evaluations over this compressed data set will be fast which opens up the possibility of performing thorough model selection studies.

3 Tests with simulated data

We have tested this method using simulated *Planck*-like data generated using various primordial power spectra including scale-invariant as well as a somewhat contrived Gaussian bump power spectra. Here we will present results assuming a scale-free input power spectrum with spectral slope $n_S - 1 = -0.03$. We integrated out the parameter space consisting of five basic cosmological parameters and a further 20 PCA mode parameters, and the results are displayed in the first panel of Figure 3. The basic result here is that this kind of analysis is indeed feasible, requiring a total of around 10^6 MCMC likelihood evaluations to relax to a good representation of the posterior peak. We exploit the fact that CosmoMC can be executed across multiple CPUs with near perfect parallelisation. In addition, CosmoMC stores the cosmological parameter transfer functions during the movement through the power spectrum parameter space, meaning that movement in the "bulk" power spectrum parameter space is fast.

Finally, the measured PCA mode amplitudes can be used to constrain the power-law slope of the initial power spectrum and we recover to within one standard deviation the input power-law slope, shown in the second panel of Figure 3.

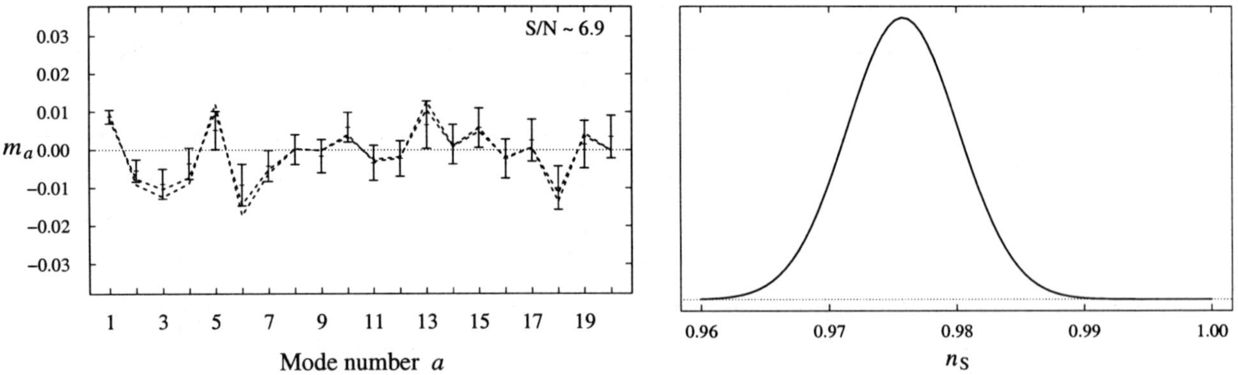

Fig. 3. Illustrating the recovery of the first 20 principal component amplitudes from simulated *Planck* data with an input $n_S = 0.97$ spectrum (left panel). The models (dashed lines) correspond to power-law spectra with $n_S(k_0 = 0.05\ \text{Mpc}^{-1}) = \{0.970, 0.985\}$ (bottom to top, mode 3). The PCA mode amplitudes can then be used to constrain more traditional power-law power spectra (right panel).

4. Final comments

The method has various extensions. For instance the PCA modes can be orthogonalised to the cosmological parameters in order to compensate for the fact that cosmological parameters degeneracies will break the desired statistical orthogonality of the recovered PCA mode amplitudes. In addition the PCA modes can be modified in order to search for deviations from scale-free spectra, not just deviations from scale-invariant spectra[2].

If one does require a (correlated) representation of the initial power spectrum and its covariance matrix in k-space given the measured PCA mode amplitudes, then an approach similiar to the "minimum variance map making" solution of Eq. (6) should be possible.

A detailed empirical approach such as the PCA method will always be an option for reconstructing some unknown function, particularly when the physics model underlying the data is thought to be basically well understood. The fact that it is derived from a Fisher matrix calculation also acts as a useful check on the consistency and scope of the final results.

Acknowledgements

The author would like to thank the organisers of PHYSTAT05, and Andrew Jaffe, Geoff Nicholls, and Roberto Trotta for their assistance and input during the conference.

References

1. S. Dodelson, *Modern Cosmology*, (Academic Press 2003).
2. W. Hu and T. Okamoto, *Phys. Rev.* D **69**, 043004 (2004).
3. S. Leach, preprint, `astro-ph/0506390`.
4. A. Lewis, A. Challinor and A. Lasenby, *ApJ* **538**, 473 (2002) (CAMB).
5. M. Tegmark, A. Taylor and A. Heavens, *ApJ* **480**, 20 (1997).
6. A. Lewis and S.L. Bridle, *Phys. Rev.* D **66**, 103511 (2002) (CosmoMC).
7. R Development Core Team, http://www.R-project.org, (2004).

TIME SERIES

ON-LINE INFERENCE FOR DATA STREAMS

PETER CLIFFORD
Statistics Department
Oxford University
E-mail: clifford@stats.ox.ac.uk

Rapid accumulation of substantial datasets is now common in many data processing applications. For example in monitoring and examining Internet traffic; analysing high-frequency financial data in market trading; voice and video capture; data logging in numerous areas of scientific enquiry. Markov Chain Monte Carlo (MCMC) methods revolutionised statistical analysis in the 1990s by providing practical, computationally-feasible access to the flexible and coherent framework of Bayesian inference. However, massive datasets have produced difficulties for these methods since, with a few simple exceptions, MCMC implementations require a complete scan of what might be several gigabytes of data at each iteration of the algorithm. For time-series data, progress is possible using modern sequential Monte Carlo methods (known as particle filters). With suitable modifications the techniques can be adapted to deal with more general data catalogues.

1. Bayesian Analysis

The basic components in the Bayesian analysis of a statistical problem are:

- Data: y
- Parameters: θ, functions of interest: $g(\theta)$
- Likelihood: $L(\theta; y)$
- Prior density: $\pi(\theta)$

The prior density and the likelihood are used to calculate integrals of the form

$$\frac{\int g(\theta)\pi(\theta)L(\theta;y)\mathrm{d}\theta}{\int \pi(\theta)L(\theta;y)\mathrm{d}\theta} = \int g(\theta)\pi(\theta|y)\mathrm{d}\theta$$

$$= E\{g(\theta)|y\} \text{ (posterior expectation)},$$

where $\pi(\theta|y)$ is the posterior density of the parameter θ given the data y. Markov chain Monte Carlo (MCMC) methods can be used to construct a chain with successive values, $\theta^1, \theta^2, \ldots, \theta^n$, simulated from the equilibrium density $\pi(\theta|y) \propto \pi(\theta)L(\theta; y)$, estimating $E\{g(\theta)|y\}$ by

$$\overline{g} = \frac{\sum_{i=1}^{n} g(\theta^i)}{n}.$$

If $\{\theta^i\}$ are independent then $\mathrm{Var}(\overline{g}) = \sigma_g^2/n$. Typically $\mathrm{Var}(\overline{g}) = \tau\sigma_g^2/n$ with 'correlation time', τ, greater than 1 and 'effective sample size' n/τ.

Suppose now that the observational framework expands, giving additional data and an expanded parameter set.

- Data: y, y^+
- Parameters: θ, θ^+, functions of interest $g(\theta, \theta^+)$
- Likelihood: $L(\theta, \theta^+; y, y^+)$
- Joint prior density: $\pi(\theta)\pi(\theta^+|\theta)$

Question: Can we use the simulations from $\pi(\theta|y)$ to simulate from $\pi(\theta, \theta^+|y, y^+)$ or do we have to start completely afresh using MCMC, for example, on the expanded problem?

2. Time-Series (Signal Processing)

In many applications, time-series data are noisy observations of an unobserved underlying process of interest (the signal). The data, $(y_1, \ldots, y_t) = \mathbf{y}_{1:t}$, expand with time, and the parameters (the history of the underlying process) expand correspondingly, $(\theta_1, \ldots, \theta_t)$.

In on-line analysis, a basic objective is to maintain knowledge about the current state θ_t, for example to allow estimation of $E\{g(\theta_t)|\mathbf{y}_{1:t}\}$. In signal processing terms, this is the *filtering* problem. Applications include: medical monitoring, robotics, finance.

For simplicity *structural assumptions* are made about the evolving data set.

- $\pi(\theta_1, \ldots, \theta_t) = \pi(\theta_1)\pi(\theta_2|\theta_1)..\pi(\theta_t|\theta_{t-1})$ (underlying state is Markov)
- $L(\theta_1, \ldots, \theta_t; \mathbf{y}_{1:t}) = \prod_{k=1}^{t} h(y_k|\theta_k)$ (current observations depend only on the current state).

In general, the underlying state process will depend on unknown (hyper)parameters that must be incorporated into a full Bayesian model. The Markov assumption is not severely restrictive and the observational assumptions can be relaxed.

With these assumptions, the current state of knowledge can be updated by

$$\pi(\theta_{t+1}|\mathbf{y}_{1:t}) = \int \pi(\theta_{t+1}|\theta_t)\pi(\theta_t|\mathbf{y}_{1:t})\mathrm{d}\theta_t \quad (1)$$

$$\pi(\theta_{t+1}|\mathbf{y}_{1:t+1}) = \frac{h(y_{t+1}|\theta_{t+1})\pi(\theta_{t+1}|\mathbf{y}_{1:t})}{p(y_{t+1}|\mathbf{y}_{1:t})} \quad (2)$$

where

$$p(y_{t+1}|\mathbf{y}_{1:t}) = \int h(y_{t+1}|\theta_{t+1})\pi(\theta_{t+1}|\mathbf{y}_{1:t})\mathrm{d}\theta_{t+1}. \quad (3)$$

The first integral is crucial. If θ_t is high-dimensional, evaluation of this integral at each stage will present problems.

When there is a linear Gaussian model for the evolution of the underlying state and when the noise is additive and Gaussian, the integrals can be evaluated explicitly. The posterior distributions then turn out to be Gaussian too. This is the basis of the *Kalman filter* (which basically just updates the means and covariances of the state θ_t). Since the posterior distributions can be obtained explicitly in the linear Gaussian model, it is comparatively straightforward to draw inferences about any unknown parameters involved in the underlying state process and the error model.

In many practical applications, these assumptions are implausible. In particular, the observation process will often be non-linear. An alternative approach in such cases is the *extended Kalman filter* (EKF), in which the updated measurements are linearised about the predicted state, permitting the Kalman filter to be applied approximately. This algorithm and its refinements have proved popular, particularly in the field of object tracking. However, the Gaussian approximation to the density of the underlying state, inherent in the EKF, will often prove to be inadequate, causing the update procedure to become unstable.

Other methods involve approximating distributions by mixtures of Gaussians (the Gaussian sum filter); approximating the first two moments of the density; evaluating the required probability density function over a grid in the state space. However, each of these techniques has to be extensively modified to tackle the particular problem in hand. For example, methods that evaluate the probability density over a grid in the state space first require the grid to be specified, which is a non-trivial problem in a multi-dimensional space. To avoid misleading results, a large number of grid points will in general be necessary. In addition, a non-trivial computation must be performed at each point.

2.1. *Sequential Monte Carlo (Particle Filters)*

Recall that the current state of knowledge is updated via equations 1, 2 and 3. We need a way of carrying out these integrals successively for $t = 1, 2, \ldots$.

Poor Man's Bayes: Rubin[15] devised a simple way of obtaining an approximate sample from a Bayesian posterior distribution.

- Simulate a sample $\tilde{\theta}^1, \tilde{\theta}^2, \ldots, \tilde{\theta}^n$ from $\pi(\theta)$.
- Calculate weights $q_i \propto L(\tilde{\theta}^i; y)$; $\sum q_i = 1$
- Sample n times (with replacement) from the discrete θ-distribution with

$$\mathrm{P}(\theta = \tilde{\theta}^i) = q_i.$$

The resulting sample $\theta^1, \theta^2, \ldots, \theta^n$ is an "approximate" sample from $\pi(\theta|y)$. The sample obtained is approximate in the sense that $n^{-1}\sum_{i=1}^n g(\theta^i)$ converges in probability to $\mathrm{E}\{g(\theta)|y\}$, as $n \to \infty$.

The *Sampling Importance Resampling (SIR/ particle filter)*[10, 6] is based on Rubin's sampler. It proceeds as follows. Assume that you have a sample $(\theta_t^i)_{i=1,\ldots,n}$ from $\pi(\theta_t|\mathbf{y}_{1:t})$:

(a) **Sampling:** Independently simulate $\tilde{\theta}_{t+1}^i$, using the state transition density $\pi(\theta_{t+1}|\theta_t^i)$, for each $i = 1, \ldots, n$,
(b) **Importance:** Upon receipt of observation y_{t+1}, for each value $\tilde{\theta}_{t+1}^i$ calculate the corresponding likelihood $h(y_{t+1}|\theta_{t+1}^i)$. Denote the set of likelihood values, normalised to sum to 1, by $(q_{t+1}^i)_{i=1,\ldots,n}$.
(c) **Resampling:** Draw a random sample of size n from the discrete distribution taking values $(\tilde{\theta}_{t+1}^i)_{i=1,\ldots,n}$ with probabilities $(q_{t+1}^i)_{i=1,\ldots,n}$. This is an approximation to a sample from $\pi(\theta_{t+1}|\mathbf{y}_{1:t+1})$.

The algorithm can be thought of as propagating a swarm of particles in the underlying state space. At time t the particles are assumed to be an approximate sample from the posterior distribution of θ_t, given the observations so far. At time $t+1$ each particle moves to a new location in the state space. The likelihood of this location given x_{t+1} is evaluated and a multinomial sample of particles is then drawn from the discrete distribution with *support points* given by the *particle locations* and *probabilities* proportional to the *likelihoods*. The process is a form of *Genetic Algorithm* where the 'fitness' of a speculative parameter value is proportional to its likelihood. In its simplest form the SIR filter has various weaknesses.

Outliers: The effect of an outlying observation is to produce a likelihood which is centered in the tail of the prior distribution. Since this tail is represented only sparsely by sample points in the SIR filter, an exceptionally large sample from the prior will be needed to yield a good support for the posterior distribution.

Sample Impoverishment: Lack of diversity: particles may be highly correlated, localised into a restricted region of parameter space, acting as one. The particle system may collapse to a singleton (extreme lack of diversity).

Track Loss: Particles become trapped in 'impossible' regions of state space (evolutionary dead-ends).

Jittering: There are various *ad hoc* fixes for these problems. In order to alleviate the problem of sample impoverishment, Gordon*et al.*[10] suggested adding a small amount of Gaussian noise, or jitter, to each sample point at each time step. If one point is replicated in the posterior r times, it is now replaced by r closely adjacent points. Jittering therefore smooths out the posterior density, using a Gaussian kernel. Choosing the jitter variance is thus equivalent to choosing the smoothing parameter in density estimation, and there is a corresponding variance/bias trade-off to be made. Standard rules of thumb can be used to choose the degree of smoothing.

Prior Boosting: This approach to sample depletion was originally proposed by Rubin[15]. At the prediction stage of the SIR filter, instead of generating the usual n points, we generate κn points. The likelihood of each of these is calculated, and then n are resampled in the update step in the usual way. Typically $\kappa = 10$.

3. Fundamentals

Particle filters work by providing a discrete approximation to the PDF which can be easily updated to incorporate new information as it arrives. More generally our interest will be in approximations which consist of a set of random locations in the state space $(s^i)_{i=1,\ldots,n}$, termed the *support*, and a set of associated weights $(m^i)_{i=1,\ldots,n}$ summing to 1. The support and the weights together form a *random measure.*

The objective is to choose measures so that

$$\sum_{i=1}^{n} g(s^i)m^i \approx \int g(\theta)\pi(\theta)\, d\mu(\theta) \qquad (4)$$

for typical functions g of the state space, in the sense that the left-hand side converges (in probability) to the right-hand side as $n \to \infty$.

The simplest example of a random measure is obtained by sampling $(s^i)_{i=1,\ldots,n}$ independently from $\pi(\theta)$, and giving equal weights $m^i = n^{-1}; i = 1,\ldots,n$. The estimate of the expected value of $g(\theta)$ is then the sample average $\sum_{i=1}^{n} g(s^i)/n$. Importance sampling provides a more general example by sampling $(s^i)_{i=1,\ldots,n}$ from another PDF $f(y)$ and attaching importance weights $m^i = A\pi(s^i)/f(s^i)$, where $A^{-1} = \sum_{i=1}^{n} \pi(s^i)/f(s^i)$.

Before attempting to improve the SIR algorithm, it is worth emphasing that our fundamental objective is to produce accurate Monte Carlo approximations to the *succession of integrals* that arise in Bayesian calculations. For accurate Monte Carlo integration, it is essential to eliminate unnecessary randomness and to make careful choices for proposals in importance sampling.

For example, the purpose of resampling is to produce a set of points with a histogram that approximates a particular probability mass function. The standard SIR algorithm achieves this with a multinomial sample $(N_i)_{i=1,\ldots,n}$. But with the following algorithm the variables N_i never differ from their required expected value by more than 1.

Algorithm: Randomised circular sampling.

$T = \mathit{unif}(0, n^{-1}); j = 1; Q = 0; i = 0$
do while $T < 1$
  if $Q > T$ *then*

```
        T = T + 1/n; output s^i
    else
        pick k in {j, ..., ℓ}
        i = s^k
        Q = Q + m^i
        switch (s^k, m^k)  with  (s^j, m^j)
        j = j + 1
    end if
end do
```

The algorithm treats the weights as contiguous intervals of $(0, 1)$. These intervals are randomly ordered, and the number of grid points $\{T + k/n\}$ in each interval is then counted. It randomly translates a 'comb' with equally spaced teeth as follows:

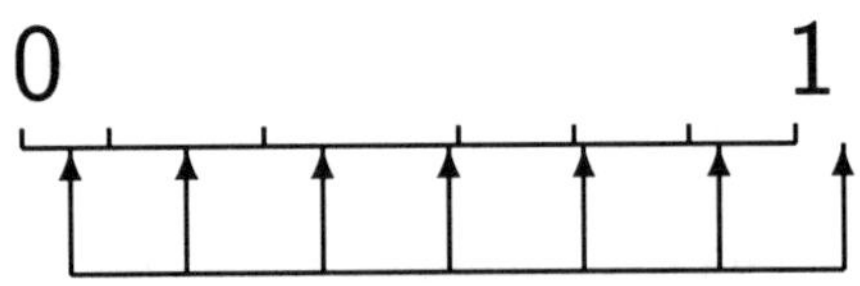

The objective of the previous sampler is to ensure that N_i has expected value nm_i for $i = 1, \ldots, \ell$, while ensuring that the variances of the N_i are as small a possible. Crisan and Lyons[5] proposed that each N_i should be chosen to be the integer part of nm_i plus a Bernoulli variable with probability equal to the fractional remainder. Liu and Chen[12] have a similar method where each N_i is again chosen to be the integer part of nm_i but with the addition of a multinomial variable based on the fractional remainders. They call their method *residual sampling.* In practice, these methods produce similar effects on sampling efficiency.

Since resampling introduces noise, this raises the question, when should we resample, and when should we carry forward the weights? The question has been addressed by Liu and Chen[12] who propose an *ad hoc* rule based on the variance of the weights $(\tilde{m}_t^i)_{i=1,\ldots,n}$. In general, if the weights are roughly even, and the system noise is small compared to the variance of the posterior at the previous time step, then it is better not to resample. In particular, if there is no system noise, resampling is always inefficient.

3.1. *Assessing Sample Depletion*

To compare refinements of the SIR algorithm, it is helpful to have a measure of the effective sample size (ESS). This is the sample size that would be required for a simple random sample from the target posterior density to achieve the same estimating precision as the random measure provided by the particle filter.

Liu[12] has suggested using ESS $= n/(1 + V)$, where V is the variance of the importance weights. The result should be used with caution, since in practice some properties of the state distribution may be estimated well, and some poorly. In general, the effective sample size will depend on the quantity being estimated and not just the weight distribution.

In principle, a Bayesian filter should be assessed by looking at its performance averaged over the population of trajectories generated by the system model. However, for non-linear problems it may happen that most of the trajectories are simple to filter and only a few are 'difficult cases'. It is therefore helpful to see how the filter performs for typical examples of these difficult cases. The integrated correlation time in MCMC calculations in non-dynamic problems and the ESS play similar roles. Neither of these diagnostics is designed to check for convergence to the *right* distribution. A noisy biased filter may have a large ESS but the sample will not have come from the correct distribution. To check for bias, the proposed particle filter will need to be compared with filters which are known to perform correctly.

We should note that there is intermediate ground between resampling and carrying forward the weights. Resampling can be carried out using modified weights: for example, using modified weights proportional to the square root of the original, i.e. $w_t^i \propto \sqrt{m_t^i}$. The resampled points are then carried forward with weights proportional to m_t^i / w_t^i. Similar techniques have been proposed in MCMC sampling to avoid problems in sampling from highly peaked densities.

Although it is unrealistic to use MCMC to sample the posterior distribution of the complete state history, under certain circumstances MCMC moves can be introduced in particle filtering. These moves may be successful in preventing sample impoverishment. In general, to accommodate arbitrary transitions it is necessary to store the whole history of the process up to time t. As we shall see in the next example, this can be avoided if the transition kernel only depends on a fixed set of summary statistics, or only upon the last τ time points.

3.2. *Example: Bearings Only Tracking*

A classic example of non-linear filtering is *bearings only tracking*. An observer (either fixed or moving) observes the bearing of a moving ship. The bearing is the angle of the observation relative to a fixed direction. The crucial problem is that we are trying to reconstruct the two-dimensional coordinates of the ship from a single non-linear observation. This type of non-linearity in tracking problems usually causes difficulties for the Extended Kalman Filter.

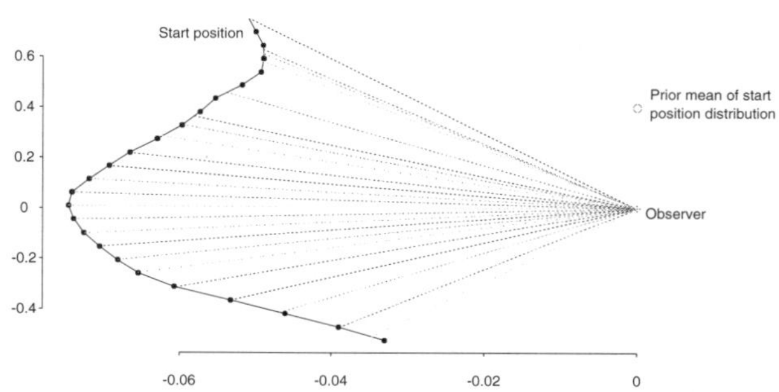

Fig. 1. Typical simulated trajectory. Dotted lines show observed bearings

We want to reconstruct the trajectory given the system model, observed bearings and a prior distribution on the initial position and velocity. In particular, suppose we observe t bearings. Notice that scaling the track toward the observer by a constant λ does not affect the likelihood since none of the angles change. It affects some of the parameters in a simple way. These factors can be incorporated into the filter by extending the *signature* of each particle. The MCMC scale move, when it is made, is a Gibbs move sampling from a truncated Gamma distribution.

3.3. *Hidden Markov Models*

By way of illustration we will work through a specific example. The problem is typical in the sense that the observation process is driven by a hidden Markov process.

Well-logs are records of the physical and mineralogical characteristics of underground rocks obtained by drilling in a region of geological interest. In traditional applications, a probe (called a sonde) is lowered into an existing well-bore by a cable, and acoustical, electrical, nuclear-magnetic or thermal properties of the surrounding rock types are recorded as the sonde descends. In this example, the measurements are of nuclear magnetic response taken at 4500 time points. The underlying signal is piecewise constant; each constant segment relating to a stratum of a single rock type with constant physical properties. The jump discontinuities in the signal occur at times when a new rock stratum is first met.

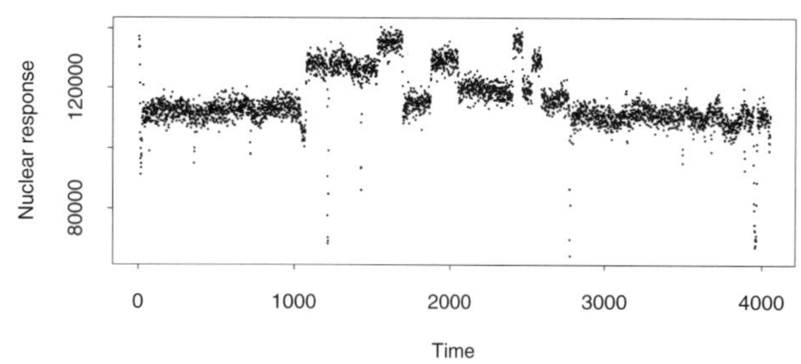

Fig. 2. The measurements of nuclear magnetic response taken at 4500 time points.

There is increasing interest in the possibility of 'measurement-while-drilling' (MWD) rather than the retrospective measurement of rock characteristics in existing boreholes. To detect changes in rock strata as drilling proceeds, data need to be collected from the vicinity of the drill-head. There are severe technical difficulties both in obtaining useful measurements and in the transmission of these data to the surface. Progress is currently been made with these problems in the gas and oil drilling industry. Other areas in which the use of traditional sondes is inadequate include the exploration of leakage below buried waste. These investigations are carried out by horizontal drilling making it impossible to lower a sonde into the borehole. Attachment of recording devices to the drill head may be the only way in which data can be collected – thus enabling drilling to be steered towards areas of high contamination.

3.3.1. *Batch processing*

The well-log data of Figure 2 have been analysed previously[13]. The whole dataset was batch-processed using a Gibbs (MCMC) sampler. Outliers were removed by hand and the number of change-points was fixed prior to the analysis. It only remained to locate the change-points as accurately as possible.

When scanning the data as a whole (by eye) the detection of change-points appears straightforward. However when the data are only available incrementally, differentiating between outliers and true change-points is difficult. Successive MCMC sampling, even when outliers have been eliminated and number of change-points is known, is too time-consuming for real-time inference. By contrast, as we shall see, particle filter methods are computationally efficient and enable uncertainty about the number of change-points and outliers to be incorporated automatically.

3.3.2. *On-line analysis*

We use a hidden Markov model to model regime switching in the well-log data. The (underlying) state is the expected nuclear magnetic response for the current rock strata. The hidden Markov chain allows for both changes in the rock strata, and the possibility that the current measurements are outliers. The conjugacy in the assumed model means that conditional on knowing the history of the hidden Markov chain, the posterior distribution of the history of the measurable state can be calculated analytically using the Kalman filter.

The posterior distribution can be written as a mixture distribution, with each term in the mixture referring to a single possible value of the history of the hidden Markov chain. Liu and Chen[12] show that for such problems, the efficiency of the particle filter can be greatly improved if, instead of each particle representing a possible value of the history of the state, each particle represents a possible history of the hidden Markov chain (or a suitable summary of that history). This technique is called *marginalisation* or *collapsing*.

With such an approach, the posterior can be calculated exactly using a finite number of particles. Unfortunately, the number of particles needs to increase exponentially with the number of measurements, and becomes unfeasibly large for even small data sets (let alone the data set shown in Figure 2, where there are 4050 measurements). To restrict the number of particles used by the particle filter, resampling must be used. At each time stage a smaller, but hopefully representative, sample of particles are chosen from the large number of current particles[16, 12, 7].

We assume a two-dimensional Hidden Markov Model, with states $I_t = (S_t, O_t)$, where S_t and O_t both taking values in $\{1, 2\}$. Conditional on I_t, the underlying state (the expected nuclear magnetic response) satisfies

$$\theta_t = \begin{cases} \theta_{t-1} & \text{if } S_t = 1, \\ \mu + \sigma Z_t & \text{if } S_t = 2. \end{cases} \tag{5}$$

and the measurements satisfy

$$Y_t = \begin{cases} \theta_t + \tau_1 Z_t^* & \text{if } O_t = 1, \\ \nu + \tau_2 Z_t^* & \text{if } O_t = 2. \end{cases} \tag{6}$$

The error terms $\{Z_t, Z_t^*\}_{t=1,\ldots}$ are uncorrelated, standard Gaussian random variables, and $\mu, \nu, \sigma, \tau_1, \tau_2$ are suitably chosen hyperparameters. The system equation (5) allows for jumps in the underlying signal, while the measurement equation (6) allows for clusters of outliers. Such a model produces the step function form for the underlying signal that is evident from the data.

A number of outliers below the main body of data are apparent. This motivated the model that we have used (see Equation 6). When the Markov chain, O_t, is in state 1, the observations will be modelled as the true state corrupted by additive noise. State 2 will represent an outlier state, and, for simplicity, the observation will be modelled as a draw from a Gaussian random variable whose parameters are independent of the true state. There are around 70 observations that appear to be outlying. These occur in 16 clusters. This suggests that suitable values of the transition probabilities would be approximately $P(O_t = 2|O_{t-1} = 1) = 0.004$ and $P(O_t = 2|O_{t-1} = 2) = 0.75$. The outlier distribution was taken to have a mean, ν, of 85000 and a standard deviation, τ_2, of 12500. The standard deviation τ_1 of non-outlying observations was taken to be 2500.

Previous analyses[13] have assumed additive Laplacian noise for the data. The analysis of the well-log data by a particle filter under such a model can be found in Fearnhead's thesis. More complicated models, which include more detailed modelling of the outliers, and allowing for correlated noise, were also considered there. For all these models the posterior distribution of the history of the state, θ_t, conditional on the history of the hidden state, I_t, could be calculated analytically.

Results: The main aim of analysing the well-log data is to detect the change-points in the signal on-line. So after processing each measurement, the probability of a jump having occurred during the last k time points was estimated. The results we present are for $k = 5$, but similar results were obtained with slightly different values of k.

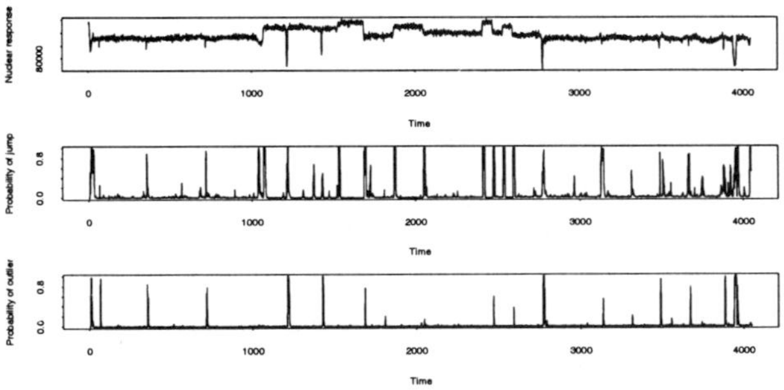

Fig. 3. Results of on-line analysis of the well-log data (top) by the new particle filter. The particle filter used 100 particles. The estimates of the probabilities of a recent change-point (middle), and the probability of the measurement being an outlier (bottom) are both shown.

The filter appears to have performed well, with all obvious change-points being given a posterior probability close to one. In a few cases, the filter appears to have misclassified outliers as change-points. An easier evaluation of the performance of the filter can be gained from looking at an estimate of the underlying signal (see Figure 4).

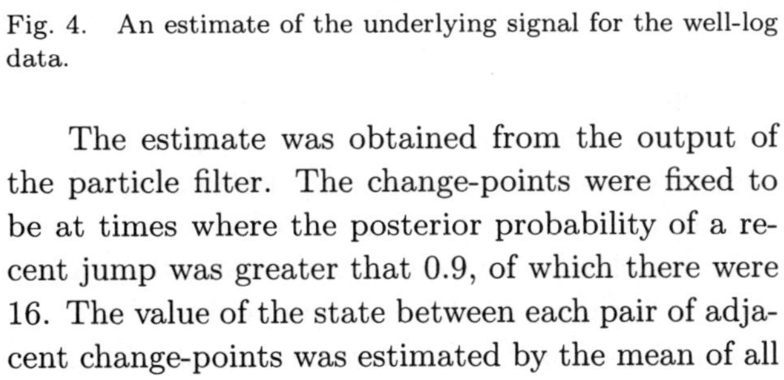

Fig. 4. An estimate of the underlying signal for the well-log data.

The estimate was obtained from the output of the particle filter. The change-points were fixed to be at times where the posterior probability of a recent jump was greater that 0.9, of which there were 16. The value of the state between each pair of adjacent change-points was estimated by the mean of all measurements in that time period which had negligible probability of being an outlier.

4. Using Particle Filters to Analyse Large Datasets

In Bayesian statistical analysis, our aim is to find the posterior density $\pi(\theta|y_{1:N})$ of the parameter given the data $y_{1:N} = \{y_1, \ldots, y_N\}$. The parameter θ may be of high dimension θ. In a standard non-dynamic (static) problem all the data $y_{1:N}$ are available at once, and we know that

$$\pi(\theta|y_{1:N}) \propto \pi(\theta)L(\theta; y_{1:N})$$

where $L(\theta; y_{1:N})$ is the likelihood. Markov chain Monte Carlo is one method of analysis.

To use MCMC we need to construct a Markov chain $(\theta^{(1)}, \theta^{(2)}, \ldots)$ on the space of possible θ values with $\pi(\theta|y_{1:N})$ as its equilibrium. By running the chain for a long period of time, values from the equilibrium can be harvested and used to summarise the target distribution.

Problems: Suppose for example that the Metropolis sampler is used. At each step r in the Markov chain the current value of $\theta^{(r)}$ is modified by proposing a new value, θ, sampled from a proposal density $g(\theta|\theta^{(r)})$. The new value θ is accepted and becomes $\theta^{(r+1)}$ with acceptance probability

$$A(\theta^{(r)}, \theta; y_{1:N}) = \min\left\{1, \frac{g(\theta^{(r)}|\theta)\pi(\theta|y_{1:N})}{g(\theta|\theta^{(r)})\pi(\theta^{(r)}|y_{1:N})}\right\}.$$

The problem is that that $A(\theta^{(r)}, \theta; y_{1:N})$ depends on the whole of $y_{1:N}$. When the data set is massive, computing the acceptance probability is a non-trivial calculation, since it involves scanning through the whole dataset. When N is of the order of millions this can be a very time-consuming task, and furthermore the task has to be repeated until the MCMC algorithm has converged, which may take several thousand steps.

There have been various attempts to use particle filters for the Bayesian analysis of large datasets. The papers by Ridgeway and Madigan[14] and Fearnhead[8] provide a simple introduction.

4.1. *Simple Use of Sub-Sampling*

The basic idea is to use a sub-sample of the data $y_{1:n}$ where $n \ll N$. If n is small enough then MCMC can be run on the subsample, to yield $\{\theta_i\}, i = 1, \ldots, M$, a sample of values from $\pi(\theta|y_{1:n})$. Each of these values then receives an importance weight w_i from the rest of the sample, given by

$$w_i = \frac{\pi(\theta_i|y_{1:N})}{\pi(\theta_i|y_{1:n})}.$$

A simplification occurs when observations are independent, since

$$\frac{\pi(\theta|y_{1:N})}{\pi(\theta|y_{1:n})} = \frac{\pi(\theta)L(\theta, y_{1:N})}{\pi(\theta)L(\theta, y_{1:n})} = L(\theta, y_{n+1:N}).$$

Similar simplifications occur when the observations have Markov dependence. The practical impact is that the remainder of the dataset only needs to be scanned once.

4.2. *Successive Sub-Samples*

Unfortunately, the set of weights produced by this procedure may be highly skewed and concentrated on only a few of the values in the set $\{\theta_i\}$. To remedy this Ridgeway and Madigan[14] consider a succession of values of n, say $n_1, n_2, \ldots N$ and apply a modified particle filter to the successively augmented datasets, proceeding as if these form a time series. The modified particle filter has two components, sampling/resampling and refreshment. MCMC transitions are introduced to refresh the particle support set. The decision on when to refresh is based on the distribution of particle weights. If the distribution is highly skewed then refreshment is carried out.

Unless the statistical model has special structure that can be exploited, these MCMC steps are computationally expensive. However, we expect that as the data are successively augmented, the distribution of particle weights will become less skewed, so moves are made less often. In Ridgeway and Madigan we see that the refresh times occur frequently at the beginning and less so toward the end of the data reading process.

4.3. *Model-Based Clustering*

Fearnhead's paper[8] is about model-based clustering. The data are assumed to come from a mixture distribution where the distributions of the mixture components have some known parametric form. For example, it could be assumed that each observation is from one of K possible multivariate normal distributions. We don't know the means and covariances of the distributions, how many different distributions there are or which distribution each observation is from.

The data are $y_{1:n} = \{y_1, \ldots, y_n\}$. Under the model each y_i comes from one of the mixture components. For any given component the observations are considered to be independent. An assignment $z_{1:n} = \{z_1, \ldots, z_n\}$ is a vector of component labels and k is the number of components identified, so $z_i \in \{1, \ldots, k\}$. The component distributions have densities $f(y; \theta)$ where θ is different for each component. The joint density of these variables is

$$p(y_{1:n}, z_{1:n}, \theta_{1:k}) \propto \pi(z_{1:n}) \prod_{j=1}^{k} \pi(\theta_j) \prod_{i=1}^{n} f(y_i; \theta_{z_i}).$$

The Dirichlet prior $\pi(z_{1:n})$ is parametrised by α, with a recursive definition:

$$\pi(z_{i+1} = j|z_{1:i}) = \begin{cases} n_j/(i+\alpha) & \text{for } j = 1, \ldots, k_i \\ \alpha/(i+\alpha) & j = k_i + 1 \end{cases}$$

where k_i is the number of clusters in the assignment $z_{1:i}$ and n_j is the number of observations that $z_{1:i}$ assigns to cluster j. With classical conjugate prior distributions for the parameters of a multivariate normal density some special simplications occur. In particular and most importantly, once the assignment vector $z_{1:n}$ is known, it is possible to evaluate the posterior distribution of the parameters explicitly. The special form of the Dirichlet prior also leads to simplications enabling the posterior probabilities of the mixture weights to be assessed when $z_{1:n}$ is known. So we can use a particle filter where each particle is tagged with its own assignment vector $z_{1:i}$ at stage i. See Fearnhead[8] for further details.

5. Data Sketching for Large Datasets

The purpose of using sequential statistical methods (particle filters) on static datasets is to reduce demands on data access. An entirely independent approach to related problems has been developed in the computer science literature. Key authors are Indyk[11], Cormode and Muthukrishnan[4] and Flagolet[9]. The first three authors exploit an ingenious combination of random projections (using stable law distributions) and universal hashing[3] to produce sketches of large datasets that enable questions concerning the distributional properties of the values in the dataset to be answered rapidly. Flagolet[9] develops an ingenious way of counting large numbers with a tiny amount of memory. This is related to the Additive-increase multiplicative-decrease processes studied by Bertoin-Biane-Yor[1].

5.1. *Projection Methods*

The data come in a stream (no particular order), $(a_1, w_1), (a_2, w_2), \ldots$ where $a_i \in A$ is the type of the ith item in the stream and w_i is the multiplicity. The problem is that the amount of data can be vast. How can you answer questions about the stream, for example, to find out how many different types there are? How many different users are there on the Internet?

We suppose that there is a pseudo-random mapping $h : A \to R$ such that

$$P(h(a) < x) = F_p(x),$$

where F_p is the distribution function of a symmetric stable distribution with parameter p, and where

$$P(|h(a) - h(b)| < \epsilon) = O(\epsilon), a \neq b.$$

(universal hash function)

Now calculate

$$S = \sum_{i=1}^{n} h(a_i) w_i = \sum_{j=1}^{m} x_j \sum_{i:h(a_i)=x_j} w_i = \sum_{j=1}^{m} x_j c_j,$$

and note that, using the property of stable distributions,

$$S \sim X \left(\sum_{j=1}^{m} |c_j|^p \right)^{1/p},$$

where X has a symmetric stable distribution with parameter p.

The *projection sketch* consists of R independent replicates of S. The median (for example) of the S values is then used to estimate the scaling term $\sum_{j=1}^{m} |c_j|^p$. When p is small this gives an estimate of the number of distinct items.

5.2. *Other Types of Data Sketches*

Sketches based on small p projections enable us to assess whether the profile of occurrences in two data streams is the same — just subtract the sketches. They also allow for removal of items (stock control). Techniques for maintaining histogram sketches are of particular interest for statistical applications.

6. Concluding Remarks

The Bayesian analysis of massive datasets remains a challenging problem. MCMC methods are not feasible for these datasets. Particle filters are promising. They are particularly effective when

- distributional conjugacy can be exploited (c.f. Section 3.3.2),
- sufficient statistics are available, permitting occasional MCMC moves to be made at low computational cost (c.f. Section 3.2).

For complex problems, particles need to be tagged with extensive information. The design of efficient database management systems for these data is an open problem.

Data sketches have the potential for summarising both the data and the particle systems that represent the posterior distribution.

References

1. J. Bertoin, P. Biane and M. Yor, *Tech. Rep. PMA-705, Lab. de Probab., Univ. Paris VI* (2002).
2. J. Carpenter *et al.*, *IEE Proc. Radar Sonar Navigation* **146**, 2 (1999).
3. T. Cormen, C.E. Leiserson and R. Rivest, *Introduction to Algorithms*. MIT Press, London (1990).
4. G. Cormode and S. Muthukrishan, *IEEE Trans. Know. Data Eng.* **15(3)**, 529 (2003).
5. D. Crisan and T. Lyons, *Probab. Th. and Rel. Fields* **109**, 217 (1997).
6. A. Doucet *et al.*, *Sequential Monte Carlo Methods in Practice*. Springer, NY.(2000).
7. P. Fearnhead and P. Clifford, *J. Royal Statist. Soc.* **65**, 887 (2003).
8. P. Fearnhead, *Statistics and Computing* **14**, 11 (2004).
9. P. Flajolet and G.N. Martin, *J. Comp. Sys. Sc.* **31**, 182 (1985).
10. N.J. Gordon *et al.*, *IEE Proc. F Radar Sig. Proc.* **140**, 107 (1993).
11. P. Indyk, *Proc. 40th Symp. Found. Comp. Sc.*189-197 (2000).
12. J. Liu and R. Chen, *J. Amer. Statist. Assoc.* **93**, 1032 (1998).
13. J.K. O'Ruanaidh *et al.*, *Numerical Bayesian Methods*. Springer, NY.(1993).
14. G. Ridgeway and D. Madigan, *KDD02 Proc. 8th ACM SIGKD* , 5 (2002).
15. D.B. Rubin, in *Bayesian Statistics*, Vol.3, Oxford University Press (1988).
16. J.K. Tugnait, *Automatica* **18**, 607 (1982).

DECONVOLUTION

SOME ASPECTS OF STATISTICAL IMAGE MODELLING AND RESTORATION

D. M. TITTERINGTON
Department of Statistics, University of Glasgow, Glasgow G12 8QQ, Scotland
E-mail: mike@stats.gla.ac.uk

A review is provided of some ways in which statistical ideas have influenced research into image analysis, in particular the problems involved in making inferences about the true scene and any parameters in the underlying model. The emphasis will be on the application of general statistical paradigms such as maximum likelihood, implemented using tools such as the EM algorithm, and Bayes' Theorem. The resulting procedures can be regarded as particular recipes for regularisation or deconvolution, according to the context.

1. Introduction

Statisticians have made a substantial impact on image modelling and analysis. The purpose of this paper is to give a quick overview of some of these contributions, to emphasise the key statistical issues, to highlight some points of contact with Physics and to mention a few applications, including one or two from contexts within Physics. The account cannot claim to cover all methodological approaches or all application areas, and there is a strong element of selectivity in the reference list! For instance, a large body of recent work in shape analysis (Dryden and Mardia [1]) is largely overlooked.

Much of the paper will concern pixellated images, and the associated notation is as follows: the true scene is denoted by x, the observed image by y, and each can be written as a vector of length N, where N denotes the number of pixels. However, it is also natural to regard the elements of both x and y as being originally arrayed as $R \times C$ matrices, where R and C denote the numbers of rows and columns of pixels and $N = RC$.

The essence of the statistical approach is to introduce probabilistic models that might plausibly represent the relationship between x and y, and possibly also the structure of x itself, and to use 'standard' statistical paradigms to make inference about the unknown x. In Section 2, we concentrate on a particularly simple model for the way in which y is created as a distorted version of x, and a notional model for x appears almost incidentally. Section 3 reviews some now-classical material in which the model for x is proposed from the outset. Section 4 returns to a starting point similar to that of Section 2 and reviews some material under the heading of 'deconvolution'.

2. A Regularization Approach to Image Restoration

The simplest and most usual model that underlies the regularization approach is that, for some square matrix H,

$$y = Hx + \epsilon. \tag{1}$$

According to this model, the true scene is deterministically blurred by H and subjected to additive noise ϵ, often assumed to be white noise, so that $\epsilon \sim N(0, \sigma^2 I)$, in which σ^2 is a variance parameter and I denotes the identity matrix. The model makes sense only if the true scene and the observed image can be regarded as continuous intensities; this might be a reasonable approximation for grey-level images, but obviously not if, for example, the true scene is binary. If H, which is a characteristic of the observing instrument, is known, then the natural estimator of the true intensities is

$$\hat{x} := H^{-1}y = x + H^{-1}\epsilon,$$

which corresponds to the least-squares estimator of x, the minimiser of

$$\Delta(x, y) := ||y - Hx||^2.$$

Since $E\hat{x} = x$, $\hat{x}$ is what is called an unbiased estimator of the true scene. The matrix H might be fairly sparse, if the amount of blurring is small. However, the high dimensionality of the problem, in imaging contexts, can lead to $\hat{x}$ being a very unstable estimator of high variability, and a common solution is to minimise instead

$$\Delta(x, y) + \beta\Phi(x), \tag{2}$$

where Φ is a measure that is intended to penalise 'roughness' in $\hat{x}$, and β is usually a positive scalar, although a version involving a high-dimensional β is described by MacKay[2] under the nomenclature of automatic relevance determination. In a computationally-convenient simplest version of this method, $\Phi(x) = x^\top Cx$, in which C is typically positive-definite, and the 'optimum' restoration is given explicitly by

$$\hat{x}_\beta := (H^\top H + \beta C)^{-1} H^\top y. \quad (3)$$

Such an estimator is biased, in that $E\hat{x}_\beta \neq x$, for any $\beta \neq 0$, but its instability, as measured by variance, is greatly reduced. A key decision is to choose β appropriately, and a variety of rationales have been investigated. The choice of C reflects the type of smoothness imposed, and is usually designed to penalise differences between intensities on neighbouring pixels or simply, with $C = I$, to penalise 'large' x.

Some of the approaches to the choice of β seek to compromise between low bias and low variance, and a natural criterion for choice is a 'minimum risk' criterion of the form

$$\min_\beta E_{y|x}\delta(x, \hat{x}_\beta),$$

where δ is a measure of distance. If $\delta(x, x') = ||x - x'||^2$ then this amounts to a minimum total mean squared error criterion. The operational difficulty with this is that the criterion and therefore the minimising β are functions of the true x, which of course is unknown! One possibility is to substitute a preliminary estimate $\tilde{x}$ for x at this point, or to apply a method of crossvalidation (CV), which leads to the use of a criterion that measures the ability to predict individual (pixel) observations, given the data on all other pixels. For example, one might define

$$CV(\beta) := N^{-1} \sum_{i=1}^{N} \{y_i - E(y_i|\hat{x}_\beta^{(i)})\}^2,$$

in which y_i is the ith element of y and $\hat{x}_\beta^{(i)}$ is the restoration computed from all observations except for y_i, and choose β to minimise $CV(\beta)$. A slight modification of this which has been very popular in practice is the generalised crossvalidation function of Golub et al.[3] When δ is a simple quadratic loss function this crossvalidation function is given by

$$GCV(\beta) := RSS(\beta)/[\text{tr}\{I - K(\beta)\}^2],$$

in which $K(\beta) = H(H^\top H + \beta C)^{-1} H^\top$, $K(\beta)y = E(y|\hat{x}_\beta)$ denotes the set of 'fitted values', and $RSS(\beta) := ||\{I - K(\beta)\}y||^2$ is the residual sum of squares. Generalised crossvalidatory choice selects $\beta = \hat{\beta}_{GCV}$ to minimise $GCV(\beta)$. Other methods exist, including the so-called 'empirical degrees of freedom' choice, $\hat{\beta}_{EDF}$, defined as the solution of the equation

$$RSS(\beta)/\text{tr}\{I - K(\beta)\} = \sigma^2,$$

that is,

$$RSS(\beta) = \{n - \text{tr}K(\beta)\}\sigma^2,$$

provided σ^2 is known or can be estimated reliably externally. The quantity $n - \text{tr}K(\beta)$ is called the equivalent degrees of freedom for error, by analogy with the corresponding version in ordinary linear models. One justification for this is that Wahba[4] recommends $RSS(\hat{\beta}_{GCV})/\text{tr}\{I - K(\hat{\beta}_{GCV})\}$ as an estimator of σ^2.

So far as the non-statistical regularisation literature is concerned, a traditional way of choosing β is to solve

$$RSS(\beta) = n\sigma^2,$$

justified on the grounds that the *true* $RSS(\beta)$ has expectation $n\sigma^2$. This method will clearly oversmooth relative to $\hat{\beta}_{EDF}$.

In its most basic form the above methodology corresponds to ridge-regression, in general it amounts to an approach to solving potentially ill-posed inverse problems, and the structure appears in the development of smoothing splines as well as in the image-analysis context. My own involvement has included investigation of the various form of choosing β in the contexts of smoothing splines (Hall and Titterington[5]) and images (Hall and Titterington[6]; Thompson et al.[7]).

3. Bayesian Image Analysis

The foundations of what became known as Bayesian image analysis are the seminal papers by Geman and Geman[8] and Besag[9]. In these papers, a 'prior' (marginal) model, $p(x|\beta)$, was assumed for the true scene, x, and a model was also assumed for the observed image, y, conditional on x, to represent the noise and/or blurring process; this model will be denoted by $p(y|x, \theta)$. Thus, for example, for the model defined in (1) $p(y|x, \theta)$ is the multivariate Gaussian density with mean vector Hx and covariance matrix $\sigma^2 I$, and $\theta = (H, \sigma^2)$. (Note that we are using 'p' generically to denote a probability density function.)

The marginal model chosen for x typically reflects local spatial correlation, and usually corresponds to a Markov random field. The simplest scenario is that of a binary (black/white) image, with each $x_i = -1$ or $+1$. In this case a possible choice is to take $p(x|\beta)$ to correspond to the Ising model:

$$p(x|\beta) = \{C(\beta)\}^{-1}\exp(-\beta \sum_{i\sim j} x_i x_j),$$

in which the sum is over neighbouring pairs of pixels and $\beta > 0$. This corresponds to a so-called first-order Markov random field, and the observed image to a hidden Markov random field. The quantity $C(\beta)$ is a normalising constant for $p(x|\beta)$, also called a partition function. Its calculation is at best a complicated computational problem and leads to difficulties in inference, as mentioned later.

The quantities β and θ represent parameters and, for brevity, we shall denote the complete set of parameters by $\psi = (\beta, \theta)$. (In practice part or all of θ may be known from the specification of the observing instrument.)

There are two other probability distributions of interest, which we shall denote by $p(x|y, \psi)$ and $p(y|\psi)$. Both of these can be expressed in terms of the joint probability function for x and y, which is given by the product of $p(y|x, \theta)$ and $p(x|\beta)$:

$$p(x|y, \psi) \propto p(x, y|\psi) = p(y|x, \theta)p(x|\beta) \qquad (4)$$

and

$$p(y|\psi) = \int p(y|x, \theta)p(x|\beta)dx, \qquad (5)$$

where, in (5), the integration is over x and represents a summation if x is discrete.

Relationship (4) is the source of 'Bayesian' inference about the underlying (hidden) true scene, whereas (5) is the likelihood function corresponding to the observed data, and is important in making inferences about the underlying, and usually unknown, parameters ψ. I have put 'Bayesian' in inverted commas in the previous sentence because it is arguably in conflict with what Bayesian inference means in statistical science. The Bayesian paradigm is characterised by the assignment of probability distributions to *parameters*, which are fixed but unknown, as well as to random variables, realised values of which represent a major component of the experimental data. Pre-experiment ideas about the parameters are summarised by the 'prior' distributions and Bayes' Theorem is used to combine the prior information with that provided by the experimental data to give the 'posterior' distribution of the parameters. In what has become known as 'Bayesian image analysis', the key unknowns are not really parameters but are the true scene, x, which are perhaps better referred to as hidden or missing values. Bayes' Theorem is used to construct the conditional distribution $p(x|y)$ from the reverse conditional distribution $p(y|x)$, corresponding to the distortion/noise model, together with the marginal model $p(x)$ for x. Of course, Bayesian inference, as statisticians know it, is one way of dealing with any unknown parameters within ψ.

If for the time being the parameters ψ are assumed known, then, ideally, one should make inferences about the true scene on the basis of $p(x|y, \psi)$. Early work concentrated on obtaining point estimates, such as the mode, using simulated annealing techniques (Geman and Geman[8]), or mode-like quantities (Besag[9]), but in principle the whole joint posterior distribution of x is available for exploitation. Just what is feasible in practice depends to some extent on what is meant by x. In low-level, pixel-based modelling, which was the case considered by Geman and Geman[8] and Besag[9] and which we are dealing with in this paper, x contained values, such as colours or intensities, associated with all individual pixels, possibly supplemented by inter-pixel edge indicators. Thus x is of extremely high dimension and it is not feasible to look at complicated features of $p(x|y, \psi)$. Modelling at a higher level is typified by the deformable-templates approach, originally conceived of by Grenander (Grenander *et al.*[10]; Grenander and Miller[11]), in which features in images are represented by skeletal frameworks summarised by a comparatively small (at least relative to the number of pixels!) number of quantities. One would also like to obtain interval estimates concerning important features of the true scene. Typically, $p(x|y, \psi)$ is not of a form that is amenable to exact analysis, but, in principle but still a daunting prospect in practice, Markov chain Monte Carlo methods allow realisations to be simulated from the distribution to be generated and quantities of interest to be estimated by empirical counterparts.

There are clear links between this formulation and Physics. As mentioned earlier, it is natural for the 'prior' $p(x|\beta)$ to reflect local association and, for a pixellated binary scene, the Ising model from statistical physics is often used as the prior; if the underlying scene is defined in terms of a known finite number of colours or land-types then a Potts model might be used. Furthermore many of the Markov chain Monte Carlo methods have their origins in physics; for example, what statisticians know as the Gibbs sampler is the same as the heat-bath method.

If the parameters ψ are unknown, then they have to be estimated from the available data, namely y. Various ad hoc methods have been used in the image-analysis context, but the statistician would prefer to implement a general paradigm, either likelihood-based or Bayesian.

In the likelihood approach, the appropriate estimator of ψ is the maximiser of $p(y|\psi)$, and the interpretation of the problem as a missing-data problem, with the true scene x being missing, makes available the general iterative EM algorithm of Dempster *et al.*[12] Let $L(x, y|\psi)$ denote the complete-data log-likelihood, given by

$$L(x, y|\psi) = \log\{p(y|x, \theta)p(x|\beta)\}.$$

Then the EM algorithm is as follows, if we envisage an iteration at stage m, with $\psi^{(m-1)}$ as the current approximation to the maximum likelihood estimate.

(1) **E-step**: calculate $Q(\psi) = E_m L(x, y|\psi)$, where the expectation is with respect to the conditional distribution represented by $p(x|y, \psi^{(m-1)})$.
(2) **M-step**: find $\psi = \psi^{(m)}$ to maximise $Q(\psi)$.

In the E-step, therefore, we evaluate the expectation of the complete-data log-likelihood, conditional on the observed data and using the model based on the current estimates of the parameters to do the averaging, and then in the M-step we maximise that expected log-likelihood in order to obtain the next set of estimates. One hopes that the sequence $\{\psi^{(m)}\}$ converges to the maximiser of $p(y|\psi)$; it is generally true that the sequence of likelihoods $\{p(y|\psi^{(m)})\}$ is monotonically increasing. As a result, convergence to at least a local maximum is ensured, except in very pathological circumstances.

For the EM algorithm to be easy, both the E-step and the M-step have to be straightforward, and unfortunately in the case of a hidden Markov random field this is true of neither step. It is not possible to obtain an explicit formula for the expectation in the E-step. One approximating alternative is to use a sample average, based on a number of realisations from the relevant distribution, but generation of each of these realisations requires a Markov chain Monte Carlo procedure. Another approach is to use an approximating measure based on so-called mean-field approximations. Here, the averaging measure is a suitably chosen fully-factorised independence model for the individual elements in x. (The mean-field approximation is another tool with its origins in Physics.) Although the use of an independence model might seem to represent a gross approximation to a typically highly complex multivariate distribution, its performance within the E-step of the EM Algorithm can be uncannily effective; see, for instance, Zhang[13,14].

Difficulties also arise in the M-step, although, maximisation with respect to θ, the parameters within the noise model, is often easy. However, this is not the case for β, the prior parameters, because $p(x|\beta)$ of the normally intractable β-dependent normalisation constant or partition function present in $p(x|\beta)$. Zhang suggests using mean-field approximations at this stage too. Other possibilities are to approximate the normalisation constant by an empirical average, as explained by Geyer and Thompson[15], or to replace $p(x|\beta)$ by Besag's[16] pseudo-likelihood, which is defined as

$$p_{PL}(x|\beta) = \prod_i p(x_i|x_{\partial i}, \beta),$$

where $x_{\partial i}$ denotes values associated with the *neighbouring* pixels to pixel i, according to the neighbourhood system defined by the Gibbs distribution $p(x|\beta)$. Thus, p_{PL} is defined by the product of the full conditional distributions of the individual x_i's, and the problem of the intractable partition function disappears. Maximum pseudo-likelihood estimators are often consistent, in that for large lattices the estimator is likely to be close to the true β, but may have rather low efficiencies. One application of the pseudo-likelihood is to use it in the M-step of the EM-algorithm as a replacement for the correct but intractable $p(x|\beta)$; see Qian and Titterington[17] for this and other ways of making the EM-algorithm practicable.

At this point we mention a few practical applications.

Qian and Titterington[17] considered four-band satellite image data of a view of the Lake of Menteith in Perthshire, the only substantial body of water in Scotland referred to as a 'Lake' rather than a 'Loch'! A six-state Potts model was assumed for the true scene and additive Gaussian noise was assumed. An ad hoc initial six-state segmentation was constructed, based on the band-3 data alone, this led to a more refined restoration, again based only on the band-3 data, and then to a number of restorations based on all the data, obtained using various versions of the above EM-type methodology.

Qian and Titterington[18] analysed magnetic induction data corresponding to cobalt-nickel evaporated tape, a high-density magnetic recording material. The image included a transition boundary and it is important to identify the boundary as precisely as possible. Altogether three 'restorations' were created, based on different ways of modelling the surfaces on either side of the boundary, with the estimated boundary identified.

Data from a transmission electron microscopy image were also examined in Qian *et al.*[19] The image depicted a magnetic domain, the ideal shape of which would be that of a tilted circle, that is, an ellipse. In the paper a number of models were proposed, and restorations were obtained. A key feature of the 'prior' model reflected the notion that there was local radial association in the true scene, bearing in mind the knowledge that the image was indeed that of a noisy ellipse.

Mean-field-like approximations have also been used in a somewhat different approach to the maximisation of a complicated likelihood such as $p(y|\psi)$, exploiting the fact that

$$\log p(y|\psi) = \log\{\sum_x p(x, y|\psi)\} \tag{6}$$

$$\geq \sum_x q(x) \log\{p(x, y|\psi)/q(x)\}, \tag{7}$$

by Jensen's inequality, where $q(x)$ is any probability distribution for x. In practice q is chosen to have a form that facilitates computation, with a fully-factorised independence model being the simplest option, and 'hyperparameters' within that form are chosen so as to maximise the lower bound to the log-likelihood given in (7). For details of this approach see Jordan *et al.*[20] Note that, so far as choice of q or its hyperparameters is concerned, maximisation of the lower bound is equivalent to minimisation of the Kullback-Leibler directed divergence between q and the 'target', $p(x|y, \psi)$, defined by

$$KL(q,p) := \sum_x q(x) \log\{q(x)/p(x|y, \psi)\}.$$

For a fully Bayesian analysis, (hyper)priors must be imposed on $\psi = (\theta, \beta)$, and inference about θ, β *and* x should be made on the basis of $p(x, \theta, \beta|y)$, and the associated marginal distributions. Needless to say, in most image-analysis contexts, and certainly in the familiar pixel-based models, there is no practically useful closed form for $p(x, \theta, \beta|y)$:

$$p(x, \theta, \beta|y) \propto p(y|x, \theta)p(x|\beta)p(\theta)p(\beta),$$

where we are assuming that θ and β are independent, a priori, with prior densities $p(\theta)$ and $p(\beta)$.

What has become the standard statistical approach is to use Markov chain Monte Carlo methods to generate a set of realisations from the above joint distribution and to make inferences about the unknown quantities, both parameters (ψ) and missing values (x), on the basis of empirical summaries of the simulated quantities. However, in the case of hidden Markov random fields, the intractable partition function within $p(x|\beta)$ once more causes problems, in that the first step in the simulation cycle is not straightforward. As a result, approximate methods have been tried. One such approach, mentioned by Heikkinen and Högmander[21] and investigated in some detail by Rydén and Titterington[22], is to replace $p(x|\beta)$ by the pseudo-likelihood function when generating the next value of β. Rydén and Titterington[22] comment that the 'Gibbs' sampling scheme that results does converge, but that it is not clear how to characterise the limiting distribution. Rydén and Titterington also report some simulation experiments involving realisations from the Ising model, corrupted by Gaussian noise. The parameters of the noise model are estimated quite well by the resulting marginal means of the simulated sample from the posterior distribution, but there can be small but perceptible biases in the corresponding estimates of the Ising parameter, β. On the other hand, their attempts at alternative ways of dealing with the partition function, in the spirit of Geyer and Thompson[15], were distinctly unsuccessful because of computational difficulties.

As in the likelihood approach, there is a technique involving deterministic variational approximations for use in the fully-Bayesian context. In this case an approximation $q(x, \theta, \beta)$ to $p(x, \theta, \beta|y)$ is sought to minimise $KL(q, p)$ subject to q having some special structure that simplifies the analysis. Generally, q is taken to have the factorised form $q(x, \psi) = q_x(x)q_\psi(\psi)$. In many specific implementations $q_\psi(\psi)$ then takes the same parametric form as is obtained in the case in which the true x is given. Since the correct, if inaccessible, $p(\psi|y)$ certainly does not take the same form, these variational approximations inevitably lead to error, but in some cases it is at least possible to show that the modes of the correct and approximate distributions are asymptotically the same; see for example Wang and Titterington[23]. For more review and references on variational Bayesian approximations see Jordan[24] and Titterington[25].

Before leaving this section about the Bayesian approach, it is appropriate to return to the models discussed in Section 2 and to note that the regularised estimator has an obvious Bayesian interpretation for grey-level images. If the noise model is given by (1), with $\epsilon \sim N(0, \sigma^2 I)$, and if the prior/marginal distribution for x is that of $N(0, \sigma^2\beta^{-1}C^{-1})$, then the negative of the logarithm of $p(x|y)$ is, apart from additive and multiplicative constants, given by

$$||y - Hx||^2 + \beta x^\top Cx,$$

so that the mode is given by $\hat{x}_\beta$ as defined in (3). This interpretation then stimulates other ways of choosing the regularisation parameter β, such as maximum likelihood, in which β is chosen to maximise

$$p(y|\beta) = \int p(y|x)p(x|\beta)dx,$$

under the assumption that H and σ^2 are known from the specification of the observing instrument. The integration can be done explicitly and the resulting $p(y|\beta)$ can be maximised numerically.

4. Deconvolution

If the noise vector is omitted from equation (1) then we are left with the problem of solving the inverse problem

$$y = Hx, \tag{8}$$

which can be thought of as a discrete deconvolution problem. With pixellated images the discreteness is achieved automatically, but it might be imposed as a way of dealing with more general scenarios governed by the integral equation

$$y(t) = \int h(s, t)x(s)ds,$$

for t and s ranging over specified domains. This corresponds to deconvolution, especially if $h(s, t)$ is a function of $s - t$. For simplicity we shall concentrate on the discrete form of the problem, although ways of dealing with the integral-equation version are covered in many of the referenced papers. If H is square and nonsingular, then the formal solution is $x = H^{-1}y$, but this may be impracticable if the original problem is ill-posed, as discussed already. Furthermore, x is likely to have to satisfy nonnegativity constraints, a fact we have not yet recognised in this paper, and typically y and H will also consist of nonnegative elements.

This type of problem is of course very well researched, and here we concentrate on just a few approaches from the statistical literature. A key source is the discussion paper of Vardi and Lee[26]. They note that, by a scaling argument, without loss of generality it can be assumed that y and x sum to 1, as do the columns of H. They derive the following iterative algorithm for obtaining a nonnegative solution for (8), starting from a positive-valued $x^{(0)}$ that satisfies the unit-sum constraint: for $m = 1, \ldots,$ and for each ith element of x, obtain

$$x_i^{(m)} = x_i^{(m-1)} \sum_j (h_{ij} / \sum_k x_k^{(m-1)} h_{kj}) y_j. \tag{9}$$

Clearly, for all m, the elements of $x^{(m)}$ are nonnegative and sum to 1. Then the algorithm converges to the probability measure x^* that maximises $\sum_i y_i \log z_i$, where $z_i = (\sum_j h_{ij} x_j)$. This is equivalent to minimising

$$\sum_i y_i \log(y_i / z_i) = KL(y, z).$$

When equation (8) has a nonnegative solution then the algorithm converges to a solution of that nature. Otherwise, it converges to the closest approximation in the above KL sense.

To statisticians, the algorithm has the appealing interpretation as a limiting version of the EM algorithm. In the context of this example, the E-step and the M-step are as follows.

- **E-step**: for each i and j calculate

$$z_{ij}^{(m-1)} = \frac{x_i^{(m-1)} h_{ij}}{\sum_k x_k^{(m-1)} h_{kj}} y_j.$$

- **M-step**: for each i, calculate

$$x_i^{(m)} = \sum_j z_{ij}^{(m-1)}.$$

The combination of these two formulae clearly amounts to equation (9). Informally, the E-step 'distributes' each y_j over the individual pixel sites and the M-step accumulates all the contributions corresponding to pixel i. That this discrete form of the algorithm was an EM algorithm was noted by Titterington and Rossi[27], stimulated by the algorithm's appearance as an ad hoc procedure in Di Gesu and Maccarone[28].

Vardi and Lee[26] list a number of disparate manifestations of the general structure, including emission tomography image reconstruction, in which x denotes pixelwise emission intensities, elements of y are event counts at a set of detectors, emissions are assumed to follow Poisson distributions and h_{ij} is the probability that a particle emitted from pixel j is picked up by detector i. The case of Poisson emissions is of course familiar in Particle Physics, although the 'image' there is typically one-dimensional, not two-dimensional, and the number of 'pixels', or rather bins, is 100 or fewer; nevertheless, the same principles are relevant. Another image-based special case is that of motion deblurring. Given the EM interpretation of the algorithm and the knowledge that unregularised maximum likelihood estimates might be ill-conditioned, it is not surprising that modified versions have been developed that involve some sort of smoothing, especially in the context of emission tomography. Such modifications include the smoothed EM algorithm of Silverman *et al.*[29], in which the $\{x_i^{(m)}\}$ obtained in the M-step are locally smoothed before being fed into the next E-step, and the modified EM algorithm of Green[30], in which a roughness penalty on the $\{x_i\}$ is included when the M-step is carried out. Hudson and Larkin[31] provide another variation of EM, applied to tomography. The papers by Green[30] and Hudson and Larkin[31] both won IEEE awards for their high levels of citation.

The algorithm in equation (9) is essentially the same as the iterative scheme described in Section 3 of the paper by D'Agostini[32]. In that paper the algorithm is not iterated till convergence, but stops early on the basis of a goodness-of-fit criterion which is somewhat arbitrary but does have a potentially helpful regularising effect.

A different algorithm for the same purpose is the so-called Iterative Image Space Restoration Algorithm, for which the iteration is

$$x_i^{(m)} = x_i^{(m-1)} (\sum_j h_{ij} y_j) / \{\sum_j h_{ij} (\sum_k x_k^{(m-1)} h_{kj})\},$$

for each i. This algorithm was introduced by Daube-Witherspoon and Muehllehner[33] and convergence properties were investigated by De Pierro[34] and Titterington[35], the latter of whom noted that the algorithm could be interpreted as an iterative approach to the calculation of least squares estimates of x. Further references and illustrations in the context of motion-blur, together with extensions to incorporate roughness penalties, thereby obtaining minimisers of (2), are available in Archer and Titterington[36]. Key references from the non-statistical literature include Byrne[37] and Eggermont[38].

Vardi and Lee[26] present a number of illustrations, one of which concerns a motion-blurred moving toy cart. Part of the image was also treated by Archer and Titterington[36]. They implemented both the EM and ISRA algorithms, running each of them for totals of 40 and 106 iterations. The restoration obtained after 40 iterations was arguably better defined, which suggests that the underlying inverse problem is somewhat ill-posed; stopping the algorithm early is one way of avoiding an ill-posed solution.

Although much of the research discussed in this section is somewhat dated, statistical research into deconvolution, with applications relevant to this Conference, is certainly continuing. For example, Hall and Yin[39] consider a model, for a signal y observed at n time-points, given by

$$y_i = g(t_i) + \epsilon_i = \mu + \sum_{j=1}^{r} g_j(t_i) + \epsilon_i,$$

for $i = 1, \ldots, n$, where the g_j are periodic components with minimal periods $0 < \theta_1 < \ldots < \theta_r$. The objective is to estimate the unknown periods $\theta = \{\theta_j\}$ and the unknown functions $\{g_j\}$, without imposing simple parametric forms on the latter. To estimate the $\{\theta_j\}$, Hall and Yin use the minimiser $\hat{\theta}$

of the residual sum of squares function

$$S(\theta) = \sum_i \{y_i - \hat{g}(t_i|\theta)\}^2,$$

in which $\hat{g}(t|\theta)$ is a (preliminary) nonparametric estimator of $g(t)$. In particular, Hall and Yin use

$$\hat{g}(t|\theta) = \{\sum_i y_i K(t, t_i)\} / \sum_i K(t, t_i),$$

in which $K(t, t')$ is a kernel function, defined as a function of θ: the kernel function is defined in such a way that $\hat{g}(t|\theta)$ is a weighted average of the $\{y_i\}$, with weights that are monotonic decreasing functions of $|t - t_i|$. Given the $\{\hat{\theta}_j\}$, the $\{g_j\}$ are estimated using orthogonal series methods. The overall response function is written as

$$g(t) = \mu + \sum_{j=1}^{r} \sum_{k=1}^{m} a_{jk} \psi_k(t/\hat{\theta}_j),$$

in which the $\{a_{jk}\}$ are generalised Fourier coefficients, μ is a constant, the $\{\psi_k\}$ are orthonormal functions and m is a truncation point. Least squares is again used, this time to estimate μ and the $\{a_{jk}\}$. Hall and Yin[39] discuss ways of choosing the $\{\hat{\theta}_j\}$, m and smoothing parameters within the kernel function, they investigate theoretical properties, and they fit the model to radiation measurements from the slowly-pulsating B-star HD 123515, showing that a multiperiodic function with $r = 4$ periods gives a good fit.

Acknowledgments

The author is grateful to Dr. Glen Cowan for his helpful comments.

References

1. I.L. Dryden and K.V. Mardia, *Statistical Shape Analysis.* (Wiley, 1998).
2. D.J.C. MacKay, *Network: Computation in Neural Systems* **6**, 469 (1995).
3. G.H. Golub *et al.*, *Technometrics* **21**, 215 (1979).
4. G. Wahba, *J. R. Statist. Soc.* B **45**, 133 (1983).
5. P. Hall and D.M. Titterington, *J. R. Statist. Soc.* B **49**, 184 (1987).
6. P. Hall and D.M. Titterington, *J. R. Statist. Soc.* B **48**, 330 (1986).
7. A.M. Thompson *et al.*, *IEEE Trans. Pattern Anal. Machine Intell.* **13**, 326 (1991).
8. S. Geman and D. Geman, D. *IEEE Trans. Pattern Anal. Machine Intell.* **6**, 721 (1984).
9. J. Besag, *J. R. Statist. Soc.* B **48**, 259 (1986).
10. U. Grenander *et al.*, *HANDS: a Pattern Theoretic Study of Biological Shapes* (Springer, 1990).
11. U. Grenander and M.I. Miller, *J. R. Statist. Soc.* B **56**, 549 (1994).
12. A.P. Dempster *et al.*, *J. R. Statist. Soc.* B **39**, 1 (1977).
13. J. Zhang, J., *IEEE Trans. Signal Proces.* **40**, 2570 (1992).
14. J. Zhang, J., *IEEE Trans. Image Proces.* **2**, 27 (1993).
15. C.J. Geyer and E.A. Thompson, *J. R. Statist. Soc.* B **54**, 657 (1992).
16. J. Besag, *Statistician* **24**, 179 (1975).
17. W. Qian, and D.M. Titterington, *Phil. Trans. R. Soc. Lond.* A **337**, 407 (1991).
18. W. Qian and D.M. Titterington, *IEEE Trans. Pattern Anal. Machine Intell.* **15**, 748 (1993).
19. W. Qian *et al.*, *J. Am. Statist. Assoc.* **91**, 944 (1996).
20. M.I. Jordan *et al.*, in *Learning in Graphical Models* ed. M.I. Jordan, 105 (MIT Press, Cambridge, MA, 1999).
21. J. Heikkinen and H. Högmander, *Appl. Statist.* **43**, 569 (1994).
22. T. Rydén and D.M. Titterington, *J. Comp. Graph. Statist.* **7**, 194 (1998).
23. B. Wang and D.M. Titterington, *Bayesian Anal.*, to appear (2006).
24. M.I. Jordan, *Statist. Sci.* **19**, 140 (2004).
25. D.M. Titterington, *Statist. Sci.* **19**, 128 (2004).
26. Y. Vardi and D. Lee, *J. R. Statist. Soc.* B **55**, (1993).
27. D.M. Titterington and C. Rossi, *Signal Proces.* **9**, 101 (1985).
28. V. Di Gesu and M.C. Maccarone, *Signal Proces.* **6**, 201 (1984).
29. B.W. Silverman *et al.*, *J. R. Statist. Soc.* B **52**, 271 (1990).
30. P.J. Green, *IEEE Trans. Med. Imaging* **9**, 84 (1990).
31. H.M. Hudson and R.S. Larkin, *IEEE Trans. Med. Imaging* **13**, 601 (1994).
32. G. D'Agostini, *Nucl. Instr. Meth. Phys. Res.* A **362**, 487 (1995).
33. M.E. Daube-Witherspoon and G. Muehllehner, G. *IEEE Trans. Med. Imaging* **5**, 61 (1986).
34. A.R. De Pierro, *IEEE Trans. Med. Imaging* **6**, 174 (1987).
35. D.M. Titterington, *IEEE Trans. Med. Imaging* **6**, 52 (1987).
36. G.E.B. Archer and D.M. Titterington, *Statist. Sinica* **5**, 77 (1995).
37. C.L. Byrne, *IEEE Trans. Image Proces.* **2**, 96 (1993).
38. P.P.B. Eggermont, *Lin. Algeb. Applics.* **130**, 25 (1990).
39. P. Hall and J. Yin, *J. R. Statist. Soc.* B **65**, 869 (2003).

ITERATIVE INVERSION METHODS FOR STATISTICAL INVERSE PROBLEMS

NICOLAI BISSANTZ
Institute for Mathematical Stochastics, University of Göttingen, Maschmühlenweg 8-10, 37083 Göttingen, Germany
E-mail: bissantz@math.uni-goettingen.de

In this paper we discuss general regularization estimators. This class includes Tikhonov type and spectral cut-off estimators as well as iterative methods, such as ν-methods and the Landweber iteration. The latter estimators achieve the same (optimal) convergence rates as spectral cut-off, but do not require explicit spectral information on the operator and are often much faster to compute than Tikhonov regularization. We demonstrate application of a ν-method by an example involving the backwards heat equation.

1. Introduction

In this paper we are concerned with Inverse Problems. Here we aim to estimate some quantity of interest, which cannot be observed directly. In more detail, suppose we want to estimate a quantity described by an element f in a Hilbert space $\mathbb{H}_1$ from indirect noisy measurements

$$Y = (Kf)(X) + \sigma \cdot \xi, \tag{1}$$

where K is a known operator $K : \mathbb{H}_1 \to \mathbb{H}_2$ mapping $\mathbb{H}_1$ to another Hilbert space $\mathbb{H}_2$. The observations Y and $\mathbb{H}_2$ are Hilbert-space-valued processes described below in more detail, and σ is the variance of the noise. We assume that K is linear, bounded and injective, but not necessarily compact.

Inverse problems are prevalent in science. Typical examples include parameter identification problems in partial differential equations, e.g. the backwards heat equation. Here K is the so-called "parameter-to-solution" operator, which simply means solving the partial differential equation for the parameter f. Another typical class of problems emerges if K is an integral operator, whence (1) may be an inverse regression or an inverse density estimation problem, e.g. estimation of the density of globular cluster luminosities in the Antennae galaxies from noisy observations[1, 2].

The organization of this paper is as follows. In section 2 we show how model (1) relates to inverse regression and inverse density (quasi-)deconvolution problems, and briefly discuss Tikhonov and spectral cut-off estimators for f, which are the most frequently used spectral regularization estimators in practical applications. Moreover, we introduce iterative spectral regularization methods, which are often computationally more feasible than the aforementioned. In section 3 we apply ν-methods to the backwards heat equation.

It is beyond the scope of this paper to discuss in detail the technical assumptions required for the results presented. Instead, we refer to Bissantz, Hohage, Munk & Ruymgaart[1].

2. Methodology

2.1. *The noise model*

In this section we discuss how model (1) is related to practically relevant statistical models. We assume that the noise ξ is a Hilbert-space valued process, which is centered and has variance 1. Important applications of model (1) are the following.

Error-in-variables, deconvolution: Suppose that the following observations are at our disposal

$$X_1, \cdots, X_n \sim X = F + W,$$

where F, W are stochastically independent, with densities $f, w \in L^2$ and w known. Our aim is to estimate f. In this case the density g of X is related to f by the convolution operator

$$g = Kf = w * f.$$

It will be shown below that estimation of f by spectral regularization methods can be achieved by estimating $q := K^*g = K^*Kf$ from the observations in the first step. In the density deconvolution or error-in-variables problem, an unbiased, $\sqrt{n}$-consistent estimator of q is

$$\hat{q}_n(\cdot) = \frac{1}{n} \sum_{i=1}^{n} w(X_j - \cdot), \tag{2}$$

and the noise process ξ is given by

$$K^*\xi = (\hat{q}_n - q)/\sigma,$$

where $\sigma = \left(\|g\|_{L^\infty} + \|g\|_{L^2}^2\right)^{1/2} / \sqrt{n}$.

Inverse regression, Fredholm equation: Next we consider the regression setting, where we want to estimate the input function f from n discrete, noisy i.i.d. observations

$$Y_i = Kf(X_i) + \epsilon_i, \; i = 1, \cdots, n,$$

where (X_i, ϵ_i) are stochastically independent design variables X_i and noise terms ε_i, and

$$\mathrm{E}[Y|X] = Kf(X)$$

for a linear integral operator K. Similarly as in the deconvolution case the generalized empirical process

$$\hat{q}_n(\cdot) = \frac{1}{n} \sum_{i=1}^{n} Y_i K(X_i, \cdot)$$

estimates $q = (K^*K)f$ in an unbiased and $\sqrt{n}$-consistent manner. Moreover, the noise process ξ can again be defined as

$$K^*\xi = (\hat{q}_n - q)/\sigma,$$

where $\sigma = \left(\mathrm{Var}\varepsilon_1 + \|Kf\|_{L^\infty}^2 + \|Kf\|_{L^2}^2\right)^{1/2} / \sqrt{n}$.

Quasi-deconvolution: Reconsider the deconvolution case, but now assume that the density g of X is given by

$$g = \int_{\mathbb{R}^d} h(\cdot - y|y) f(y) dy =: Kf,$$

where $h(\cdot|y)$ is the conditional density of W given $Y = y$. Note that K is a convolution operator if Y and W would be stochastically independent. However, in many practical applications the variance of the noise term W depends on F. For a typical example consider observations of the brightness of a globular cluster belonging to some remote galaxy. Here the measurement gets increasingly difficult with fainter cluster brightness, and the measurement noise increases. For quasi-deconvolution we replace the estimator (2) of $q := K^*g$ by

$$\hat{q}_n(z) := \frac{1}{n} \sum_{j=1}^{n} h(X_j - z|z).$$

2.2. *Inverse estimators*

We assume that the operator $K : \mathbb{H}_1 \to \mathbb{H}_2$ is bounded and injective. Therefore its generalized (Moore-Penrose) inverse

$$K^\dagger = (K^*K)^{-1}K^* : R(K) \oplus R(K)^\perp \to \mathbb{H}_1$$

is in general unbounded, and noise in the measurements Y is blown up by the inversion $\hat{f} := (K^*K)^{-1}\hat{q}_n$, which, in general, yields useless results $\hat{f}$. A possible solution to this problem consists in *regularization*, i.e. to replace $K^\dagger$ by a sequence of bounded operators R_α with *regularization parameter* α, such that $R_\alpha \to K^\dagger$ for $\alpha \searrow 0$ (pointwise).

How can we construct such regularization estimators for general inverse problems? The fundamental tool is *Halmos' spectral theorem*[5]: Let $A : \mathbb{H} \to \mathbb{H}$ be a bounded, self-adjoint operator defined on a separable Hilbert space $\mathbb{H}$. Then there exists a σ-compact space $\mathbb{S}$, a Borel measure Σ on $\mathbb{S}$, a unitary operator $U : \mathbb{H} \to L^2(\Sigma)$, and a measurable function $\rho : \mathbb{S} \to \mathbb{R}$ such that

$$UAf = \rho \cdot Uf, \quad \Sigma - \text{almost everywhere},$$

for all $f \in \mathbb{H}$. The spectral theorem justifies the *functional calculus*, which will be used to define general spectral regularization estimators. Let $\Phi : \sigma(A) \to \mathbb{R}$ a bounded function on the spectrum $\sigma(A)$ of A. Then

$$\Phi(A) = U^* M_{\Phi(\rho)} U,$$

where $M_{\Phi(\rho)}$ is the operator given by multiplication with $\Phi(\rho)$, and U^* the adjoint of U. For example, if K is the operator generated by convolution with some (known) density w on $\mathbb{R}$, K^* its adjoint and $A := K^*K$, then the unitary transform U which appears in the spectral theorem and in functional calculus are the Fourier transformation $\mathcal{F}$, and ρ is the Fourier transform of w.

We now define the regularized inverse of $K^\dagger$ as

$$\Phi_\alpha(K^*K)K^*, \tag{3}$$

where $\Phi_\alpha : \sigma(A) \to \mathbb{R}$, $\alpha > 0$ are bounded functions which satisfy

$$\lim_{\alpha \searrow 0} \Phi_\alpha(t) = \frac{1}{t}, \quad \text{for all } t \in \sigma(A),$$

in particular. The (generalized) spectral regularization estimator for f is given by

$$\hat{f}_{\alpha,n} := \Phi_\alpha(K^*K)\hat{q}_n.$$

For a detailed discussion on admissible collections of functions Φ_α, and the necessary regularity properties, we refer to Engl, Hanke & Neubauer[4] and Bissantz, Hohage, Munk & Ruymgaart[1].

In practical applications, Tikhonov regularization type methods and spectral cut-off are the most frequently used methods. Tikhonov regularization results if $(K^*K)^{-1}$ is replaced by $(K^*K + \alpha I)^{-1}$, and can therefore be computed without refering to the spectral information $\sigma(A), U$. However, it can also be defined as the regularization estimator (3) for $\Phi_\alpha^{\mathrm{Tik}}(t) := 1/(t+\alpha)$. In the case of spectral cut-off methods we have

$$\Phi_\alpha^{\mathrm{SC}} := \begin{cases} t^{-1}, & t \geq \alpha \\ 0, & t < \alpha \end{cases}$$

To provide a specific example, consider density deconvolution on $\mathbb{R}$ (cf. Section 2.1). Then the spectral cut-off estimator of f reads

$$\hat{f}_{\alpha,n} = \mathcal{F}^{-1} \int_{\rho(\omega)\geq\alpha} \frac{\mathcal{F}\hat{q}_n(\omega)}{\rho(\omega)} d\omega, \qquad (4)$$

where $\rho = |\mathcal{F}w|^2$. Note from eq. (4) that the regularization property of spectral cut-off is achieved by neglecting the high-frequency information in the observations $\hat{q}_n$. This is because $\rho(\omega) \searrow 0$ for $\omega \to \infty$, and division by ρ would blow up the measurement noise by an arbitrarily large amount for increasing frequency $|\omega|$ if no regularization is performed.

Both Tikhonov regularization and spectral cut-off methods require setting up a matrix representing the operator K, and moreover a matrix inversion or eigenvalue decomposition. This can be computationally very costly, e.g. in the case of parameter identification problems in partial differential equations. Another reason can be that estimates $\hat{f}_{\alpha,n}$ are computed for many different values of the regularization parameter α in the case of data-driven regularization parameter selection methods such as cross-validation.

On the other hand, iterative methods can be defined for suitable collections of functions Φ_α, which require for their computation only to apply the matrix representing the operator K and its transpose to a solution vector. This is an important advantage since for many problems there exist algorithms to apply the matrix to a given vector at a much smaller computational cost than the cost of setting up the matrix.

Important iterative spectral regularization methods are Landweber iterations and ν-methods. For these methods the regularization parameter α is given by the stopping index k of the iterations. The more iterations are performed, the less regularization is imposed on the solution.

For *Landweber iterations* we have $\Phi_{1/(k+1)}(t) := \sum_{j=0}^{k-1}(1-t)^j$, but the method can be implemented by the recursion formula

$$\hat{f}_{0,\sigma} = 0, \quad \hat{f}_{k+1,\sigma} = \hat{f}_{k,\sigma} - A\hat{f}_{k,\sigma} + K^*Y, \quad k = 0, 1, \ldots,$$

i.e. we do not require the spectral information U, ρ of $A = K^*K$. Here, the regularization parameter can be identified as $\alpha = 1/(k+1)$, and the norms on $\mathbb{H}_1$ and $\mathbb{H}_2$ have to be scaled such that $\|A\| \leq 1$.

Better numerical convergence than for Landweber iterations can be achieved by $\nu-$methods[3]. For a given parameter $\nu > 0$, the estimator $\hat{f}_{k,\sigma}$ can be computed by the three-term recursion

$$\begin{aligned} \hat{f}_{k,\sigma} &= \hat{f}_{k-1,\sigma} + \theta_k \left(\hat{f}_{k-1,\sigma} - \hat{f}_{k-2,\sigma} \right) \\ &\quad + \omega_k K^* \left(Y - K\hat{f}_{k-1,\sigma} \right), \quad k \geq 2, \end{aligned}$$

with starting values $\hat{f}_{0,\sigma} := 0$, $\hat{f}_{1,\sigma} = \omega_1 K^* Y$, coefficients $\theta_1 = 0$, $\omega_1 = (4\nu+2)/(4\nu+1)$ and

$$\theta_k = \frac{(k-1)(2k-3)(2k+2\nu-1)}{(k+2\nu-1)(2k+4\nu-1)(2k+2\nu-3)},$$

$$\omega_k = 4\frac{(2k+2\nu-1)(k+\nu-1)}{(k+2\nu-1)(2k+4\nu-1)},$$

for $k \geq 2$. Now the regularization parameter can be identified by $\alpha = (1+k)^{-2}$, which implies that the number of iterations required for ν-methods typically are of order square root the number of required Landweber iterations.

Bissantz, Hohage, Munk & Ruymgaart[1] analyzed the convergence of general spectral regularization methods of the form (3), and determined their rates of convergence, which depend on the smoothness properties of the input function f. It turns out that all methods defined in (3) converge with the same rates of convergence as spectral cut-off, which are in many cases optimal (cf. Mair & Ruymgaart[6]). However, this only holds true as long as the smoothness of f is within the *qualification* of the respective method. Spectral cut-off and Landweber iterations have infinite qualification, and $\nu-$methods are available for arbitrary qualification, but Tikhonov regularization has small qualification 1. This implies that in many cases Tikhonov methods cannot converge with optimal order. For details we refer to Engl, Hanke & Neubauer[4] and Bissantz, Hohage, Munk & Ruymgaart[1].

3. The backwards heat equation

Finally, we briefly discuss an application of ν-methods to the backwards heat equation. To this end consider the inverse problem of reconstructing the temperature distribution at time $t = 0$ on some compact domain $H \subset \mathbb{R}^2$ from discrete, noisy observations

$$Y_i = g(X_i) + \varepsilon_i$$

of the temperature distribution at time $t = T$, where the design points X_i form a regular mesh on H, and ε_i is a centered, i.i.d. noise term with standard deviation σ.

For the backwards heat equation, the forward "parameter-to-solution" problem is described by the partial differential equation of parabolic type

$$\begin{aligned} &\partial_t u(x,t) = \Delta u(x,t), && x \in H, t \in (0,T) \\ &u(x,t) = 0, && x \in \partial H, t \in (0,T] \\ &u(x,0) = f(x), && x \in H, \end{aligned} \qquad (5)$$

with an initial temperature distribution $f \in L^2(H)$ and the final temperature distribution $g(x) := u(x,T)$, $x \in H$.

We have implemented the backwards heat equation for a two-dimensional, approximately heart-shaped, smooth domain H and defined the operator K as the evolution of the heat equation from time $t = 0$ to $T = 0.001$. For the observations the sample size is $n = 200$ and $\sigma = 0.001$. Moreover, the Laplace operator on the domain H was discretized by a finite difference scheme using 16038 unknowns. The matrix representing the forward solution operator K is therefore a dense 16038×16038 matrix, which would require a huge amount of computation time to be set up. However, the application of the operator K to a vector f can be implemented efficiently by time stepping methods. We have used a BDF multistep method.

To apply a ν-method to this problem we first have to estimate $q = K^*g$. To this end we first estimate $g := Kf$ with a locally linear estimator $\hat{g}$ from the observations Y_i. In the second step we compute $\hat{q} := K^*\hat{g}$. This estimator of q is not unbiased because the local polynomial estimator $\hat{g}$ used in the first step is not either. However, in numerical simulations this approach turned out to be very stable. For a discussion of local polynomial estimators cf. Wand & Jones[7].

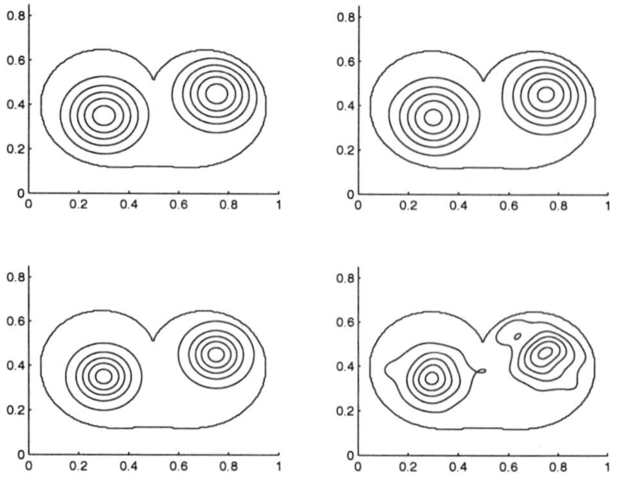

Fig. 1. A typical simulation of the backwards heat equation. Upper row (from left to right): True q and estimate $\hat{q}$ from $n = 200$ observations Y_i. Contour levels are $0.1, 0.2, 0.3, \ldots$ Lower row (from left to right): True f and estimate $\hat{f}$. Here the contour levels are $0.1, 0.3, 0.5, \ldots$ The outer, heart-shaped contour indicates the boundary of the domain under consideration.

Fig. 1 shows a typical example of a simulation, where a ν-method was used with $\nu = 1$ and 8 iterations, which amounts to approximately 5 minutes of CPU time on a Pentium IV 1.7 Ghz processor.

Acknowledgments

The author would like to thank A. Blümel and T. Hohage for help and interesting discussions.

References

1. N. Bissantz, T. Hohage, A. Munk and F. Ruymgaart, Convergence rates of general regularization methods for statistical inverse problems and applications, submitted.
2. P. Anders, N. Bissantz, L. Boysen, U. F. v. Alvensleben and R. de Grijs, The luminosity distribution of young massive clusters in the Antennae galaxies, in preparation.
3. P. R. Halmos, *Amer. Math. Monthly* **70**, 241 (1963).
4. H. W. Engl, M. Hanke and A. Neubauer, *Regularization of Inverse Problems* (Kluwer Academic Publisher, Dordrecht, Boston, London, 1996).
5. H. Brakhage, On ill-posed problems and the method of conjugate gradients, in *Inverse and Ill-Posed Problems*, eds. H. W. Engl and C. W. Groetsch (Academic Press, Orlando, 1987) pp. 191–205.
6. B. A. Mair and F. Ruymgaart, *SIAM J. Appl. Math.* **56**, 1424 (1996).
7. M. P. Wand and M. C. Jones, *Kernel Smoothing* (Chapman & Hall, London, 1995).

UNFOLDING WITH SYSTEM IDENTIFICATION

N.D. GAGUNASHVILI

University of Akureyri, Faculty of Information Technology, Borgir, v/Nordurslód, IS-600 Akureyri, Iceland
E-mail: nikolai@unak.is

A procedure for unfolding the true distribution from experimental data is presented. Methods of system identification are applied for the creation of a model of a transformation of the true distribution to the experimentally measured distribution. A priori information about the true distribution which is known from theory or previous experiments is used. The stability of the result of the unfolding is obtained by a sensible binning and by application of D-optimization. In this paper it is shown how to decrease the bias of the unfolded distribution by introducing the X^2 selection criteria for distributions used for system identification. Application of the D-optimization and the Least Squares Method allow us to minimize the statistical errors of the unfolded distribution. The unfolding procedure may be applied for detectors with a linear or nonlinear transformation of a true distribution into the experimentally measured one. The dimensionality of the solved problem can be arbitrary. The procedure can be applied for solving the unfolding problem with both smooth and non-smooth solutions. This method does not require a large amount of Monte-Carlo simulations of the experiment.

1. Introduction

An experimentally measured distribution differs from the true physical distribution due to limited acceptance and finite resolution of a set-up. To obtain a physical distribution an unfolding procedure is applied [1 2 3 4]. The unfolding problem is an under-specified problem. Any approach to solve the problem requires a priori information about the solution. Different methods of unfolding differ, directly or indirectly, through the use of this a priori information.

In [5] an approach to an unfolding problem related to methods of system identification is presented. To obtain a stable solution of an unfolding problem, information about the shape of the distribution to be measured is used for system identification. This paper further develops the ideas presented in [5]. D-optimization which is used in the theory of experimental design is applied to minimize the statistical errors of the unfolded distribution. The X^2 selection criterion is introduced for a set of distributions used for system identification; this criteria minimizes the bias of the solution.

2. Main equation

In this work we will use the linear model for a transformation of a true distribution to the measured one

$$\boldsymbol{f} = \mathsf{P}\boldsymbol{\phi} + \boldsymbol{\epsilon}, \tag{1}$$

where $\boldsymbol{f}$ is an m-component column vector of an experimentally measured histogram content, P is an $m \times n$ matrix, with $m \geq n$, $\boldsymbol{\phi}$ is an n-component vector of some true histogram content and $\boldsymbol{\epsilon}$ is an m-component vector of random residuals with expectation value $\mathrm{E}\,\boldsymbol{\epsilon} = \mathbf{0}$ and a diagonal variance matrix $\Sigma = \mathrm{Var}\,\boldsymbol{\epsilon} = \mathrm{diag}(\sigma_1^2, \cdots, \sigma_m^2)$, where σ_i is the statistical error of the measured distribution for the ith bin. The linear model (1) is reasonable for the majority of set-ups. It is only an approximate model for set-ups with non-linear transformation from a true distribution into the measured distribution.

A Least Squares Method can give an estimator for the true distribution $\boldsymbol{\phi}$,

$$\hat{\boldsymbol{\phi}} = (\mathsf{P}'\Sigma^{-1}\mathsf{P})^{-1}\mathsf{P}'\Sigma^{-1}\boldsymbol{f} \tag{2}$$

where $\hat{\boldsymbol{\phi}}$, the estimator, is the *unfolded distribution*, and the full matrix of errors of the unfolded distribution is given by

$$\mathrm{Var}\,\hat{\boldsymbol{\phi}} = (\mathsf{P}'\Sigma^{-1}\mathsf{P})^{-1}\,. \tag{3}$$

3. System identification and regularization

To realize the scheme described in the previous section, the matrix P must be defined. This problem can be solved using system identification methods [7]. System identification may be defined as the process of determining a model of a dynamic system using observed input-output data. In our case it is the model of transformation of a true physical distribution into the experimentally measured distribution, represented by the matrix P. The Monte-Carlo simulation of a set-up can be used to get input-output data. Control input signals are used for system identification. The most popular choice is to use impulse control signals.

An impulse input control signal is a generated distribution in which the histogram has only one bin with non-zero content. For the model (1) there are n different impulse inputs that can be presented as the diagonal matrix $\Phi^{c} = \mathrm{diag}(\phi^{c}_{11}, ..., \phi^{c}_{nn})$, where each row contains the content from a generated histogram. Let us denote corresponding values of ith component of the reconstructed vector (output) as $\boldsymbol{f^c_i} = (f^c_{i1} f^c_{i2} \cdots f^c_{im})'$. Each element of the ith row of the matrix

$$\mathsf{P} = \begin{pmatrix} p_{11} & p_{12} & \cdots & p_{1n} \\ \cdots & \cdots & \cdots & \cdots \\ p_{i1} & p_{i2} & \cdots & p_{in} \\ \cdots & \cdots & \cdots & \cdots \\ p_{m1} & p_{m2} & \cdots & p_{mn} \end{pmatrix},$$

can be found from the equation

$$\boldsymbol{f^c_i} = \Phi^{c} \boldsymbol{p_i}\,, \tag{4}$$

where $\boldsymbol{p_i} = (p_{i1}\, p_{i2} \cdots p_{in})'$, and $p_{ij} = f^c_{ij}/\phi_{jj}$. Equation (2), with the matrix P calculated this way, gives a highly fluctuating unfolded function with large statistical errors. Also, it is possible that the matrix $\mathsf{P}'\Sigma^{-1}\mathsf{P}$ is singular, in which case a solution does not exist.

To regularize the solution of the unfolding problem, let us use for system identification not an impulse control distribution, but rather a priori distributions that may be known from theory, or from some other experimental data.

Assume we have q control generated distributions, and now present them as a $q \times n$ matrix

$$\Phi^{c} = \begin{pmatrix} \phi^c_{11} & \phi^c_{12} & \cdots & \phi^c_{1n} \\ \phi^c_{21} & \phi^c_{22} & \cdots & \phi^c_{2n} \\ \cdots & \cdots & \cdots & \cdots \\ \phi^c_{q1} & \phi^c_{q2} & \cdots & \phi^c_{qn} \end{pmatrix},$$

where each row represents a generated histogram content. For each ith row of the matrix P we can write the equation

$$\boldsymbol{f^c_i} = \Phi^{c} \boldsymbol{p_i} + \boldsymbol{\xi_i}\,, \tag{5}$$

where $\boldsymbol{p_i} = (p_{i1}\, p_{i2} \cdots p_{in})'$, $\boldsymbol{f^c_i}$ is a q-component vector of reconstructed ith bin content for different generated control distributions, and $\boldsymbol{\xi_i}$ is a q-component vector of random residuals with expectation value $\mathrm{E}\,\boldsymbol{\xi_i} = \boldsymbol{0}$ and a diagonal variance matrix $\Delta_i = \mathrm{Var}\,\boldsymbol{\xi_i} = \mathrm{diag}(\delta^2_{i1}, \cdots, \delta^2_{iq})$, where δ_{ij} is the statistical error of the reconstructed distribution for the ith bin and the jth control generated distribution. A Least Squares Method gives an estimator for $\boldsymbol{p_i}, i = 1, ..., m$

$$\hat{\boldsymbol{p_i}} = (\Phi^{c\prime}\Delta_i^{-1}\Phi^{c})^{-1}\Phi^{c\prime}\Delta_i^{-1}\boldsymbol{f^c_i}\,. \tag{6}$$

Columns of the matrix Φ^{c} can correlate with each other. This means that transformation of the control generated distribution to the ith bin of reconstructed distribution can be parametrized by the subset of elements of the row $\boldsymbol{p_i}$. Elements of the row that do not belong to the subset are set to 0. Moreover, there can be more than one subset that describes this transformation in a sufficiently good manner. Thus for each ith reconstructed bin we will have the set of N_i candidate rows, and for all reconstructed bins the set of $N_1 \times N_2 \times \cdots \times N_m$ candidate matrices P. We need to choose a matrix P that is good, or optimal, in some sense. The most convenient criterion in our case is D-optimality [8] that is related to the minimization of

$$\det(\mathsf{P}'\Sigma^{-1}\mathsf{P})^{-1} = \det(\mathrm{Var}\,(\boldsymbol{\phi}))\,. \tag{7}$$

There are many algorithms and programs of minimization (7). The matrix P that minimizes function (7) gives us a stable solution of the unfolding problem (2) with minimal volume of the confidence ellipsoid. Further improvement of the quality of the solution can be achieved by introducing the selection criteria described below.

A control generated distribution has a corresponding reconstructed control distribution that can be compared with the experimentally measured distribution using a χ^2 test [6]. Let us take for identification a generated control distribution that has a corresponding reconstructed distribution satisfying a $X^2 < a$ selection criteria. The statistic X^2 is calculated to test the compatibility of the experimental distribution with the reconstructed control distribution [6]. The parameter a defines how close the set of reconstructed control distributions is to the experimental distribution. A decrease in parameter a represents a decrease in systematic and statistical errors of the solution.

4. The unfolding procedure

In this section a description of the complete unfolding procedure is presented. The procedure can be divided into four parts: initialization, system identification, solution of the basic equation, and test of goodness-of-fit.

Initialization

Define the binning for the experimental data. The strategy in selecting the size of the bins is to start with large bin sizes, then increase the number of bins incrementally and stop the process when the value of the determinant of the complete matrix of errors of unfolded distribution stops decreasing.

Define the binning for the unfolded distribution. The way to choose the bin size is to pick a reasonably large size of bin for the first step, then decrease the size of bins on further steps and stop this process before the correlation between adjusted bins becomes too big. The number of bins of an unfolded distribution, n, must be lower then the number of bins for the experimentally measured distribution, m, due to the fact that we use the Least Squares Method for the solution of the main equation.

System identification

Choose a set of control generated distributions. Control generated distributions for the set must be chosen with lowest possible values of the X^2 statistics. A second iteration can be made to find a better set of control distributions. The method of the reweighting of events can be used in this case. The number of generated distributions must be greater than the expected number of non-zero elements in any row of matrix P (for reasons related to the use of the Least Squares Method).

Calculate the set of candidates for the matrix P. A stepwise regression algorithm can be used for this calculation [9]. The first element in the stepwise algorithm can define a candidate row. To obtain as many variants as possible of each row, each element of the matrix is used as a first element in the stepwise algorithm.

Calculate the D-optimal matrix P. On the first step, matrix P is chosen randomly from the set of candidates. After this, optimization can be done by Fedorov's reliable EA algorithm [8]. In the majority of cases this algorithm finds a matrix that has a global minimum for $\det(\mathsf{P}'\Sigma^{-1}\mathsf{P})^{-1}$. The optimization procedure can be repeated with another randomly chosen matrix to be sure that the minimum is global.

Solution of the basic equation

Calculate the unfolded distribution Eq. (2) with the full matrix of errors Eq. (3). The correlation matrix calculated from the full matrix of errors can give hints for an improved binning of the unfolding distribution. For example, if the correlation between two adjacent bins is high, they should be combined.

Test of goodness-of-fit

Fit unfolded distribution, and then use the fit to generate a new 'experimental' distribution (including effects of resolution and acceptance), to compare with the real data. This is the only objective test of goodness-of-fit of the unfolding procedure and it should be done with an analysis of the studentised residuals [6].

5. A numerical example

The method described above is now illustrated with an example taken from [1]. We take a true distribution

$$\phi(x) = A_1 \frac{C_1^2}{(x-B_1)^2 + C_1^2} + A_2 \frac{C_2^2}{(x-B_2)^2 + C_2^2} \quad (8)$$

with parameters $A_1 = 2, A_2 = 1, B_1 = 10, B_2 = 14, C_1 = C_2 = 1$; x is defined on the interval $[4,16]$. An experimentally measured distribution is defined as

$$f(x) = \int_4^{16} \phi(x')A(x')R(x,x')dx' \quad (9)$$

where the acceptance function $A(x)$ is

$$A(x) = 1 - \frac{(x-10)^2}{36} \quad (10)$$

and

$$R(x,x') = \frac{1}{\sqrt{2\pi}\sigma}\exp(-(\frac{(x-x')^2}{2\sigma^2})) \quad (11)$$

is the detector resolution function with $\sigma = 1.5$. The true distribution, acceptance and the resolution functions are shown in Fig. 1a. A histogram of the measured distribution $\boldsymbol{f}$ was obtained by simulating 10^4 events with $m = 90$ bins, and is shown in Fig. 1b.

For the true distribution histogram we choose thirty equal size bins, i.e. $n = 30$. We use for the detector identification 100 distributions defined by formula (8) with parameters simulated according to uniform distributions on the intervals :
$[1,3]$ for A_1; $[8,12]$ for B_1; $[0.5,1.5]$ for C_1;
$[0.5,1.5]$ for A_2; $[10,18]$ for B_2; $[0.5,1.5]$ for C_2.

Each distribution is represented by a histogram with 10^4 events. The first example is calculated without X^2 selection cut. Fig. 2a shows 30 of the 100 control distributions used for identification, and Fig. 2b shows the unfolded distribution and the true distribution as a solid line.

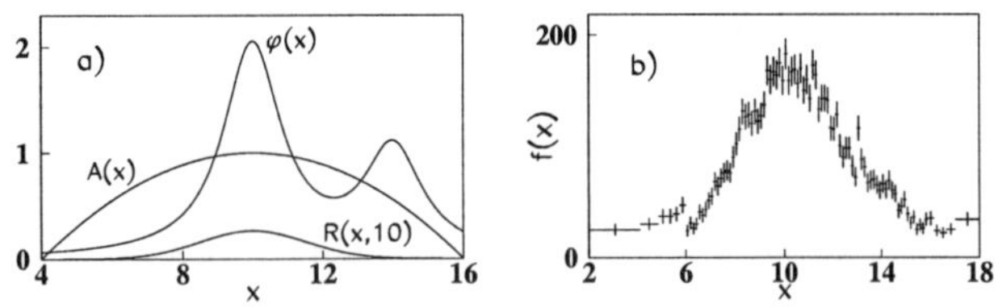

Fig. 1. An example of the true distribution $\phi(x)$, the acceptance function $A(x)$, the resolution function $R(x, 10)$ and of the measured distribution $\boldsymbol{f}$ (number of events per bin).

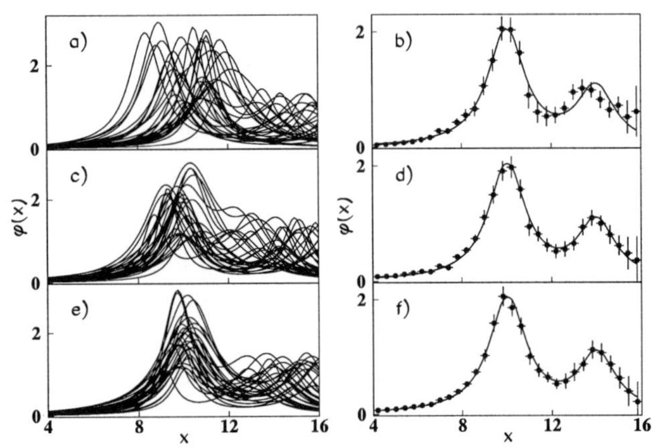

Fig. 2. The first 30 control distribution generated for system identification and an unfolded distribution a),b) without X^2 cut; c),d) with $X^2 < 200$ cut; e),f) with $X^2 < 124$ cut.

Two other examples are calculated with $X^2 < 200$ selection criteria (significance level $p = 2.5 \cdot 10^{-10}$) (see Figs. 2c,d) and with $X^2 < 124$ cut (significance level $p = 10^{-2}$) (see Figs. 2e,f). The D-optimal determinant in the first case is equal to $2117^{-1/30}$, in the second case equal to $4445^{-1/30}$ and in the third one $4719^{-1/30}$. The average number of non-zero elements in the rows of matrix P in the last two cases is 4 and for first case 5. Notice that for a lower values of the X^2 cut we have a lower values of the determinant of the full matrix of errors and a lower systematic deviation of the unfolded distribution from true one. The errors seem a bit large in comparison with fluctuations of the unfolded distribution. The reason for these errors is the positive correlation between adjacent bins.

6. Conclusion

The main idea of the method presented in this paper is the use of a set of a priori distributions for system identification, i.e. the construction of the transformation matrix. For this set of distributions, we introduce a X^2 selection criteria, which permits us to decrease the possible bias of the procedure. D-optimization and the application of the Least Squares Method gives the stable solution with minimal statistical errors. The method of identification provides a linear approximation of the transformation of the true distribution into the measured distribution in case this transformation is nonlinear. The method does not require a large amount of Monte-Carlo simulation of the experiment, because of the relatively low number of non-zero elements in the transformation matrix and because a re-weighting technique is used. The procedure has no restrictions due to the dimensionality of the problem. The method can be applied for solving the unfolding problem with a non-smooth solution. Based only on a statistical approach, the method has a good statistical interpretation.

Acknowledgments

The author would like to thank Dr. Mark O'Brien and Dr. J. Nystrom for a critical reading of this paper and making comments that improved the text.

References

1. V.P. Zhigunov, *Nucl. Instrum. Meth.* **216**, 183 (1983).
2. V. Blobel, *CERN 85-02*, 1985.
3. G. Zech, *DESY 95-113*, 1995.
4. A. Höcker, V.Kartvelishvili, *Nucl. Instrum. Meth.* A **372**, 469 (1996)
5. N. Gagunashvili, *Nucl. Instrum. Meth.* A **451**, 657 (1994).
6. N. Gagunashvili, χ^2 Test for the Comparison of Weighted and Unweighted Histograms, in these proceedings.
7. L. Ljung, *System Identification: Theory for User*, (Prentice Hall, 1999).
8. V.V. Fedorov, *Theory of Optimal Experiments*, (Academic Press, New York, 1972).
9. G.A.F. Seber and A.J. Lee, *Linear Regression Analysis*, (John Wiley & Sons Inc, 2003).

SUMMARIES

STATISTICS IN ASTROPHYSICS AND COSMOLOGY: PHYSTAT05

ANDREW H. JAFFE
Blackett Laboratory, Imperial College London
E-mail: a.jaffe@imperial.ac.uk

In this conference summary on Astrophysics talks at PhyStat05, I will discuss the various philosophical and pragmatic approaches to problems of statistics in astrophysics (and cosmology in particular), and their application to a few modern problems discussed at this meeting. In particular, I will develop a Bayesian formalism for the analysis of data from Cosmic Microwave Background (CMB) experiments.

In his PhyStat talk on the Transit of Venus, Johnston said "Precision astronomy depends on an individual's judgment" and that remains as true today as in the Enlightenment.

1. Philosophy

One remarkable difference between the members of the astrophysics and particle physics communities present at this meeting was the relative prevalence of Bayesian and Frequentist methods in the two communities. In particle physics, the prevailing methods are strictly frequentist, while astrophysics (and especially cosmology) has become increasingly Bayesian in its outlook in recent years. This philosophical distinction is grounded in the very different practical realities of the two fields: particle physicists can usually run their experiments for longer and longer, "building up statistics", and making the underlying asymptotic assumptions of a frequentist approach more valid. In cosmology, on the other hand, "there is only one Universe" and there are some experiments that can never be re-run. Moreover, as we will see below in the discussion of CMB data analysis, many of the predictions of cosmological theory are inherently statistical, so we must infer the properties of a correlated multivariate probability distribution from a *single* realization.

2. Case Study: The Cosmic Microwave Background

The Cosmic Microwave Background provides perhaps a rich example of data analysis in a cosmological setting. (Let me emphasize several points at the outset. First, this is a personal view of the CMB data analysis process. Second, this has become quite a large sub-field of cosmology, and I have been quite spare in my use of references, for which I apologize to my many colleagues whose work has not been cited herein despite important contributions to the field.)

The raw data – voltages output by some sort of antenna or temperature sensor – bear no simple relationship to the ultimate parameters to be measured, cosmological parameters such as the curvature of the Universe and the spectrum of primordial perturbations. Moreover, those raw voltage data are dominated by the noise properties of the measuring instrument. Yet somehow we must find an algorithm to "radically compress"[2] the millions or billions or more of raw data to just a few cosmological parameters. We start by writing down a simple model for the data from a single detector (the generalization to multiple detectors is straightforward; the data can just be appended as a single very long vector):

$$d_t = A_{tp} s_p + n_t = As + n \tag{1}$$

where d_t is the data taken at time t, s_p is the signal in pixel p (i.e., the CMB *map*), n_t is the noise, and the "pointing matrix", A_{tp} gives the response of the instrument at time t to pixel $p = 1 \ldots \#_p$. Note that *pixel* here refers to a finite area of sky. For simplicity, we can assume that the signal already contains the action of the experimental beam and any pixelization scheme we impose on the sky, in which case $A_{tp} = 1$ when pixel p is being observed at time t, and 0 otherwise. Finally, in the first equality we assume the Einstein summation convention, and in the second use matrix notation, so $As = A_{tp}s_p \equiv \sum_p A_{tp}s_p$. In general there will also be terms representing various other effects that may be present in the data, such as foreground contamination, instrumental systematics, etc. By a suitable generalization of the pixel domain and the pointing matrix, we can in fact estimate (and marginalize over) such effects.[19]

To proceed further, we need a model for the noise. We will assume that it can be represented by a stationary zero-mean Gaussian process, with correlations given by

$$\langle n_t n_{t'} \rangle \equiv N_{T,tt'} = N_T(t-t') \ . \tag{2}$$

More generally, we may subdivide the timestream into individual "stationary periods" within which this equation holds, and between which we assume zero correlation. This assignment is *conservative*, at least in the sense that the Gaussian is the maximum-entropy distribution with a given correlation structure. (This fact will have further implications later on when we discuss the correlations of the underlying signal, that is, the power spectrum, C_ℓ.)

2.1. *Mapmaking*

The first step, then, is to estimate the map, s_p given Eq. 1. This is a fairly standard inverse problem, but we choose to address it from a Bayesian standpoint. For these purposes, Bayes' theorem states

$$P(s_p|d_t I) \propto P(s_p|I) \times P(d_t|s_p I) \ , \tag{3}$$

where the left hand side is the *posterior* probability, the first factor on the right is the *prior* probability for the signal, and the final factor is the *likelihood*, the probability of the data given the signal. We write all probabilities as conditional upon some background information, I; in this case I encodes our knowledge of the noise correlation function, $N(t)$, the fact that we are imposing a Gaussian distribution upon the noise, etc. With this setup, the likelihood is just a multivariate Gaussian:

$$P(d_t|s_p I) = \frac{1}{|2\pi N|^{1/2}} \exp -\frac{1}{2}\left(d - As\right)^\dagger N_T^{-1} \left(d - As\right) \ , \tag{4}$$

where the superscript † means matrix transpose. Finally, we impose a uniform (albeit improper) prior on the signal $P(s_p|I) \propto$ const. As we shall see, this prior is actually irrelevant to the ultimate determination of power spectra and cosmological parameters.

By completing the square in the exponential (or taking derivatives, etc.) we see that the likelihood (and the posterior with our constant prior) is proportional to a Gaussian distribution in $\bar{s}_p = s_p + n_p$ with

$$\bar{s}_p = \left(A^\dagger N_T^{-1} A\right)^{-1} A^\dagger N_T^{-1} d \tag{5}$$

(the overbar denotes a generalization of the mean over all observations of a single pixel for correlated noise) and variance

$$\langle n_p n_{p'} \rangle = N_{P,pp'} = \left(A^\dagger N_T^{-1} A\right)^{-1}_{pp'} \tag{6}$$

which is just the usual Generalized Least Squares (GLS) solution.

[In fact with complex data like that expected from the Planck Surveyor, we cannot always calculate the full Bayesian map, Eq. 5 because of the complicated matrix manipulations involved. However, even in the case of some more general approximation to the map, we can still calculate its full noise correlation structure, replacing Eq. 6, as long as the operations are linear in the data and unbiased — a word not usually associated with Bayesian methods! — with respect to the signal.]

The output of this procedure is represented by the quantities $\bar{s}_p$ and $N_{P,pp'}$, our estimate of the map and its noise correlation structure. Specifically, $\bar{s}_p$ is an estimate of the beam-smoothed and pixelized sky in the pixels labelled by p. We take the beam to be circularly symmetric, with spherical harmonic transform, B_ℓ.

2.2. *Power Spectrum Estimation*

Next we must estimate the power spectrum which, by hypothesis, is responsible for realizing the map. Conventionally, we assume a zero-mean Gaussian process with covariance given by

$$\langle s_p s_{p'} \rangle = S_{P,pp'}(C_\ell) = \sum_\ell \frac{2\ell+1}{4\pi} C_\ell B_\ell^2 P_\ell(\hat{x}_p \cdot \hat{x}_{p'}) \ , \tag{7}$$

where C_ℓ is the cosmological power spectrum, $\hat{x}_p \cdot \hat{x}_{p'}$ is the cosine of the angular distance between the pixels p and p', and the P_ℓ are the Legendre polynomials. Now, the parameter we wish to estimate is C_ℓ; we can use the posterior of the previous step as the effective likelihood for the signal, so the model for the data, now just the map $\bar{s}_p$, is simply

$$\bar{s}_p = s_p + n_p \ , \tag{8}$$

where pixel noise correlations are given by Eq. 6. Alternately, we can start with the full likelihood, Eq. 4, and jointly estimate s_p and C_ℓ, with prior

$$\begin{aligned} P(s_p, C_\ell|I) &= P(C_\ell|I) P(s_p|C_\ell I) \\ &= P(C_\ell|I) \frac{1}{|2\pi S|^{1/2}} \exp -\frac{1}{2} s^\dagger S^{-1} s \ . \end{aligned} \tag{9}$$

We can then marginalize over s_p giving us the posterior for C_ℓ alone. It turns out that these approaches are mathematically equivalent, showing that indeed the mean and variance of Eqns. 5–6 are *sufficient statistics* for any further calculations. At this point, then, we have the following likelihood function:

$$\begin{aligned} P(d_t|C_\ell I) &= P(\bar{s}_p|C_\ell I) \\ &= \frac{\exp -\frac{1}{2}\bar{s}^\dagger (S_P + N_P)^{-1}\bar{s}}{|2\pi(S_P + N_P)|^{1/2}} \end{aligned} \qquad (10)$$

where now the parameter of interest, C_ℓ, appears in the covariance matrix, $S_P(C_\ell) + N_P$. Because of this, there is no simple analytic description of the posterior probability, or indeed for the shape of the likelihood considered as a function of C_ℓ. However, we can relatively easily use techniques like Newton-Raphson iteration to find the peak of the likelihood and calculate its curvature about the maximum[1].

Unfortunately, these techniques are prohibitively expensive for data from upcoming experiments such as the Planck Surveyor, scaling as $O(\#_p^3)$ in time and $O(\#_p^2)$ in storage; indeed the latter implies that, for coming megapixel experiments, the covariance matrix is likely too large to store, much less calculate in full generality. There have also been efforts to develop so-called Gibbs Sampler Monte Carlo techniques to calculate the full posterior for the power spectrum.[6]

Thus, we must be practical. Even if we are philosophically disposed to Bayesianism (as are many in the cosmology community), we may need to consider other techniques, although I, in particular, take the rather unorthodox view that these methods are useful as approximations to the Bayesian result. This stands in contrast to many of the Bayesian approaches in particle physics discussed at this meeting, in which the analysts try to find Bayesian techniques which give the same answer as the orthodox frequentist techniques already in use.

The most common of these techniques for estimating C_ℓ are the so-called unbiased pseudo-C_ℓ quadratic estimators, in which some approximation to the spherical harmonic transform of the full sky is calculated, and squared to give the "pseudo-C_ℓ" spectrum, $\widehat{C}_\ell$, which is then corrected to give an unbiased estimate of the true spectrum by inverting the relation

$$\langle \widehat{C}_\ell \rangle = \sum_{\ell'} M_{\ell\ell'} C_{\ell'} + N_\ell \qquad (11)$$

where the ensemble average is taken over Gaussian realizations of the signal and noise with the variances given above. We know that in the limit of a full sky and uniform noise the estimator thus derived is exactly the same as the Bayesian maximum likelihood, with the variance the same as the curvature about the maximum, and indeed the usual relations from the theory of probability and statistics state that these hold "asymptotically", which is usually understood to mean that they will hold for high ℓ, where very many modes contribute to the measurement. (We do know from experience that the results do differ in detail for realistic experiments, such as BOOMERANG[18].)

2.3. *Cosmological Parameter Estimation*

Finally, we must use these C_ℓ to determine the underlying parameters, θ_i, (e.g., the densities, Ω_i; the Hubble Constant, H_0, etc.). Unlike in previous steps, there is a direct relationship between the parameters and the power spectrum, simply $C_\ell = C_\ell(\theta)$, i.e., we have a delta-function prior $P(C_\ell, \theta_i) = P(\theta_i)\delta[C_\ell - C_\ell(\theta)]$. So the likelihood function remains as before, but we wish to determine its parameters as we vary θ. The calculation of $C_\ell(\theta)$ for standard models requires the solution of coupled Einstein-Boltzmann linear differential equations describing the distribution of matter and radiation in the expanding universe, and has been in implemented in publicly available codes such as CMBFAST[20] and CAMB[14]. For realistic models, this is straightforward but relatively time-consuming, and moreover the general exploration of a multi-parameter space is a difficult task. In recent years, the favored technique for this exploration has been Markov Chain Monte Carlo (MCMC).[3, 15]

In this meeting, MCMC in a cosmological context was discussed in the talks of Leach, Nicholls and Trotta (this volume).

A very different way of exploring CMB power spectra was discussed by Nichol at this meeting: he used a (very non-Bayesian!) non-parametric smoothing technique to examine the overall shape of the spectrum, and answer some very basic questions: does the data in Figure 1 show a series of definite peaks? He then extended these methods to the Cosmological Parameter Estimation problem itself: do the parameters predict the correct overall smoothed shape? Qualitatively his results agree largely with the consensus discussed in the following, although there are detailed — and not unexpected — differences.

3. Results: Cosmology in 2005

In Figure 1 we show the results of various calculations of C_ℓ along these lines, each from a different dataset; the results are in quite good agreement overall. In general, these results are either the frequentist mean and variance, or the maximum likelihood and curvature; as emphasized above, the prior does not really matter at this point.

Where the prior does enter, however, is in the calculation of the cosmological parameters from these spectra. As was emphasized at this meeting by Cox and Le Diberder, flat priors are dangerous. Indeed, in cosmology, there is no one set of natural parameters on which to impose flat priors. For example, would we want a flat prior on the density relative to the critical density, Ω? However, that critical density itself depends on the *a priori* unknown Hubble Constant, $H_0 \equiv 100h$ km/s/Mpc, so perhaps a more physical quantity would be Ωh^2? There is no "correct" answer to this question; rather we take the advice of Cousins at this meeting: we must perform sensivity analyses to determine the effect of our priors upon the analysis. In effect, by comparing the work of different authors, we can do just this sort of meta-analysis. If we estimate the cosmological parameters from the data of Figure 1 (from WMAP[10], ACBAR[11], BOOMERANG (B03)[9, 16], CBI[17], DASI[7], MAXIMA[13] and VSA[5]) in various combinations, we see that many features are robust to these changes, and we can highlight a few here:

- The Universe is flat: $\Omega_{\rm tot} = 1$;
- The primordial perturbations are well described by a nearly scale-invariant power spectrum ($n_s \simeq 1$); and
- The Hubble Constant is approximately $H_0 = 72$ km/s/Mpc.

Perhaps startlingly, the first two are just the predictions of the inflationary theory of the early Universe! So finally, the discussion of statistics leads us, as it should, to the underlying physics.

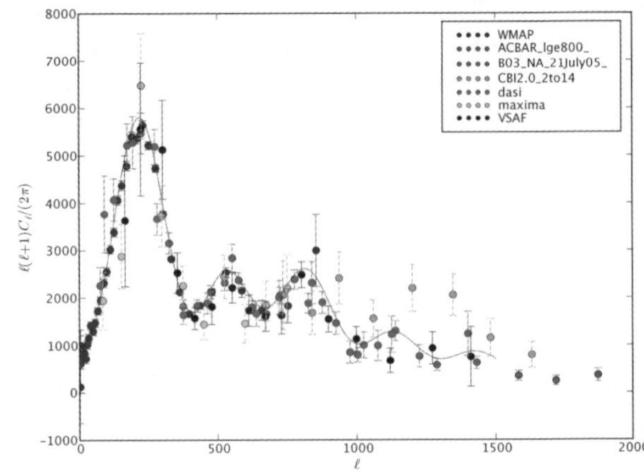

Fig. 1. Recent CMB Power spectrum data; publications are cited in the text.

4. Complications: Non-Gaussianity

The model for CMB data in the previous section becomes more restrictive in each step of the process. When making the map, we first assume a model for the *noise*, that it is stationary over some period of time, and a Gaussian of the noise power spectrum. When calculating the power spectrum, we assume that the signal is isotropic on the sky, and a Gaussian realization from the cosmological power spectrum, C_ℓ. Finally, we assume the cosmological parameters directly determine the power spectrum. Much effort has been put into going beyond these assumptions, specifically the Gaussianity of the signal. As the saying goes, "non-Gaussian distributions" are like "non-elephant animals", and it is very hard to describe an arbitrary distribution without just giving all possible detailed information (e.g., the functional form of the distribution or its moments). Methods must be tuned to find specific "sorts" of non-Gaussianity, such as the existence of higher-order connected moments of the data. Even there, of course, the task immediately becomes very difficult: are the higher moments best described in Fourier (spherical harmonic) space, or spatially on the sky? Which moments are we searching for? What counts as a detection (if you look hard enough, you will always find something!)? For a Bayesian, these problems seem even worse: how do you even write down the distribution in the absence of a very concrete and calculable model?

With the actual WMAP data, we also may be encountering a related problem: There is some evidence that, on the largest scales, the WMAP data do not seem to obey statistical isotropy. That is, statistical quantities may not just be functions of the distance between points, but may depend on where you are on the sky, as has been discussed in a number of recent works[4, 8, 12] (and others). This could be evidence of foreground contamination, or, more excitingly, of some underlying misunderstanding of the physics of the universe on large scales. In the absence of a physical model, it is prima facie impossible to distinguish a non-Gaussian distribution from an anisotropic distribution or from a combination of the two effects.

Nonetheless, various techniques have been proposed and applied to tease out non-Gaussianities from current data. At this meeting, they were discussed by Jin and Starck (and by Digel and by Bissantz in more traditional astronomical contexts). But a word of caution is in order: most of these methods are derived assuming some sort of independent and identically distributed (iid) random variable is responsible for the non-Gaussianity, but in real-world astrophysics, nothing is ever iid!

5. Other Problems

I have concentrated on my speciality, cosmology in general and the CMB in particular, but of course statistics plays a paramount role throughout astrophysics. Indeed, as emphasized in the presentations of Cox and Johnston, astronomy has played a leading role in the development of statistics since the beginning. Other exciting developments discussed at the Conference include:

- Time-series analysis [Clifford];
- Image processing/reconstruction/restoration [Titterington]. For some applications, it is crucial to be able get full error information (i.e., the posterior distribution) of the reconstructed image, and this restricts the possible algorithms;
- Classification problems [e.g., Gray]: finding unusual objects (or usual ones: photometric redshift; galaxy classification from pictures). Note that these tasks usually have vastly different kinds of prior information: physics vs. training sets vs. "experience".

5.1. *Virtual Observatories*

As discussed at this meeting by Alex Szalay, we in cosmology and astrophysics are beginning to deal with the massive, heterogeneous datasets covering a variety of instruments, wavebands, areas of the sky, etc. The community is attempting to build tools for uniform and distributed access to and analysis of these data under the rubric of Virtual Observatories. Bob Nichol discussed searching through massive astronomical datasets using KD-trees and the plans to finally move large-scale astronomical data-processing from the desktop to the grid.

6. Conclusions

In his opening talk at this meeting, Sir David Cox said that "We're eclectic"; this perfectly captures the pragmatism of astronomers and astrophysicists confronting our data, the need to find tools to handle its complexity and volume. Indeed, astronomers have always had to deal with data just beyond the ability of obvious current techniques, and therefore have always been avid consumers — if not developers — of cutting-edge statistical techniques. As we saw throughout this meeting, this fruitful confluence of fields continues to this day.

Acknowledgments

I would like to thank Louis Lyons and the other Phystat05 conference organizers for this great opportunity, and the many wonderful speakers at the meeting on whose work I based this review. I'd like also to thank my collaborators in the MAXIMA, BOOMERANG and COMBAT collaborations who participated in the aspects of my own work that I have discussed, as well as PPARC for their financial support.

References

1. J. R. Bond, A. H. Jaffe, and L. Knox. Estimating the power spectrum of the cosmic microwave background. *Phys. Rev. D*, 57:2117–2137, 1998.
2. J. R. Bond, A. H. Jaffe, and L. Knox. Radical Compression of Cosmic Microwave Background Data. *Astrophys. J.*, 533:19–37, 2000.
3. N. Christensen, R. Meyer, L. Knox, and B. Luey. Bayesian methods for cosmological parameter estimation from cosmic microwave background measurements. *Classical and Quantum Gravity*, 18:2677–2688, July 2001.

4. A. de Oliveira-Costa, M. Tegmark, M. Zaldarriaga, and A. Hamilton. Significance of the largest scale CMB fluctuations in WMAP. *Phys. Rev. D*, 69(6):063516–, Mar. 2004.
5. C. Dickinson et al. High sensitivity measurements of the CMB power spectrum with the extended very small array. astro-ph/0205436.
6. H. K. Eriksen et al. Power Spectrum Estimation from High-Resolution Maps by Gibbs Sampling. *Astrophys. J. Supp.*, 155:227–241, Dec. 2004.
7. N. W. Halverson et al. Degree Angular Scale Interferometer First Results: A Measurement of the Cosmic Microwave Background Angular Power Spectrum. *Astrophys. J.*, 568:38–45, 2002.
8. F. K. Hansen, A. J. Banday, and K. M. Górski. Testing the cosmological principle of isotropy: local power-spectrum estimates of the WMAP data. *Mon. Not. R. Astr. Soc.*, 354:641–665, Nov. 2004.
9. W. C. Jones et al. A measurement of the angular power spectrum of the CMB temperature anisotropy from the 2003 flight of BOOMERANG. astro-ph/0507494. 2005.
10. A. Kogut et al. Wilkinson Microwave Anisotropy Probe (WMAP) first year observations: TE polarization. *Astrophys. J. Suppl.*, 148:161, 2003.
11. C.-l. Kuo et al. High resolution observations of the CMB power spectrum with ACBAR. *Astrophys. J.*, 600:32–51, 2004.
12. K. Land and J. Magueijo. Examination of Evidence for a Preferred Axis in the Cosmic Radiation Anisotropy. *Physical Review Letters*, 95(7):071301–, Aug. 2005.
13. A. T. Lee et al. A High Spatial Resolution Analysis of the MAXIMA-1 Cosmic Microwave Background Anisotropy Data. *Astrophys. J. Lett.*, 561:L1–L5, 2001.
14. A. Lewis. http://camb.info/.
15. A. Lewis and S. Bridle. Cosmological parameters from VSA, CBI and other data: a Monte-Carlo approach. *Phys. Rev. D*, 66:103511, 2002.
16. F. Piacentini et al. A measurement of the polarization-temperature angular cross power spectrum of the cosmic microwave background from the 2003 flight of BOOMERANG. astro-ph/0507507.
17. A. C. S. Readhead et al. Extended mosaic observations with the cosmic background imager. *Astrophys. J.*, 609:498–512, 2004.
18. J. E. Ruhl et al. Improved Measurement of the Angular Power Spectrum of Temperature Anisotropy in the Cosmic Microwave Background from Two New Analyses of BOOMERANG Observations. *Astrophys. J.*, 599:786–805, 2003.
19. R. Stompor et al. Making maps of the cosmic microwave background: The MAXIMA example. *Phys. Rev. D*, 65(2):022003–, 2002.
20. M. Zaldarriaga and U. Seljak. CMBFAST for Spatially Closed Universes. *Astrophys. J. Supp.*, 129:431–434, 2000.

SUMMARY OF SOME STATISTICAL ISSUES

N. REID

Department of Statistics, University of Toronto, 100 St. George St., Toronto Canada M53 3G3
E-mail: reid@utstat.utoronto.ca

A brief summary of some statistical issues that arose during the conference is presented.

In terms of statistical ideas, I would make a very broad distinction between two prominent sets of problems at this conference. First there are a number of problems in which the main feature is a very large amount of data, requiring new methods and considerable computing power. An example that has already been used with success in astronomy is the use of false discovery rates in problems involving a great number of tests, and we heard here about new adaptations of wavelet and ridgelet techniques for identifying structure in images, about smoothing methods in multi-dimensional image processing, and new methods for on-line data mining. I won't attempt to summarize this class of problems, although it is clearly very important, not only in physics and astronomy but in a number of scientific problems, especially including genomics, where there is very active development of statistical techniques.

Another class of problems seems simpler (to a statistician) on a first reading. An example is independent Poisson counts from background events and possible signal events. We should not forget, though, that elaborate experimental techniques and considerable ingenuity in data processing, have preceded the presentation of a small amount of data. For this setting one would expect that standard statistical methods would provide a simple, and even a best, answer, but as we have seen even in this context this is not always the case. Certainly inference about the ratio of Poisson mean parameters is satisfactorily solved using the binomial likelihood. Statistical inference for the difference between two Poisson means is somewhat more difficult, as we have to rely on some approximate argument, and with small counts the usual normal approximations will not be reliable. As Sir David Cox stressed in the panel discussion, the science of statistics develops most fruitfully in close collaboration with applications, and this problem is a good example of something that is indeed sufficiently specialized to the HEP context that it is not in the repertoire of 'off-the-shelf' statistical methods.

Some general ideas which should inform the solution include the very important notion that confidence intervals, however developed, should have good properties in repeated observation of the same experimental system, even if these repetitions are hypothetical. In my view the definition of 'same experimental system' needs great care, in order to avoid difficulties similar to, but more subtle than, the problem of two measuring instruments discussed in Cox[1] and mentioned in Cousins[2]. Unfortunately it seems extremely difficult to 'mathematize' this notion; statisticians have spent many years of effort on the topic, and a single widely accepted solution has not emerged. At this time the best we can advise is to look at problems on a case by case basis.

Likelihood methods are well accepted in the HEP community, but not always used in quite the same manner as used by statisticians. To clarify, suppose we have a single parameter model $f(x;\theta)$ and observe a sample $\underline{x} = (x_1, \ldots, x_n)$ of independent observations from this model. The log-likelihood function $\ell(\theta;\underline{x}) = \log \Pi f(x_i;\theta)$ is a sum of n terms, and we can apply the central limit theorem to $\partial\ell(\theta;\underline{x})/\partial\theta$ to derive the following approximations:

$$
\begin{aligned}
(\hat{\theta} - \theta) i^{1/2}(\theta) &\;\dot\sim\; N(0,1) \\
\ell'(\theta) i^{-1/2}(\theta) &\;\dot\sim\; N(0,1) \\
\pm\sqrt{[2\{\ell(\hat{\theta}) - \ell(\theta)\}]} &\;\dot\sim\; N(0,1)
\end{aligned}
$$

where $\hat{\theta}$ is the maximum likelihood estimate and $i(\theta) = E\{-\partial^2\ell(\theta)/\partial\theta^2\}$ is the expected information. Barlow[3] described the second of these as Bartlett's statistic and the third as $-2\ln L$ (although I have here taken the square root, since the parameter is scalar). Each approximation provides a different way to compare the expected value to the observed value, but each is a so-called 'first order approximation', because the error in the approximation is $O(n^{-1/2})$. In the limit when the log-likelihood function becomes quadratic, with second derivative equal in the limit to its expectation, they all lead to the same measure. To these three approximations we can further confuse things by adding standardization by observed information:

$$(\hat{\theta} - \theta)j^{1/2}(\hat{\theta}) \;\dot{\sim}\; N(0,1)$$
$$\ell'(\theta)j^{-1/2}(\hat{\theta}) \;\dot{\sim}\; N(0,1)$$

where $j(\hat{\theta}) = -\ell''(\hat{\theta})$ is the curvature of the log-likelihood function at the maximum.

A very natural question is which of these approximations is to be preferred in finite samples, and some reasons for expecting the log-likelihood ratio to be preferred are that it is invariant to reparametrization, and that it preserves the asymmetry in the log-likelihood function. It is also the leading term in a higher order expansion, the correction term of which uses one or other of the two j-standardized statistics. Indeed the statistical literature has since Efron & Hinkley[4] preferred the j-standardization for $\hat{\theta}$, and later somewhat technical development of improved approximations to the distribution of $\hat{\theta}$ have confirmed this preference. It is related to conditioning on ancillary statistics, i.e. functions of the data that have a distribution exactly or approximately free of θ.

Unfortunately however there are no general results on rates of convergence or other properties that could lead to a definitive conclusion about which departure measure to use, and case by case studies are thus needed. Barlow[3] showed that for the exponential mean, Bartlett's statistic, i.e. the score function using the i-standardization (which coincidentally is the same in this example as the i-standardized maximum likelihood estimate), is better approximated by a standard normal than the log-likelihood ratio. This I found quite surprising, given my 'prior belief' in the log-likelihood ratio statistic. The explanation is that Bartlett's statistic has exact mean 0 and exact variance 1, these moments coinciding with those of the normal approximation, which is therefore reasonably accurate for moderate deviations. However if we move out to the tails, the likelihood ratio statistic is more accurately approximated by a standard normal than Bartlett's statistic. Figure 1 compares the p-values, as functions of the mean parameter, to the exact p-value based on the gamma distribution, for a sample of size 5 and an observed sample mean of 1, first in the '1-sigma' range and then in the '4-sigma' range. This example is also treated in Barndorff-Nielsen & Cox[5].

It does seem very difficult to draw any general conclusions about the first order approximations, although for most examples I have looked at the normal approximation to the square root of the likelihood ratio has been the most accurate in the tails. A relatively simple combination of this with the Bartlett score statistic, as outlined in Reid & Fraser[6] gives essentially exact results for the exponential example.

As has been mentioned several times during this workshop, adding nuisance parameters further complicates the issues. There are a number of somewhat different lines of argument in the statistical literature leading to the idea of improving the profile likelihood by adding a term to allow for the estimation of the nuisance parameters. The simplest motivation is from a Bayesian argument. We can get an approximation for the marginal posterior distribution of the parameter of interest as follows:

$$\begin{aligned}
\pi_m(\psi \mid \underline{x}) &= \int \pi(\psi, \nu \mid \underline{x}) d\nu \\
&\propto \int \exp\{\ell(\psi,\nu)\}\pi(\psi,\nu)d\nu \\
&= \int \exp\{\ell(\psi,\nu)\}\pi(\nu \mid \psi)d\nu\pi(\psi) \\
&\doteq \exp\{\ell(\psi,\hat{\nu}_\psi)\}|j_{\nu\nu}(\psi,\hat{\nu}_\psi)|^{-1/2} \cdot \\
&\qquad \pi(\hat{\nu}_\psi \mid \psi)\sqrt{(2\pi)^{k-1}}\pi(\psi)
\end{aligned}$$

where $\theta = (\psi, \nu)$ has been partitioned into a parameter of interest ψ and a $k-1$-dimensional nuisance parameter ν, and $j_{\nu\nu}(\psi,\hat{\nu}_\psi) = -\partial^2\ell(\psi,\nu)/\partial\nu\partial\nu^T$ is the portion of the observed information matrix related to the nuisance parameter. The last approximation comes from a Laplace approximation of the integral defining the marginal posterior.

Now it can be shown that if ψ and ν are orthogonal parameters, in the sense that the (ψ,ν) components of the expected Fisher information matrix are 0, then $\hat{\nu}_\psi = \hat{\nu} + O_p(1/n)$; in the absence of parameter orthogonality the error would be $O_p(1/\sqrt{n})$. Sweeting[7] in the discussion of Cox & Reid[8] argued that if ψ and ν are orthogonal then it would make sense to assign independent priors to them, in which case the term involving the prior on ν vanishes (to $O(n^{-1})$) and the log of the posterior marginal

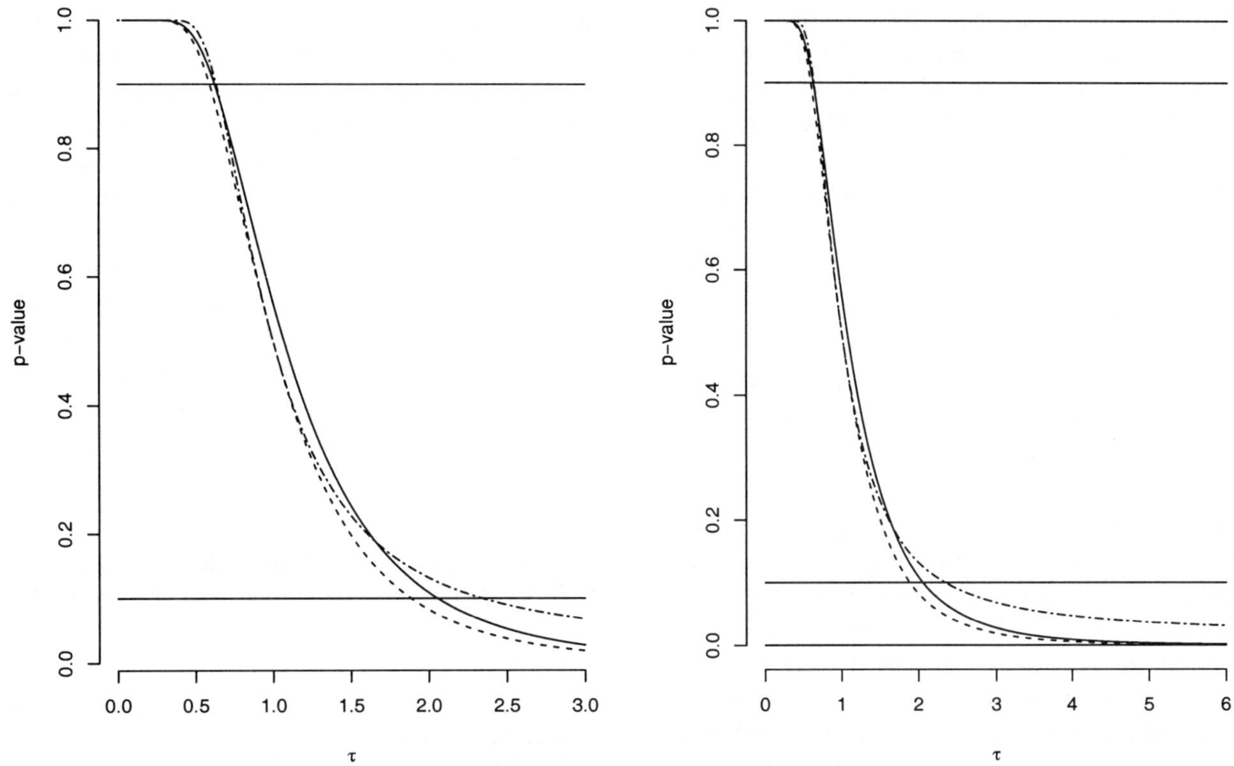

Fig. 1. Plots of p-value functions for exponential mean parameter τ, computed using the exact distribution (solid), the normal approximation to the square root of the log-likelihood difference (dashed), and the normal approximation to the standardized score (dot-dash). Horizontal lines show the 0.10 and 0.90 limits (left), as well as the 0.0001 and 0.9999 limits (right). The sample size n is 5 and the sample mean is 1. Very similar results are obtained for both smaller and larger values of n.

density is

$$\log \pi_m(\psi \mid \underline{x}) \doteq \ell(\psi, \hat{\nu}_\psi) - \frac{1}{2}\log|j_{\nu\nu}(\psi, \hat{\nu})| + \log \pi(\psi);$$

this is one way to motivate the so-called "adjusted" or "modified" profile log-likelihood

$$\ell_a(\psi) = \ell(\psi, \hat{\nu}_\psi) - \frac{1}{2}\log|j_{\nu\nu}(\psi, \hat{\nu})|.$$

If ψ is scalar then a transformation from some original parameterization (ψ, ϕ) to (ψ, ν) where ν is orthogonal to ψ can always be found; Cox & Reid[9] indicate how to compute the adjusted profile without explicitly reparameterizing the model. The term "modified profile likelihood" is usually used for one of a family of adjusted profile log-likelihoods of the form

$$\ell(\psi, \hat{\nu}_\psi) - \frac{1}{2}\log|j_{\nu\nu}(\psi, \hat{\nu})| + B(\psi)$$

where $B(\psi)$ is to be specified, but is always $O(1)$, i.e. the same order as the $\log j$ term, and serves among other things to make the result parameterization invariant, which the simple version ℓ_a is not.

Although motivated by higher order asymptotic arguments, only first order asymptotics apply to ℓ_a and its variants. In particular we have, in analogy to the results for a scalar parameter

$$\begin{aligned}(\hat{\psi}_a - \psi)\{-\ell_a''(\hat{\psi}_a)\}^{1/2} &\;\dot{\sim}\; N(0,1)\\ \ell_a'(\psi)\{-\ell_a''(\hat{\psi}_a)\}^{-1/2} &\;\dot{\sim}\; N(0,1)\\ \pm\sqrt{2\{\ell_a(\hat{\psi}_a) - \ell_a(\psi)\}} &\;\dot{\sim}\; N(0,1)\end{aligned}$$

where $\hat{\psi}_a$ is the maximum likelihood estimate from $\ell_a(\psi)$. These approximations are no more accurate in asymptotic theory than those based on the profile likelihood but in practice the adjustment for nuisance parameters seems to lead to better approximations, especially when the number of nuisance parameters is large.

There are two classes of models where, at least for some of their parametrizations, exact elimination of nuisance parameters is possible: exponential family models and non-normal linear regression models. Some examples are given in Reid & Fraser[6]. In these two classes the adjusted profile likelihood ℓ_a arises quite naturally as a kind of 'leading term'.

In models with a single scalar parameter, there is a uniquely determined, albeit improper, prior for which Bayesian posterior upper limits are guaranteed to have frequentist coverage to high accuracy: more precisely we have

$$\Pr(\theta \leq \theta^{(1-\alpha)}(\underline{x}) \mid \underline{x}) = \Pr_\theta(\theta^{(1-\alpha)}(\underline{X}) \geq \theta) + O(1/n)$$

if and only if the prior is proportional to $i^{1/2}(\theta)$; the first probability above is calculated under the posterior distribution, and defines $\theta^{(1-\alpha)}(x)$ by the requirement that this probability equal α, and the second probability is calculated under the sampling model $f(\underline{x};\theta)$. In multiparameter problems, matching priors do not exist in general, but there is an important exception. In statistical models whose mathematical structure is generated by a group of transformations, then it is possible to obtain the exact distribution of the maximum likelihood estimator by conditioning, and this is identical to the Bayesian posterior distribution for a special choice of prior measure related to the group structure; see Fraser[10], Barndorff-Nielsen[11] and also Podobnik & Zivko[12]. These arguments do not apply however to models for discrete data.

A recurring theme in this meeting has been the possible dangers in using flat priors in multiparameter problems. An early and compelling example is described in Example 10.6 of Cox & Hinkley[13]. Suppose $X_1, \ldots, X_n$ are independent normal random variables with mean μ_i and variance σ^2, and that

$$\mu_i = EX_i = \gamma + \beta\rho^{x_0+ia}, \quad 0 \leq \rho \leq 1$$

where x_0 and a are known, and $\theta = (\gamma, \beta, \rho, \sigma)$. In a linear regression model, the matching prior and most usual prior is proportional to $d\beta d\sigma/\sigma$, so a very natural 'flat' extension of this is to choose the prior

$$\pi(\theta) \propto d\gamma d\beta d\sigma/\sigma d\rho, \quad 0 \leq \rho \leq 1;$$

however the marginal posterior for ρ concentrates on the points $\rho = 0$ and $\rho = 1$. I don't know if this phenomenon is widespread or not, but the fact that one can so easily get into trouble in a relatively simple model with a seemingly vague choice of prior is somewhat worrying. Heinrich[14] also raises several issues with flat priors. There is an active research effort in the statistics community to investigate what have come to be called 'objective' priors; the most recent conference was 'OBayes5', held in June, 2005.

Speaking as a statistician who has been largely involved with theoretical issues, it is exciting to discuss these issues in the context of applications to high energy physics, and I look forward to further fruitful collaborations between the two disciplines.

References

1. D.R. Cox, *Ann. Math. Statist.* **29**, 257 (1958).
2. R. Cousins, *this volume* (Treatment of Nuisance Parameters in High Energy Physics, and Possible Justifications and Improvements in the Statistics Literature), 2005.
3. R. Barlow, *this volume* (Errors from the likelihood function), 2005.
4. B. Efron and D.V. Hinkley, *Biometrika* **65**, 457 (1978).
5. O.E. Barndorff-Nielsen and D.R. Cox, *Inference and Asymptotics.* (Chapman & Hall, London, p.83, 1994).
6. N. Reid and D.A.S. Fraser, in *Proceedings of PHYSTAT2003*, L. Lyons, R. Mount, R. Reitmeyer, eds. SLAC e-Conf C030908, 265 (2003).
7. T.J. Sweeting, *J. R. Statist. Soc.* B **49**, 20 (1987).
8. D.R. Cox and N. Reid, *J. R. Statist. Soc.* B **49**, 1 (1987).
9. D.R. Cox and N. Reid, *J. R. Statist. Soc.* B **55**, 467 (1993).
10. D.A.S. Fraser, *Inference and Linear Models.*, Ch. 7, McGraw-Hill, New York.
11. O.E. Barndorff-Nielsen, *Biometrika* **67**, 293 (1980).
12. T. Podobnik and T. Zivko, *this volume* (Towards reconciliation between Bayesian and Frequentist reasooning), 2005.
13. D.R. Cox and D.V. Hinkley, *Theoretical Statistics.* (Chapman & Hall, London, 1974).
14. J. Heinrich, *this volume* (The Bayesian approach to setting limits: what to avoid), 2005.

CONCLUDING TALK: PHYSICS

GARY J. FELDMAN
Department of Physics
Harvard University
Cambridge, MA 02138
E-mail: feldman@physics.harvard.edu

This concluding talk by a particle physicist reviews two topics, event classification and nuisance parameters.

1. Introduction

I will just cover two topics in this talk, both of which I am interested in and which have received a fair amount of attention at this conference, event classification and nuisance parameters. I apologize to other speakers whose work I will not have time to cover.

2. Event Classification

2.1. *Introduction*

The general problem that we wish to solve is given a measurement of an event $\mathbf{X} = (x_1, x_2, \ldots x_n)$, find the function $\mathrm{F}(\mathbf{X})$ which returns 1 if the event is signal (s) and 0 if the event is background (b) to optimize a figure of merit, say $s/\sqrt{b}$ for a discovery or $s/\sqrt{s+b}$ for an established signal.

In principle the solution is straightforward. Use a Monte Carlo simulation to calculate the likelihood ratio $L_s(\mathbf{X})/L_b(\mathbf{X})$ and derive $\mathrm{F}(\mathbf{X})$ from it. This just amounts to counting the number of signal and background at each point in the parameter space. Then F can be calculated by ordering the likelihood ratios and accepting them in decreasing order until the figure of merit starts to decrease. Further, by the Neyman-Pearson Theorem, this is the optimum solution. Of course, this does not work due to the "curse of dimensionality." In a high-dimension space, even the largest data set is sparse, with the distance between neighboring events comparable to the radius of the space. Thus, we are forced to substitute cleverness for brute force. In recent years, physicists have come to learn that computers may be cleverer than they are; they have turned to machine learning. One gives the computer samples of signal and background events and lets the computer figure out what $\mathrm{F}(\mathbf{X})$ is.

2.2. *Artificial Neural Networks*

Originally most of the machine learning effort was in artificial neural networks (ANN). Although used successfully in many experiments, ANNs tend to be finicky and often require real cleverness from their creators. At this conference, there was an advance in ANNs reported by Harrison Prosper.[1] The technique is to average over a collection of networks. Each network is constructed by sampling the weight probability density constructed from the training sample. Prosper notes that this Bayesian technique "takes us another step closer to realizing optimal results in classification (or density estimation) problems. It allows a fully probabilistic approach with proper treatment of uncertainties." He states that the "the initial results are promising, though computationally challenging."

2.3. *Rules and Trees*

In the past couple of years, interest has started to shift to other techniques, such as decision trees, at least partially sparked by Jerry Friedman's talk at PHYSTAT2003.[2] A cartoon of a decision tree is shown in Fig. 1. The best variable and value to separate signal and background are chosen and the sample is divided into two branches. For each branch, the process is repeated until a leaf is found with a preset minimum of either signal or background events. Each leaf is then labeled as either signal or background.

A single decision tree has limited power, but its power can be increased by techniques that effectively sum many trees. Several approaches were presented at this conference. Jerry Friedman discussed a technique based on rules, which effectively combines a series of trees.[4]

Harrison Prosper presented a talk for Ilya

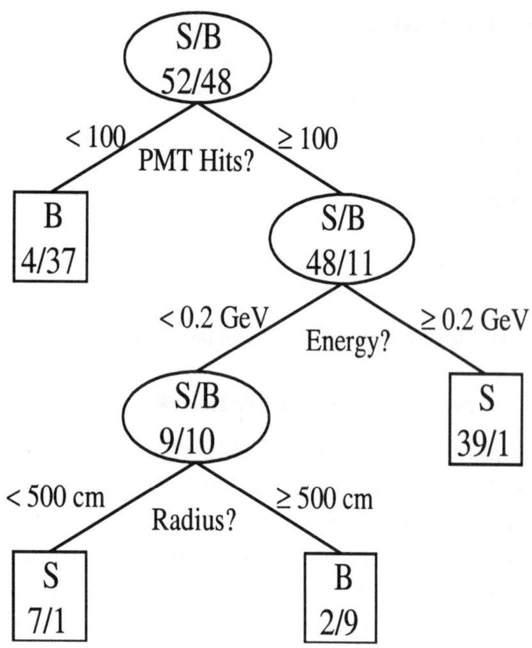

Fig. 1. A cartoon of a decision tree. From Ref. 3.

Narsky on bagging decision trees.[5] "Bagging" stands for Bootstrap AGGregatING. In this technique, one builds a collection of trees by selecting a sample of the training data and, optionally, a subset of the variables. Although boosted trees (to be discussed next) are generally more sensitive than bagging trees, Narsky reported the use of bagging trees in the study of the decay $B \rightarrow \gamma e \nu$, in which bagging trees gave the most significant results. A single decision tree gave a 2.16 σ significance; boosted decision trees gave a 2.62 σ significance; and bagging decison trees gave a 2.99 σ significance. However, about half of the difference between the bagging and boosted trees was not fully optimizing the boosted trees.

Byron Roe gave a talk on the use of boosted trees in MiniBooNE.[3] The boosted tree technique is to give misclassified events in one tree a higher weight in the generation of a new tree. In the MiniBooNE analysis, this process is repeated to generate 1000 trees. The final classifier is a weighted sum of all of the trees. Roe found that the boosted trees are about a factor of 1.8 more sensitive than an ANN technique and that it is more robust.

Other talks on the subject of event classification were given by Puneet Sarda[6] and Alex Gray.[7] Unfortunately, I was unable to attend these talks as I was chairing another session at the time.

3. Nuisance Parameters

3.1. *Introduction*

Nuisance parameters are parameters with unknown true values for which coverage is required in a frequentist analysis. They may be statistical, such as number of background events in a sideband used for estimating the background under a peak. Or they may be systematic, such as the shape of the background under the peak, or the error caused by the uncertainty of the hadronic fragmentation model in the Monte Carlo. Most experiments have a large number of systematic uncertainties, so having an efficient way of dealing with them is an important issue in the statistical analysis of the results. The single statistical issue that I have been asked most about is probably the proper treatment of these uncertainties.

In his talk, Kyle Cranmer has pointed out that these issues will be even more important at the LHC.[8] Cranmer's argument has to do with the size of the systematic uncertainties. For example, a typical negative search at LEP might have had no signal events. In this case the statistical error is of order 1 and the systematic error might have been of order 0.1. The contribution of the systematic error would then have been of order 0.01, and the details of its treatment would have been largely irrelevant. However, Cranmer expects that at the LHC it might be typical to have a signal of 100 events and systematic uncertainties of 10%. And to make the issue more critical, Cranmer notes that the standard for discovery in bump hunting has come to be five standard deviations. (See Appendix A for an aside on this issue.)

At this meeting we have seen a wide series of techniques discussed for constructing confidence intervals in the presence of nuisance parameters. The one unifying aspect of these techniques is that everyone has expressed a concern that their methods cover, at least approximately. This appears to be important for LHC physics in the light of Cranmer's concerns.

3.2. *Bayesian Treatment*

Joel Heinrich presented a recommendation for the CDF collaboration to do Bayesian analyses with the

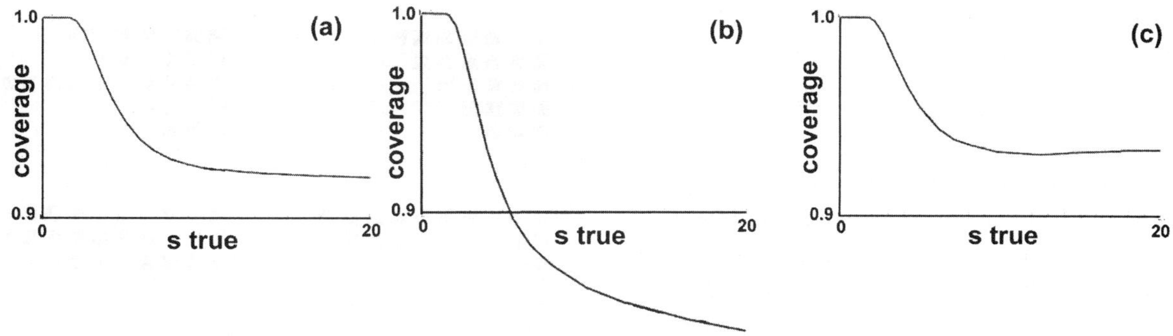

Fig. 2. Examples of coverage for various Bayesian priors on the nuisance parameters. See the text for parameters.

requirement that non-subjective priors are chosen to provide coverage.[9] The perceived advantage is Bayesian conditioning with frequentist coverage.

Figure 2 shows some examples of coverage studies from Heinrich's talk. Figure 2(a) shows the case of a single Poisson signal distribution with flat priors on the signal (s), the background (b), and the normalization (ϵ). There is modest overcoverage. Figure 2(b) shows the case of four independent Poisson signal distributions with flat priors, as in (a). There is now substantial undercoverage. Coverage is recovered by switching to $1/b$ and $1/\epsilon$ priors, shown in Fig. 2(c).

This example illustrates that the choice of priors for nuisance parameters is important and that flat priors, particularly in multiple dimensions, are likely to lead to undesirable results. The problem here is that the volume of a hypersphere is increasingly concentrated at large radius as the dimension of the hypersphere increases. Thus, the nuisance parameters are pushed to unreasonably large values, depressing the signal and causing undercoverage.

3.3. *Frequentist-Bayesian Hybrids*

Fredrik Tegenfeldt presented a likelihood-ratio ordered Neyman construction[10] after integrating out the nuisance parameters with flat priors.[11] In a single channel test, there was no undercoverage.

What would happen for a multi-channel case? Would it fall prey to the same difficulties illustrated in Heinrich's talk?[9] I am not sure. It is likely that the confidence belt will be distorted by the use of flat priors, but it is possible that the method will still cover due to the Neyman construction.

Cranmer considered a similar technique, as used for LEP Higgs searches.[8] Both techniques are referred to as Cousins-Highland, from their 1992 paper which used essentially the same technique.[12]

3.4. *Profile Likelihood*

Forty-four years ago, Kendall and Stuart suggested how to eliminate nuisance parameters in a likelihood-ratio ordered Neyman construction.[13] They first defined

x	the vector of measurements,
θ_r	the vector of unknown parameters with θ_{r_0} representing the parameters of the null hypothesis H_0,
θ_s	the vector of nuisance parameters,
$\hat{\theta}_r$, $\hat{\theta}_s$	the values which unconditionally maximize $L(x\|\hat{\theta}_r, \hat{\theta}_s)$, and
$\hat{\hat{\theta}}_s$	the values which conditionally maximize $L(x\|\hat{\theta}_{r_0}, \hat{\hat{\theta}}_s)$,

and then defined the likelihood ratio l,

$$l = \frac{L(x|\hat{\theta}_{r_0}, \hat{\hat{\theta}}_s)}{L(x|\hat{\theta}_r, \hat{\theta}_s)}.$$

They then presented a charmingly simple argument for this approach for eliminating nuisance parameters, which we now refer to as "profile likelihood": "Intuitively, l is a reasonable test statistic for H_0: it is the maximum likelihood under H_0 as a fraction of its largest possible value, and large values of l signify that H_0 is reasonably acceptable."

They concluded with the prescription for the Neyman construction: "The critical region for the test statistic is therefore

$$l \leq c_\alpha,$$

where c_α is determined from the distribution $g(l)$ of l to give a size-α test, *i.e.*

$$\int_0^{c_\alpha} g(l)dl = \alpha.\text{"}$$

These terse passages from Kendall and Stuart contain the full content of what is commonly known as "Feldman-Cousins"[10] with the addition of the suggestion of how to eliminate the nuisance parameters.

Coverage is not guaranteed with this technique since the treatment of the nuisance parameters involves an approximation. However, coverage is excellent in all cases which have been studied. This is pointed out in Giovanni Punzi's talk[14] and it agrees with my experience.

There is, however, a minor problem with this technique when used with a Poisson distribution. This is illustrated in Fig. 3, which comes from Punzi's talk. As the nuisance parameter is better and better known, the confidence intervals do not converge to the limit of the nuisance parameter being perfectly known. The reason is that the introduction of a nuisance parameter breaks the discreteness of the Poisson distribution.

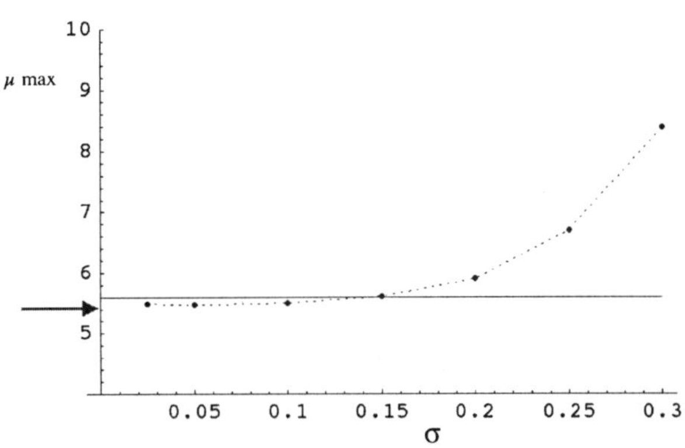

Fig. 3. The upper limit from a Poisson distribution as a function of the normalization error σ. As σ approaches 0, the upper limit falls below the value for $\sigma = 0$, shown by the horizontal line. From Ref. 14.

I discussed this problem and its solution at the 2000 Confidence Level Workshop at Fermilab.[15] However, since two plenary speakers[8, 16] indicated that the explanation was incomprehensible from those slides, I will try again here. (Also see Punzi's talk[14] for another solution.)

Consider the case of a Poisson signal of mean μ where n events have been measured for the sum

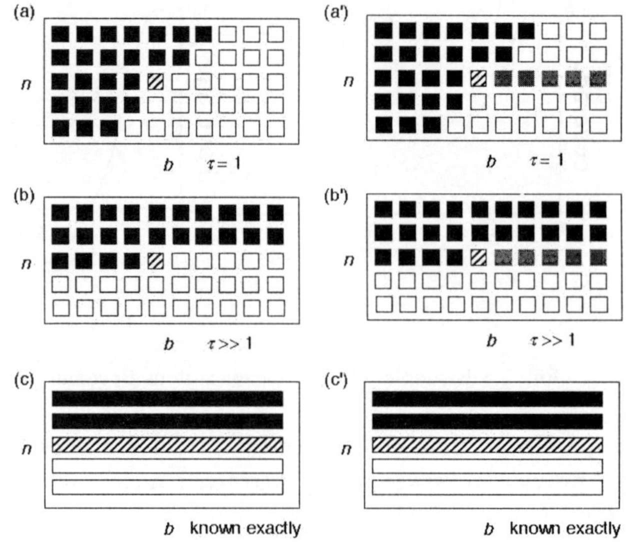

Fig. 4. Diagrams showing the reason for the confidence limits for a very well known background not approaching the limit of a perfectly known background for a likelihood-ratio ordered Neyman construction, and the solution to this problem. See the text for the explanation. Redrawn from Ref. 15 for black and white printing.

of the signal and background. The background is estimated by a measurement of b events from a sideband of τ times the size of the signal region. Figure 4 shows a fragment of the construction of the confidence belt for the value of μ that is the upper limit for the measurement of n_0 and b_0, which are shown by the diagonally-lined box. The confidence belt consists of this box and the black boxes. Figure 4(a) shows the case for an equal size sideband and signal region, $\tau = 1$. The confidence belt includes the region in the n-b space that is bounded by a 45° line since each unit of b reduces n by one unit for constant μ. Figure 4(b) shows the case for a very large sideband compared to the signal region. The angle becomes very shallow, but even in the limit of a very well known background only the lower b values are included in the confidence belt for $n = n0$. The probability for the diagonally lined square always exceeds those to its right, even if only by an infinitesimal amount. This gives greater granularity in the probabilities that can be achieved by the confidence belt construction, compared to the case of a perfectly known background, shown in Fig. 4(c). Here all the squares for $n = n0$ have the same probability and must all be included. Less integrated probability is included in Fig. 4(b) than in Fig. 4(c), allowing a lower upper limit for μ.

A solution is shown in the right half of Fig. 4. To restore the overcoverage due to the discreteness of the Poisson distribution it is necessary to add the boxes shown in gray shading. Figures 4(b′) and 4(c′) are now identical and Fig. 4(a′) will approach the limit smoothly.

The Cousins-Highland paper[12] indicated that it was using the hybrid frequentist-Bayesian technique discussed earlier because the pure frequentist technique gave a lower upper bound with a small normalization uncertainty than with no uncertainty. The reason for this is the same as shown above, and the same solution yields a reasonable frequentist upper limit.

3.5. *Hill Climbing*

Wolfgang Rolke presented a talk on eliminating the nuisance parameters via profile likelihood, but with the Neyman construction replaced by the $-\Delta \ln L$ hill-climbing approximation.[17] This is also what the popular MINUIT program does.[18] The coverage is good with some minor undercoverage. Cranmer[8] also discusses this method in his talk.

3.6. *Full Neyman Constructions*

In principle a frequentist should provide coverage for nuisance parameters by doing a full Neyman construction for both signal and nuisance parameters. Both Punzi[14] and Cranmer[8] attempt this in their talks. I do not recommend this procedure for a number of reasons.

(1) The ordering principle is not unique. Both Punzi and Cranmer ran into some problems.
(2) Unless great care is taken, they are likely to substantially overcover.[16]
(3) The technique is not feasible for more than a few nuisance parameters.
(4) It is unnecessary since removing the nuisance parameters through profile likelihood works quite well.

Appendix A. Why 5 σ?

We noted above that the standard for discovery in bump hunting has been established at five standard deviations. Given that the probability of a statistical fluctuation giving a 5 σ effect is 3×10^{-7}, is this a reasonable requirement?

There are two major high energy experiments at the LHC, but let us first consider this requirement if there were only one experiment. A reasonable expectation is that in the analysis of the experiment, there may 500 searches, each of which has 100 resolution elements (mass, angle bins, *etc.*), which yields 5×10^4 chances to find something. Thus, the chance of a false positive would be $(5 \times 10^4)(3 \times 10^{-7}) = 0.015$, an acceptably low number.

Now consider the situation with two experiments. First, we need to consider the number of allowable false positives in either experiment. False positives have some cost — they generate unnecessary experimental and theoretical activity (often a hundred or so theoretical papers) and if they are too frequent, they give the impression that results in the field cannot be relied on. I would guess that perhaps ten false positives over the course of the LHC would be acceptable. Then, we can solve for the significance level Σ that would yield ten events:

$$2(5\times 10^4)\mathrm{P}(\Sigma) = 10 \Rightarrow \mathrm{P}(\Sigma) = 1\times 10^{-4} \Rightarrow \Sigma = 3.7\sigma$$

If we take 0.01 as the acceptable probability of a false positive after verification by the other experiment, then the required significance level Σ' is given by

$$10\mathrm{P}(\Sigma') = 0.01 \Rightarrow \mathrm{P}(\Sigma') = 0.001 \Rightarrow \Sigma' = 3.1\sigma$$

Thus, it would appear from this analysis that a 3.7σ discovery by one experiment followed by a 3.1σ conformation by the other would be satisfactory. However, great care needs to be taken to consider to what extent common systematic uncertainties exist. For example, both experiments are likely to rely on common hadronic production and fragmentation models.

Acknowledgment

I would like to thank Roger Barlow for comments on the first draft of this paper. This work was supported by a grant from the U. S. Department of Energy.

References

1. P. Bhat and H. Prosper, "Bayesian Neural Networks," these proceedings.
2. J. H. Friedman, `http://www.slac.stanford.edu/econf/C030908/proceedings.html`.

3. B. Roe, H. Yang, J. Zhu, "Boosted Decision Trees, a Powerful Event Classifier," these proceedings.
4. J. H. Friedman, "Seperating Signal from Background Using Ensembles of Rules," these proceedings.
5. I. Narsky, "Optimization of Signal Significance by Bagging Decision Trees," these proceedings.
6. R. Vilalta, P. Sarda, G. Mutchler, P. Padley, "Signal Enhancement Using Multivariate Classification Techniques and Physical Constraints," these proceedings.
7. A. Gray, "How to do Bayes-Optimal Classification with Massive Datasets: Large-scale Quasar Discovery," these proceedings.
8. K. Cranmer, "Statistical Challenges for Searches for New Physics at the LHC," these proceedings.
9. J. Heinrich, "The Bayesian Approach to Setting Limits: What to Avoid," these proceedings.
10. G. J. Feldman and R. D. Cousins, *Phys. Rev. D* **57**, 3873 (1998).
11. J. Conrad and F. Tegenfeldt, "Likelihood Ratio Intervals with Bayesian Treatment of Systematic Uncertainties," these proceedings.
12. R. D. Cousins and V. L.. Highland, *Nucl. Inst. and Meth. A* **320**, 331 (1992).
13. M. Kendall and A. Stuart, *The Advanced Theory of Statistics*, Volume 2: Inference and Relationship (Hafner, 1961).
14. G. Punzi, "Ordering Algorithms and Confidence Intervals in the Presence of Nuisance Parameters," these proceedings.
15. G. J. Feldman, "Multiple Measurements and Parameters in the Unified Approach," `http://conferences.fnal.gov/cl2k/`.
16. R. D. Cousins, "Treatment of Nuisance Parameters in High Energy Physics, and Possible Justifications and Improvements in the Statistics Literature," these proceedings.
17. W. A. Rolke, A. M. López, and Jan Conrad, "Limits and Confidence Intervals in the Presence of Nuisance Parameters," these proceedings.
18. F. James, "MINUIT, Function Minimization and Error Analysis Reference Manual," `http://wwwasdoc.web.cern.ch/wwwasdoc/minuit/minmain.html`.

APPENDICES

QUESTIONS FOR PANEL DISCUSSION

The following questions were submitted by PHYSTAT2005 participants for the Panel Discussion session.

1. **GENERAL :** Do the statisticians on the Panel consider there are any techniques we should be using, but currently we are not?

2. **PARAMETER INTERVALS :** What are the properties that confidence intervals should respect? e.g. Frequentist coverage? Short but not too short? No empty intervals? Invariance with respect to reparametrisation? Robustness? Consistent approach to incorporating nuisance parameters?

3. **FREQUENTIST PARAMETER DETERMINATION :** Particle Physicists like to use frequentist approaches. Is it practically possible to use a Neyman construction in several dimensions i.e. of data and/or parameters? And if only one of the parameters is a physics parameter and the rest are nuisance parameters, so that it is necessary to project the multi-parameter confidence region in order to obtain that for the single physics parameter, what ordering rule will give optimal behaviour for the resulting one-dimensional intervals?

4. **P-VALUES WITH NUISANCE PARAMETERS :** An experiment is looking for some rare or perhaps non-existent process. It involves simply counting events. There is an uninteresting background b which also contributes to the counting rate. The number of observed events N is expected to be Poisson distributed with mean b, which has been measured in a subsidiary experiment as $b_0 \pm \sigma_b$. (This can be thought of as being determined as c/r, where c is the number of events in a situation which is sensitive only to the background, and r is a scale factor which typically could be 5, and is accurately specified with zero uncertainty. Larger r results in a smaller error σ_b). We want to calculate the p-value for the null hypothesis (only background), for observing at least N events. To be specific, we could take $N = 9$ and $b = 3.1 \pm 0.4$. What is the recommended statistical technique? It is desirable that it could easily be extended to a larger number of nuisance parameters.

5. **PROFILE LIKELIHOOD AND BAYES PRIORS :** If I understand correctly, eliminating a nuisance parameter by using the profile likelihood is equivalent to using a delta function prior for the nuisance parameter in the Bayesian philosophy. More refined likelihood methods can probably also be interpreted in the Bayesian way. If this is correct, why doesn't one use directly Bayesian methods?

6. **BAYESIAN TREATMENT OF SYSTEMATIC UNCERTAINTIES :** In a Bayesian approach to parameter estimation, it is straightforward to include nuisance parameters in a Monte Carlo Markov Chain, then marginalize over them in order to recover high probability regions for the parameters of interest which include the effect of our imperfect knowledge of the nuisance parameters. I am uncertain about the proper way to treat systematical errors in this context: an example could be the uncertainty associated with numerical inaccuracies of the code used, or the error induced by the fact that some second-order physical processes have been neglected in the code. This results in an uncertainty associated with the output of the code itself (which I'd classify as "systematical"), rather than a statistical uncertainty, associated with the data used. A common way to deal with this is to add the statistical and the estimated systematical errors in quadrature (or linearly if one wants to be conservative), then use this new artificial error on the data at hand. I would like to know whether there are more satisfactory ways of dealing with systematical errors of the kind described above, and in particular methods which recognise the different nature of the statistical and systematical errors.

7. **MAXIMUM ENTROPY PRIORS :** Setting the prior correctly by making use of all available information is a central problem of Bayesian model selection, where the result is strongly dependent on the prior scale and does not disappear with better data (as it is the case for parameter estimation). I am getting interested in ways of setting the prior by maximum entropy arguments. I would like to hear the opinion of the Panel regarding this method, and in particular whether this way of determining priors is now well accepted in the Bayesian community. I would be interested in comments about the applicability and limitations of maximum entropy priors, if possible with examples illustrating situations where that kind of argument has been proven successful (or has failed for a clear reason).

8. **PARAMETER DETERMINATION/HYPOTHESIS TESTING :** To what extent are hypothesis testing and parameter determination equivalent? Are there simple examples to illustrate when they are equivalent and when they are not?

9. **MULTI-DIMENSIONAL CLASSIFICATION WITH VERY MANY VARIABLES :** Analyzing HEP data, physicists more and more often deploy multivariate classification methods. We have seen a bunch of HEP publications where analysts separate signal and background by training a neural net on 10 or more input variables. Byron Roe and his associates in their recent work on PID at MiniBoone used 100 input variables for classification by boosted decision trees; this seems to set a record on the dimensionality used in HEP analysis. At the same time, there is a number of conservative physicists who refuse to adopt such multivariate methods with many input dimensions. They argue that it is very hard to assess how well Monte Carlo models data in so many dimensions, especially if one needs to take various systematic effects into account. Is there a generic prescription that relates the maximal reasonable dimensionality to the size of available Monte Carlo and data samples, in the context of a specific classification method? Can professional statisticians recommend good literature on variable selection, perhaps for two different problems: a) statistics- and b) systematics-dominated analysis? Should we attempt multidimensional analysis with dozens of input variables only when systematic effects do not matter much - for example, in rare signal searches - and stick to more robust and simple-minded techniques when systematics are important?

10. **MULTI-DIMENSIONAL CLASSIFICATION WITH VERY MANY VARIABLES :** I have a data set where each entry is described by 18 variables. I want to reduce the dimensionality from 18 to, say, 4. Are there well-understood techniques for doing this, so that some measure of information loss is minimised?

11. **NUMBER OF VARIABLES IN MULTIVARIATE PROBLEMS :** I assume we have 11 variables per event. The reason this is of interest is the observation that the D0 experiment obtained its most precise measurement of the top quark mass using the so-called matrix element method, a method that CDF is working on also. In the matrix element method one writes an explicit formula for the N-dimensional differential density, based on one's knowledge of the matrix element squared and the mapping from partons to observed objects. The premise is that, if one knew this N-dimensional density, one need look no further in terms of the search for new variables – one would just use the density directly. So the question is this: Given $p(x)$ where x is N-dimensional and given $q(y)$ where y is M-dimensional and $y = f(x)$ (and perhaps $M > N!$), can the use of $q(y)$ yield better signal/background discrimination than the use of $p(x)$? We spend a lot of time constructing $y = f(x)$, by hand! Is this necessary, if we have $p(x)$?

12. **KOLMOGOROV-SMIRNOV :** Is there a good method of using the Kolmogorov-Smirnov goodness of fit test with multi-dimensional data?

13. BLIND ANALYSES : What do you do when you unblind your analysis and find something in there that wasn't predicted by either the background or signal estimates? This is usually the case where there's no data off source region to estimate the background and the background estimate is purely simulation based. You predict the background, open the box and find a big excess but then see that the excess lies in an event-observable region that doesn't look like either signal or background predictions. You guess it's an unsimulated background, but then what? I would have thought that maybe you could possibly cut it away and recalculate. But then I learned from Kath Rawlins that LIGO saw a similar thing in their analysis and she had showed that coverage got all screwed up if one removes events post-unblinding, even if one could tell for sure that the extra events were a background from some source you hadn't considered and that, had you known about this class in the first place, you would have designed cuts to never allow them into the final analysis.

14. PULLS : If we have a complicated parameter-estimation technique, we may want to use Monte Carlo simulation to check whether the procedure is behaving sensibly. One way to do this is to look at the distribution of

$$\text{pull} = (p_f - p_t)/{\sigma_f}^2$$

where p_f is the fitted parameter, p_t is its true value, and σ_f is the estimated error in p_f. The distribution is calculated for repeated simulations. Asymptotically and if all is well, we expect the distribution of pull to be Gaussian centered on zero with unit width. However there are simple non-asymptotic examples where this is not so. How do I know for small sample simulations whether deviations from standard Gaussianity is a cause for worry or not?

15. ASYMMETRIC ERRORS : Are there recognised methods for dealing with asymmetric errors? These often arise when estimating parameters in low statistics experiments? For example, we may measure a lifetime as $1.6^{+0.6}_{-0.3}$ picoseconds. Then we might be interested in comparing it with another measurement (e.g. taking the ratio of it with another lifetime); incorporating another contribution to the error, possibly also with asymmetric errors; or combining this result with another to obtain a weighted average.

16. ASYMPTOTICS : Can there be or has there been progress to improve asymptotic techniques' convergence in the tails? In particular the modified and adjusted profile likelihood techniques attempt to improve convergence of the first and second moments, but for a 5σ test we are more interested in describing the tails.

17. IMPROVED LIKELIHOOD TECHNIQUES : Under what circumstances do improved profile likelihood methods help? We have a range of N from 10-10,000 events, we have a range of likelihoods from Gaussian to highly-non Gaussian, and we are interested in a range of significance levels from $2 - 5\sigma$. Under the circumstances that they do help, how much do we stand to gain?

18. JAMES-STEIN ESTIMATOR : There exist other shrinkage estimators that could be helpful for high-dimensional problems like Supersymmetry. The estimators are biased, which will alarm most physicists and provide an obstacle for the methods to be accepted; however, these estimators can significantly improve the mean-squared error. Do you have any words of wisdom regarding these estimators: when are they a good idea when are they a bad idea?

PANEL DISCUSSION

Panel Members: Bernard Silverman, David Cox, Jerry Friedman and Bob Cousins
Chairman: Louis Lyons

Louis Lyons I think members of the panel need no introduction whatsoever: Bernard Silverman, David Cox, Jerry Friedman and Bob Cousins.

What I was going to do was to ask the members of the panel if there are any of the questions they would particularly like to talk about. I told David I would give him the first go, so David let's start with you.

David Cox My impression is that most advances in the use of statistical methods come not from looking through a library of techniques which are available but by those with a primary interest in statistics and with some knowledge of the subject matter sitting down over a period with a group in the subject field who have some knowledge of statistics. Issues of formulation are crucial. Eventually, one hopes, new ideas will emerge that both address the subject-matter questions at issue and which maybe will be more widely useful.

Most of the really interesting and important developments in statistical methods have come that way. Direct transplantation of ideas from one field into another are less commonly totally successful. Now collaboration with physicists and statisticians is going to be particularly difficult for statisticians because you know so much already and also of course you have an enormously strong tradition of independent mathematical thought.

To address Question 1 I feel it is helpful to think of statistical methods in four chapters and two of the chapters have been quite strongly represented in this meeting, which incidentally I have highly enjoyed; it has been much more interesting than most statistics meetings.

The four chapters, not in any particular order, are first of all likelihood, Bayes, Neyman, Fisher, confidence interval calculations and so forth; clearly there is a lot of interest in that. It may be there are things in the statistical literature, and matters that could be absorbed into the thinking, perhaps particularly the notion of taking profile likelihoods and modifying them to make them perform better, particularly when there are a large number of nuisance parameters which is the critical issue in many contexts.

In a sense at the other extreme, there's the enormous collection of particular statistical methods that had been found useful somewhere or other, some highly exploratory, some partially exploratory, some graphical, some numerical. You might say that that's statistics as covered or potentially covered by R.

But I mentioned there are two chapters that have been hardly represented here. Now one would be what we call roughly applied stochastic processes. I know in some contexts this might not be regarded as a part of statistics at all. I mean constructing particular probabilistic models often dynamic, but not necessarily so, and studying their properties as issues in applied mathematics rather than in the pure mathematics of probability theory. It may be you regard all that as part of theoretical physics anyway and not part of statistics. So there is the issue of constructing new models that incorporate the physics and the probability, not quantum probability, but the physics of observational probability into new models.

Now the fourth chapter that is scarcely represented is the issue of the design of investigations and there are two major sections of statistical theory here, one to do with sampling and I point out that includes stereology. The other aspect is the design of experiments and I think this particularly relevant here to the design of computer experiments, including systematic sensitivity studies of models with many adjustable input features. Notions of Latin hypercube sampling and fractional replication which come originally largely out of technological industrial statistics could be useful.

Louis Lyons What exactly is stereology?

David Cox The study of the properties of objects in three dimensions by two or one dimensional probes or in general k dimensions, by appropriately sampling in lower numbers of dimensions.

Louis Lyons Well I'm sure that your first remark, about getting statisticians involved in analyses, will be welcome news to members of the physics department here and we'll be knocking at your door with problems that we would like to have you help us solve. Do any other members of the panel want to add anything to this?

Jerry Friedman In terms of the machine learning component, you seem to be coming up to speed rather fast. If you want to see if there are some things you have overlooked that might be useful, I'll shamelessly recommend our book. Get it from the library and go over the table of contents and see if you find some things in there that look a little strange that you haven't seen before and go to the relevant chapter. But as I said, I think that in the machine learning area you have got up to speed rather fast.

Bernard Silverman A very interesting aspect of this Conference is that the main impetus for the development of statistics in the twentieth century was its relevance to agriculture and biological and later medical applications. To see it in a physics context is fascinating because one of the things you seem to be doing, quite reasonably, is to recapitulate a lot of the discussions that went on in statistics in other contexts previously and to appropriate those discussions to the physics context. I suspect you need to think more about Bayesian methods, and to be more comfortable about using Bayesian techniques. While not everyone on the panel would necessarily approve, that has been a major shift in statistics in recent decades. There are many issues in these questions where essentially frequentist methods are more problematic, and the Bayesian approach might be more natural.

But I would really like to stress what David Cox said, which is that you should not think that statistics is about techniques. You would never think physics was about techniques, would you? Physics is about ideas and understanding and intuition and so on, and the techniques are just on top of that. Astronomy is not about techniques, is it? It's not, because it is about ideas, and the techniques are just a way through to the ideas. It's the same with statistics. Statistics is a developed scientific field and it is not a collection of techniques. It's a way of thinking which then gives rise to techniques and it is important to get into that way of thinking, rather than to say "Wouldn't such and such a technique be more appropriate for this problem?"

Louis Lyons Let's move on to Bernard on Question 4.

Bernard Silverman Now you'll probably discover when we do this that when you have two statisticians you have three opinions. I was fascinated by Question 4 in a way because the real issue to Question 4 is to try and formulate the problem in such a way that you can actually see what's going on. This problem seems to have arisen from someone saying : "I've got an experiment and I have observed the data point x which has a Poisson distribution with parameter λ and I want to know if λ is bigger than λ_0". Then they said "By the way I know something about λ, I know that λ is $b_0 \pm \sigma_b$." Then you realise that that's not what they know at all; in the question we were posed we were told it could be thought of as being determined in a subsidiary experiment. So what I actually had to do on reading this was to re-work out the original experiment. There seems to be another experiment going on where there is another data point y which has Poisson distribution with parameter $r\lambda_0$. So it isn't only the main experiment that is conducted, but this other one as well, and really you just need to make sure you have written both of them down. Once you do that, you can see that to do any inference at all you want to know what the distribution of x is, conditional on y.

One way to approach this is to take a Bayesian viewpoint, putting a prior on λ and then calculating $p(x|y) = p(x|\lambda) * p(\lambda|y)$, using Bayes theorem to find the latter conditional probability.

The key point, however, is not the use of a Bayesian vs frequentist approach, but the need to go right back to the experiment that was conducted, to look at all the experiments at once and write everything down about it. That's better than saying that b has been measured in a subsidiary experiment as something or other.

Louis Lyons We've got two more statisticians here so I expect four more opinions! Jerry, David do either of you want to say something about that?

David Cox And the audience could have opinions too, maybe an infinite number!

Jerry Friedman I'm going to disappoint you, I don't have another opinion.

David Cox You have to write down the likelihood. It's the key and then the issue is what do you do with the likelihood. If you have a prior that is evidence-based, I don't think anyone would dispute that you should use it. If you've not got such a prior, you're into this issue of reference priors and flat priors, and there's been a lot of work about that, in a way for two hundred years, and certainly for the last two days. It's a minefield and done properly the reference prior would give a beautiful answer that we are all satisfied with, but what does a beautiful answer mean? Well if I wanted to be argumentative, and I'm totally not argumentative, I would say if it gives something that has at least tolerable frequentist properties. If it gives an answer that has very bad frequentist properties, I can't think anyone would defend it. So in the end I would be very happy to use the Bayesian formalism as a way of getting an answer. I have done so and have no qualms - if that's the way to get an answer, I'll do it. But in the last analysis, I have to say, "I doing something that is going to produce answers close to the right answer most of the time, in other words is it calibrated properly?" which is much about what that means. In some sense I would see it as a hypothetical frequentist interpretation.

Steffen Lauritzen Just to keep things down to earth, in this example when you formulated it the way that Bernard did, if r is known, then $x + y$ is sufficient for λ and it becomes a straightforward simple hypothesis in the conditional binomial, just with the scale factor $1 + r$ entering, whether you are Bayesian or not.

Louis Lyons Bob do you want to say anything?

Bob Cousins Well I had my 40 minutes on exactly this problem so I won't repeat too much. I'll just say what Sir David said, that at the end of the day you want to say that some probability P equals some number. If a student asks "How do you define probability P in that statement?" you would like to be able to give them an answer. I think that in our field the frequentist answer is the easiest one to explain, but I'm also perfectly happy with the subjective answer. What I'm unhappy with are the ones in between, those priors which are used in a Bayesian machinery but unless the probability P comes out with the right frequency I don't know how to interpret it. Since this particular problem is so interesting in our field, it's been studied a lot and for the variety of methods that I talked about, their frequentist coverage has been studied. I refer, for example, to the papers by Conrad and collaborators on integrating out the nuisance parameter formally and seeing how it works. What we call the MINUIT MINOS method is profile likelihood, and is discussed in a recent paper by Rolke, Conrad and Lopez. Those papers contain a lot of information, and they also tell you what sort of study you can do with whatever technique you use. Then if you're really up to it you can do a full-blown frequentist construction the way Kyle Cranmer and Giovanni Punzi talked about. Then you don't have to check if it covers because it does so by construction, but it likely overcovers. So that is interesting to study as well.

Luc Demortier I just thought it would be useful to clarify that when statisticians say flat priors they don't mean uniform priors, they just mean objective priors. Am I correct in that because there might be some confusion? We try to discourage the use of flat uniform priors but not necessarily the use of objective priors or reference priors. But the statisticians sometimes use the terminology flat prior to mean objective prior in general.

David Cox I certainly had in mind either the Jeffreys prior in simple cases or something like Bernardo's reference priors in more complicated ones.

Bob Cousins Well it's OK because you did not say in what metric your prior was flat! It will be flat in some metric!

Tomi Zivko I would like to give a short comment about priors. Yesterday I gave a talk in which I presented work of my colleague and myself. In the talk I claimed that we started from Jaynes, Polya and Cox's desiderata. That means that we started from a purely Bayesian point of view, and using only those assumptions which are found in the desiderata, we obtained calibrated solutions, which is a purely frequentist result. But there were no comments after the talk, no objections.

Louis Lyons OK so I guess people need to have a chance to read it and absorb the ideas there.

Maybe we should move on to another question. I think Bob hasn't had a chance to choose anything yet.

Bob Cousins So I've thrown out all the ones about nuisance parameters and Bayesian analysis, and I've instead chosen the one about blind analysis (Question 13). I have been involved in three experiments that performed blind analyses, including the BNL E791 rare kaon decay experiment, where the blind analysis led by Bill Molzen seems to have led to widespread use of blind analyses in HEP, and also one including Josh Klein, who has recently written a review with Aaron Roodman [Ann. Rev. Nucl. Sci. 55 (2005) 141]. So the question is "What do you do if you open the box and despite of all your due diligence, you see that you've been stupid and there is an obvious background that you did not anticipate?" This actually happened to me. We opened up the box and looked at the events in the signal region, and found that two events had all their ADC (analogue-to-digital converter) readouts zero – not even at the pedestal value, but really zero. So we came up with the criterion that it is OK to throw away an event after you open the box, if you would look foolish by not throwing it away. You should feel free to throw away an event, rather than go to a conference and stand up and say "I'm going to stick to my principle of blind analysis and keep this event, even though my read-out was not working."

In our case, the effect on acceptance was completely negligible when we added the 'ADC read-out was working' cut to all events, and it was such a clear-cut case that we did not take any further action. If this was not the case, however, as the question points out, this can introduce a bias in a subtle way. Suppose there are twelve possible backgrounds that you haven't thought about, and if you cut on them each would introduce a 5% inefficiency. Then you open up your box you find there is only one of them that actually appears that you haven't thought about. You add this one cut and take a 5% hit on efficiency. But if you had really thought about all your backgrounds in advance and decided to eliminate them, you would have had twelve hits on efficiency, each of 5%. So this is a source of bias one is left with, but probably it is even worse for an unblind analysis. In practice I have found that people doing blind analyses are very good about thinking hard about potential background sources precisely because they want no surprises when they open the box.

There is another principle I'm convinced of from seeing all these blind analyses. You should freely look at 10% of the data inside the blind box, especially if it's a new experiment and it's not the third data set or what not. If you are going to tune your cuts on 10% of data, even if you do a bad job, the bias is not very strong. So I would first of all try to prevent this problem of unexpected backgrounds by looking at 10% of the data; and second of all I think if you really can improve your analysis after you open the box (for example by better calibration), then go ahead and improve the analysis.

Louis Lyons You are assuming you keep that 10% in the final analysis?

Bob Cousins Yes, keep the 10%. There are plenty of people in the world who think you can tune on 100% of the data and still have valid answers, so I'm perfectly happy to tune on 10% of the data and keeping that 10% in the final sample.

Gary Feldman I just wanted to add to Bob's comment that Josh Klein and Aaron Roodman have recently written a very nice paper on blind analysis and in it they have a great line which says "Doing a blind analysis is not an excuse for publishing a wrong result." Their recommendation is if you find a problem once you've opened the box, fix it and then just explain in your paper what you did.

Louis Lyons There's one blind analysis that I recently heard about and that's from the TWIST experiment at TRIUMF. They are doing a precision measurement to determine the value of a parameter by comparing their data with Monte Carlo simulations with different values of the parameter. They blind the value of the parameter used in the Monte Carlo experiment. So they can look at their data as much as they like and see if there any problems there because it's the Monte Carlo parameter that's blind. That seems quite a nice technique.

Bangalore Sathyaprakash I'm part of the LIGO scientific collaboration looking for rare events in our gravitational wave detector. To cut down the background, we look for coincidences within a certain time window - when there is an event in one of the instruments, we look for an event in the other instrument. After opening the box, we later discovered that one particular event was associated with an aeroplane flying over the instrument. Now, who could have thought an aeroplane-veto beforehand? So it was very hard. So I'm very heartened to hear that the advice that we should not just do blind analysis blindly. When you open the box, if you find something funny, just go ahead and allow for it in your further analysis.

Rajendran Raja The dangers of a blind analysis far outweigh its benefits since while you are blinded you are not monitoring the data in the blind box. Any loss in objectivity in an unblind analysis can be overcome by having simulators that model the data well and using the simulators to set the cuts. If D0 had been looking for the top quark blindly, we'd still be looking for a 65 GeV top quark.

Byron Roe I agree with Raja that one should be cautious about using blind analyses. I know that we have a minority opinion on this, but to me blind analyses are very useful when you don't know what to do with the data to give you a positive result for the parameters of interest. That is, an unconscious bias will not be able to bias the data to give you a positive result. The blind experiment I'm in is MiniBooNe, and in that experiment, you do know what to do. I think that diminishes greatly the value of blindness.

Now at breakfast Gary Feldman was pointing out there is a second problem called the stopping problem where you keep going until you've got an answer you like and then you stop looking for corrections. That certainly is a point to be worried about but you have to balance against that the problems that you introduce in your analysis by having things blind. Surely you can always correct it afterwards, but it certainly is not such a great idea if you can do the analyses correctly in the first place.

Louis Lyons OK, so maybe we move on to another question. Jerry would you like to choose a topic?

Jerry Friedman I guess the message I get from this discussion of blind analysis that it's good to think outside the box!

There were a number of questions about machine learning, mainly about variable selection that can be dealt with rather quickly so I'll just try to give short answers to many of them, and some of them were the same. There were a couple of questions about variable selection and whether there are well understood techniques for selecting subsets of variables. The answer to that question is 'yes' but the question should be whether there are good techniques for variable selection. Variable selection techniques fall into two categories of filters and integral. With filters, before you apply whatever machine learning procedure you are going to use, you apply a different procedure to filter out bad variables before you run the machine learning procedure. This is often fast, but the problem is that the criterion you are using to select the variables is not the machine learning procedure that's going to use them, and so you may filter out useful variables. An example I have seen here is where you look at the power of each variable one at a time and filter out those appearing to be weakly related to the outcome variable. The problem is that a variable may not be strongly related by itself, but in combination with others it may have quite an effect. So the best way to do variable selection is in the context of the procedure that is going to be using the variables. I discussed this a bit in the techniques that I talked about during my talk where you actually get the relevance of the variables as used by the procedure that was trying to do the learning. Then you could filter out the ones that the algorithm said it didn't need.

There was another question about using many variables when you are suspicious that your Monte Carlo may in fact not describe the experiment; so maybe you should use less variables rather than try to use more variables that attempt to capture more features to do the discrimination between signal and background. I'm certainly sympathetic to the fact that the Monte Carlo may not be exactly correct. This is a common problem even when you have actual data. We call it non-stationarity in statistics and concept drift in the machine-learning literature. You take data at some time, you build a model that can describe that data rather well and you can cross-validate it. But then you apply it in the future and it doesn't work very well because the relationship between the variables has simply changed – you don't have the same system any more. There's been some work on that, but it's very hard and nothing is really satisfactory. In all these machine learning techniques, the presumption is that your training data is a random sample from the population of the future predictions. If that's not the case there's really not a lot you can do. You can try and over-regularise, because regularisation implies not fitting your data as well as you can. There are two reasons for doing this. One is because the data is random and randomness can lead you astray. That's the kind of thing statisticians deal with. Then there is the situation where the data has simply changed and again a solution to that is not to trust it too much, don't fit it quite so strongly. But systematic ways of doing that in the presence of concept drift are really not well developed in the machinery described in the literature.

There was one question on the Kolmogorov-Smirnov method for the goodness of fit test for multi-dimensional data. I was amused by that, well actually nostalgic, because as a young physicist it was that problem that got me interested in statistics. I realised back then that it was an important problem so I started to try and solve it. That led me into the statistics literature and I said "Hey this is pretty interesting". So that's how I got into statistics and I never left. At the last meeting I talked about general multi-variable goodness of fit testing and a procedure for doing that based on machine learning procedures. That's in the PHYSTAT2003 proceedings. Actually, the first statistics paper I ever wrote came out of this problem. It is published in the Annals of Statistics, and had in its title "Kolmogorov-Smirnov test in high dimensional data". It's around 1989-1990 Annals and if you look plus or minus a year in those you'll find a multi-variable generalisation of the K-S test and multi-variable generalisations for other goodness of fit tests like the Wald-Wolfowitz. You cannot straightforwardly extend K-S tests in high dimensions in the obvious way using the multi dimensional CDF. That doesn't work at all because of the curse of dimensionality. The CDF of the joint variables would tend to realise only two values (either 0 or 1) because it's the number of observations that are dominated by whatever point you are considering in the high dimensional space. You tend to dominate very few points simultaneously in all of the dimensions.

Are there any other machine learning questions? I thought Question 11 was kind of interesting. You have a system where you know that the target function can only be a function of a certain set of variables, like the matrix element technique talked about in the question. Since you know all of the variables that the target function could possibly depend upon, is there any value in constructing new variables that are functions of those variables and extending the variable set? The answer is 'yes'. The reason for this is that all machine learning procedures have some functions that they are good at learning and some functions that they are bad at learning. When you add the new variables, you change the function. I tend to try and understand things sometimes by considering extreme cases. As an example, suppose that the background was on a two-dimensional ball and the signal is on a larger ball surrounding it. It's only a function of those two variables. Given a perfect procedure those two variables would separate perfectly. But given finite data and an imperfect procedure (and all procedures are imperfect), if you added the variable which is just the sum of the squares of the two and put that into your machine learning algorithm you'd do a whole lot better. It would ignore x_1 and x_2 and just use $x_1^2 + x_2^2$. So the answer to that is 'yes' and the best way to do it is to use knowledge about the problem. If you use a technique based on trees that are not sensitive to lots of irrelevant variables, then you can feel free to add many derived variables if you have any suspicion at all that they might be useful. Then the natural variable selection technique of tree-based procedures will weed them out if they are not good, but include them if they are.

Finally, concerning the question on the James-Stein estimator. I guess I would answer that with a question: "What on earth is wrong with biased estimators?" Accuracy is an important thing, and if the lack of accuracy comes not so much from bias but from the variance, it is still lack of accuracy. If you can get a much more accurate answer by allowing a little bit of bias, I just don't see the downside.

Louis Lyons I was just going to say, maybe not every member of the audience knows about James-Stein estimators. So could somebody provide us with a two sentence explanation of James-Stein estimators?

Bernard Silverman James-Stein simply says this: If you are estimating a parameter of a large number of dimensions, then shrinking it back towards zero will give you a more accurate estimate than simply making it equal to the data. James-Stein in fact works with about four or more dimensions, but if you had thousands of observations - suppose you had a vector of a thousand parameters and you had one observation on each parameter - it's pretty obvious that it's much better to shrink the observations than simply to let them be equal to the parameters.

I want to add something to what Jerry said about Question 18 which is very interesting. It is not about James-Stein estimators as such, but to note that the discussion around this question is an example of recapitulating discussions that have gone on in statistics before. The objections raised in Question 18 are what people said when the James-Stein estimator was originally suggested, and so don't be surprised or worried if you're alarmed by issues which statisticians have been alarmed at before. The thing to bear in mind is that we may have thought through some of them already.

Jerry Friedman There are lots of other shrinkage estimators that you can probably use, such as ridge regression, and lasso regression that I mentioned in my talk. Those are all shrinkage estimators and in general for prediction as opposed to estimation, shrinking is almost always a lot better than selecting variables.

Variable selection is also a shrinkage estimator by the way; you just shrink some of the coefficients of the variables to zero.

David Cox The mathematically peculiar thing about the James-Stein estimator is that there is an apparent gain if you shrink towards anywhere. You've talked about shrinking towards the origin, which is natural, or generally to some linear space represented by a regression. If you shrink towards that, you can view that as a kind of empirical Bayes, even though it is not explicitly formulated in that way. The mathematical paradox or semi-paradox about it is that you may shrink towards anywhere you like, but you won't gain very much.

Going to the unbiased estimates, I can see only one situation in which unbiased estimates are particularly compelling. That would be if you had a lot of data in sections and you had a parameter that you were interested in for each section. You analyse each section of data separately and you get an estimate of that parameter and then you put those estimates into some linear representation. Then biasing the estimates would be a systematic error that persisted through the whole analysis.

Jerry Friedman But could you combine the data and do a general analysis with the shrinkage?

David Cox Yes, but in some contexts it is both easier but also I think more insightful to proceed stage by stage and you can see, as it were, what's happening in each bit of the data first before you put it into some big system.

Bob Cousins I'll just make one point that the way bias is usually expressed is in terms of a mean, and that is a metric-dependent statement. The most common bias we learn about in freshman's physics class is to take a sample variance and correct by $n/(n-1)$, which is to correct for the known bias in the maximum likelihood estimate. This gives a non-biased estimate for the variance but the RMS is biased. I think that for historic reasons it was probably defined that way. Fred James has suggested that if you are going to worry about bias, you should consider trying the median instead of the mean, because the median is independent of metric.

Jerry Friedman One last thing is that Bayesian techniques seem to be popular in Particle Physics, and all Bayesian estimates are shrinkage estimates.

Steffen Lauritzen Just to add to the number of opinions, what would worry me about the James-Stein estimate is certainly not the bias (and in that sense I agree completely with the rest of the statisticians) but rather the lack of invariance on changes of units and scale. I think that if I was a physicist, this would send a chill down my spine.

Kyle Cranmer I'd like to address this question and the physical context in which I looked at it, and I ran into exactly these changes of scale issues and funny things like that. We are looking at super-symmetry and what we would do in that case is that we'd measure masses of 10 or 12 particles, something like that, and the idea that we'd then shift all the masses from what we actually measured to stick into some other calculation seems really bizarre at first.

I'd like to briefly extend the question as if what we have really is some fundamental theory that might have a lot of parameters like a super-symmetry with 105 parameters and they predict the masses of these particles. Then we'd measure the masses of those particles. We'd be using the James-Stein part to try and improve our estimator of the masses but then back propagate that to try and get the parameters of the more fundamental theory. I'm wondering in that more extended context – if that made any sense – are there any more things to worry about? It's a sort of two step procedure.

Jerry Friedman There are always plenty of things to worry about.

David Cox Could I make a couple of comments about the bias issue? One is the issue of $n-1$. Of course it doesn't matter in one set of data but if you have variance being built up from various different sources which you think have about the same variability, then it would matter for the same reason that I indicated before.

The other point is that empirical Bayes estimates are typically shrinkage estimates, but not necessarily. If you do empirical Bayes and allow priors with very long tails, highly non-Gaussian, then under some circumstances empirical Bayes can be anti-shrinkage. It can take relatively extreme observations and push them further out rather than pull them closer in. One doesn't often see that, but certainly mathematically that's the situation.

Harrison Prosper I think our obsession with bias is really historical. In the old days it was much simpler to have each experiment do their analysis internally and then provide some summaries of what they have done. In that circumstance of course you'd like to have the summaries be such that you can combine them linearly and have an unbiased answer. Today we have 2, 3 Gigahertz machines and thousands and thousands of CPUs all over the planet. If ever we got to the point where we'd be willing to publish our data in some form that could be usable by other people, we could then do what Jerry suggests, which is that you do a large analysis of all these data and the whole issue of whether the thing should be unbiased becomes moot.

David Cox There's also of course the connection indirectly with Question 6. It seems to me a particularly important question and that's headed "Bayesian treatment of systematic uncertainties" because the terminology bias tends to suggest that a biased estimate is the same as one with a systematic error and that in some sense is misleading. Systematic errors are surely very important and there is a lot of concern in many fields that conventional statistical analyses, largely whether they are Bayesian or frequentist, deal with the errors that arise out of the random variations in the data, not out of any systematic errors in measurement. Bias from them is assumed eliminated by design. It is a strength of the Bayesian treatment that if you can put a reasonable prior on these systematic errors, then of course they can be incorporated into a fuller assessment of error. One danger there concerns independence assumptions.

Louis Lyons In particle physics some of the systematic errors come from trying to correct for biases and then the contribution to the systematic error will be how uncertain we were that we have allowed for this bias correctly.

Bernard Silverman The question posed is quite interesting. It uses the words 'systematical errors' but it's trying to get at some other kind of error which is an error which is in some sense unknown. But we know roughly what it might be and so if I were the prophet Dennis Lindley I would say the problem with the way the question is posed is that the last line isn't particularly correct. It says "I'd like to know whether there are methods that recognise the different nature of the statistical and systematical errors". Within a pure Bayesian way of thinking there is no different nature between the systematical and statistical errors. That's the whole point. For a Bayesian there is only one kind of randomness. Errors do not have different natures, they may have different origins physically, but in terms of how you model them they do not have different natures. That's the strength and weakness if you like of the Bayesian approach.

Geoff Nicholls I agree that in the Bayesian inference there is just one kind of error – the modelled error, errors that you've accommodated correctly, the statistical error from the fluctuations, that is to say uncertainties due to randomness in the realisation of the data. Our Bayesian error bars measure this error very well. The focus in Bayesian inference is on fitting a parametric model – so model misspecification errors, biases caused by fitting the wrong model, are not expressed in the error bars we report. One of the things I've found interesting about the physicists' contribution to systematic errors is that they attempt to report them in a rather explicit way. I like the way you often see in physics papers $\pm x$ followed by $\pm y$, the second being an attempt to quantify uncertainty due to variation in the model. You can formalise this model-error by fitting a larger class of models, but that often isn't computationally feasible. So the physicists' approach of simply having a go – considering at least the obvious and physically important modes of model variation – is a lot better than simply ignoring the problem.

Rajendran Raja I would like to speak about the distinction between the systematic errors and the statistical errors. In an experiment there are quantities which have a certain frequency of occurrence. Some of them are longer lived than others. So depending on how long the experiment lasts, some things will be systematic. For example, if CDF lasted a hundred years the luminosity error will be statistical because the luminosity errors would change many times during that time, but if it lasted a few years the error on the luminosity will be systematic because it will have one value during that interval of time. That's something that I haven't seen discussed, as to when errors become systematic as opposed to statistical, depending on the timescales involved.

Louis Lyons Can I encourage members of the panel to express a view on Question 2, about parameter intervals? When we estimate some parameter and get some range for the parameter, what properties would we like these intervals to have? Has anybody got an opinion on that?

Bernard Silverman They should have the properties they are claimed to have! If they are confidence intervals, they should have frequentist coverage.

Louis Lyons OK but we could widen the question a little bit so it didn't necessary say confidence intervals but rather any intervals. Is coverage an overriding feature? How unhappy would we be about the method that gave empty intervals or sometimes very very short intervals? What should we aim to do when we are investigating methods for producing confidence intervals?

David Cox Assuming you are using a reasonably high level of confidence or posterior probability or whatever, any true value should lie within the interval most of the time – that would be my answer. Secondly there is the issue that in most cases, I think one wants not intervals but upper limits and lower limits separately. For instance the Poisson problem is very clearly a situation where you can formally do a good job for the upper limit, but perhaps all you could possibly say about a lower limit is that it could perfectly well be zero. So there's that aspect.

No empty intervals? Well think one must accept empty intervals, in certain situations, because if the confidence interval, or a posterior interval, is a list of those parameter points that are reasonably consistent with the data and the model, the answer may be that no value is consistent with the data.

Then there is a complementary problem where the confidence set is the whole space, and any parameter value is consistent with the data. Again you are making a statement that's trivially true. I don't see the difficulty with that as a formalization of what the data imply.

Bernard Silverman Would you be worried by a very short interval? There is a danger here; you could have a very short interval because the model didn't really work and there was only a very small parameter set that fitted, so it's a sort of limiting case of the empty interval. That's scary because people would interpret that to mean we have estimated this parameter with very great accuracy, where what we've actually said is the model doesn't really fit but there is a very small range where it just about OK.

David Cox Yes but there's qualitative prior knowledge involved in judging what is small and that prior knowledge has to be used, if only informally.

Nancy Reid I think there's been a little bit more emphasis than necessary in the physics literature on exact coverage, that seemed to be bordering on an obsession! Coverage refers to this property in the long run over a whole lot of experiments and something that's been argued about in statistics over many many years is how many of those conceptual experiments are relevant to the one you have. That was laid out most clearly by David in 1958 with the two measuring instruments problem. I think that, in some cases, you are almost duplicating the two measuring instruments problem, by using the Neyman construction to get intervals that are guaranteed to cover in such a wide variety of situations that you're losing for the particular situation that you are going to use it. There's lots and lots of literature on this and it's not an easy literature to study but I think you'd be well advised to consider a little bit more the more pragmatic view that's been expressed by David and Bernard. It came up in the question on blinding: "If I look at the data then I'm going to ruin my coverage." But that coverage refers to a whole lot of perfectly carried out experiments where nothing weird happened, and you have a different experiment where something weird has happened, so that coverage is not really relevant in that situation. There's obviously a tension because every experiment is in some sense unique but we are talking about statistics so we have to average over something. So there is a tension between the two and it's not easy to resolve, but I don't think the right resolution is to average over everything.

Sergei Bityukov I want to express my opinion about confidence intervals. If we have a procedure which allows us to construct intervals, we can also construct the confidence density, and that contains more information.

Bruce Yabsley Just on the question of empty intervals and why we might be concerned about them: It's my impression part of our problem with statistical methods is that we use them for non-statistical purposes as well as for statistical purposes. You put a statement in a paper, and rather than just saying "Our confidence interval is such and such" or "Our upper limit is such and such", it tends to be overloaded with what I'm going to call (in a non-technical sense) sociological claims like "We have observed something" or "It's not there". This creates a serious problem with upper limits [in the case where there's a small excess over the expectation for background only] because if we are not absolutely confident that we've seen a signal, people want to quote an upper limit, even if they're quoting that limit at 90% confidence, which is the convention in the field. So you might have a weak signal (where a 90% interval in a unified approach excludes zero, but (say) a 99.7% interval includes it) but still want to quote a 90% upper limit: we return to this business of flip-flopping that Gary Feldman and Bob Cousins fixed. People throw away the solution, i.e. a unified approach to interval-setting, because they're nervous about what a two-sided interval would imply. And so you get a situation where something has a perfectly clear statistical meaning — 90% of the time the real value will be in the interval and 10% of the time it won't — but people aren't willing to stop there. So returning to the case of an empty interval, it will be taken as saying "We're confused, maybe our model was wrong or maybe the data is discrepant or we just don't know." I think it's very hard to imagine a physics collaboration actually writing that in a paper, even though it might be perfectly valid as it stands.

Bob Cousins I think my opinions are pretty well advertised on most of these so I'll just comment on the shortness issue. For two-sided intervals, 'shortness' of course is a metric-dependent statement. I think the way you want to look at it is that the coverage of confidence intervals is just a statement about one type of error and you do need to worry about the other type of error. So you want the most powerful intervals against alternative hypotheses. People know that this is not generally possible for all alternative hypotheses.

For a discrete observable you can make the acceptance region shortest in the construction direction, but that is not necessarily the same as in the parameter direction. Crow and Gardner did this many years ago, but it seems not to be popular.

Louis Lyons I think that actually points to a difficulty if you are trying to decide between which of two methods you are going to use, before you look at the data of course, and you want a method that gives intervals which are not too long but not too short. It's not quite obvious what criterion you would use to optimise the choice of interval.

David Cox In the case of empty intervals the data are sending the message that something is wrong and of course if at all possible the source of the confusion must be identified.

Bob Cousins I'll just repeat something I said at the first Confidence Limits Workshop at CERN in 2000. I am not sure that statisticians are aware of it, but in Particle Physics we have a sort of thing that goes under the name of robustness. The Particle Data Group does lots of averaging of several measurements of the same parameter. When these are inconsistent with each other and have a large χ^2, then the error on the weighted average is not determined simply by error propagation, but is scaled until the χ^2 per degree of freedom becomes something reasonable like unity. Of course it's like all robustness things in that it's kind of a black box that may or may not fix the actual problem, because you don't know what the actual problem is, or otherwise you'd fix it. In a 1999 paper Mike Chanowitz wrote on Higgs mass constraints, the input data from LEP and SLAC had some discrepancies, so he suggested blowing up the errors in an analogous way before producing a combined constraint on the Higgs mass.

One thing I thought was at least worth exploring five years ago, and I don't know it has ever been done, is to take that error on the background we were talking about and blow it up until there is a reasonable probability that you got the data you got. Then you quote an upper limit, including that larger error on the background. As with the PDG method, you don't really know if this is solving the problem, but I think it is worth exploring.

Louis Lyons Maybe at this stage we could ask members of the audience if any of them wanted to comment on any of the issues here.

Bangalore Sathyaprakash This is really a question rather than a comment. We are looking for signals that are weak and rare. When you're really not expecting very strong signals there is probably not much point in debating about which methods we follow; frequentist or Bayesian, does it really matter? We would certainly learn a lot from particle physicists who have had this experience for generations so I would like to see some discussion of that.

What I'm asking is this: let's suppose we are looking for a specific type of signal in a time series. We could follow either a frequentist approach and do an analysis wherein we try to evaluate the likelihood and either claim a detection or set an upper limit. Alternatively, you could follow a Bayesian approach and assume a prior – we know nothing about the prior – and then follow that procedure and get an upper limit. These two upper limits are different but does it really matter? Should we really quarrel about it? Should we not worry about detecting rather than setting upper limits?

Bob Cousins I've been talking about this with my colleagues at the Large Hadron Collider. There has been an enormous amount of interest over the last fifteen years or so, including the Confidence Limits Workshops, on how we measure upper limits. I do look forward to having signals to worry about. We can take as one example the statistics issues at the TeVatron with the discovery of the top quark, and how CDF and D0 measure its mass. I'll also just mention that Gary Hill gave a talk here on the difference between optimising for upper limits and for discovery.

Louis Lyons I wanted to add that at the Fermilab conference in 2000, Ilya Narsky investigated what you get for upper limits by analysing the same data, using the number of events seen and the expected background, with a whole series of different methods. He's got an interesting plot that compares all these different methods. What becomes clear is that you can get very different upper limits by varying your technique and in some cases you can get variations by a factor of 10 or more, especially when the number of observed events is less than the estimated background. So it's a very good idea to choose your method before you look at the data, rather than tuning up the method that gives you the tightest or the weakest upper limit, depending on your feelings about things.

Harrison Prosper I'm one physicist who is not quite so obsessed with coverage. Certainly I'm very happy, and I also insist, that if one invents a method one should at least see whether it works well on average in some ensemble. But the crucial thing to realise is that this is a hypothetical ensemble. Our real experiment presumably is embedded in some ensemble, but we don't know what it is. We make some assumptions, for example we assume that things are perfectly Poisson, but presumably that's not exactly true; things are Poisson to some degree. So I'm happy to have approximate coverage in an ensemble, which is, after all, hypothetical. In fact such a situation arose in a collaboration of which I'm a member. We had two analyses measuring the top quark mass. The question arose as to whether it was sensible to choose the better of the two answers. We looked at the error that was computed for each analysis and asked whether or not we should use the analysis that gave the smaller error. As I noted then, that means inventing some ensemble in which we decide to toss a coin and choose which answer to report. The point is that you can invent any number of ensembles that are plausible and for each of these you'll get different coverage. This is why I'm not quite so obsessed with exact coverage.

Jeremy Lys We're missing out a lot here on the sociology as Bruce referred to before, when Harrison asks whether it is better to use one method rather than the other. You can ask "Better in what sense?" Many of us are experimental physicists, we know we are in competition with other people with other groups, we are in competition for our livelihoods in a sense, we are in competition for grants and so on and it's essential that we appear to be doing good physics and hence it's better to get what we call an accurate answer rather than an inaccurate one, so it's clearly tempting for us to change the way we define coverage for example. If you release the conditions of coverage and get a better answer then you might say that you should publish that result, maybe not. There are sociology points that arise here, that's all.

Rajendran Raja I would like to harp back to the Durham conference which was what got me into this whole thing. It's not just coverage that's important, but the stability of the limit in an ensemble of similar experiments is also important. The degree of fluctuation of a one-sided limit is important to compute and quote. We had this problem when we were looking for the top quark, with CDF publishing one limit and D0 was publishing another limit. The question was which was better. They were within 20 GeV of each other, but if you change the accepted sample by one event by changing the cuts slightly, the limit fluctuates by about 20 GeV. So it's important not only what the actual value of the limit is but also what the band of fluctuation is. Until we see an analysis along those directions, we'll be preoccupied with the coverage, and we won't get the full picture.

Bob Cousins Back to these criteria for parameter intervals, maybe I've said it 20 times but I think it's important that, when an interval is associated with some probability P, there exists a definition of what P is. For confidence intervals P is defined as frequentist coverage. Harrison's point is that you need a well-defined ensemble in order to define frequentist coverage, and that brings us back to points both of us made at the first Confidence Limits Workshop. Professional statisticians seem to go in the direction of conditioning, and that is something we should look at.

The other way of defining P which I think is quite useful is subjective degree of belief. Real Bayesians, like Michael Goldstein at the Durham Conference, are comfortable with that P. What I have trouble with are intervals in between which come out of so-called objective Bayes, where you don't know what the P is: it's not subjective degree of belief and it's not necessarily frequentist coverage. Therefore I think that when we use the machinery of objective Bayes, the P we want to teach our students to apply is the frequentist P, and then check that it works. Joel Heinrich gave an outstanding talk on this in one of the parallel sessions yesterday, using the Bayesian machinery to obtain intervals which undercovered by the criterion of frequentist P, and he figured out how to change priors so that the result was more consistent with frequentist expectations. This is quite useful, but we should make sure that our students understand that the result of all this Bayesian machinery is not a P which is subjective degree of belief.

Byron Roe Improbable events do happen and sometimes your empty interval or almost empty interval tells you have an improbable event, even if you are confident in your parameterization. There are certainly famous examples, as Bob well knows. My question is really for the statisticians. Suppose you have an improbable distribution and by various means you really know it's an improbable distribution. Do you have any general ideas for what should be done to set limits on parameters?

For example, suppose you are measuring signal plus background and you know the mean value of your background well, and you get an observation which is much lower than the background. You know something strange has happened to you. My question is: "In this kind of situation can you make suggestions as to what one might want to do?"

Unknown Get a better estimate of your background!

Bernard Silverman This is a bit like the story about Mendel and the beans, and counting number of plants of different kinds in one particular experiment. If the theory is true, then you expect three of one sort to one of the other. Mendel's published data is at the wrong end of the χ^2 distribution. In other words it fits far better than you would expect at random. So you could then, I suppose, reject random selection. The interesting thing is that you get results which appear to challenge the very randomness assumption sometimes. There was an explanation: it was suggested that Mendel had a gardener who was adjusting the results to what his boss wanted to get! Fortunately, the experiment was correct anyway but there was a little fiddling going on. But I think that you might look for some reason for that low value. In other words, interact with your data intelligently and say "Maybe we're spotting something that we weren't expecting to see, perhaps it's some other phenomenon". I don't think you can get a statistical answer to the question; you just need to know it will happen sometimes.

Louis Lyons That brings us to coffee time, so let's close the session. Thanks to all of you who contributed to this session, and especially to our Panel members: David, Bob, Jerry and Bernard.

Committees

International Scientific Committee:

Roger Barlow (Manchester)
Peter Clifford (Oxford)
Bob Cousins (UCLA)
Glen Cowan (Royal Holloway, London)
Sir David Cox (Oxford)
Roger Davies (Oxford)
Luc Demortier (Rockefeller)
Bradley Efron (Stanford)
Eric Feigelson (Penn State)
Pedro Ferreira (Oxford)
Jerome Friedman (Stanford)
Fred James (CERN)
Steffen Lauritzen (Oxford)
Jim Linnemann (Michigan State)
Tom Loredo (Cornell)
Louis Lyons (Oxford)
Bill Murray (Rutherford Appleton Lab)
Jeffrey Scargle (NASA)
Joe Silk (Oxford)
Gunter Zech (Siegen)

Local Organising Committee:

Andy Carslaw
John Cobb
Pete Gronbech
Louis Lyons
Sue Geddes

List of Participants

Seyed Ali Asgha Alavi	Sabzevar University	alavi@sttu.ac.ir
Roger Barlow	Manchester University	roger.barlow@man.ac.uk
Alan Barr	UCL	alan.barr@cern.ch
Pushpalatha Bhat	Fermi National Accelerator Laboratory	pushpa@fnal.gov
Nicolai Bissantz	University of Goettingen	bissantz@math.uni-goettingen.de
Sergey Bityukov	Institute for High Energy Physics - Protvino	Serguei.Bitioukov@cern.ch
Volker Blobel	Institut fuer Experimentalphysik - Hamburg	volker.blobel@desy.de
Martin Block	Northwestern University	mblock@northwestern.edu
Giovanni Bonvicini	Wayne State University	giovanni@physics.wayne.edu
Pawel Bruckman de Renstrom	University of Oxford	p.bruckman1@physics.ox.ac.uk
Rene Brun	CERN	Rene.Brun@cern.ch
Andy Buckley	IPPP - Durham University	andy.buckley@durham.ac.uk
Pierluigi Catastini	Universita' di Siena - INFN Pisa	pierluigi.catastini@pi.infn.it
Peter Clifford	Oxford University	clifford@stats.ox.ac.uk
John Cobb	Oxford University	j.cobb@physics.ox.ac.uk
Jan Conrad	CERN	Jan.Conrad@cern.ch
Robert Cousins	Univ. of California - Los Angeles	cousins@physics.ucla.edu
Glen Cowan	Royal Holloway - University of London	g.cowan@rhul.ac.uk
David Cox	Nuffield College - Oxford	david.cox@nuf.ox.ac.uk
Kyle Cranmer	Brookhaven National Laboratory	cranmer@cern.ch
Luc Demortier	Rockefeller University	luc@fnal.gov
Seth Digel	Stanford Linear Accelerator Center	digel@slac.stanford.edu
Cristina Espana-Bonet	Universitat de Barcelona	cespana@am.ub.es
Sebastien Fabbro	CENTRA Lisbon	seb@ist.utl.pt
Gary Feldman	Harvard University	feldman@physics.harvard.edu
Pedro Ferreira	University of Oxford	p.ferreira1@physics.ox.ac.uk
Jerome Friedman	Stanford University	jhf@stanford.edu
Nikolai Gagunashvili	University of Akureyri - Iceland	nikolai@unak.is
Alexander Gray	Georgia Institute of Technology	agray@cs.cmu.edu
Joel Heinrich	University of Pennsylvania	heinrich@hep.upenn.edu
Gary Hill	University of Wisconsin - Madison	ghill@icecube.wisc.edu
Susan Holmes	Stanford University	susan@stat.stanford.edu
Aldo Ianni	Gran Sasso Laboratory	aldo.ianni@lngs.infn.it
Andrew Jaffe	Imperial College	a.jaffe@imperial.ac.uk
Jiashun Jin	Purdue University	jinj@stat.purdue.edu
Stephen Johnston	Museum of the History of Science - Oxford	stephen.johnston@mhs.ox.ac.uk
Muge Karagoz Unel	Oxford University	karagozm@cern.ch
Anna Kreshuk	CERN	anna.krechtchouk@cern.ch
Steffen Lauritzen	Dept. of Statistics - University of Oxford	steffen@stats.ox.ac.uk
Francois Le Diberder	CNRS/IN2P3	diberder@admin.in2p3.fr
Samuel Leach	SISSA - Trieste	leach@sissa.it
James Linnemann	Michigan State University	linnemann@pa.msu.edu

Sotiris Loucatos	DAPNIA-SPP CEA-Saclay	sloucatos@cea.fr
Louis Lyons	Oxford University - Particle Physics	l.lyons@physics.ox.ac.uk
Jeremy Lys	Lawrence Berkeley National Lab.	lys@lbl.gov
John Magorrian	Oxford Physics	j.magorrian1@physics.ox.ac.uk
David Maurin	SAp-CEA/Saclay	dmaurin@cea.fr
Lorenzo Moneta	CERN	Lorenzo.Moneta@cern.ch
Robin Morris	USRA	rdm@email.arc.nasa.gov
William Murray	RAL	w.j.murray@rl.ac.uk
Bob Nichol	ICG Portsmouth	bob.nichol@port.ac.uk
Geoff Nicholls	Statistics Department - Oxford University	nicholls@math.auckland.ac.nz
Andrei Nomerotski	University of Oxford	A.Nomerotski@physics.ox.ac.uk
Andres Osorio	The University of Manchester	andres@hep.man.ac.uk
Popat M. Patel	McGill University - Montreal	patel@hep.physics.mcgill.ca
Marc Paterno	Fermilab	paterno@fnal.gov
Simon JM Peeters	University of Oxford	s.peeters1@physics.ox.ac.uk
Troels Petersen	Niels Bohr Institute	petersen@nbi.dk
Maria Grazia Pia	INFN Sezione di Genova	MariaGrazia.Pia@ge.infn.it
Harrison Prosper	Florida State University	harry@hep.fsu.edu
Giovanni Punzi	INFN - Pisa Italy	giovanni.punzi@pi.infn.it
Rajendran Raja	Fermilab	raja@fnal.gov
Nancy Reid	University of Toronto	reid@utstat.utoronto.ca
Byron Roe	University of Michigan	byronroe@umich.edu
Wolfgang Rolke	University of Puerto Rico - Mayaguez	wolfgang@puerto-rico.net
Joseph Romano	Cardiff University	Joseph.Romano@astro.cf.ac.uk
Pilar Ruiz-Lapuente	University of Barcelona	pilar@am.ub.es
Puneet Sarda	University of Houston	puneetsarda@gmail.com
Bangalore Sathyaprakash	Cardiff University	B.Sathyaprakash@astro.cf.ac.uk
Anand Sankar Sengupta	Cardiff University	Anand.Sengupta@astro.cf.ac.uk
Joe Silk	Univ. of Oxford	silk@astro.ox.ac.uk
Bernard Silverman	St Peter's College - Oxford	bernard.silverman@spc.ox.ac.uk
Patrick Sizun	CEA-Saclay - Gif-sur-Yvette France	sizun@discovery.saclay.cea.fr
Chihwa Song	University of Wisconsin - Madison	csong@icecube.wisc.edu
Jean-Luc Starck	CEA-Saclay	jstarck@cea.fr
Alexander Szalay	Johns Hopkins University	szalay@jhu.edu
Fredrik Tegenfeldt	Iowa State University	fredrik.tegenfeldt@cern.ch
Mike Titterington	University of Glasgow	mike@stats.gla.ac.uk
Roberto Trotta	Oxford University - Astrophysics	rxt@astro.ox.ac.uk
Wouter Verkerke	NIKHEF	verkerke@nikhef.nl
Stefania Xella	University of Zurich	xella@physik.unizh.ch
Bruce Yabsley	University of Sydney	yabsley@bmail.kek.jp
Guenter Zech	Universitaet Siegen	zech@physik.uni-siegen.de
Tomi Zivko	Jozef Stefan Institute	tomi.zivko@ijs.si